CLIMATE CHANGE AND BIOTIC FACTORS

A Molecular Approach

CLIMATE CHANGE AND BIOTIC FACTORS

A Molecular Approach

Edited by

Ashutosh Singh, PhD

Saurabh Pandey, PhD

Amarjeet Kumar, PhD

AAP | APPLE ACADEMIC PRESS

First edition published 2025

Apple Academic Press Inc.
1265 Goldenrod Circle, NE,
Palm Bay, FL 32905 USA

760 Laurentian Drive, Unit 19,
Burlington, ON L7N 0A4, CANADA

CRC Press
2385 NW Executive Center Drive,
Suite 320, Boca Raton FL 33431

4 Park Square, Milton Park,
Abingdon, Oxon, OX14 4RN UK

Library and Archives Canada Cataloguing in Publication

Title: Climate change and biotic factors : a molecular approach / edited by Ashutosh Singh, PhD, Saurabh Pandey, PhD, Amarjeet Kumar, PhD.
Names: Singh, Ashutosh (Assistant professor at the Centre for Advanced Studies on Climate Change), editor. | Pandey, Saurabh (Lecturer in agriculture), editor. | Kumar, Amarjeet, editor.
Description: First edition. | Includes bibliographical references and index.
Identifiers: Canadiana (print) 20240479696 | Canadiana (ebook) 20240479785 | ISBN 9781774918487 (hardcover) | ISBN 9781774918494 (softcover) | ISBN 9781003568704 (ebook)
Subjects: LCSH: Crops—Genetic engineering. | LCSH: Crops and climate. | LCSH: Crops—Effect of stress on. | LCSH: Agricultural pests—Control. | LCSH: Agricultural innovations. | LCSH: Plant breeding.
Classification: LCC SB123.57 .C55 2024 | DDC 631.5/233—dc23

Library of Congress Cataloging-in-Publication Data

CIP data on file with US Library of Congress

ISBN: 978-1-77491-848-7 (hbk)
ISBN: 978-1-77491-849-4 (pbk)
ISBN: 978-1-00356-870-4 (ebk)

About the Editors

Ashutosh Singh, PhD
Assistant Professor, Center for Advanced Studies on Climate Change, Dr. Rajendra Prasad Central Agricultural University, PUSA, Samastipur, Bihar, India

Ashutosh Singh, PhD, is an Assistant Professor at the Center for Advanced Studies on Climate Change at Dr. Rajendra Prasad Central Agricultural University, Pusa, India. Dr. Singh has been awarded a PhD degree from the University of Agricultural Sciences Bangalore, GKVK, Karnataka (Plant Biotechnology) in 2018. He has more than eight years of experience in different areas of plant biotechnology, including plant stress biology, pollen-based selection, molecular marker studies, etc. Currently, he is working on developmental aspects of male gametes under stressful environmental conditions. He has published more than a dozen research papers in various journals of international repute. Dr. Singh has also published one book and more than 20 popular articles in many magazines and periodicals.

Saurabh Pandey, PhD
Assistant Professor, Department of Agriculture, Guru Nanak Dev University, Amritsar, Punjab, India

Saurabh Pandey, PhD, is an Assistant Professor at the Department of Agriculture at Guru Nanak Dev University, Amritsar, Punjab, India. He earned an MSc (Agriculture) with a specialization in Plant Biotechnology from the University of Agricultural Sciences, Bangalore, and a PhD in Plant Virology from the National Institute of Plant Genome Research, New Delhi. With eight years of experience in plant biotechnology, his expertise includes plant stress biology, plant virus interactions, abiotic stresses, bio-fortification, molecular marker studies, and metabolism related to tomato and rice crops. Currently, he is focused on biotic stress management of unconventional crops like banana and apple in the Punjab region, India. Dr. Pandey has published over 20 research and review papers in reputable international journals such as *Frontiers in Plant Science, Environmental and Experimental Botany, Plant Science, Journal of Plant Growth and Regulation, Plant Breeding, Advances in Genetics, Physiology, and Molecular Biology of Plants, Plant Stress,* and *South African Journal of Botany.* He has also contributed articles to various magazines and periodicals. Dr. Pandey has authored numerous book chapters on molecular markers, genomics, plant breeding, and plant molecular virology to support modern agricultural research. He is an active member of the Indian Virological Society, the Indian Academy of Horticultural Sciences, and the Society of Plant Biochemistry and Biotechnology. Dr. Pandey was awarded a gold medal for his MSc degree from UAS, Bangalore, in 2016, and has received a best postgraduate thesis award for his MSc thesis, along with other awards for his research work. Additionally, he has received NTS, JRF, and SRF awards from the Indian Council of Agriculture Research (ICAR) India and the Department of Biotechnology, India during his degree program.

Amarjeet Kumar, PhD
Assistant Professor, Department of Genetics and Plant Breeding, College of Agriculture, Central Agricultural University (Imphal), Kyrdemkulai, Meghalaya, India

Amarjeet Kumar, PhD, is currently an Assistant Professor in Genetics and Plant Breeding at the College of Agriculture, Central Agriculture University (Imphal) Kyrdemkulai, Meghalaya, India. He embarked on his professional journey as an Assistant Professor cum Junior Scientist specializing in Genetics and Plant Breeding at Agriculture College Garhwa, Birsa Agricultural University, Ranchi, Jharkhand, India.. Dr. Kumar earned his BSc honors degree in Agriculture at Visva-Bharati University, West Bengal, and pursued his MSc (Ag.) and PhD degrees in Genetics and Plant Breeding at G. B. Pant University of Agriculture and Technology, Pantnagar, India. During his doctoral research, titled "Mapping drought tolerance using teosinte-derived population of maize," Dr. Kumar exhibited exceptional academic prowess, consistently achieving first-class distinctions throughout his career. Dr. Kumar has an impressive publication record, with over 50 contributions in renowned national and international publications. This includes 27 research papers, two lead papers, 24 book chapters, five review papers, four popular articles, and more than 2o research abstracts presented atnational and international seminars and symposia. Additionally, he has authored two books: *Wheat: A Premier Food Crop and Classical* and *Molecular Approaches in Plant Breeding*. Throughout his academic journey, Dr. Kumar has received prestigious scholarships and fellowships. ICAR awarded him a National Talent Scholarship (NTS) during his BSc Agriculture degree program and a Junior Research Fellowship (JRF) in Plant Science for his MSc (Ag.) degree program. He also received a INSPIRE Fellowship from the Department of Science and Technology (DST), Government of India, for his PhD studies. He has also twice qualified the ICAR-National Eligibility Test (NET) in Genetics and Plant Breeding. Recognized internationally, Dr. Kumar was honored with a Young Scientist Associate Award in Genetics and Plant Breeding by the Biologix Research and Innovation Center Pvt. Ltd. (BRICPL) in 2016 at Kathmandu, Nepal. He has been awarded a Student Excellence Award-2022 by the College of Agriculture, G. B. Pant University of Agriculture and Technology, Pantnagar. Furthermore, his master's research work on "Morphological, Physiological, and Molecular Characterization for Heat Tolerance Traits in Bread Wheat (*Triticum aestivum* em. Thell.)" was awarded an Outstanding MSc Thesis Award by the Society for Agriculture Innovation and Development (SAID), Ranchi, India, in 2017. Dr. Kumar's contributions extend beyond academia. He has delivered talks on radio and television, sharing his expertise with a broader audience. He holds memberships in esteemed professional societies such as the Indian Society of Genetics and Plant Breeding, New Delhi, the Indian Society of Plant Genetic Resources, New Delhi, and the Society for Agriculture Innovation and Development (SAID), Ranchi, Jharkhand, India. These affiliations reflect his active involvement and commitment to advancing his field.

Contents

Contributors

R. S. Afiya
Department of Horticulture, School of Life Sciences, Central University of Tamil Nadu, Thiruvarur, Tamil Nadu, India

Priyanka Agarwal
ICAR–Indian Agricultural Research Institute, New Delhi, India

Mohd. Ali
Department of Zoology, Guru Nanak Dev University, Amritsar, Punjab, India

Partha Pratim Behera
Department of Plant Breeding and Genetics, Assam Agricultural University, Jorhat, Assam, India

Alka Bharati
Scientist, ICAR–Central Agroforestry Research Institute, Jhansi, Uttar Pradesh, India

Tanmaya Kumar Bhoi
Forest Protection Division, ICFRE–Arid Forest Research Institute (AFRI), Jodhpur, Rajasthan, India

Ajay Kumar Chandra
ICAR–Indian Agricultural Research Institute, New Delhi, India

Manisha Chaudhary
Department of Entomology, College of Agriculture, Chaudhary Sarwan Kumar Himachal Pradesh Krishi Vishwavidyalaya, Palampur, Kangra, Himachal Pradesh, India

Aman Chauhan
Dr. Yashwant Singh Parmar University of Horticulture and Forestry, Nauni, Solan, Himachal Pradesh, India

M. Chithra
Department of Horticulture, School of Life Sciences, Central University of Tamil Nadu, Thiruvarur, Tamil Nadu, India

Vinay Kumar Dhiman
Department of Basic Sciences, Dr. Yashwant Singh Parmar University of Horticulture and Forestry, Solan, Himachal Pradesh, India

Vivek Kumar Dhiman
Department of Biotechnology, Dr. Yashwant Singh Parmar University of Horticulture and Forestry, Solan, Himachal Pradesh, India

Vishal Dinkar
ICAR–Central Institute of Temperate Horticulture, Rangreth, Srinagar, Jammu and Kashmir, India

K. Jagadeesh
Department of Genetics and Plant Breeding, Professor Jayashankar Telangana State Agricultural University, Hyderabad, Telangana, India

Kritika Kaushal
Department of Zoology, Guru Nanak Dev University, Amritsar, Punjab, India

Ram Keval
Department of Entomology and Agriculture Zoology, Institute of Agriculture Sciences, Banaras Hindu University, Varanasi, Uttar Pradesh, India

Gopi Kishan
ICAR–Indian Institute of Seed Science, Kushmaur, Mau, Uttar Pradesh, India

J. Komal
Department of Entomology, Navsari Agricultural University, Navsari, Gujarat, India

Charishma Krishnappa
Division of Plant Pathology, ICAR–Indian Agricultural Research Institute, New Delhi, India

Amarjeet Kumar
Assistant Professor, Genetics and Plant Breeding, College of Agriculture, Central Agricultural University (Imphal), Kyrdemkulai, Meghalaya, India

Deepak Kumar
Department of Zoology, Guru Nanak Dev University, Amritsar, Punjab, India

Ravi Kumar
Research Scholar, Department of Plant Pathology, Chandra Shekhar Azad University of Agriculture and Technology, Kanpur, Uttar Pradesh, India

Radheshyam Kumawat
Research Scholar, College of Post-Graduate Studies in Agricultural Sciences, Central Agricultural University (Imphal), Umiam, Meghalaya, India

Kanupriya Kuniyal
Department of Biochemistry and Biotechnology, Sardar Bhagwan Singh University, Balawala, Dehradun, Uttarakhand, India

Deepak Kumar Mahanta
Department of Entomology, Dr. Rajendra Prasad Central Agricultural University, Pusa, Bihar, India

Prasanta Kumar Majhi
Department of Plant Breeding and Genetics, Regional Research and Technology Transfer Station (RRTTS), Odisha University of Agriculture and Technology (OUAT), Keonjhar, Odisha, India

S. Manivannan
Department of Horticulture, School of Life Sciences, Central University of Tamil Nadu, Thiruvarur, Tamil Nadu, India

Sahil Mehta
Department of Botany, Hansraj College, University of Delhi, New Delhi, India

Varsha Mishra
Sir J.C. Bose Technical Campus, Department of Biotechnology, Kumaun University, Bhimtal, Nainital, Uttarakhand, India

Sajad Un Nabi
ICAR–Central Institute of Temperate Horticulture, Srinagar, Jammu and Kashmir, India

R. M. Nagalakshmi
Sethu Bhaskara Agricultural College and Research Foundation, Karaikudi, Tamil Nadu, India

Niki Nautiyal
Department of Biochemistry and Biotechnology, Sardar Bhagwan Singh University, Balawala, Dehradun, Uttarakhand, India

Puja Ohri
Department of Zoology, Guru Nanak Dev University, Amritsar, Punjab, India

Himanshu Pandey
Department of Biotechnology, Dr. Yashwant Singh Parmar University of Horticulture and Forestry, Solan, Himachal Pradesh, India

Pramod Kumar Pandey
Assistant Maize Breeder, AICRP on Maize, College of Agriculture, Central Agricultural University, Kyrdemkulai, Meghalaya, India

Rishi Nath Pandey
Department of Plant Pathology, Banda University of Agriculture and Technology, Banda, Uttar Pradesh, India

Supriya Pandey
Govind Ballabh Pant National Institute of Himalayan Environment, Kosi-Katarmal, Almora, Uttarakhand, India

Parkirti
Department of Zoology, Guru Nanak Dev University, Amritsar, Punjab, India

Asharani Patel
School of Agricultural Sciences, GD Goenka University, Sohna, Gurugram, Haryana, India

R. Ponmani
Mother Teresa College of Agriculture, Pudukkottai, Tamil Nadu, India

R. Sangeetha Vishnu Prabha
Vanavarayar Institute of Agriculture, Pollachi, Tamil Nadu, India

Umara Rahmani
Department of Plant Pathology, College of Agriculture, Sardar Vallabhbhai Patel University of Agriculture and Technology, Modipuram, Meerut, Uttar Pradesh, India

Sumit Rai
Govind Ballabh Pant National Institute of Himalayan Environment, Kosi-Katarmal, Almora, Uttarakhand, India

Ramkumar
Department of Entomology and Agriculture Zoology, Institute of Agriculture Sciences, Banaras Hindu University, Varanasi, Uttar Pradesh, India

Swati Ray
Department of Food Technology, Haldia Institute of Technology, ICARE Complex, Hatiberia, Haldia, West Bengal, India

Kuleshwar Prasad Sahu
School of Agricultural Sciences, GD Goenka University, Sohna, Gurugram, Haryana, India

Prince Sahu
Department of Entomology, Chandra Shekhar Azad University of Agriculture and Technology, Kanpur, Uttar Pradesh, India

Ipsita Samal
Department of Entomology, Faculty of Agriculture, Sri Sri University, Cuttack, Odisha, India

Anshuman Semwal
Department of Entomology, Dr. Yashwant Singh Parmar University of Horticulture and Forestry, Nauni, Solan, Himachal Pradesh, India

Manaranjan Senapati
Division of Genetics, ICAR–Indian Agricultural Research Institute (IARI), Pusa Campus, New Delhi, India

Subrat Kumar Senapati
Department of Floriculture and Landscaping, Odisha University of Agriculture and Technology (OUAT), Bhubaneswar, Odisha, India

Yagnayani Senapati
Department of Plant Breeding and Genetics, College of Agriculture, Odisha University of Agriculture and Technology (OUAT), Bhubaneswar, Odisha, India

S. Senthilkumar
Department of Horticulture, School of Life Sciences, Central University of Tamil Nadu, Thiruvarur, Tamil Nadu, India

K. Shamini
Center for Plant Breeding and Genetics, Tamil Nadu Agricultural University, Coimbatore, Tamil Nadu, India

Roohi Sharma
Department of Zoology, Guru Nanak Dev University, Amritsar, Punjab, India

Anju Shukla
Research Scholar, Department of Plant Pathology, Chandra Shekhar Azad University of Agriculture and Technology, Kanpur, Uttar Pradesh, India

Ashish Kumar Singh
Crop Protection Division, ICAR-VPKAS, Almora, Uttarakhand, India

Chander Singh
Department of Entomology, Dr. Yashwant Singh Parmar University of Horticulture and Forestry, Nauni, Solan, Himachal Pradesh, India

Devendra Singh
Faculty of Biotechnology, Institute of Biological Sciences and Technology, Shri Ramswaroop Memorial University, Barabanki, Uttar Pradesh, India

Jagmohan Singh
Department of Vegetable Science, CCS Haryana Agricultural University, Hisar, Haryana, India

Kartikey Singh
Department of Entomology, Chandra Shekhar Azad University of Agriculture and Technology, Kanpur, Uttar Pradesh, India

Niharika Singh
Acharya Narendra Dev University of Agriculture and Technology, Ayodhya, Uttar Pradesh, India

Yamini Tak
Agricultural Research Station, Agriculture University, Kota, Rajasthan, India

Swagata Thakur
Division of Plant Pathology, ICAR–Indian Agricultural Research Institute, New Delhi, India

Raman Tikoria
Department of Zoology, Guru Nanak Dev University, Amritsar, Punjab, India

Sunil Kumar Verma
Faculty of Biotechnology, Institute of Biological Sciences and Technology, Shri Ramswaroop Memorial University, Barabanki, Uttar Pradesh, India

Foreword

With great pleasure and anticipation, I write the foreword for the book *Climate Change and Biotic Factors: A Molecular Approach*. Climate change has emerged as one of the most pressing challenges of our time, with far-reaching implications for the environment, ecosystems, and the well-being of humanity. Understanding the intricate interactions between climate change and biotic factors is crucial for developing effective strategies to mitigate the negative impacts and ensure the resilience of our planet's diverse ecosystems. This book brings together an impressive collection of research and knowledge, focusing on the molecular approach to unraveling the complexities of the relationship between climate change and biotic factors. The contributors, experts in their respective fields, explore the fundamental molecular mechanisms that govern the responses of organisms to changing climatic conditions and the subsequent interactions with biotic stresses such as pests, diseases, and invasive species. By adopting a molecular lens, this book goes beyond traditional ecological approaches, delving into the intricate workings of genes, proteins, and signaling pathways that dictate the responses of organisms to climate change and biotic factors. It offers a comprehensive exploration of the molecular mechanisms underlying organisms' adaptation, resistance, and resilience in the face of shifting environmental conditions. The chapters in this book cover a wide range of topics, including the molecular basis of plant responses to climate change, the dynamics of disease under changing climatic conditions, and molecular tools for enhancing the resilience of crops and ecosystems. Each chapter provides valuable insights into the intricate molecular processes that shape the interactions between climate change and biotic factors, fostering a deeper understanding of the challenges and opportunities our changing world presents. What sets this book apart is its emphasis on the practical application of molecular approaches to address the challenges posed by climate change and biotic factors. The authors highlight innovative strategies, techniques, and tools that can be employed to develop climate-resilient crops, manage invasive species, and mitigate the impact of diseases.

I commend the editors and contributors for their dedication, expertise, and commitment to advancing our understanding of this crucial field. Their collective efforts have resulted in a comprehensive and insightful compilation that will inspire further research, collaboration, and action. I am confident that this book will serve as a valuable resource for researchers, students, policymakers, and anyone interested in unraveling the mysteries of climate change and its intricate interactions with biotic factors. As we face the urgent need to address climate change and safeguard the biodiversity and sustainability of our planet, this book serves as a beacon of knowledge, illuminating the path towards a resilient and harmonious coexistence between nature and humanity. May the insights within these pages inspire transformative actions that will shape a brighter future for generations to come.

—R. L. Ravikumar
Emeritus Scientist,
Department of Plant Biotechnology,
UAS, GKVK, Bangalore, Karnataka, India

Preface

In recent decades, the scientific community and the global population have become increasingly aware of the profound impacts of climate change on our planet. Rising temperatures, changing precipitation patterns, and extreme weather events alter ecosystems and challenge the delicate balance of life on Earth. Alongside these climatic shifts, the intricate interactions between organisms, including pests, pathogens, and other biotic factors, have emerged as critical components in understanding the complex dynamics of our changing environment. This book, *Climate Change and Biotic Factors: A Molecular Approach,* delves into the intricate relationship between climate change and the interplay of biotic factors, shedding light on the molecular mechanisms underlying these interactions. It explores how climate change influences the dynamics of pests, diseases, and other biotic stresses, while also examining how these factors, in turn, shape the response of plants and microorganisms to the changing climate. The molecular approach adopted in this book provides a deeper understanding of the underlying mechanisms through which climate change affects biotic factors. By exploring organisms' intricate molecular pathways and genetic responses, we gain valuable insights into their ability to adapt, survive, and interact in changing environmental conditions. Furthermore, we explore the potential for utilizing molecular tools to develop innovative strategies for mitigating the negative impacts of climate change on biotic factors and promoting sustainable ecosystems.

This book brings together an interdisciplinary group of experts who have significantly contributed to climate change and molecular biology. Through their collective knowledge, research, and experiences, they provide a comprehensive overview of the current state of understanding in this field and highlight the potential for advancements in molecular approaches to address the challenges posed by climate change and biotic factors. The chapters within this book cover a wide range of topics, including the molecular responses of plants to biotic stresses, the role of microorganisms in mediating climate change impacts, the impact of changing climate on disease dynamics, and molecular tools for enhancing plant resistance to biotic stresses. Each chapter delves into the fascinating molecular intricacies and interactions within these systems, offering valuable insights for researchers, students, policymakers, and practitioners interested in climate change, biotic factors, and molecular biology. We hope this book will serve as a valuable resource for those seeking a deeper understanding of the molecular mechanisms driving the complex interplay between climate change and biotic factors. We envision it as a catalyst for further research, dialogue, and collaboration to address the challenges of climate change and promote sustainable management of our ecosystems. Together, through a molecular lens, we can unlock the secrets of our changing world and work towards a resilient future for generations.

—Ashutosh Singh, PhD
Saurabh Pandey, PhD
Amarjeet Kumar, PhD

Introduction

The world population is expected to reach over 9.8 billion by 2050. Consequently, increasing pressure to feed a growing population in the future would need additional production of major crops. On the contrary, the limiting resources and challenges associated with changing climate and the unpredictability of pest and disease incidents pose a severe threat to global food security. The effect of climate change on insect pests and diseases of crops is multi-dimensional. The magnitude of this impact could vary with the species and their growth patterns. Climate change, such as elevated temperature, humidity, and increased concentration of CO_2, can result in an expansion of their geographic distribution, increased survival during overwintering, increased number of generations, altered synchrony between plants and pests, limited interspecific interaction, increased risk of invasion by migratory pests, increased incidence of insect-transmitted plant diseases, and reduced effectiveness of biological control, especially natural enemies. As a result, there is a severe risk of crop economic losses and a challenge to human food security. Therefore, changing disease and insect pest scenarios due to climate change have warranted the need to understand the molecular mechanism of plant interaction.

This book systematically studies the effects of climate change on pest and disease incidence of crop plants. The book has four parts; where Part I introduces the readers to climate change leads to the unpredictability of rainfall and temperature. These two factors are essential drivers of the spreading of pests and diseases. Part I deals with the impact of temperature, rainfall, and humidity affects the adaptation of different biotic agents such as insects, pests, and diseases. Part II deals with breeding and molecular approaches to tackling pest disease in changing climate. Each chapter deals with different components that affect the pest incident, which could be targeted by biotechnological tools to tackle the biotic factors under unpredictable climate change. Part III summarizes the advanced tools, including genome editing and lipoxygenase, etc., which directly affect target pest and disease incidence.

PART I

Climate Change and Biotic Factors: Introduction

CHAPTER 1

Effect of Humidity on Pest and Disease Incidence in Crops

RAMKUMAR,[1] MANISHA CHAUDHARY,[2] PRINCE SAHU,[3] KARTIKEY SINGH,[3] RAVI KUMAR,[4] ANJU SHUKLA,[4] CHANDER SINGH,[5] VISHAL DINKAR,[6] ASHISH KUMAR SINGH,[7] NIHARIKA SINGH,[8] RAM KEVAL,[1] ANSHUMAN SEMWAL,[5] and RISHI NATH PANDEY[9]

[1]*Department of Entomology and Agriculture Zoology, Institute of Agriculture Sciences, Banaras Hindu University, Varanasi, Uttar Pradesh, India*

[2]*Department of Entomology, Chaudhary Sarwan Kumar Himachal Pradesh Krishi Vishwavidyalaya, Palampur, Kangra, Himachal Pradesh, India*

[3]*Department of Entomology, Chandra Shekhar Azad University of Agriculture and Technology, Kanpur, Uttar Pradesh, India*

[4]*Research Scholar, Department of Plant Pathology, Chandra Shekhar Azad University of Agriculture and Technology, Kanpur, Uttar Pradesh, India*

[5]*Department of Entomology, Dr. Yashwant Singh Parmar University of Horticulture and Forestry, Nauni, Solan, Himachal Pradesh, India*

[6]*ICAR–Central Institute of Temperate Horticulture, Rangreth, Srinagar, Jammu and Kashmir, India*

[7]*Crop Protection Division, ICAR-VPKAS, Almora, Uttarakhand, India*

[8]*Acharya Narendra Dev University of Agriculture and Technology, Ayodhya, Uttar Pradesh, India*

[9]*Department of Plant Pathology, Banda University of Agriculture and Technology, Banda, Uttar Pradesh, India*

Climate Change and Biotic Factors: A Molecular Approach.
Ashutosh Singh, Saurabh Pandey, and Amarjeet Kumar, (Eds.)

ABSTRACT

Humidity is a term used to describe how much moisture or water vapor is present in an atmosphere. It is a crucial environmental factor in the emergence, growth, and development of diseases and pests. Insects have a greater ability for reproduction, and high humidity has an impact on their physiology. There are several differences in how ambient moisture affects an insect's metabolic rate. It may have an impact on insect behavior such as post-diapause egg hatching, molting, mating, and pest movement. Dry environments trigger diapause, but moist ones promote the growth and pupation of insect pests. With a maximal rate of disease progress at intermediate RHs (50–56%), disease development is often faster at close to room temperature. Low RH levels speed up host tissue death, inhibit disease progression, and limit spore germination and lesion growth. Within a range of suitable temperatures, intermediate RH levels enhance disease progression by increasing spore germination. Although prolonged exposure to these conditions seems to be detrimental for the development of disease, high RH enhances spore germination. Epidemics of disease, the prevalence of pests, and the use of pesticides are all influenced by weather conditions, especially high ambient humidity. Soil moisture and temperature have a considerable impact on the emergence and spread of nematodes and plant diseases. The population growth, survival, and incidence of nematodes are influenced by humidity. When egg-laying in a dry environment, mites produce more eggs more quickly and live longer than when doing so in a nearly saturated environment. In a moist environment, newly hatched mites have a limited chance of survival. At higher RH levels (>96%) with a well-defined incubation period, entomopathogenic agents exhibit their highest infection potential and mortality rates.

1.1 INTRODUCTION

Climate change and global warming are two of the most talked-about topics right now, and both are quite concerning for agriculture. Agricultural productivity, as well as insects and pests in agriculture, are significantly affected by climate variables such as temperature rises, changes in relative humidity (RH), and changes in precipitation patterns. Pests are a significant biotic component that is influenced by climate change and other weather-related phenomena (Prakash et al., 2014). Climate change impacts pest insects in several ways such as the range of geographic spread, their multiplication via overwintering, required generations to increase, the synchrony of pests with plants, interspecific interactions, invasive migratory pests, insect-transmitted plant diseases, and their biological control such as natural enemies becoming a greater risk (Skendžić et al., 2021).

It is anticipated that pest infestations shall become more unpredictable and they can spread over more areas under changing climates. The insect-plant connections in ecosystems are yet to be known comprehensively, which is added to the ambiguity around how climate change may directly affect agricultural harvests (Gregory et al., 2009). It may be suggested that climate modification adaptation is a continuous process of lowering the potential risks associated with climate change impacts by using current risk management techniques (Howden et al., 2007). Due to significant environmental changes in both natural and artificial habitats, several developmental changes in behavioral and physiological reactions could have occurred in insects (Jain, 2019). Certain insects can alter their preferences or leave their current site in reaction to these alterations in the physical and chemical environment such as pesticides applied and changed atmosphere (Guedes et al., 2009).

Humidity is a term used to describe how much moisture or water vapor is present in an atmosphere. It is a crucial environmental factor in insect evolution, the emergence of virulent biotypes, proliferation, growth, and development of diseases and pests (Table 1.1). Humidity affects insect physiology and reproductive ability. However, the effect of atmospheric moisture on insect metabolic speed seems extremely variable. The presence of floods and high humidity can threaten the development of offspring and delay the time when the eggs are laid. The distribution and regularity of rainfall may have an impact on the prevalence of pests both directly and indirectly through variations in humidity levels. It has been shown in various insect species that rising humidity slows down metabolism. The hatching period of eggs varies, normally 3–8 days depending upon RH and temperature. RH and temperature have an adverse effect on the egg period (Srivastava, 2003). The spotted stem borer's larval period was influenced by both RH and temperature, and it was much shorter at 30°C and 40% RH (Tamiru et al., 2012). The pupal period was reduced at 30°C temperature and 40% RH. The duration of the insect life cycle (egg to adult) was increased at 40%, 60%, and 80% RH under 22°C, 26°C, and 30°C temperatures. According to Ouedraogoa et al. (1996), high relative humidity increases the number of stored grain pests, and *Dinoderus minutus* eggs hatch with a high mortality rate at high relative humidity levels. Certain species, such as mites, are typically restricted to environments above 65% RH (Beckett, 2011). The initial developmental stages of *Acarus farris* and *Tyrophagus putrescentiae* were unable to reach the protonymph stage at 70% RH because of 100% larval mortality, while *Tyrophagus neiswanderi* was able to complete their development.

At a particular period, the combination of three key variables – host's sensitivity, pathogen's virulence, and favorable environment – determines how a disease develops in a host plant. The development of tomato powdery mildew is clearly influenced by humidity. The optimal RH range for tomato powdery mildew development is between 60% and 80% (Corbaz, 1990; Hannig, 1996). An optimal temperature of 25°C and relative humidity levels in the test range of 39–98% are most optimum for powdery mildew development on grape seedlings and the germination of *Uncinula necator* conidia in vitro. For most fungi to propagate and function effectively, they require >75% humidity. Juveniles of entomopathogenic nematodes (*Rhabditia Steinernematidae* and *Heterorhabditidae*) under infection may endure unfavorable environmental circumstances by staying in the host carcass for up to 50 days. *Galleria mellonella* larvae were subjected to RH levels of 75, 85, 96, or 100% after being infested at 25°C with one of four nematode species (Brown and Gaugler, 1997). A low relative humidity reduced *Heterorhabditis bacteriophora* and *Steinernema feltiae* emergence, but unchanged *S. glaseri* and *S. carpocapsae* emergences. *H. bacteriophora* and *S. feltiae* infection was considerably reduced in cadavers with low relative humidity, but no effect was observed in cadavers infected with *S. glaseri* or *S. carpocapsae* (Brown and Gaugler, 1997). Lacey and Unruh (1998) found that codling moth larvae exposed to *S. carpocapsae, S. riobrave,* and *H. bacteriophora* at 25% relative humidity had a lower mortality rate compared to larvae exposed to 75% relative humidity. Thus, humidity is a vital factor for insects and pests' growth and development and affects each stage of development. Therefore, it is possible to use RH as an effective tool in integrated pest management (IPM).

1.2 EFFECT OF HUMIDITY ON INSECT PESTS

Insects are poikilothermic (cold-blooded) organisms and thus very sensitive to changing climate conditions which can have significant effects on the biology and behavior of insects. Climate conditions affect invertebrate populations including insects, mites, and other invertebrates as well as their reproduction, development, and dispersal. Many abiotic factors such as relative humidity, rainfall, temperature, etc., affect the plants and their products and favor insect pests for infestation and establishment. Insect pests can infest economic parts of plants at both pre- and post-harvest stages, resulting in significant losses (Kodjo et al., 2011). Moisture is lost easily from the insects through the wax layer of the external cuticle via respiration, evaporation, and defecation processes (Busvine, 1980). Although certain insects

TABLE 1.1 Optimum Relative Humidity and Temperature for Insect Pests' Proliferation

Insect	Scientific Name	RH (%)	Temperature (in °C)	Reference
Top Borer	*Scirpophaga excerptalis*	86	33.8	Kurmi et al., 2022
Early Shoot Borer	*Chilo infuscatellus*	80	33.9	Kurmi et al., 2022
Brown Plant Hopper	*Nilaparvata lugens*	70	28.9	Patil et al., 2020
Yellow Stem Borer	*Scirpophaga incertulas*	99	23.7	Samrit et al., 2019; Patil et al., 2020
Rice Leaf Folder	*Cnaphalocrocis medinalis*	95	28.5	Patil et al., 2020
Gram Pod Borer	*Helicoverpa armigera*	81	30.5	Meena et al., 2021; Ramkumar et al., 2023
Tur Pod Fly	*Melanogromyza obtusa*	79	30.5	Ramkumar et al., 2023
Tur Pod Bug	*Clavigralla gibbosa*	79	30.5	Ramkumar et al., 2023
Spotted Pod Borer	*Maruca testulalis*	81	30.5	Meena et al., 2021
Hairy Caterpillar	*Psalis pennatula*	92	27.3	Patil et al., 2020

may directly take moisture from the air when the relative humidity reaches 45%, many insects rely on their food source. While many insects have enormous populations and short generation periods, climate change may cause significant changes in their selection, fecundity, phenology, survival, and habitat usage (Shrestha, 2019). Insect abundance and distribution may alter quickly as a result of these changes to their life histories. It has been revealed that widespread generalist species have expanded their range at their cool margins, whereas localized species specialized in habitats have narrowed their ranges (Konvicka et al., 2003; Hill et al., 2002).

Rainfall distribution and frequency affect pest incidence either directly or via altering the humidity levels. According to projections of the climate, rainfall would be less frequent but might be greater in intensity (Skendžić et al., 2021). On one side, this would cause floods, while on the other, it would cause droughts. However, due to the high rains washing away minor pests like whiteflies, jassids, aphids, mites, etc., on crops, the occurrence of these pests may be decreased (Skendžić et al., 2021). For example, the *Mythimna separata* (Armyworm) reaches large epidemic measurements following prolonged rainfall and floods, whereas drought years correspond with the least severe outbreaks (Wang et al., 2015). A study by Lever (1969) found that all but three outbreaks of armyworm and *Spodoptera mauritia* occurred when rainfall exceeded 89 cm on average from 1938 to 1965. Study the rainfall effect on the insect's sprinkler system was used for supplying continuous moisture. The population of aphids in wheat and other crops is influenced by the rainfall and sprinkler system of irrigation used for experiments (Chander, 1998; Daebeler and Hinz, 1977).

As a major component of grasslands, insect fauna homopteran insects are influenced by both wet and dry summers as studied by Masters et al. (1998). Due to the modification of climatic conditions, it directly and indirectly influences the insects' growth and development. An increase in vegetation cover was observed due to the supplemented summer rainfall resulting in an increase in insect abundance. While summer droughts decreased vegetation cover, they did not affect insect abundance. A nymphal development rate that remained unaffected by temperature was found in the winter-warmed plots when egg hatching and hibernation occurred earlier. Manipulation in RH can also play a vital role in integrated pest management programs (IPM). Entomopathogenic nematodes (EPNs) act as an alternative or as one of the components of IPM (Abd-Elgawad, 2017; Grewal et al., 2012). EPNs are affected by many factors that determine their survival and effectiveness including temperature, RH, the time required for them to enter a leaf, and the number of applications on the host. RH is the most vital factor for the EPNs' survival and efficacy; if RH decreases, the efficiency and survivability of EPNs decrease (Husin and Port, 2021). The importance of environmental parameters on the growth and development of insect pests can be understood by taking the example of *Hermetia illucens*, the black soldier fly (BSF). The life cycle of BSF and factors that directly and indirectly affect its growth and development are illustrated in Figure 1.1.

FIGURE 1.1 Black soldier fly (BSF) life cycle and factors affecting different stages.

Source: Adapted from Kim et al., 2021

Optimal ranges are indicated below the factors: (i) RH: decreases the survival rate when fluctuation in the optimal range of RH (Holmes et al., 2012; Barragan-Fonseca et al., 2017); (ii) larval density: in the presence of a high larval density, decreases survival rate due to the ratio of substrate weight (g) to the number of larvae (Pastor et al., 2015); (iii) salinity of food material: decrease the growth and development rate of larvae when the salinity of food material will be increased (Pastor et al., 2015); (iv) moisture of food material: may decrease the degradation rate when moisture increases from the optimal range (Diener et al., 2011; Barragan-Fonseca et al., 2017); (v) temperature: above the optimal range has a negative effect on the lifespan and pupal development time of adults (Shumo et al., 2019; Barragan-Fonseca et al., 2017); (vi) light intensity: effect on the mating behavior depending on the light source (Barragan-Fonseca et al., 2017; Zhang et al., 2010).

1.2.1 EFFECT OF HUMIDITY ON INSECT BEHAVIOR

Behavioral actions include any activity taken in response to a stimulus or its environment, especially ones that can be observed and characterized. But insects exhibit spontaneous behavior even in the absence of a distinct signal (Hoy, 2019). To control insect pests, it is essential to understand insect behavior, especially in the present situation where IPM programs are becoming more focused. In order to implement IPM successfully, one must understand the behavioral characteristics of pests and how they affect ecological factors like population dynamics and geographic distribution (Campbell, 2006). Programs for pest management might be improved by better understanding the behavior of both harmful and useful insects. Environmental instability, as well as an increase in extreme weather events, may reduce the effectiveness of pesticides to kill targeted pests or harm non-target organisms. Both positive and negative effects can occur on the biological control system. Agriculture will probably continue to face difficulties due to pests (Patterson et al., 1999).

Relative humidity affects the oviposition of insects. At 70% RH, rice moth/rice meal moth lay more eggs than at 50% RH. At 70% RH, larvae develop more quickly than at lower RH, taking 34 and 43 days respectively (Reddy and Reddy, 2016). Eggs are laid spontaneously by many insects in humid areas; most insects are not affected by being around saturated air; however, certain insects are known to be. Sometimes the death of the embryo results in the death of the egg, and other times the eggshell becomes very hard, that time hatching is restricted (Buxton, 1932). Some eggs withstand dryness, so it turns out to be dormant. Egg stage duration is prolonged in many insects when they are exposed to low levels of humidity that are not deadly.

Population buildup studies of different insect pests vary with the combined effect of weather parameters. One of the crucial considerations is the relative humidity. The mustard aphid population and the afternoon RH exhibited a strong positive correlation. The leaf miner *Aproaerema modicella* was able to multiply on soybean under dry weather conditions, low RH, and an increase in maximum temperature on a clear sunny day. According to the multiple linear regression model, maximum temperature and sunshine would greatly reduce leafhopper populations, whereas a rise in rainfall, morning RH, and minimum temperature would significantly increase the population. Correlation studies between weather parameters and the incidence of the entomophagous fungus *Nomuraea rileyi* on *Spodoptera litura* indicated positive significant effects of afternoon RH in groundnut, soybean, and potato during the kharif season.

1.2.2 EFFECT OF HUMIDITY ON THE INCIDENCE OF INSECT PESTS ON CROPS

There seems to be a wide range of how ambient moisture affects the rate of insect metabolism. Relative humidity (RH) can affect various developmental stages differently in different insect species. For instance, moist air retards the egg development of *Dendrolimus pini* in some cases while in other cases it may hasten the development (Serebrjanikaw et al., 1901). Moist air favors the larval development (larvae to pupae) of *Mayetiola destructor*, and dry air significantly retards it (Headlee et al., 1917). Dry air retards the pupal development in *Lygellus epilachna* (Urech, 1890); however, it favors pupal development in *Pieris brassicae* (Bachmetjew, 1907a). Lethargy is caused by dry air even when it is warm in *Phytodecta viminalis* and numerous other species (Bachmetjew, 1907b). *Toxoptera graminum* does not appear to be affected by dry or humid air when consuming young and succulent wheat (Headlee, 1914). The larval population of *Helicoverpa armigera* (gram pod borer) increases as RH declines (Bisane et al., 2013; Ramkumar et al., 2023).

High humidity does reduce the production of eggs in the *Metatetranychus ulmi* (Hueck et al., 1952). In Germany, Linke (1953) found that rainfall can explain some of *Tetranychus telurius'* decline, but it cannot entirely explain the population decline. It has been found that constant high RH causes spider mites to die during the chrysalis stage (the period of inactivity prior to molt).

Understanding how humidity affects the onset of diapause seems to be a challenging problem. The development of diapause can be influenced both positively and negatively by water and moisture. In times of flooding and high humidity, offspring development can be negatively affected, and oviposition can be delayed until after the rainy season. When female Mexican bruchid beetles (*Acanthoscelides obtectus*) are in reproductive diapause, they hang dead leaves over plants from April to November. When rainfall is under 50 mm per month, these beetles delay their oviposition until mid-November to late December (Hodek et al., 1981). Diapause termination by rains in tropical borers is one of the well-known examples. Upon the onset of rain, the larval stage of the sorghum stem borer diapauses in dry stalks for six to eight months, pupating in spring. There have been numerous experiments showing that the stimulatory effect of moisture requires previous exposure to drought (Gill et al., 2017).

1.2.3 EFFECT OF HUMIDITY ON INSECT METABOLISM

The sum of all internal chemical and physical processes in an organism is called metabolism. Basic climatic factors like humidity and temperature impact directly or indirectly on the insect's metabolism, growth, and development. They have a direct effect on the larvae and adults' activities, dispersion of insects in new places and environments, effect on phenology and duration of lifespan along with the survivability of insects. They also affect the distribution, fecundity, population dynamics, abundance, feeding behavior, physiological mechanisms, and intensity of insect pests (Fränzle, 2003; Overgaard and Sorenson, 2008; Ayres and Schneider, 2009; Karl et al., 2011; Khaliq et al., 2014). Humidity has an indirect impact on the habitat where insects appear, including effects on plant development, food quality, plant phenology, parasitoids, predators, and entomopathogen activity (Jaworski and Hilszczański, 2013). Climatic factors affecting the crop plant's physiology and metabolism indirectly affect the insects (Netherer and Schopf, 2010; Ayres and Lombardero, 2000; Moore and Allard, 2008; Rouault et al., 2006).

The effect of climatic factors like humidity and temperature on mulberry silkworm *Bombyx mori* influences the physiology, developmental stages, quality of silk, and productivity (Singh et al., 2009). Those components are responsible for dispersal (Genkai et al., 2005; Bale, 2002; Brower, 1995), habitat modification and selection of habitat (Powell et al., 2006), low-temperature resistance (Ramlov, 2000), dry conditions (Dautel, 1999), and limitation of food (Anderson, 1974), diapause (Ando, 1974; Nakamura and Numata, 2000), the developmental rate also alteration (Honek and Kocourek, 1990), sensitivity to environmental signals (Danks, 2005), life-cycle patterns (Danks, 2013), variation in phenology (Chen et al., 2003), and development (Bennett et al., 2005; Danks, 2007) and all they are the outcome of external pressures. For healthy larval growth and the production of high-quality cocoons, humidity levels throughout the rearing period of 1st, 2nd, 3rd, 4th, and 5th instars should be kept at 85%, 85%, 80%, 75%, and 70% respectively. For the embryo to develop normally during egg incubation, humidity must be kept at an average of 80%. Low-quality eggs always hatch if humidity levels during incubation fall below 70% (Singh et al., 2009).

According to entomologists, in dry air at the same temperature, some insects may have a faster metabolism than in moist air. They propose that such enhanced metabolism should generate more metabolic water as a result of enhanced metabolism in order to maintain a constant ratio of water to dry matter in various atmospheric humidities. Insects at rest are not actively moving and experience an increase in metabolic rate (Mellanby, 1936; Moore and Allard, 2008; Rouault et al., 2006). *Tenebrio molitor*, a mealworm beetle, is presently regarded as an economically damaging insect as it attacks grains or their products during storage (Vigneron et al., 2019). *Ophraella communa*, a ragweed pest with good control over *Ambrosia artemisiifolia*, was analyzed for effects of three different RH levels (i.e., 60, 75, and 90% RHs) (Zhou et al., 2010; Table 1.2). The time of all developmental stages of *O. communa* is shortening with rising RH except for the pupal stage. *O. communa* can grow best in RH levels between 75% and 90% with the highest values of *rm* (intrinsic rate of increase), *R0* (net reproduction rate), and λ (finite rate of increase) found at 75% RH, i.e., most conducive for *O. communa* population expansion. Clearly, *O. communa* favors environments with moist microclimates and wet seasons. *Beauveria bassiana* is a potential and effective biological control agent for reducing the increasing population of various economically significant pest species. The highest infection potential and mortality rates of entomopathogenic agents are observed at higher RH values (>96%) with a clearly defined incubation period. The germination and sporulation of *B. bassiana* are highly dependent on humidity. In petri dish experiments at 97.5% RH, a significantly greater number of pest insects were infected (60–88.8%) while at 75 and 80% RH only 15.3–43.9% of pest

TABLE 1.2 Humidity Requirement of Various Insect Pests for Their Different Growth Stages

Insect Pest	Scientific Name	Host	Biological Processes	RH (%)	References
Spotted stem borer	*Chilo partellus*	Maize and Sorghum	Egg, larva, pupa, and adult	40–80	Omkar and Pervez, 2004; Srivastava and Omkar, 2003; Mbapila et al., 2002; Yonow et al., 2017
Gram pod borer	*Helicoverpa armigera*	Pigeonpea, chickpea, pea, tomato	Larva	81	Meena et al., 2021; Ramkumar et al., 2022, 2023
Leaf beetle	*Ophraella communa*	Ragweeds	Egg, larvae, and entire immature stage, development	75–90	Zhou et al., 2010; Fukano et al., 2016
Codling moth	*Cydia pomonella*	Apple, pear, and walnuts	Larvae	75	Lacey and Unruh, 1998; Salinas-Castro et al., 2014
Rice weevil	*Sitophilus oryzae*	Wheat, rice, barley, and maize	Growth and development	65	Beckett et al., 1994; Deepthi and Manjunatha, 2015
Red flour beetle	*Tribolium castaneum*	Black and green gram, cereals, flours, and dry fruits	Growth	65	Beckett et al., 1994; Kayode et al., 2014; Falah and Azher, 2020
Saw-toothed grain beetle	*Oryzaephilus surinamensis*	Legumes, oilseeds, and dry fruits	Growth and development	70	Beckett and Evans, 1994; Awadalla et al., 2021
Confused flour beetle	*T. confusum*	Cereals, flours, and dry fruits	Eggs	50	Falah and Azher, 2020; Kayode et al., 2014; Athanassiou et al., 2008
Flat grain beetle	*Cryptolestes pusillus*	Stored grain, beans, nuts, and oil seeds	Growth	90	Currie, 1967; Hou et al., 2019
Synanthropic mold mite	*Tyrophagus putrescentiae*	Stored products, plant seeds, and mushroom beds	Growth and development	85	Aspaly et al., 2007; Kheradmand et al., 2007; Malik et al., 2018
Capsanem nematode	*Steinernema carpocapsae*	EPNs (*Galleria mellonella*, *Agrotis ipsilon*, *Schistocerca gregaria*)	Juveniles	95	Lacey and Unruh, 1998; Lewis et al., 1996; San-Blas et al., 2012

TABLE 1.2 *(Continued)*

Insect Pest	Scientific Name	Host	Biological Processes	RH (%)	References
NemAttack Beneficial Nematode	*S. riobrave*	EPNs (*T. castaneum*, *Plodia interpunctella*)	Juveniles	95	Lacey and Unruh, 1998; Ramos-Rodríguez et al., 2007; Bastidas et al., 2014
Beneficial nematodes	*Heterorhabditis bacteriophora*	EPNs	Juveniles	95	Lacey and Unruh, 1998; Stefanovska et al., 2008; Bhat et al., 2019
Black soldier flies	*Hermetia illucens*	Organic waste material	Egg and adults	70	Tomberlin et al., 2002; Holmes et al., 2012; Joosten et al., 2020

insects become infested (Shipp et al., 2003). *B. bassiana* produces mycosis in *Hippodamia convergens*, a coccinellid, indicating that high humidity is necessary for conidial germination (James et al., 1998). An intensive period of exposure to high humidity is not required for the development of mycosis in this insect (Luz and Fargues, 1999).

1.2.4 EFFECT OF HUMIDITY ON FECUNDITY AND DEVELOPMENT OF INSECTS

The immediate impact of humidity can be observed in the regulation of the larvae and adults' activities, dispersal of the insects in a particular environment, how they behave, and the duration of growth, their ability to survive inclement weather, population genetics, etc. Climate has an indirect impact on the habitat where insects occur, including effects on plant growth, development, phenology, food quality, predators and parasitoids, and entomopathogen activity (Jaworski & Hilszczański, 2013). The effect of RH on various biological processes of pests is given in Table 1.2. In stored grain pests, *Trogoderma granarium*, *Tribolium castaneum*, and *Tribolium confusum* females laid the highest number of eggs at an RH of 50% as opposed to 20%, 30%, and 40% RHs. At an RH of 50%, the highest proportion of eggs hatched as well as the fastest rate of larval metamorphosis into pupa. However, lower RH (i.e., 20%) hinders the transition of larvae into pupae (Falah & Azher, 2020).

Humidity's effect on insect growth and development can be best understood by taking the stem borer example. One of the most significant insect pests damaging fields of maize and sorghum is *Chilo partellus* (Spotted stem borer). The present research has revealed its expansion to new areas where it was previously unreported (Tamiru et al., 2012). *Chilo partellus* reproduction fitness and population expansion are influenced by RH, temperature, and their interactions (Zulfiqar et al., 2010; Yadav & Chaudhary, 1987; Mbapila et al., 2002). However, Sithole (1987) showed that the distribution of the stem borer is influenced by humidity, rainfall, and temperature, with temperature being the most significant impact. The impact of RH on spotted stem borer development and fecundity revealed that RH is essential for life cycle development and environmental adaptation. All of *C. partellus'* developmental stages, including the egg, larva, pupa, and egg to adult phases, are impacted by RH levels (Tamiru et al., 2012).

RH and temperature have an adverse effect on the egg period of spotted stem borer (Srivastava, 2003). The period of eggs was significantly increased by 8 days at 40% RH at 22°C temperature, but due to a change in the RH to 60% and 80% at 22°C and 26°C, there is no significant difference in the egg period (Tamiru et al., 2012). The hatching period of eggs depended upon the RH and temperature, normally 3–8 days. The correlation between RH and temperature and the spotted stem borer's larval time revealed that at 40% RH and 30°C, the larval period was much shorter. However, at 40% RH and 22°C, the larval duration was greatly increased. There is no discernible difference between the larval phase at 80% and 60% RH and 30°C and 26°C (Pervez, 2004; Tamiru et al., 2012). The pupal period reduced at 40% relative humidity and 30°C temperature. However, between 60% to 80% RH, the pupal period did not change much within the temperature range of 26°C to 30°C. The pupal period was substantially longer at 22°C than it was at 26°C and 30°C (Tamiru et al., 2012). The duration of the cycle (egg to adult) will be increased at 22°C temperature compared to 26°C to 30°C temperature under 80%, 60%, and 40% RH. The developmental period (egg to adult) shortened from 60–62 days at 60% RH and 22°C to 26.5–26.7 days at 40% RH and 30°C temperature. The egg-adult period did not change considerably between the 26°C and 30°C temperature regimes at 60% and 80% RH. Except for the immature larval and pupal stages at 26°C, the developmental period of *C. partellus* was affected at 80% RH distinct from those of 40% and 60% RH at the other temperatures. At 30°C and 40% RH, the life cycle of *C. partellus* was shortened from 59.9 days at 22°C to 26.5 days. At 80% and 60% RH, a comparable shortening of the life cycle was seen when the temperature was raised. RH also affects the adult longevity of *C. partellus* (Yadav & Chaudhary, 1987; Rahim et al., 1991; Muegge & Lambdin, 1989; Ponsonby & Copland, 1996; Mbapila et al., 2002; Levesque et al., 2002; Emana, 2007).

1.2.5 EFFECT OF HUMIDITY ON STORED GRAIN PESTS

In *Dinoderus minutus* (Bamboo borer), RH has an effect on egg-hatching ability. High hatch rates at 40% and 75% RH and low hatch rates at 25% and 85% RH were observed. Moreover, RH influences the mortality of stored grain pests and increases their infestation. Low reproductivity and incubation period are caused by high RH (Ouedraogo et al., 1996). Optimum reproduction and growth conditions for pests are given in Tables 1.1 and 1.2.

In *D. minutus*, increasing the number of eggs from 3.5 to 57.70 with increasing RH from 20–85%. There is a correlation between the number of eggs laid and RH for female beetles but an inverse correlation between the lay period and RH for male beetles. Increasing RH in adult females is associated with better reproductive traits and since breathing is increased with movement, spiracles will lose water as well (Zachariassen, 1991). Emad et al. (2004) showed that 20–40% RH leads to dehydration and a lack of hatching, whereas 60% RH causes the eggs to hatch and to last longer during the larval stage in *Sitophilus oryzae*, a primary internal feeder of maize, wheat, barley, and rice grains. It is also possible that soil moisture may influence the time of egg hatching through direct effects on eggs or by altering the behavior of fertilized females during spawning. Before hatching, mole cricket eggs grow by absorbing water and enlarging by 25% (Regniere et al., 1981). Woodman (2017) determined the effect of soil, as well as the salts in the female fertilized insects, affected the laying of eggs, egg development, survival, egg hatching ratios, body weight, and survival. As sodium chloride concentrations increased above this limit, the eggs' size was reduced as a result of the water balance between the egg and the impurities associated with the aqueous activity. In *Trogoderma granarium*, *Tribolium confusum*, and *Tribolium castaneum*, RH levels affect all growth stages, with each female depositing 23.3, 14.0, and 13.0 eggs at 50% RH (Falah & Azher, 2020). At 20% and 30% RH, *T. confusum* and *T. castaneum* could not lay eggs. Larval metamorphosis into pupa could be prevented at 20% RH. Among the three insects, the greatest larval to pupal transition rates were 40%, 30%, and 26.6%, respectively (Falah & Azher, 2020).

By controlling insect damage on cereals with adequate drying and cooling, the market will be able to store commodities without pesticides and will be able to control insect resistance, especially to phosphine. The response of insects to temperatures at various RH within the range of 9–55°C has been investigated, including those that induce individual mortality, those at the threshold for population growth, and those where growth rates are slow. Drying is examined mainly in terms of improving the detrimental effects of temperature. About 10°C is the minimum threshold temperature for population growth in all of the insect species studied. For the better growth and development of *C. pusilloides*, 90% RH at 11°C temperature is required, but at 50% RH it can only complete development (Lefkovitch, 1964). Some other species, *Cryptolestes capensis* and *C. turcicus*, required the same RH (90%) and temperature (11–15°C) (Lefkovitch, 1962a).

Unlike *C. turcicus*, which cannot develop at 40% RH and only at 27.5°C at 50% RH (Lefkovitch, 1962b), *C. capensis* is capable of surviving at 30°C and 10% RH (Lefkovitch, 1962a), so drying or cooling might not be enough to control it. The major stored product beetles *C. ferrugineus*, *O. surinamensis*,

R. dominica, *Sitophilus granarius*, *S. oryzae*, and *T. castaneum* were wiped out at 13.5°C with 45% RH. 100% RH was killed at 9°C with 45% relative humidity. Aside from *R. dominica*, over 95% of immature stages died after 6 months at 13.5°C and 45% relative humidity. In addition, over 97% died at 9°C with 45% RH. Levels of immature mortality over 99% occurred at both temperatures after 3 months for *C. ferrugineus*, *O. surinamensis*, and *T. castaneum*. Observations made after 6 months at 13.5°C with 45% RH suggested that adults were more tolerant of moderately low temperatures than immatures. Nevertheless, after 3 and 6 months at 9°C with 45% RH, mortality levels reached 99% or more for all species. Adult mortality was also recorded at 70%. Relative humidity and moist conditions increased survival. *Sitophilus* species benefit from higher RH.

A substantial level of protection was seen at temperatures just above these thresholds. At conditions roughly 6°C below the threshold for population growth, over 99% mortality of major coleopteran species is possible after 9 months at 45% RH. Insect mortality at moderately elevated grain temperatures (35–55°C) is examined as an opportunity to disinfest grain during drying. Several coleopteran and psocopteran species suffered at least 99% mortality at 50°C after 2.5 hours. Bell (1983) suggested 100% mortality of diapausing larvae of *Ephestia calutella* (Hübner) is achieved at 60–70% RH within 96 hours at 40°C, 24 hours at 43°C, and 16 hours at 45°C. At 60% RH, total mortality of *Ephestia cautella* (Walker) and *Plodia interpunctella* (Hübner) pupae can be achieved within 2 hours at 50°C and 45°C, respectively (Arbogast, 1981). Beckett and Evans (1994) observed that the immature mortality of *O. surinamensis* was 46% at 20°C, 70% RH, but 99% at 20°C, 30% RH. However, Eaton and Kells (2009) showed that *Tyrophagus putrescentiae* can survive at a relative humidity as low as 45% up to 20°C with minimal mortality. This could allow them to migrate to more favorable conditions. The psocids as a group can be controlled at all temperatures if humidity is kept low. *Liposcelis brunnea* Motschulsky and *Liposcelis rufa* Broadhead cannot grow at 43% RH (Opit & Throne, 2009; Gautam et al., 2010).

1.2.6 EFFECT OF HUMIDITY ON DISEASE DEVELOPMENT

Plant diseases do not occur spontaneously. Successful manifestation of pathogens on plants requires three key components of the disease cycle to coincide at a specific time: a virulent pathogen, a susceptible host plant, and a conducive environment. In the absence of suitable environmental conditions, an otherwise susceptible plant species appears immune even in the presence of a virulent pathogen, which is actually termed an escape. The presence of moisture, temperature, light, oxygen, carbon dioxide levels, soil pH, and nutrients all play a significant role in the development of disease. Among environmental factors, humidity or moisture availability is the most important factor for pathogen contact, infection, proliferation, and spread. Environmental conditions affect the pathogen's ability to survive, as well as the density and capacity of inoculum, and the speed and distance of dissemination. Similar to other climatic variables, the quantity of moisture (humidity) has a significant influence on the occurrence and intensity of plant diseases in all their forms, including precipitation, soil moisture, and atmospheric relative humidity. Humidity directly affects sporulation by many fungi. There are a number of diseases caused by pathogens depending on the availability of moisture content in the microclimate, e.g., late blight, downy mildew, blast, powdery mildew, smut, etc.

1.2.7 IMPACT OF HUMIDITY ON OCCURRENCE AND DEVELOPMENT OF PLANT DISEASES

Environmental factors such as relative humidity have a broad impact on the survival, germination, expression, and delivery of virulence factors in pathogens. Relative humidity plays a critical role in driving actual outbreaks of fungal plant diseases. Additionally, leaf wetness can affect the production and transportation of fungal inoculum. High atmospheric humidity levels, typically above 85%, are conducive to disease development and infection. From all incidence reports of fungal outbreaks during the period of 2014–2019, 36.4% indicated that an increase in relative humidity may have provoked or facilitated the outbreak. Fungal genera like *Puccinia* and most species within the *Fusarium* genus, as well as the fungal-like oomycete genus *Phytophthora*, are frequently cited as causative agents of plant diseases that are influenced by moisture (Romero et al., 2022). Weather conditions have a significant impact on disease development and have been extensively researched as predictors of disease outbreaks. Among the various environmental factors, moisture (water availability) is considered the most crucial factor influencing disease outbreaks caused by fungi and bacteria. The term "moisture" encompasses rainfall (intensity and duration), relative humidity (RH), dew, and leaf wetness (intensity and duration). Most airborne pathogens, such as rusts and late blight of potatoes, require a period of free water on the leaf surface for spore germination and infection. The devastating epidemics of late blight of potatoes in Ireland during the 1840s were due to a combination of a susceptible host, an aggressive pathogen, and prolonged periods of warm, moist weather that kept leaf wetness at optimal conditions for pathogen infection. In Australia, wet spring weather significantly favors the development of cereal rusts. The period of leaf wetness, which varies depending on the temperature, can be utilized to predict outbreaks of brown rot of stone fruit caused by *Monilinia fructicola* in the spring (Keane and Kerr, 1997). Moisture affects disease development in the following ways: (i) As an important factor to make germination of resting spores of pathogens. (ii) Confers a turgid force that allows pathogens to enter host cells. (iii) Rain-water or irrigation makes it easier for inoculums to spread. (iv) Activates microbial pathogens so that they can infect the host by imparting a succulent surface to the host surfaces, thus making plants more susceptible to pathogen attack. (v) Rhizospheric moisture solubilizes dry salts and under moisture deficit, salt builds up, turning roots toxic and more prone to infection.

1.2.8 EFFECT OF HUMIDITY ON INCIDENCE AND SEVERITY OF DISEASE

Humidity plays a crucial role in both the incidence (i.e., the number or proportion of diseased plants in a population) and severity (i.e., the area or proportion of plant tissue that shows symptoms) of diseases. It mainly triggers disease progression by altering the infection process, spore germination, and spore dissemination (Damiri, 2011). For example, humidity appears to have a significant impact on the epidemiology of grapevine powdery mildew, with incidence and severity increasing as humidity approaches an optimum of around 85% RH. Additionally, conidial density and chain length were found to be proportional to humidity. Carroll and Wilcox (2003) observed a strong positive linear relationship between humidity levels and the frequency of conidium germination, with RH treatments of 84%. In the case of rice blast, there is a significant correlation between disease severity and RH levels recorded on varieties/lines such as F7, Basmati 2000, Basmati C622, M3, and KS 282. The severity of rice blast increases with an increase in RH (72–95%) (Shafaullah et al., 2011). The outbreak of powdery mildew in tomato (*Leveillula taurica*) epidemics is significantly impacted by the interplay of temperature and humidity. These environmental conditions can affect the germination, formation, release, and survival of spores, as well as mycelial development. Additionally, each process may require a unique optimal combination of temperature and humidity to occur (Reuveni and Rotem, 1974; Xu and Butt, 1998). Humidity clearly influences the development of tomato powdery mildew. Lower humidity (80% RH) is more conducive to tomato powdery mildew development. Optimum RH conditions for tomato powdery mildew development have been reported to lie between 60% and 80% RH (Corbaz, 1990; Hannig, 1996). Conidia of powdery mildew fungi contain about 72% moisture, which helps in the germination of conidiospores. The impact of humidity on the estimated number of conidia produced per leaf seems to depend more on its influence on the sporulating area rather than the sporulation density. Disease development and pathogen sporulation measures were found to be lowest in the lowest humidity treatments and increased proportionally with increasing humidity levels. The study on the effect of humidity on the development of powdery mildew in grape seedlings and the germination of *Uncinula necator* conidia *in vitro* revealed that an optimum temperature of 25 ± 2°C and an RH level ranging from 39 to 98% were most conducive for growth. The incidence and severity of powdery mildew increased with rising humidity, peaking at around 85% RH, and then appeared to level off or slightly decrease at higher values. All measures of humidity were equally effective in predicting germination responses, and conidium germination frequency responded to increased humidity, showing a modest but distinct linear increase from the minimum test level of approximately 10% RH to an optimum level of approximately 87% RH, beyond which the germination frequency dropped sharply (Carroll and Wilcox, 2003).

As late blight affects potatoes (*Phytophthora infestans*), it is most prevalent at RH levels of or close to 100%; sporangia lose viability after a few hours if RH is lower than 80% (Crosier, 1934). When dry conditions prevail and a water film is present on the surface of the leaves, sporangia grow into zoospores rather than germ tubes. If the weather is humid after infection, blighting occurs swiftly and results in the destruction of the entire foliage within two to three days. Blighting of leaves is limited to their margins if the RH is low. Most fungi require >75% humidity for their efficient functioning. At constant humidities ranging from 75% to 95.5% and a temperature of 25°C, mycosis mortality caused by the mycoinsecticide *Beauveria bassiana* on first instar *Rhodnius prolixus* nymphs did not exceed 50% within ten days of treatment. However, at higher humidities of 96.5%, 98.5%, and 100%, all insects died during a 10-day incubation period (Luz and Fargues, 1999). Rice false smut (RFS), caused by *Villosiclava virens* (Anamorph: *Ustilaginoidea virens*), is prevalent on rice panicles worldwide, and primary infection success is dependent on growth stage, temperature, RH, and wetness duration, which have been extensively studied. A temperature of 16°C was found to induce higher disease incidence compared to other temperatures, and 5 days of wetness at 100% RH was determined to be the optimal wetness duration for infection. The highest RFS incidence occurred at 25°C and 95% RH for 120 hours of wetness, while the lowest incidence (5%) was observed in plants exposed to 35°C for 120 hours of wetness (Jia et al., 2014).

The significance of leaf wetness for ascospore germination and infection has been demonstrated through experiments on the incidence of Sclerotinia disease in field-grown lettuce caused by *Sclerotinia sclerotiorum* (Abawi and Grogan, 1979; Tu, 1989), and the relationship between leaf wetness duration and infection levels has been demonstrated for soybean, white bean, and dry bean (Boland and Hall, 1988; Phillips, 1994). Eggplant and oilseed rape have also been found to be susceptible to infection and disease development due to prolonged exposure to high relative humidity (Heran et al., 1999; Tores and Moreno, 1991). The impact of varying levels of RH on Sclerotinia disease development was evident, as the initial symptoms were observed much quicker at 100% RH, appearing within 5 to 7 days across the three experiments, followed by symptoms at 80–90% RH after 5–17 days, and at 50–70% RH after 8–21 days, with a final disease incidence of 33–76%. Additionally, disease progress was faster at 15–27°C compared to 5–10°C (with 85% RH) (Clarkson et al., 2014). Purple blotch (*Alternaria porri*) can cause significant damage to onion crops in many countries, such as India, and the disease development is influenced by the components of the host, pathogen, and environment; a decrease in the density of inoculum led to a late appearance and poor development of the disease, while an increase in relative humidity resulted in an increase in disease incidence and severity, and a decrease in the incubation period, with the highest disease development (75.1%) and shortest incubation period (4.93 days) observed in plants kept under high RH for 120 hours (Gupta and Pathak, 1986). Phoma black stem and leaf spot (*Phoma medicaginis*) and leaf spot (*Leptosphaerulina trifolti*) are commonly found in lucerne crops, and there is a strong increasing trend in leaf disease severity with an increase in periods of high humidity. Significantly higher disease severity can be observed at a 168-hour period of high humidity compared to 96 hr, which in turn had more severe disease compared to 48 hours, as reported by Barbetti in 1991. Similarly, *Cercosporidium personatum*, the causal agent of leaf spot in peanuts, is an important disease of the crop. Controlled experimental conditions have revealed that *C. personatum* requires a minimum of 4 hr of RH ≥95% per day for conidial production, and the highest numbers of conidia were produced when lesions were subjected to daily periods of ≥16hr of RH ≥95%. However, the optimum temperature for the production of spores of *C. personatum* is around 20°C (Alderman et al., 1994).

1.2.9 EFFECT OF HUMIDITY ON MITES' DEVELOPMENT, OCCURRENCE AND THEIR INTERACTION WITH CROPS

Mites are the smallest organisms that belong to the phylum Arthropoda, class Arachnida, subclass Acari, and order Acarina. Worldwide, mites have an impact on human existence, either directly or indirectly. Mites can survive in every environment on Earth (Shukla, 2021). The productivity of cultivated crops is substantially reduced by plant-feeding mites, also known as phytophagous mites; they are vectors of several plant diseases. Phytophagous mites are significant pests in a variety of agricultural crops all over the world (Jeppson et al., 1975; Bolland et al., 1998; Torres et al., 2023). They affect nearly all cultivated crops, decorative crops, and wild plants. After the development and wide utilization of toxic chemicals (pesticides) in agriculture during the past few decades, the problem of phytophagous mites grew more severe (Shukla, 2021). Mites cause severe economic damage to food or feed commodities, and fiber, if not kept below the damage threshold level. There are several distinct families of mite species, i.e., Tetranychidae (spider mites), Eriophyidae (rust mites and gall), Tenuipalpidae (false spider mites), and Tarsonemidae (tarsonemid mites), which are economically important because they cause damage to crops directly by feeding or by transmitting disease-causing agents (fungi, bacteria, and viruses) (Van Leeuwen et al., 2010a; Jeppson et al., 1975; Van Leeuwen et al., 2010b). The red spider mite species *Panonychus citri*, *Panonychus ulmi*, and *Tetranychus urticae* are measured as major infesting pests of agricultural crops. Whereas *P. citri* and *P. ulmi* are pests of apple and citrus, *T. urticae* has an extremely broad host range with 1000 or more plant species from over 250 families of plants (Migeon and Dorkeld, 2014). The host plants include major field and legume crops, vegetables, fruits, as well as corn, soybean, and cotton (Karban and Carey, 1984; Ryan, 1990; Kant et al., 2004; Ament et al., 2004; Sarmento et al., 2011; Migeon and Dorkeld, 2014). Among invertebrate pests, red spider mites are infamous for having a high potential for developing pesticide resistance. Because of their short lifespan and high fertility, controlling them necessitates using acaricides often, which quickly builds resistance (Van Leeuwen et al., 2010b, 2015).

Eriophyoid mites are obligatory parasites that feed on a variety of plant organs and can occasionally mimic viral diseases, nutritional deficiencies, and physiological problems in their symptoms (Van Leeuwen et al., 2010a). Several of these mites are considered a quarantine issue in many nations because they can spread viruses and other plant infections that have major effects on agricultural productivity (Lindquist et al., 1996; Oldfield and Proeseler, 1996; Duso et al., 2010). Therefore, the second most economically significant mite species are eriophyoid mites following tetranychid mites (Royalty and Perring, 1996; Van Leeuwen et al., 2010a). Crop pests include eriophyid mites. Across the world, seven species of eriophyoid mites are associated with citrus, with the citrus bud mite *Aceria sheldoni*, pink citrus rust mite *Aculus pelekassi* (Keifer), and citrus rust mite *Phyllocoptruta oleivora* (Ashmead) being the most damaging (McCoy, 1996; Bergh et al., 1999; Hall et al., 2007, 2005). The apple rust mite *Aculus schlechtendali* also belongs to the eriophyid mite group, causing significant economic damage (Ciampolini et al., 1976; Jeppson et al., 1975; Easterbrook and Fuller, 1986; Li and Cai, 1996; Easterbrook, 1996).

The mites are economically important to the agriculturist because they are used as biocontrol agents for phytophagous mites and other insect pests (Banks, 1917; Evans, 1952; Koehler, 1999; Knapp et al., 2018). Predatory mites are the main biological control agents used in commercial augmentation. They are mostly employed to manage phytophagous insects (thrips and whiteflies) and mites in protected vegetable and ornamental production systems. The mite family Phytoseiidae is the most commercially adopted biocontrol agent with 20 species worldwide. Among them, some important species are *Amblyseius swirskii*, *Neoseiulus californicus*, *Neoseiulus cucumeris*, and *Phytoseiulus persimilis*, collectively accounting for around two-thirds of the market for arthropod biocontrol agents (Bolckmans, 1999; Nomikou et al., 2001; Calvo et al., 2015). Less often utilized are soil predatory mites such as *Macrocheles robustulus* (Macrochelidae) or *Stratiolaelaps scimitus* (Laelapidae) for the control of thrips pupae and sciarid fly larvae (De Azevedo et al., 2015; Moreira and Moraes, 2015). Predatory mites, such as the chicken red mite, *Dermanyssus gallinae*, can be utilized to manage animal parasites in addition to plant pests. The predatory mites *Cheyletus eruditus* (Cheyletidae) and *Androlaelaps casalis* (Laelapidae) have been successfully applied in conjunction with laying hen stables with positive results (Knapp et al., 2018). Certain species, such as mites, which are typically restricted to environments above 65% RH, and numerous important Psocid species above 60% RH, are particularly sensitive to lower amounts of environmental moisture (Beckett, 2011). Mites produce more eggs and live longer when they oviposit in a dry environment compared to a nearly saturated one. In a moist environment, newly hatched mites have a limited chance of survival. In a dry environment, an organism can consume and utilize more food by eliminating body fluids through evaporation from the cuticle. If a high RH is maintained in the greenhouse, the possibility to manage the spider mites increases. Ovipositing mites are unable to feed when body fluids do not evaporate, which is caused by high RH. As a result, maintaining extremely high relative humidity can prevent spider mite damage in greenhouse crops tolerant of water-saturated conditions, especially in areas where spider mites are resistant to acaricides (Boudreaux, 1958).

Sanchez-Ramos and coworkers (2007) examined the three species of storage mites – *Acarus farris*, *Tyrophagus putrescentiae*, and *T. neiswanderi*, at 70%, 80%, and 90% RH under a similar temperature of 25°C to determine how quickly immature stages developed, how quickly adults reproduced, and for the development of management practices. The immature stage does not reach the protonymph stage at 70% RH, with 100% larval mortality of *A. farris* and *T. putrescentiae*. *T. neiswanderi* could not fully develop. In all three species, there was a significant increase in developmental time with lower RH. These three storage mites are highly vulnerable to RH reductions, to the point that no mite development could occur at 70% RH. *Phyllocoptes fructiphilus*, a small eriophyid mite, transmits rose rosette disease, a major disease of roses caused by the rose rosette virus (Laney et al., 2011; Amrine et al., 1988; Allington et al., 1968). Arthropod biological control processes, including oviposition, reproduction, survival, and other processes, are affected by variations in RH. When RH was high (95%) compared to low (20%), the frequency and severity of symptoms associated with rose rosette disease differed significantly. There were no signs of rose rosette disease on rose plants grown at 21°C and 40% relative humidity in the greenhouse without exposure to *P. fructiphilus* or the virus that causes rose rosette disease.

1.2.10 EFFECT OF HUMIDITY ON NEMATODES

Nematodes have the highest population density of any group of multicellular creatures on Earth (number of individuals), even though there are more than 4,100 species of plant-parasitic nematodes that are known (Mitiku, 2018; Decraemer and Hunt, 2006; Elling, 2013). When cropping patterns change, formerly considered non-damaging species are turning into major pests while others are constantly being described (Nicol, 2002). The economically significant phytoparasitic nematodes may be divided into a small number of highly specialized subgroups that either directly harm their hosts or serve as virus carriers. There are around 50 million nematodes per square meter of 30 cm deep soil that is moderately productive. Approximately 50% of nematode species are marine, 25% are soil-dwelling free-living nematodes, 15% are parasites of animals and people, and 10% are parasites of plants (Mitiku, 2018). Nematode-related agricultural production losses were projected to be 14.6% in subtropical and tropical climates, compared to 8.8% in developed nations (Sasser and Carter, 1985).

Plant parasitic nematodes are small organisms that severely infest the root system of the plant by creating specific feeding sites on the vascular tissue of the plants (e.g., syncytia by cyst nematodes, giant cells by root-knot nematodes). Migratory endoparasites feed on cortical tissues, while reniform nematodes feed on nurse cells, putting the soil's ability to provide nutrients at risk and leading to restricted growth (Jones et al., 2013). Due to their status as obligatory biotrophs, PPNs promote a strong host-pathogen relationship in which their esophageal gland secretions play a crucial role. This allows PPNs to significantly alter cells of the host plant for their own benefit (Berg et al., 2009). PPNs must be exposed to several biotic and abiotic factors in order to survive, reproduce, and increase their populations. These factors include a continuous availability of favorable hosts with little genetic variation, minimal exposure to extreme environmental conditions, and a predominance of favorable temperature and moisture (Greco and Esmenjaud, 2004). PPNs that typically experience diapause in natural settings adapt to the greenhouse environment to the extent that they no longer respond to triggers that induce diapause (Phani et al., 2021; Sabir and Walia, 2017; Walia and Walia, 2014).

Juveniles of entomopathogenic nematodes (*Rhabditia*, *Steinernematidae*, and *Heterorhabditidae*) that are infected may endure unfavorable environmental circumstances by staying in the host carcass for up to 50 days. Depending on the environmental factors that the cadaver is exposed to, different species have different chances of surviving. *Galleria mellonella* larvae were subjected to RH levels of 75, 85, 96, or 100% after being infested at 25°C with more than one nematode (Brown and Gaugler, 1997). *Steinernema carpocapsae* and *S. glaseri* were examined for survival and pathogenicity under laboratory conditions that were impacted by RH, soil temperature, and moisture (Kung et al., 1991). In a controlled setting, RH is a crucial element for nematode persistence (Kamionek et al., 1974), is additionally significant in soil conditions, and has an impact on nematode survival (Simons, 1973). It is difficult, however, to control these two steinernematid nematodes due to the high relative humidity in soil pores, which is generally close to 100% (Croney et al., 1952).

1.2.11 EFFECT OF HUMIDITY ON SURVIVAL AND PATHOGENICITY OF NEMATODES

The pathogenicity and survival of *S. glaseri* and *S. carpocapsae* will be affected by the RH increase or decrease. They survived 32 days at 100% RH, but at 25% RH, *S. glaseri* will survive only 4 hours and *S. carpocapsae* for 2 days. Over the course of the 32-day test period, there will be noticeable variations in the survival of *S. glaseri* and *S. carpocapsae* at the six ranges of relative humidity (RH). During the first 24 hours, *S. carpocapsae* survival reduced when RH dropped from 100 to 25%. After 32 days at 100% RH, the survival rate stayed at 50%. Nematodes were still alive after 2 days at 50% and 25% RH, 16 days at 97% RH, 4 days at 75% RH, and 8 days at 90% RH. Likewise, as RH dropped from 100 to 90% in the first 24 hours, *S. glaseri* survival was reduced. Nematodes endured at 75% RH for 12 hours and at 50% and 25% RH for 4 hours. Nematodes endured at 97% RH for 16 days and at 90% for 4 days. While *S. glaseri* lived for at least 4 hours, *S. carpocapsae* survived for at least 2 days in all tested RH. Again, the pathogenicity outcomes and nematode survival showed a strong correlation. *S. carpocapsae* completely eliminated larval mortality at 100% RH for over 32 days, while nematodes still exhibited 80% *G. mellonella* mortality. Larval mortality varies with percentage RH; at 97% RH, it takes 32 days, and at 90% RH, it takes 16 days. Studies on quick and progressive desiccation under controlled environmental circumstances have mostly focused on *S. carpocapsae* survival. The *S. carpocapsae* DD-136 strain did not survive three hours at 26% RH (Schmiege, 1963), and *S. carpocapsae* perished after 102 hours at 30°C and after 36 hours at 5°C when the RH was 85% (Kamionek et al., 1974). Simons and Poinar (1973) studied *S. carpocapsae* survival by predicting the effects of progressive desiccation and found that > 90% survived after 12 days at 79.5% RH and after 4 days at 48.4% RH. *S. carpocapsae* survived longer under progressive desiccation than under fast desiccation, perhaps because cellular integrities are not compromised. *S. glaseri* and *S. carpocapsae*'s survival and pathogenicity declined as the RH decreased. *S. glaseri* was more susceptible to low RH conditions and less persistent than *S. carpocapsae*. *S. glaseri* tends to travel lower, which suggests that it lives in the deeper soil profile (Georgis and Poinar, 1983). As a result, infectious juveniles are more likely to endure progressive desiccation and adapt to it.

1.2.12 EFFECT OF HUMIDITY ON THE EMERGENCE OF NEMATODES

Low RH resulted in noticeable support for *H. bacteriophora* and *S. feltiae* emergences, while *S. glaseri* and *S. carpocapsae* emergences were unaffected.

In a second experiment, *G. mellonella* were given nematode infections at 25°C and then either remained at that temperature or were moved to 5, 10, or 15 °C, all while maintaining 100% RH (Brown and Gaugler, 1997). Reduced RH had no impact on the emergence of cadavers infected with *S. glaseri* or *S. carpocapsae* but significantly reduced the emergence of cadavers infected with *H. bacteriophora* and *S. feltiae*. When RH varied from 100 to 85%, the proportion of cadavers from which *H. bacteriophora*, *S. carpocapsae*, and *S. glaseri* emerged did not significantly differ across species. At 96% RH but not at 85% RH, *S. feltiae* developed from many fewer cadavers than the other species. When *S. feltiae* from cadavers that had not emerged at 96% and 85% RH were dissected, it was found that all but one of them contained nematodes and that all but one had less than 5% nematode survival. While *S. glaseri* and *S. carpocapsae* do not exhibit a substantial drop in emergence at 75% RH, *S. feltiae* and *H. bacteriophora* do. More than 75% of the nematodes in the *S. carpocapsae* cadavers had survived, compared to fewer than 25% in *S. glaseri*, less than 5% in *S. feltiae*, and none in *H. bacteriophora* (Brown and Gaugler, 1997). In the cadaver at 75% RH and 15 °C, all strains of *S. carpocapsae* lasted approximately 50 days. Other experiments found that just 50% of infective free juveniles of all strains could survive 8 days of desiccation on 1% agarose with 80% RH outside the corpse (Womersley, 1990). The *S. carpocapsae* agriotos strain will survive at 79.5% RH and 20°C for 28 days outside the cadaver (Simons and Poinar, 1973; Moore, 1965).

1.2.13 EFFECT OF HUMIDITY ON MORTALITY OF THE EPNS

1.2.13.1 INFECTED LARVAL OF CODLING MOTH

An experiment was conducted to test the effect of RH on codling moth mortality by EPNs. Each nematode species' suspensions were applied to surfaces before being incubated at 25°C and 20–25% or 75% RH in uncovered Petri dishes. Larvae exposed to 10 *S. carpocapsae* infective juveniles/cm^2 reacted with 4.7 to 76.4% mortality when incubated at 95% RH for 0–8 hours. Mortality and the amount of time an infection was incubated at 95% RH were highly correlated. The death rate increased most dramatically between 1 and 2 hours of incubation, although there was a constant rise in mortality for each hour added to the incubation time at 95% RH. After 8 hours of incubation, the death rate was approaching that of 24 hours at 95% RH. The average mortality rate for codling moth larvae over 11 test dates with *S. carpocapsae* at 25°C, 10 infected juveniles/cm^2, and 24 hours of incubation in 95% RH was 83.6% compared to 2.6% of controls.

The suppression of codling moth diapausing larvae was substantially more successful with *Steinernema carpocapsae* than with the other two nematode species. When compared to control emergence, the emergence of codling moth adults was reduced by 83.4%, 31.2, and 43.4%, respectively, in the *S. carpocapsae*, *S. riobrave*, and *H. bacteriophora* treatments (Lacey and Unruh, 1998).

1.3 CONCLUSION

Humidity is an essential climatic factor for the growth and development of diseases and insect pests. Every insect and its species will require specific related humidity for its proper function. RH can affect the metabolism, fecundity, incidence, feeding habits, and the emergence of different pests and diseases on crops. It affects all living organisms on Earth. RH may change to alter an organism's life cycle; it varies from insect to insect or organism to organism. Nowadays, humidity can be used as a tool in IPM programs because modification in the RH at particular developmental stages (susceptible stage) can play a role in the management of insect pests and diseases.

KEYWORDS

- **atmosphere**
- **entomopathogenic agents**
- **environmental factor**
- **humidity**
- **incubation period**
- **lesion growth**
- **mating**
- **moisture**
- **molting**
- **mortality rates**
- **Rh levels**
- **spore germination**
- **water vapor**

REFERENCES

Abawi, G. S., & Grogan, R. G. (1979). Epidemiology of diseases caused by Sclerotinia species. *Phytopathology, 69*, 899–904.

Abd Elgawad, M. M. M. (2017). Status of entomopathogenic nematodes in integrated pest management strategies in Egypt. In M. M. M. Abd-Elgawad, T. H. Askary, & J. Coupland (Eds.), *Biocontrol Agents: Entomopathogenic and Slug Parasitic Nematodes* (pp. 473–501). CAB International.

Alderman, S. C., & Nutter Jr, F. W. (1994). Effect of temperature and relative humidity on development of *Cercosporidium personatum* on peanut in Georgia. *Plant Disease, 78*, 690–694.

Allington, W. B., Staples, R., & Viehmeyer, G. (1968). Transmission of rose rosette virus by the eriophyid mite *Phyllocoptes fructiphilus*. *Journal of Economic Entomology, 61*, 1137–1140.

Ament, K., Kant, M. R., Sabelis, M. W., Haring, M. A., & Schuurink, R. C. (2004). Jasmonic acid is a key regulator of spider mite-induced volatile terpenoid and methyl salicylate emission in tomato. *Plant Physiology, 135*(4), 2025–2037.

Amrine Jr, J. W., Hindal, D. F., Stasny, T. A., Williams, R. L., & Coffman, C. C. (1988). Transmission of the rose rosette disease agent to *Rosa multiflora* by *Phyllocoptes fructiphilus* (Acari: Eriophyidae). *Entomological News, 99*, 239–252.

Anderson, J. F. (1974). Responses to starvation in the spiders Lycosa lenta Hentz and *Filistata hibernalis* (Hentz). *Ecology, 55*(3), 576–585.

Ando, Y. (1974). Egg diapause and ambient oxygen tension in the false melon beetle, *Atrachya menetriesi* Faldermann (Coleoptera: Chrysomelidae). *Applied Entomology and Zoology, 9*(4), 261–270.

Arbogast, R. L. (1981). Mortality and reproduction of *Ephestia cautella* and *Plodia interpunctella* exposed as pupae to high temperatures. *Environmental Entomology, 10*, 708–711.

Aspaly, G., Stejskal, V., Pekár, S., & Hubert, J. (2007). Temperature-dependent population growth of three species of stored product mites (Acari: Acaridida). *Experimental and Applied Acarology, 42*, 37–46.

Athanassiou, C. G., Palyvos, N. E., & Kakouli-Duarte, T. (2008). Insecticidal effect of (Filipjev) (Nematoda: Steinernematidae) against *Tribolium confusum* du Val (Coleoptera: Tenebrionidae) and *Ephestia kuehniella* (Zeller) (Lepidoptera: Pyralidae) in stored wheat. *Journal of Stored Products Research, 44*(1), 52–57.

Awadalla, H. S., Guedes, R. N. C., & Hashem, A. S. (2021). Feeding and egg-laying preferences of the sawtoothed grain beetle *Oryzaephilus surinamensis*: Beyond cereals and cereal products. *Journal of Stored Products Research, 93*, 101841.

Ayres, J. S., & Schneider, D. S. (2009). The role of anorexia in resistance and tolerance to infections in Drosophila. *PLoS Biology, 7*(7), 1000150.

Ayres, M. P., & Lombardero, M. J. (2000). Assessing the consequences of global change for forest disturbance from herbivores and pathogens. *Science of the Total Environment, 262*(3), 263–286.

Bachmetjew, P. (1907a). Experimentelle entomologische Studien vom physikalisch-chemischen Standpunkt aus. *Wilhelm Engelmann, 2*, 574.

Bachmetjew, P. (1907b). Experimentelle entomologische Studien vom physikalisch-chemischen Standpunkt aus. *Wilhelm Engelmann, 2*, 688.

Bale, J. S. (2002). Insects and low temperatures: From molecular biology to distributions and abundance. *Philosophical Transactions of the Royal Society of London. Series B: Biological Sciences, 357*(1423), 849–862.

Banks, N. (1917). New mites, mostly economic (Araeh., Acar.). *Entomological News, 28*(5), 193–199.

Barbetti, M. J. (1991). Effects of temperature and humidity on diseases caused by *Phoma medicaginis* and *Leptosphaerulina trifolii* in lucerne (*Medicago sativa*). *Plant Pathology, 40*, 296–301.

Barragan-Fonseca, K. B., Dicke, M., & van Loon, J. J. (2017). Nutritional value of the black soldier fly (*Hermetia illucens* L.) and its suitability as animal feed–A review. *Journal of Insects as Food and Feed, 3*(2), 105–120.

Bastidas, B., Portillo, E., & San-Blas, E. (2014). Size does matter: The life cycle of Steinernema spp. in micro-insect hosts. *Journal of Invertebrate Pathology, 121*, 46–55.

Beckett, S. J. (2011). Insect and mite control by manipulating temperature and moisture before and during chemical-free storage. *Journal of Stored Products Research, 47*, 284–292.

Beckett, S. J., & Evans, D. E. (1994). The demography of *Oryzaephilus surinamensis* (L.) (*Coleoptera: Silvanidae*) on kibbled wheat. *Journal of Stored Products Research, 30*(2), 121–137.

Beckett, S. J., Longstaff, B. C., & Evans, D. E. (1994). A comparison of the demography of four major stored grain coleopteran pest species and its implications for pest management. In E. Highley, E. J. Wright, H. J. Banks, & B. R. Champ (Eds.), *Proceedings of the 6th International Working Conference on Stored-Product Protection, Vol. 1* (pp. 491–497). CAB International.

Bell, C. H. (1983). Effect of high temperatures on larvae of *Ephestia cautella* (Lepidoptera: Pyralidae) in diapause. *Journal of Stored Products Research, 19*, 153–157.

Bennett, V. A., Sformo, T., Walters, K., Toien, O., Jeannet, K., Hochstrasser, R., Pan, Q., Serianni, A. S., Barnes, B. M., & Duman, J. G. (2005). Comparative overwintering physiology of Alaska and Indiana populations of the beetle *Cucujus clavipes* (Fabricius): Roles of antifreeze proteins, polyols, dehydration, and diapause. *Journal of Experimental Biology, 208*(23), 4467–4477.

Berg, R. H., Fester, T., & Taylor, C. G. (2008). Development of the root-knot nematode feeding cell. In *Plant Cell Monographs* (Vol. 15, pp. 115–152).

Bergh, J. C., Rugg, D., Jansson, R. K., McCoy, C. W., & Robertson, J. L. (1999). Monitoring the susceptibility of citrus rust mite (Acari: Eriophyidae) populations to abamectin. *Journal of Economic Entomology, 92*(4), 781–787.

Bhat, A. H., Askary, T. H., Ahmad, M. J., & Chaubey, A. K. (2019). Description of *Heterorhabditis bacteriophora* (Nematoda: Heterorhabditidae) isolated from hilly areas of Kashmir Valley. *Egyptian Journal of Biological Pest Control, 29*(1), 1–7.

Bisane, K. D., Wadaskar, R. M., & Deotale, R. O. (2013). Tritrophic interaction of *Helicoverpa armigera* (Hubner) in major pulses ecosystem. *Legume Research, 36*(2), 116–124.

Boland, G. J., & Hall, R. (1988). Numbers and distribution of apothecia of *Sclerotinia sclerotiorum* in relation to white mold of white bean (*Phaseolus vulgaris*). *Canadian Journal of Botany, 66*(2), 247–252.

Bolckmans, K. J. F. (1999). Commercial aspects of biological pest control in greenhouses. In R. Albajes, M. Lodovica Gullino, J. C. van Lenteren, & Y. Elad (Eds.), *Integrated Pest and Disease Management in Greenhouse Crops* (pp. 310–318). Springer. (Developments in Plant Pathology, Vol. 14).

Bolland, H. R., Gutierrez, J., & Flechtmann, C. H. (1998). *World Catalogue of the Spider Mite Family (Acari: Tetranychidae)*. Brill.

Boudreaux, H. B. (1958). The effect of relative humidity on egg-laying, hatching, and survival in various spider mites. *Journal of Insect Physiology, 2*(1), 65–72.

Brower, L. P. (1995). Understanding and misunderstanding the migration of the monarch butterfly (Nymphalidae) in North America: 1857–1995. *Journal of the Lepidopterists' Society, 49*(4), 304–385.

Brown, I. M., & Gaugler, R. (1997). Temperature and humidity influence emergence and survival of entomopathogenic nematodes. *Nematologica, 43*(5), 363–375.

Busvine, J. R. (1966). *Insects and Hygiene: The Biology and Control of Insect Pests of Medical and Domestic Importance* (2nd ed.). Methuen.

Buxton, P. A. (1930). Evaporation from the meal-worm (*Tenebrio*: Coleoptera) and atmospheric humidity. *Proceedings of the Royal Society of London. Series B, Biological Sciences, 106*(747), 560–577.

Buxton, P. A. (1932). Terrestrial insects and the humidity of the environment. *Biological Reviews, 7*(4), 275–320.

Calvo, F. J., Knapp, M., van Houten, Y. M., Hoogerbrugge, H., & Belda, J. E. (2015). *Amblyseius swirskii*: What made this predatory mite such a successful biocontrol agent. *Experimental and Applied Acarology, 65*(4), 419–433.

Campbell, J. F. (2006). Stored-product insect behavior. In *Insect Management for Food Storage and Processing* (2nd ed.). American Association of Cereal Chemists Press.

Carroll, J. E., & Wilcox, W. F. (2003). Effects of humidity on the development of grapevine powdery mildew. *Phytopathology, 93*(9), 1137–1144.

Chander, S. (1998). Infestation of root and foliage/ear head aphids on wheat in relation to predators. *Indian Journal of Agricultural Sciences, 68*(11), 754–755.

Chen, Z., Clancy, K. M., & Kolb, T. E. (2003). Variation in budburst phenology of Douglas-fir related to western spruce budworm (Lepidoptera: Tortricidae) fitness. *Journal of Economic Entomology, 96*(2), 377–387.

Ciampolini, M., Rota, P. A., & Schulthaus, S. D. (1976). Apple rust caused by the eriophyid *Aculus schlechtendali*. *Informatore Agrario, 32*(40), 24243–24245.

Clarkson, J. P., Fawcett, L., Anthony, S. G., & Young, C. (2014). A model for *Sclerotinia sclerotiorum* infection and disease development in lettuce, based on the effects of temperature, relative humidity and ascospore density. *PLoS One, 9*(4), e94049.

Corbaz, R. (1990). L'oidium de la tomate, une maladie nouvelle en Suisse. *Revue Suisse de Viticulture, Arboriculture, Horticulture, 22*(3), 159–161.

Croney, D., Colman, J. D., & Bridge, P. M. (1952). The suction of moisture held in soil and other porous materials. *Technical Paper Road Research Laboratory, 24*, 1–40.

Crosier, W. (1934). Studies in the biology of *Phytophthora infestans* (Mont) de Bary. *Studies in the Biology of Phytophthora Infestans (Mont) de Bary, 155*.

Currie, J. E. (1967). Some effects of temperature and humidity on the rates of development, mortality and oviposition of *Cryptolestes pusillus* (Schönherr) (Coleoptera, Cucujidae). *Journal of Stored Products Research, 3*(2), 97–108.

Daebeler, F., & Hinz, B. (1977). Compensation of aphid injury to sugar beet by means of supplementary rain. *Archiv für Phytopathologie und Pflanzenschutz, 13*(3), 199–205.

Damiri, N. (2011). Climate change, environment and plant diseases development. *Plant Archives, 11*(1), 200–205.

Danks, H. V. (2005). Key themes in the study of seasonal adaptations in insects I. Patterns of cold hardiness. *Applied Entomology and Zoology, 40*(2), 199–211.

Danks, H. V. (2007). The elements of seasonal adaptations in insects. *The Canadian Entomologist, 139*(1), 1–44.

Danks, H. V. (2013). Measuring and reporting life-cycle duration in insects and arachnids. *European Journal of Entomology, 97*(3), 285–303.

Dautel, H. (1999). Water loss and metabolic water in starving *Argas reflexus* nymphs (Acari: Argasidae). *Journal of Insect Physiology, 45*(1), 55–63.

De Azevedo, L. H., Emberson, R. M., Esteca, F. D. C., & de Moraes, G. J. (2015). Macrochelid mites (Mesostigmata: Macrochelidae) as biological control agents. In D. Carrillo, G. J. de Moraes, & J. E. Peña (Eds.), *Prospects for Biological Control of Plant Feeding Mites and Other Harmful Organisms* (pp. 103–132). Springer. (Progress in Biological Control, Vol. 19).

Decraemer, W., & Hunt, D. J. (2006). Structure and classification. In *Plant Nematology* (pp. 3–32). CABI.

Deepthi, N., & Manjunatha, M. (2015). Comparative development of rice weevil, *Sitophilus oryzae* (L.) in stored split legumes. *Eco-Friendly Agriculture Journal, 10*(2), 180–183.

Diener, S., Zurbrügg, C., Gutiérrez, F. R., Nguyen, D. H., Morel, A., Koottatep, T., & Tockner, K. (2011). Black soldier fly larvae for organic waste treatment-prospects and constraints. In *Proceedings of 2nd International Conference on Solid Waste Management in Developing Countries, WasteSafe* (Vol. 2, pp. 13–15). Khulna, Bangladesh.

Duso, C., Castagnoli, M., Simoni, S., & Angeli, G. (2010). The impact of eriophyoids on crops: Recent issues on *Aculus schlechtendali, Calepitrimerus vitis* and *Aculops lycopersici*. *Experimental and Applied Acarology, 51*(1–2), 151–168.

Easterbrook, M. A. (1996). Damage and control of eriophyoid mites in apple and pear. In E. E. Lindquist, M. W. Sabelis, & J. Bruin (Eds.), *World Crop Pests* (Vol. 6, pp. 527–541).

Easterbrook, M. A., & Fuller, M. M. (1986). Russeting of apples caused by apple rust mite *Aculus schlechtendali* (Acarina: Eriophyidae). *Annals of Applied Biology, 109*(1), 1–9.

Eaton, M., & Kells, S. A. (2009). Use of vapor pressure deficit to predict humidity and temperature effects on the mortality of mold mites, *Tyrophagus putrescentiae*. *Experimental and Applied Acarology, 47*(3), 201–213.

Elling, A. A. (2013). Major emerging problems with minor *Meloidogyne* species. *Phytopathology, 103*(11), 1092–1102.

Emad, A., Sulaf, A., & Muhammad, P. (2004). The effect of low temperatures and different levels of relative humidity on the life of *Chrysomya bezziana* (Villeneuve) (Diptera: Calliphoridae). *Umm Salamah Journal of Science, 1*(1), 2004.

Emana, G. D. (2007). Comparative studies of the influence of relative humidity and temperature on the longevity and fecundity of the parasitoid, *Cotesia flavipes*. *Journal of Insect Science, 7*(1), 19.

Evans, G. O. (1952). On a new predatory mite of economic importance. *Bulletin of Entomological Research, 43*(2), 397–401.

Falah, A. S., & Azher, M. A. (2020). Effect of different levels of relative humidity and impurities in three stored insects. *Plant Archives, 20*(1), 257–261.

Fränzle, O. (2003). Bioindicators and environmental stress assessment. In B. A. Markert, A. M. Breure, & H. G. Zechmeister (Eds.), *Trace Metals and Other Contaminants in the Environment* (Vol. 6, pp. 41–84). Elsevier.

Fukano, Y., Doi, H., Thomas, C. E., Takata, M., Koyama, S., & Satoh, T. (2016). Contemporary evolution of host plant range expansion in an introduced herbivorous beetle *Ophraella communa*. *Journal of Evolutionary Biology, 29*(4), 757–765.

Gautam, S. G., Opit, G. P., & Giles, K. L. (2010). Population growth and development of the psocid *Liposcelis rufa* (Psocoptera: Liposcelididae) at constant temperatures and relative humidities. *Journal of Economic Entomology, 103*(5), 1920–1928.

Genkai-Kato, M., Mitsuhashi, H., Kohmatsu, Y., Miyasaka, H., Nozaki, K., & Nakanishi, M. (2005). A seasonal change in the distribution of a stream-dwelling stonefly nymph reflects oxygen supply and water flow. *Ecological Research, 20*(2), 223–226.

Georgis, R., & Poinar, G. O., Jr. (1983). Effect of soil texture on the distribution and infectivity of *Neoaplectana glaseri* (Nematoda: Steinernematidae). *Journal of Nematology, 15*(3), 329–332.

Gill, K. H., Goyal, G., & Chahil, G. (2017). Insect diapause: A review. *Journal of Agricultural Science and Technology, 7*, 454–473.

Greco, N., & Esmenjaud, D. (2004). Management strategies for nematode control in Europe. In *Proceedings of the Fourth International Congress of Nematology, 8–13 June 2002, Tenerife, Spain* (pp. 33–43). Brill.

Gregory, P. J., Johnson, S. N., Newton, A. C., & Ingram, J. S. (2009). Integrating pests and pathogens into the climate change/food security debate. *Journal of Experimental Botany, 60*(10), 2827–2838.

Grewal, P. S., Abrol, D. P., & Shankar, U. (2012). Entomopathogenic nematodes as tools in integrated pest management. In *Integrated Pest Management: Principles and Practice* (pp. 162–236). CABI Books. CABI International.

Guedes, N. M. P., Guedes, R. N. C., Ferreira, G. H., & Silva, L. B. (2009). Flight take-off and walking behavior of insecticide-susceptible and-resistant strains of *Sitophilus zeamais* exposed to deltamethrin. *Bulletin of Entomological Research, 99*(4), 393–400.

Gupta, R. B. L., & Pathak, V. N. (1986). Effect of age of host, inoculum density and duration of high relative humidity on development of purple blotch of onion. *Phytophylactica, 18*(3), 151.

Hall, D. G., Childers, C. C., & Eger, J. E. (2005). Effects of reducing sample size on density estimates of citrus rust mite (Acari: Eriophyidae) on citrus fruit: Simulated sampling. *Journal of Economic Entomology, 98*(3), 1048–1057.

Hall, D. G., Childers, C. C., & Eger, J. E. (2007). Binomial sampling to estimate rust mite (Acari: Eriophyidae) densities on orange fruit. *Journal of Economic Entomology, 100*(1), 233–240.

Hannig, C. (1996). *Untersuchungen zu Oidium lycopersicum, dem Echten Mehltau an Tomate* (Doctoral dissertation, Technische Universität München).

Headlee, T. J. (1914). Some data on the effect of temperature and moisture on the rate of insect metabolism. *Journal of Economic Entomology, 7*(6), 413–417.

Headlee, T. J. (1917). Some facts relative to the influence of atmospheric humidity on insect metabolism. *Journal of Economic Entomology, 10*, 31–38.

Heran, A., McCartney, H. A., & Li, Q. (1999). The effect of petal characteristics, inoculum density and environmental factors on infection of oilseed rape by Sclerotinia sclerotiorum. In *Proceedings 10th International Rapeseed Congress: New Horizons for an Old Crop, Canberra, 26–29 September 1999*. (CD-ROM).

Hill, J. K., Thomas, C. D., Fox, R., Telfer, M. G., Willis, S. G., Asher, J., & Huntley, B. (2002). Responses of butterflies to twentieth century climate warming: Implications for future ranges. *Proceedings of the Royal Society of London. Series B: Biological Sciences, 269*(1505), 2163–2171.

Hodek, I., Bonet, A., & Hodkova, M. (1981). Some ecological factors affecting diapause in adults of *Acanthoscelides obtectus* from Mexican mountains. In *The Ecology of Bruchids Attacking Legumes (Pulses): Proceedings of the International Symposium Held at Tours (France), April 16–19, 1980* (pp. 43–55). Springer Netherlands.

Holmes, L. A., Vanlaerhoven, S. L., & Tomberlin, J. K. (2012). Relative humidity effects on the life history of *Hermetia illucens* (Diptera: Stratiomyidae). *Environmental Entomology, 41*(4), 971–978.

Honěk, A., & Kocourek, F. (1990). Temperature and development time in insects: A general relationship between thermal constants. *Zoologische Jahrbücher, Abteilung für Systematik, Ökologie und Geographie der Tiere, 117*(4), 401–439.

Hou, L., Wu, Y., & Wang, S. (2019). Thermal death kinetics of *Cryptolestes pusillus* (Schonherr), *Rhyzopertha dominica* (Fabricius), and *Tribolium confusum* (Jacquelin du Val) using a heating block system. *Insects, 10*(5), 119.

Howden, S. M., Soussana, J. F., Tubiello, F. N., Chhetri, N., Dunlop, M., & Meinke, H. (2007). Adapting agriculture to climate change. *Proceedings of the National Academy of Sciences, 104*(50), 19691–19696.

Hoy, M. A. (2019). Molecular genetics of insect behavior. In *Insect Molecular Genetics* (4th ed., pp. 413–461). Academic Press.

Hueck, H. J., Kuenen, D. J., Denboer, P. J., & Jaegerdraafsel, E. (1952). The increase of egg production of the fruit tree red spider mite (*Metatetranychus ulmi* Koch) under influence of DDT. *Physiological and Comparative Oecology, 2*(4), 371–377.

Husin, O. T., & Port, R. G. (2021). Efficacy of entomopathogenic nematodes against *Tuta absoluta*. *Biological Control, 60*, 104699.

James, R. R., Croft, B. A., Shaffer, B. T., & Lighthart, B. (1998). Impact of temperature and humidity on host-pathogen interactions between *Beauveria bassiana* and a coccinellid. *Environmental Entomology, 27*(6), 1506–1513.

Jaworski, T., & Hilszczański, J. (2013). The effect of temperature and humidity changes on insects' development their impact on forest ecosystems in the context of expected climate change. *Leśne Prace Badawcze, 74*(4), 345–355.

Jeppson, L. R., Keifer, H. H., & Baker, E. W. (1975). *Mites Injurious to Economic Plants*. University of California Press.

Jia, Q., Lv, B., Guo, M., Luo, C., Zheng, L., Hsiang, T., & Huang, J. (2015). Effect of rice growth stage, temperature, relative humidity and wetness duration on infection of rice panicles by *Villosiclava virens*. *European Journal of Plant Pathology, 141*(1), 15–25.

Jones, J. T., Haegeman, A., Danchin, E. G., Gaur, H. S., Helder, J., Jones, M. G., & Perry, R. N. (2013). Top 10 plant-parasitic nematodes in molecular plant pathology. *Molecular Plant Pathology, 14*(9), 946–961.

Joosten, L., Lecocq, A., Jensen, A. B., Haenen, O., Schmitt, E., & Eilenberg, J. (2020). Review of insect pathogen risks for the black soldier fly (*Hermetia illucens*) and guidelines for reliable production. *Entomologia Experimentalis et Applicata, 168*(6–7), 432–447.

Kamionek, M., Maslana, I., & Sandner, H. (1974). The survival of invasive larvae of *Neoaplectana carpocapsae* Weiser in a waterless environment under various conditions of temperature and humidity. *Zesz. Probl. Postepow Nauk Roln. 154*, 409–412.

Kant, M. R., Ament, K., Sabelis, M. W., Haring, M. A., & Schuurink, R. C. (2004). Differential timing of spider mite-induced direct and indirect defenses in tomato plants. *Plant Physiology, 135*(1), 483–495.

Karban, R., & Carey, J. R. (1984). Induced resistance of cotton seedlings to mites. *Science, 225*(4657), 53–54.

Karl, I., Stoks, R., De Block, M., Janowitz, S. A., & Fischer, K. (2011). Temperature extremes and butterfly fitness: Conflicting evidence from life history and immune function. *Global Change Biology, 17*(2), 676–687.

Kayode, O. Y., Adedire, C. O., & Akinkurolere, R. O. (2014). Influence of four cereal flours on the growth of *Tribolium castaneum* Herbst (Coleoptera: Tenebrionidae). *Ife Journal of Science, 16*(3), 505–516.

Keane, P., & Kerr, A. (1997). Factors affecting disease development. In *Plant Pathogens and Plant Diseases* (pp. 287–298).

Khaliq, A. M., Javed, M., Sohail, M., & Sagheer, M. (2014). Environmental effects on insects and their population dynamics. *Journal of Entomology and Zoology Studies, 2*(2), 1–7.

Kheradmand, K., Kamali, K., Fathipour, Y., & Goltapeh, E. M. (2007). Development, life table and thermal requirement of *Tyrophagus putrescentiae* (Astigmata: Acaridae) on mushrooms. *Journal of Stored Products Research, 43*(3), 276–281.

Kim, C. H., Ryu, J., Lee, J., Ko, K., Lee, J. Y., Park, K. Y., & Chung, H. (2021). Use of black soldier fly larvae for food waste treatment and energy production in Asian countries: A review. *Processes, 9*(1), 161.

Knapp, M., van Houten, Y., van Baal, E., & Groot, T. (2018). Use of predatory mites in commercial biocontrol: Current status and future prospects. *Acarologia, 58*, 72–82.

Kodjo, T. A., Gbénonchi, M., Sadate, A., Komi, A., Yaovi, G., Dieudonné, M., & Komla, S. (2011). Bio-insecticidal effects of plant extracts and oil emulsions of *Ricinus communis* L. (Malpighiales: Euphorbiaceae) on the diamondback, *Plutella xylostella* L. (Lepidoptera: Plutellidae) under laboratory and semi-field conditions. *Journal of Applied Biosciences, 43*, 2899–2914.

Koehler, H. H. (1999). Predatory mites (Gamasina, Mesostigmata). *Agriculture, Ecosystems & Environment, 74*(1–3), 395–410.

Konvicka, M., Maradova, M., Benes, J., Fric, Z., & Kepka, P. (2003). Uphill shifts in distribution of butterflies in the Czech Republic: Effects of changing climate detected on a regional scale. *Global Ecology and Biogeography, 12*(5), 403–410.

Kung, S. P., Gaugler, R., & Kaya, H. K. (1991). Effects of soil temperature, moisture, and relative humidity on entomopathogenic nematode persistence. *Journal of Invertebrate Pathology, 57*(2), 242–249.

Kurmi, D. K. S., Meena, R. S., Gupta, A. K., & S.T.A.K. (2022). Studies on Seasonal Incidence and Management of Early Shoot Borer and Top Shoot Borers using New Insecticides in Sugarcane (*Saccharum officinarum* L.). *International Journal of Environmental and Agriculture Research, 8*(10), 16–27.

Lacey, L. A., & Unruh, T. R. (1998). Entomopathogenic nematodes for control of codling moth, *Cydia pomonella* (Lepidoptera: Tortricidae): effect of nematode species, concentration, temperature, and humidity. *Biological Control, 13*(3), 190–197.

Laney, A. G., Keller, K. E., Martin, R. R., & Tzanetakis, I. E. (2011). A discovery 70 years in the making: Characterization of the Rose rosette virus. *Journal of General Virology, 92*, 1727–1732.

Lefkovitch, L. P. (1962a). The biology of *Cryptolestes capensis* (Waltl) (Coleoptera, Cucujidae). *Bulletin of Entomological Research, 53*, 529–535.

Lefkovitch, L. P. (1962b). The biology of *Cryptolestes turcicus* (Grouvelle) (Coleoptera: Cucujidae), a pest of stored and processed cereals. *Proceedings of the Zoological Society of London, 138*, 23–35.

Lefkovitch, L. P. (1964). The biology of *Cryptolestes pusilloides* (Steel and Howe) (Coleoptera, Cucujidae), a pest of stored cereals in the Southern Hemisphere. *Bulletin of Entomological Research, 54*, 649–656.

Lever, R. J. W. (1969). Do armyworm follow the rain. *Wild Crops, 21*, 351–352.

Levesque, K. R., Fortin, M., & Mauffette, Y. (2002). Temperature and food quality effects on growth, consumption and post-ingestive utilization efficiencies of the forest tent caterpillar *Malacosoma disstria* (Lepidoptera: Lasiocampidae). *Bulletin of Entomological Research, 92*(2), 127–136.

Lewis, E. E., Ricci, M., & Gaugler, R. (1996). Host recognition behavior predicts host suitability in the entomopathogenic nematode *Steinernema carpocapsae* (Rhabditida: Steinernematidae). *Parasitology, 113*(6), 573–579.

Li, Q. X., & Cai, R. X. (1996). Occurrence and damage of the apple rust mite *Aculus schlechtendali*. *Plant Protection, 22*(3), 16–17.

Lindquist, E. E., Bruin, J., & Sabelis, M. W. (1996). *Eriophyoid Mites: Their Biology, Natural Enemies and Control*. Elsevier.

Linke, W. (1953). Investigation of the biology and epidemiology of the common spider mite, *Tetranychus althaeae* v. Hanst. With consideration of the hop as the host. *Hoefchen-Briefe, 6*, 181–232.

Luz, C., & Fargues, J. (1999). Dependence of the entomopathogenic fungus, *Beauveria bassiana*, on high humidity for infection of *Rhodnius prolixus*. *Mycopathologia, 146*(1), 33–40.

Malik, A., Gulati, R., Duhan, K., & Poonia, A. (2018). *Tyrophagus putrescentiae* (Schrank) (Acari: Acaridae) as a pest of grains: A review. *Journal of Entomology and Zoology Studies, 6*(2543), 2550–2555.

Masters, G. J., Brown, V. K., Clarke, I. P., Whittaker, J. B., & Hollier, J. A. (1998). Direct and indirect effects of climate change on insect herbivores: Auchenorrhyncha (Homoptera). *Ecological Entomology, 23*(1), 45–52.

Mbapila, J. C., Overholt, W. A., & Kayumbo, H. Y. (2002). Comparative development and population growth of an exotic stemborer, *Chilo partellus* (Swinhoe), and an ecologically similar congener, *C. orichalcociliellus* (Strand) (Lepidoptera: Crambidae). *International Journal of Tropical Insect Science, 22*(1), 21–27.

McCoy, C. W. (1996). Damage and control of eriophyoid mites in crops 3.2. 1 Stylar feeding injury and control of eriophyoid mites in citrus. In *World Crop Pests* (Vol. 6, pp. 513–526).

Meena, V. P., Khinchi, S. K., Kumawat, K. C., & Choudhary, S. (2021). Seasonal incidence of gram pod borer, *Helicoverpa armigera* (Hubner) and spotted pod borer, *Maruca testulalis* (Geyer) on green gram in relation to weather parameters. *Pharma Innovation Journal, 10*(10), 696–699.

Mellanby, K. (1936). Humidity and insect metabolism. *Nature, 138*(3481), 124–125.

Migeon, A., Nouguier, E., & Dorkeld, F. (2010). Spider Mites Web: a comprehensive database for the Tetranychidae. In *Trends in Acarology: Proceedings of the 12th International Congress* (pp. 557–560). Springer: Netherlands.

Mitiku, M. (2018). Plant-parasitic nematodes and their management: A review. *Agriculture Research & Technology, 8*, 30–38.

Moore, B. A., & Allard, G. B. (2008). Climate change impacts on forest health. *Forest Health and Biosecurity Working Paper FBS/34E*. Forest assessment, management and conservation department, FAO, Rome, Italy. 38.

Moore, G. E. (1965). The bionomics of an insect-parasitic nematode. *Journal of the Kansas Entomological Society, 38*(2), 101–105.

Moreira, G. F., & de Moraes, G. J. (2015). The potential of free-living laelapid mites (Mesostigmata: Laelapidae) as biological control agents. In *Prospects for Biological Control of Plant Feeding Mites and Other Harmful Organisms* (pp. 77–102).

Muegge, M. A., & Lambdin, P. L. (1989). Longevity and fecundity of *Coccophagus lycimnia* (Walker) (Hymenoptera: Aphelinidae), a primary parasitoid of *Coccus hesperidum* (Homoptera: Coccidae). *Journal of Agricultural Entomology, 6*, 169–174.

Nakamura, K., & Numata, H. (2000). Photoperiodic control of the intensity of diapause and diapause development in the bean bug, *Riptortus clavatus* (Heteroptera: Alydidae). *European Journal of Entomology, 97*(1), 19–24.

Netherer, S., & Schopf, A. (2010). Potential effects of climate change on insect herbivores in European forests—General aspects and the pine processionary moth as specific example. *Forest Ecology and Management, 259*(4), 831–838.

Nomikou, M., Janssen, A., Schraag, R., & Sabelis, M. W. (2001). Phytoseiid predators as potential biological control agents for *Bemisia tabaci*. *Experimental & Applied Acarology, 25*, 271–291.

Omkar, & Pervez, A. (2004). Temperature-dependent development and immature survival of an aphidophagous ladybeetle, *Propylea dissecta* (Mulsant). *Journal of Applied Entomology, 128*(7), 510–514.

Opit, G. P., & Throne, J. E. (2009). Population growth and development of the Psocid *Liposcelis brunnea* (Psocoptera: Liposcelididae) at constant temperatures and relative humidities. *Journal of Economic Entomology, 102*, 1360–1368.

Ouedraogo, P. A., Sou, S., Sanon, A., Monge, J. P., Huignard, J., Tran, B. E., & Credland, P. F. (1996). Influence of temperature and humidity on populations of *Callosobruchus maculatus* (Coleoptera: Bruchidae) and its parasitoid *Dinarmus basalis* (Pteromalidae) in two climatic zones of Burkina Faso. *Bulletin of Entomological Research, 86*(6), 695–702.

Overgaard, J., & Sørensen, J. G. (2008). Rapid thermal adaptation during field temperature variations in *Drosophila melanogaster*. *Cryobiology, 56*(2), 159–162.

Pastor, B., Velasquez, Y., Gobbi, P., & Rojo, S. (2015). Conversion of organic wastes into fly larval biomass: Bottlenecks and challenges. *Journal of Insects as Food and Feed, 1*(3), 179–193.

Patil, S. D., Kusalkar, D. V., Patil, H. M., & Bhoite, K. D. (2020). Seasonal incidence of insect pests on rice and impact of various abiotic factors on their incidence. *Journal of Pharmacognosy and Phytochemistry, 9*(2), 1869–1872.

Patterson, D. T., Westbrook, J. K., Joyce, R., Lingren, P. D., & Rogasik, J. (1999). Weeds, Insects, and Diseases. *Climate Change, 43*, 711–727.

Pervez, A. (2004). Temperature-dependent development and immature survival of an aphidophagous ladybeetle, *Propylea dissecta* (Mulsant). *Journal of Applied Entomology, 128*(7), 510–514.

Phani, V., Khan, M. R., & Dutta, T. K. (2021). Plant-parasitic nematodes as a potential threat to protected agriculture: Current status and management options. *Crop Protection, 144*, 105573.

Phillips, A. J. L. (1994). Influence of fluctuating temperatures and interrupted periods of plant surface wetness on infection of bean leaves by ascospores of *Sclerotinia sclerotiorum*. *Annals of Applied Biology, 124*(3), 413–427.

Ponsonby, D. J., & Copland, M. J. W. (1996). Effect of temperature on development and immature survival in the scale insect predator, *Chilocorus nigritus* (F.) (Coleoptera: Coccinellidae). *Biocontrol Science and Technology, 6*(1), 101–110.

Powell, G., Tosh, C. R., & Hardie, J. (2006). Host plant selection by aphids: Behavioral, evolutionary, and applied perspectives. *Annual Review of Entomology, 51*, 309–330.

Prakash, A., Rao, J., Mukherjee, A. K., Berliner, J., Pokhare, S. S., Adak, T., & Shashank, P. R. (2014). Climate change: Impact on crop pests. *Applied Zoologists Research Association (AZRA), Central Rice Research Institute, Cuttack, Odisha.*

Rahim, A., Hashmi, A. A., & Khan, N. A. (1991). Effects of temperature and relative humidity on longevity and development of *Ooencyrtus papilionis* Ashmead (Hymenoptera: Eulophidae), a parasite of the sugarcane pest, *Pyrilla perpusilla* Walker (Homoptera: Cicadellidae). *Environmental Entomology, 20*(3), 774–775.

Ramkumar, Prajapati, K. K., Keval, R., & Srivastava, C. P. (2022). Bioefficacy of some insecticides against pod pest complex on pigeonpea (*Cajanus cajan* (l.) mill sp.). *Journal of Experimental Zoology, 25*, 2515–2520.

Ramkumar, Prajapati, K. K., Keval, R., Padwal, G. K., & Srivastava, C. P. (2023). Seasonal incidence of pod pests complex of pigeonpea, *Cajanus cajan* (l.) Mill sp. *Journal of Experimental Zoology, 26*(1), 1183–1186.

Ramlov, U. B. (2000). Aspects of natural cold tolerance in ectothermic animals. *Human Reproduction, 15*(5), 26–46.

Ramos-Rodríguez, O., Campbell, J. F., & Ramaswamy, S. B. (2007). Efficacy of the entomopathogenic nematode *Steinernema riobrave* against the stored-product insect pests *Tribolium castaneum* and *Plodia interpunctella. Biological Control, 40*(1), 15–21.

Reddy, T. Y., & Reddy, G. H. S. (2016). *Principles of Agronomy*, 5th revised ed. Kalyani Publishers, 94–96.

Regniere, J., Rabb, R. L., & Stinner, R. E. (1981). *Popillia japonica*: Effect of soil moisture and texture on survival and development of eggs and first instar grubs. *Environmental Entomology, 10*(5), 654–660.

Reuveni, R., & Rotem, J. (1974). Effect of humidity on epidemiological patterns of the powdery mildew (*Sphaerotheca fuliginea*) on squash. *Phytoparasitica, 2*, 25–33.

Romero, F., Cazzato, S., Walder, F., Vogelgsang, S., Bender, S. F., & van der Heijden, M. G. (2022). Humidity and high temperature are important for predicting fungal disease outbreaks worldwide. *New Phytologist, 234*(5), 1553–1556.

Rouault, G., Candau, J. N., Lieutier, F., Nageleisen, L. M., Martin, J. C., & Warzée, N. (2006). Effects of drought and heat on forest insect populations in relation to the 2003 drought in Western Europe. *Annals of Forest Science, 63*(6), 613–624.

Ryan, C. A. (1990). Protease inhibitors in plants: Genes for improving defenses against insects and pathogens. *Annual Review of Phytopathology, 28*(1), 425–449.

Sabir, N., & Walia, R. K. (2017). *Management of Nematodes in Protected Cultivation with Short Notes on Key Pests*. Project coordinating cell all India coordinated research project on nematodes in cropping systems, ICAR-Indian Agricultural Research Institute.

Salinas-Castro, A., Sandi, M. T. M. P., Ramírez-Reyes, T., Luna-Rodríguez, M., & Trigos, Á. (2014). An unusual food plant for *Cydia pomonella* (Linnaeus) (Lepidoptera, Tortricidae) in Mexico. *Revista Brasileira de Entomologia, 58*, 261–264.

Samrit, R. M., Chaudhari, B. N., & Gahane, K. D. (2019). Seasonal incidence of rice yellow stem borer, *Scirpophaga incertulas* (Walk.) and its correlation with weather parameters and natural enemies. *Journal of Pharmacognosy and Phytochemistry, 8*(5), 740–742.

San-Blas, E., Pembroke, B., & Gowen, S. R. (2012). Scavenging and infection of different hosts by *Steinernema carpocapsae* [carroñerismo e infección de diferentes hospedadores por *Steinernema carpocapsae*]. *Nematropica, 42*(1), 123–130.

Sarmento, R. A., Lemos, F., Dias, C. R., Kikuchi, W. T., Rodrigues, J. C., Pallini, A., & Janssen, A. (2011). A herbivorous mite down-regulates plant defense and produces web to exclude competitors. *PLoS One, 6*(8), 23757.

Sasser, J. N., & Carter, C. C. (1985). Overview of the International *Meloidogyne* Project 1975–1984. In J. N. Sasser & C. C. Carter (Eds.), *An Advanced Treatise on Meloidogyne* (pp. 19–24). North Carolina State University Graphics.

Schmiege, D. C. (1963). The feasibility of using a neo-aplectanid nematode for control of some forest insect pests. *Journal of Economic Entomology, 56*, 427–431.

Serebrjanikaw, A. W. (1901). *Nachr. Des Forst. Instit. zu St. Petersburg, VII*, 29–102.

Shafaullah, M. A. K., Khan, N. A., & Mahmood, Y. (2011). Effect of epidemiological factors on the incidence of paddy blast (*Pyricularia oryzae*) disease. *Pakistan Journal of Phytopathology, 23*(2), 108–111).

Shipp, J. L., Zhang, Y., Hunt, D. W. A., & Ferguson, G. (2003). Influence of humidity and greenhouse microclimate on the efficacy of *Beauveria bassiana* (Balsamo) for control of greenhouse arthropod pests. *Environmental Entomology, 32*(5), 1154–1163.

Shrestha, S. (2019). Effects of climate change in agricultural insect pest. *Acta Scientiarum Agronomy, 3*(12), 74–80.

Shumo, M., Khamis, F. M., Tanga, C. M., Fiaboe, K. K., Subramanian, S., Ekesi, S., & Borgemeister, C. (2019). Influence of temperature on selected life-history traits of black soldier fly (*Hermetia illucens*) reared on two common urban organic waste streams in Kenya. *Animals, 9*(3), 1–14.

Simons, W. R. (1973). Nematode survival in relation to soil moisture. *Wageningen University and Research.*

Simons, W. R., & Poinar Jr, G. O. (1973). The ability of *Neoaplectana carpocapsae* (Steinernematidae: Nematodea) to survive extended periods of desiccation. *Journal of Invertebrate Pathology, 22*(2), 228–230).

Singh, T., Bhat, M. M., & Khan, M. A. (2009). Insect adaptations to changing environments-temperature and humidity. *International Journal of Industrial Entomology, 19*(1), 155—164.

Sithole, S. Z. (1987). The effect of date of planting on shoot fly and stem borer infestations on sorghum. In *Proceedings of the 3rd Regional Workshop on Sorghum and Millets for Southern Africa* (pp. 6–10).

Skendžić, S., Zovko, M., Živković, I. P., Lešić, V., & Lemić, D. (2021). The impact of climate change on agricultural insect pests. *Insects, 12*(5), 440.

Srivastava, S. (2003). Influence of temperature on certain biological attributes of a ladybeetle *Coccinella septempunctata* Linnaeus. *Insect Science, 10*(3), 185–193.

Srivastava, S., & Omkar. (2003). Influence of temperature on certain biological attributes of a ladybeetle *Coccinella septempunctata* Linnaeus. *Entomologia Sinica, 10*(3), 185–193.

Stefanovska, T., Pidlishyuk, V., & Kaya, H. (2008). Host range and infectivity of *Heterorhabditis bacteriophora* (Heterorhabditidae) from Ukraine. *Communications in Agricultural and Applied Biological Sciences, 73*(4), 693–698.

Tamiru, A., Getu, E., Jembere, B., & Bruce, T. (2012). Effect of temperature and relative humidity on the development and fecundity of *Chilo partellus* (Swinhoe) (Lepidoptera: Crambidae). *Bulletin of Entomological Research, 102*(1), 9–15.

Tomberlin, J. K., & Sheppard, D. C. (2002). Factors influencing mating and oviposition of black soldier flies (Diptera: Stratiomyidae) in a colony. *Journal of Entomological Science, 37*(4), 345–352.

Tores, J. A., & Moreno, R. (1991). *Sclerotinia sclerotiorum*: Epidemiological factors affecting infection of greenhouse aubergine crops. *Journal of Phytopathology, 132*(1), 65–74.

Torres, E., Álvarez-Acosta, C., del Pino, M., Wong, M. E., Boyero, J. R., Hernández-Suárez, E., & Vela, J. M. (2023). Economic impact of the Persea mite in Spanish avocado crops. *Agronomy, 13*(3), 668. https://doi.org/10.3390/agronomy13030668.

Tu, J. C. (1989). Modes of primary infection caused by *Sclerotinia sclerotiorum* in navy bean. *Microbios, 57*, 85–91.

Urech, F. (1890). Chemisch-analytische Untersuchungen an lebenden Raupen, Puppen und Schmetterlingen und an ihren Secreten. *Microbios, 57*, 335–345.

Van Leeuwen, T., Tirry, L., Yamamoto, A., Nauen, R., & Dermauw, W. (2015). The economic importance of acaricides in the control of phytophagous mites and an update on recent acaricide mode of action research. *Pesticide Biochemistry and Physiology, 121*, 12–21. https://doi.org/10.1016/j.pestbp.2014.12.016.

Van Leeuwen, T., Vontas, J., Tsagkarakou, A., Dermauw, W., & Tirry, L. (2010). Acaricide resistance mechanisms in the two-spotted spider mite *Tetranychus urticae* and other important Acari: A review. *Insect Biochemistry and Molecular Biology, 40*(8), 563–572. https://doi.org/10.1016/j.ibmb.2010.04.006.

Van Leeuwen, T., Witters, J., Nauen, R., Duso, C., & Tirry, L. (2010). The control of eriophyoid mites: State of the art and future challenges. *Experimental and Applied Acarology, 51*, 205–224. https://doi.org/10.1007/s10493-009-9308-3.

Vigneron, A., Jehan, C., Rigaud, T., & Moret, Y. (2019). Immune defenses of a beneficial pest: The mealworm beetle, *Tenebrio molitor*. *Frontiers in Physiology, 10*, Article 138, 1–17. https://doi.org/10.3389/fphys.2019.00138.

Wang, L., Hui, C., Sandhu, H. S., Li, Z., & Zhao, Z. (2015). Population dynamics and associated factors of cereal aphids and armyworms under global change. *Scientific Reports, 5*, Article 18801. https://doi.org/10.1038/srep18801.

Womersley, C. Z. (1990). Dehydration survival and anhydrobiotic potential. In J. A. Glazer & H. K. Kaya (Eds.), *Entomopathogenic Nematodes in Biological Control* (pp. 117–137). CRC Press.

Woodman, J. D. (2017). Effects of substrate salinity on oviposition, embryonic development and survival in the Australian plague locust, *Chortoicetes terminifera* (Walker). *Journal of Insect Physiology, 96*, 9–13. https://doi.org/10.1016/j.jinsphys.2016.12.008.

Xu, X. M., & Butt, D. J. (1998). Effects of temperature and atmospheric moisture on the early growth of apple powdery mildew (*Podosphaera leucotricha*) colonies. *European Journal of Plant Pathology, 104*, 133–140.

Yadav, R. P., & Chaudhary, J. P. (1987). Effect of temperature and relative humidity on longevity and development of the parasitoid, *Tetrastichus pyrillae* Crawford (Hymenoptera: Eulophidae). *International Journal of Tropical Insect Science, 8*(1), 39–41.

Yonow, T., Kriticos, D. J., Ota, N., Van Den Berg, J., & Hutchison, W. D. (2017). The potential global distribution of *Chilo partellus*, including consideration of irrigation and cropping patterns. *Journal of Pest Science, 90*, 459–477. https://doi.org/10.1007/s10340-016-0773-6.

Zachariassen, K. E. (1991). Routes of transpiratory water loss in a dry habitat tenebrionid beetle. *Journal of Experimental Biology, 157*(1), 425–437.

Zhang, J., Huang, L., He, J., Tomberlin, J. K., Li, J., Lei, C., Sun, M., Liu, Z., & Yu, Z. (2010). An artificial light source influences mating and oviposition of black soldier flies, *Hermetia illucens*. *Journal of Insect Science, 10*(1), Article 202. https://doi.org/10.1673/031.010.20201.

Zhou, Z. S., Guo, J. Y., Chen, H. S., & Wan, F. H. (2010). Effect of humidity on the development and fecundity of *Ophraella communa* (Coleoptera: Chrysomelidae). *BioControl, 55*, 313–319. https://doi.org/10.1007/s10526-009-9252-7.

Zulfiqar, M. A., Sabri, M. A., Raza, M. A., Hamza, A., Hayat, A., & Khan, A. (2010). Effect of temperature and relative humidity on the population dynamics of some insect pests of maize. *Pakistan Journal of Life and Social Sciences, 8*, 16–18.

PART II

Molecular Mechanism of Stress Tolerance Under Climate Change

CHAPTER 2

Molecular Mechanism of Host-Pathogen Interaction in Crop Plants

KANUPRIYA KUNIYAL and NIKI NAUTIYAL

Department of Biochemistry and Biotechnology, Sardar Bhagwan Singh University, Balawala, Dehradun, Uttarakhand, India

ABSTRACT

Crop plants are of enormous importance for human survival. However, they are under constant threat from a wide range of pathogens causing a variety of diseases. Over the course of their long co-evolution, plants have created a highly powerful defense mechanism against microbial pathogens. Plants have two types of defense mechanisms: primary (preformed) and secondary (induced). The first step in activating the defense response is to recognize the presence of microorganisms through elicitors, which initiate basal resistance. A variety of signaling processes are involved in this defense response, which results in the control of defense-related genes and the generation of defense responses. Additionally, plants have an innate immune system that detects pathogens via MAMPs, PAMPs, and effector proteins. PTI and ETI are two of the many reactions that are sparked by this recognition. With the help of their effectors, pathogens can avoid detection and decrease host immunity, leading to ETS. R proteins can identify some effectors known as Avr proteins, which give the pathogen an advantage over PTI. The rapid advancement of genome editing technologies has made them one of the most crucial genetic tools for implementing disease resistance in plants. Meganucleases, zinc finger nucleases (ZFNs), transcription activator-like effector nucleases (TALENs), and clustered regularly interspaced short palindromic repeats (CRISPR) have all been used as site-directed modification techniques in recent years (Cas9).

Climate Change and Biotic Factors: A Molecular Approach.
Ashutosh Singh, Saurabh Pandey, and Amarjeet Kumar, (Eds.)

2.1 INTRODUCTION

Crop plants are of enormous importance for human survival, providing most of our food and other agricultural products. They are the basis of global food security and are critical for sustaining human populations and economic development. However, crop plants are under constant threat from a wide range of pathogens, causing a variety of diseases in crop plants, including wilts, rots, blights, and viruses, which can cause significant yield losses and economic damage.

Plant disease poses a constant threat to agriculture globally, and developing sustainable agriculture relies heavily on comprehending the interactions between plants and pathogens. Plants have developed a highly effective defense mechanism over the course of their long co-evolution against microbial pathogens (Attard et al., 2008) as they are devoid of mobile defense cells and somatic adaptive systems. Instead, each cell relies on innate immunity, generating systemic signals from infection sites. Receptors are essential for the awareness of pathogens because they can identify conserved chemical patterns and particular effector proteins, which activate related responses. The first immune response is known as PTI, which has evolved to recognize common features of microbial pathogens. However, to inhibit PTI and facilitate their growth, pathogens have also developed the capacity to release effector proteins. To detect the presence of pathogen effector proteins, plants have also developed surveillance proteins (R proteins) in response, which activate a chain of defense (Mahadevakumar et al., 2020).

The study of interactions of host-pathogen lies at the core of our comprehension of infectious diseases in plants, encompassing their transmission, progression, diagnosis, and treatment. Thus, it is imperative to focus on investigating these interactions, particularly considering the growing challenge of antibiotic resistance for modern technologies to advance effectively (Swain et al., 2023). Furthermore, the threats to crop plants by pathogens are constantly evolving. New strains of pathogens can emerge, and existing strains can evolve resistance to pesticides and other control measures. Climate change and global trade can also increase the spread of pathogens across borders and into new areas, posing new challenges for crop protection.

2.1.1 PLANT PATHOGEN INTERACTIONS

Plants have two types of defense mechanisms against pathogens: preformed (primary) and induced (secondary). Detecting microbes is the first step in activating the defense response. Elicitors (molecules that induce a plant's defense response at low concentrations) start the plant's basal resistance (Thakur et al., 2013). Ion fluxes, protein kinases, phosphatases, and the creation of signaling molecules are just a few of the processes that make up this defensive reaction. These actions trigger the development of defense-related responses, including the strengthening of cell walls, the accumulation of phytoalexins and pathogenesis-related (PR) proteins, and localized programmed cell death (Dangl et al., 2001).

In addition to elicitors, plants have an innate immune system that recognizes MAMPs or PAMPs or by sensing effector proteins released by the host during plant-pathogen interactions to identify the presence of pathogens. By activating multicomponent and multilayered responses, PAMP/MAMP interactions with cell surface receptors lead to appropriate defenses. Several pathways have been identified, including Ca2+ influx, the production of reactive oxygen species (ROS) and reactive nitrogen species (RNS), and the synthesis of phytohormones like jasmonic acid (JA), salicylic acid (SA), and ethylene (ET), which act as signal molecules that can all lead to the establishment of defense (Kunz et al., 2013).

Plant immunity can be divided into two levels (Jones et al., 2006). At first, PAMPs are detected by pattern recognition receptors (PRRs), which then trigger PTI. The second involves NB-LRR proteins, which detect pathogen effectors and activate a powerful immune response known as ETI. ETI typically results in local cell death, the hypersensitive response (HR), which prevents the spread of the infection. ETI is quicker, longer, and stronger than PTI (Jones et al., 2006). However, some pathogens can avoid recognition and reduce host immunity with effectors, leading to ETS. Avr proteins are effectors that are recognized by R proteins and allow the pathogen to bypass PTI (Jones et al., 2006).

2.2 TYPES OF PATHOGENS

The plant kingdom is under attack from a variety of pathogens, including fungi, bacteria, and viruses. Different diseases have developed unique methods of invading the host plants, feeding on them, and reproducing there. Plant pathogens can be categorized as biotrophs, hemibiotrophs, or necrotrophs based on their mode of invasion and infection of the plant (Oliver et al., 2006). Biotrophic pathogens rely on living host tissue for their growth and reproduction, while hemibiotrophic pathogens initially require living tissue and later infect and grow on dead tissue. Necrotrophic pathogens infect and kill the host tissue immediately, feeding on the dead tissue. Biotrophic pathogens include viruses, while bacteria and fungi can adopt both biotrophic and necrotrophic strategies. Plants mount distinct responses to different types of pathogens.

2.2.1 BIOTROPHS

Biotrophic pathogens are a type of plant pathogen that requires living host tissue in order to grow and reproduce. The formation of specialized structures like haustoria, appressoria, and hyphae for infecting the host plant and taking nutrients distinguishes them due to their complex way of invasion (Schulze-Lefert et al., 2003). Biotrophs sustain the host's viability and cause relatively little damage to the host tissue, making them stealthy invaders. As they rely on the host plant for nutrition, they can also decrease the HR, which is localized programmed cell death of the host plant. Examples of biotrophic pathogens are rust, powdery mildew, and some other types of viruses.

2.2.2 NECROTROPHS

Necrotrophs are characterized by their less sophisticated and more destructive virulence methods for killing and obtaining nutrients from host plant cells. They cause tissue maceration or soft rots and can induce host cell death like HR. "Necrotrophs can be further classified into two subcategories based on their toxin secretion: host-specific necrotrophs (HSNs) and broad host range necrotrophs (BHNs)" (Wolpert et al., 2002). HSNs produce host-specific toxins to enhance their virulence, while BHNs have a wider range of hosts they can infect without requiring host-specific toxins.

2.2.3 HEMIBIOTROPHS

Hemibiotrophs are a type of pathogen that initially invades plant tissues as a biotroph but later switches to a necrotrophic lifestyle. This means they need living host tissue to reproduce and colonize first, but then they kill the host cells and feed on the dead tissue. Hemibiotrophic infections can cause both localized programmed cell death (PCD) and tissue necrosis, resulting in a mixed symptomology of biotrophic and necrotrophic infections (Vleeshouwers and Oliver, 2014). Examples of hemibiotrophic pathogens are the fungus *Colletotrichum* spp. and the oomycete *Phytophthora* spp. Regardless of the classification of the pathogens, during the early phases of invasion, the pathogen produces effector proteins to start colonizing, and the host plant activates defense systems to stop the spread of the infection. Necrotrophic invasion causes the host plant to experience cell death, release hormones and secondary metabolites like ET, abscisic acid (ABA), SA, and JA, as well as an increase in ROS and callose (Mengiste, 2012).

2.3 PLANT DEFENSE RESPONSES AGAINST PATHOGEN

Necrotrophic and biotrophic pathogens face similar initial defenses in plants, including the waxy cuticle layer and rigid cell walls (Delventhal et al., 2007). To detect pathogens, plants use PAMPs, which are conserved molecules that are crucial for pathogens and classify a group of microbes. PAMPs include things like chitin, xylanase from fungi, and flagellin from bacteria. These PAMPs are detected by PRRs, which trigger a PTI (Bent et al., 2007). This response involves the induction of callose deposition, the formation of ROS, PR proteins, and defensins, as well as the activation of mitogen-activated protein (MAP) kinases and WRKY transcription factor phosphorylation (Silverstein et al., 2005). Along with PAMPs, plants can also identify damaged host cells through damage-associated molecular patterns (DAMPs) such as polysaccharides (Lotze et al., 2007). ETI response occurs when a plant *R* gene recognizes an *Avr* gene (Mishra and Grover, 2018). ETI is linked to the formation of ROS, which causes a HR at the infection site (Torres et al., 2005).

2.4 PLANT IMMUNITY

Plants have developed a number of defense systems to defend against invasive pathogens, including a range of pre-existing, inducible, and systemic immune responses (Figure 2.1). There are two major types of plant immunity: basal immunity and specific immunity.

2.4.1 BASAL IMMUNITY

The first line of defense against pathogens is basal immunity, which includes pre-existing physical and chemical barriers such as the cell wall, cuticle, and secondary metabolites that prevent pathogen entry into the plant cell. Basal immunity leads to the activation of PTI by recognizing conserved PAMPs by plant PRRs. PTI triggers a range of defense responses, including producing

ROS, callose deposition, and synthesizing defense-related proteins and phytohormones.

2.4.2 SPECIFIC IMMUNITY

Specific immunity involves using plant R proteins to recognize pathogen effectors, activating ETI. ETI involves a fast and robust activation of defense responses, including the production of ROS, programmed cell death, and the production of PR proteins and phytohormones. ETI frequently results in localized HR, a type of programmed cell death that prevents pathogen growth and expansion at the site of infection. Both basal immunity and specific immunity can also trigger systemic acquired resistance (SAR), a broad-spectrum defense response that provides long-lasting protection against a wide range of pathogens. SAR involves the accumulation of systemic signals such as SA that activate defense responses in distal plant tissues. When sufficiently stimulated, a plant might acquire an improved defensive capacity known as induced resistance. The two types of induced resistance, known as SAR and induced systemic resistance (ISR), involve preconditioning plant defenses through prior infection or treatment, which confers resistance to a pathogen or parasite's subsequent challenge (Choudhary et al., 2007).

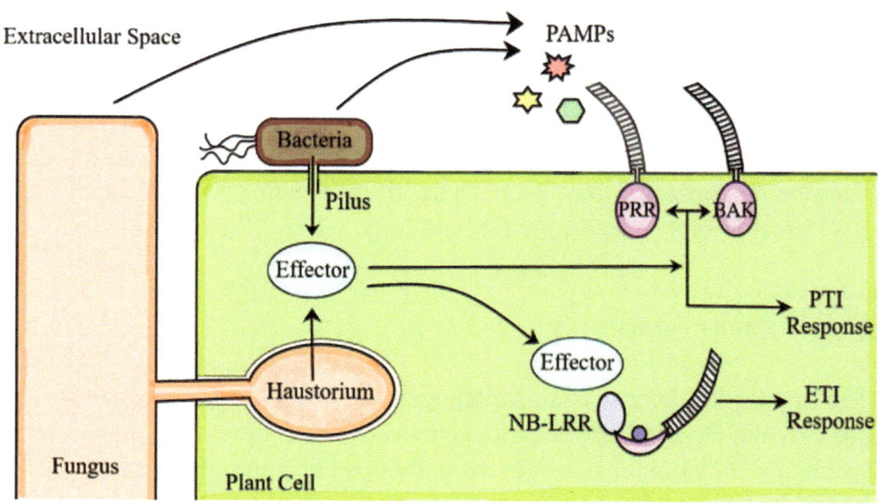

FIGURE 2.1 Plant defense responses.

2.4.3 PATTERN-TRIGGERED IMMUNITY (PTI)

PTI is a type of defense mechanism that is activated when a plant detects molecules known as PAMPs that are unique to pathogens. It is a form of basal immunity and is the first line of defense. When a plant recognizes a PAMP, it activates a cascade of biochemical signaling pathways, which produce defense compounds, including ROS, phytohormones, and antimicrobial peptides (Bigeard et al., 2015). PTI also involves the activation of genes in the plant cell wall reinforcement, which can further prevent pathogen invasion. The identification of PAMPs by PRRs, which are found on the surface of plant cells, is one of the fundamental elements of PTI. PRRs are receptor-like proteins that bind to specific PAMPs and initiate downstream signaling pathways that activate defense mechanisms. PTI is effective against many types of pathogens, including bacteria, fungi, and viruses. However, some pathogens have evolved to evade or repress PTI by secreting effector proteins, which can disrupt plant defense responses. In response, plants have created a second line of defense called ETI, activated when plants recognize specific effector proteins produced by pathogens.

2.4.4 EFFECTOR-TRIGGERED IMMUNITY (ETI)

ETI is a second line of defense that is activated in plants when they recognize specific effector proteins produced by pathogens. ETI is a more specialized and powerful defense response than PTI, the first line of defense against invading pathogens. When a pathogen secretes effector proteins, they can interfere with the plant's normal physiological processes and suppress PTI. However, if the plant has a specific receptor recognizing the effector protein, it will trigger a strong immune response. The "gene-for-gene" concept, which states that a plant's resistance gene (*R* gene) identifies the equivalent *Avr* gene in the pathogen, provides the foundation for the theory that plants can detect effectors.

ROS, phytohormones, antimicrobial peptides, and plant cell wall reinforcement are defensive chemicals produced when ETI is activated. The HR, a consequence of ETI, can cause programmed cell death at the site of infection and halt the spread of the pathogen to other areas of the plant. ETI is a highly specific defense response, as it requires the recognition of a specific effector protein by the plant's *R* gene. This specificity can limit the effectiveness of ETI, as pathogens can evolve to produce new effector proteins that are not recognizable by the R genes of the plant. Nonetheless,

ETI is a crucial defense mechanism that works in conjunction with PTI to defend plants against a variety of pathogens.

2.5 PLANT DEFENSE RESPONSE

Plants exhibit a phenomenon known as non-host resistance, which means they are typically able to resist most pathogens. This resistance is initially provided by pre-existing physical barriers such as the cuticle and cell wall, as well as naturally produced antimicrobial compounds. The protective layer cuticle on the plant surface is made up of cutin and waxes (Yeats and Rose, 2013), which can act as a barrier against pathogens. However, some fungi can penetrate the cuticle by secreting enzymes that break down cutin. The plant cell wall, on the other hand, surrounds the plant cell and provides structural support and protection against biotic and abiotic stressors (Hamann, 2012). Even with the cuticle and cell wall in place, some infections can still enter through natural openings like stomata. Plant wounds are also potential sites of infection. To further protect themselves, plants produce antimicrobial compounds (Osborn et al., 1996), including proteins and secondary metabolites, which can inhibit pathogen growth. If pathogens successfully breach the physical barriers, plants have a complex immune system that can recognize and defend against them. Plants have a variety of defense mechanisms that can be activated upon detecting pathogens. These include stomatal closure to restrict bacterial entry (Melotto et al., 2008), limiting nutrient transfer to prevent bacterial multiplication (Chen et al., 2010), producing and secreting antimicrobial compounds and defense-related proteins/peptides, generating ROS that are toxic to pathogens, and inducing programmed cell death at the site of infection to halt pathogen progression (Mur et al., 2008). An intricate web of signaling channels is involved in activating various defense mechanisms.

2.5.1 R GENES

In plants, resistance genes enable plants to resist or defend themselves against diseases, pests, and environmental stressors. These genes are typically involved in the plant's immune response and recognize and respond to specific pathogenic agents. *R* genes are those that encode the specificity elements of ETI. There are two main types of resistance genes in plants: those that confer resistance through a HR and those that confer resistance through SAR. HR-type resistance genes typically work by causing a rapid and localized cell death response in the infected area, which prevents the pathogen from spreading. On the other hand, SAR-type resistance genes work by inducing a broad-based systemic response throughout the plant, making it more resistant to a wide range of pathogens. Plant resistance genes can be naturally occurring or introduced through genetic engineering techniques. They are an important tool for sustainable agriculture, as they reduce the need for harmful pesticides and herbicides and can help to ensure food security by increasing crop yields and protecting crops from disease and pest damage. Without knowing the specific loci involved, *R* genes are genetically defined as the variable element between susceptible and resistant plant genotypes. One of the most fascinating discoveries in plant pathology is the discovery and characterization of *R* genes. Over the past 15 years, more than 60 plant R genes have been cloned, which play an important role in activating disease resistance to various pathogens, including nematodes, viruses, bacteria, and fungi. While Avr-encoded proteins are highly diverse, the cloned R genes share a common feature with 40 of them encoding proteins with predicted extracellular or intracellular leucine-rich repeat (LRR) motifs. The suggested receptor role of R proteins is compatible with the ability of LRRs to mediate protein-protein interactions. A putative nucleotide-binding (NB) domain that is either connected to an amino-terminal Toll/interleukin-1-receptor (TIR) homologous region or a coiled-coil (CC) domain is seen in R proteins with intracellular LRRs (Mahadevakumar et al., 2020).

2.5.2 FUNCTION OF NBS-LRR PROTEINS IN PLANT DEFENSE

Numerous R genes produce nucleotide-binding site leucine-rich repeat (NBS-LRR) proteins, which have both a nucleotide-binding site (NBS) and leucine-rich repeats (LRRs). They have a crucial role in recognizing pathogen effectors, also known as Avr proteins, which could act as virulence factors in the absence of the corresponding *R* gene (Chisholm et al., 2006). Unlike animals, where NBS-LRR proteins seem to recognize PAMPs, in plants these proteins are specialized in recognizing pathogen effectors (Inohara et al., 2005). Depending on the sequence of the domain before the NBS domain, plant NBS-LRR proteins can be divided into TIR and non-TIR classes. While the non-TIR class has fewer well-defined characteristics, it frequently contains α-helical coiled-coil-like sequences in its N-terminal region, whereas the TIR class comprises a TIR homologous domain in the

N-terminal region (Pan et al., 2000). About 40 of the more than 60 plant R genes that have been cloned and described encode proteins with anticipated extracellular or intracellular LRR motifs.

2.5.3 PR PROTEINS

A plant's response to these dangers involves the activation of a series of genes that code for various effectors, receptors, signaling molecules, and defensive chemicals. The production and accumulation of PR proteins in plants as a result of these unfavorable circumstances are particularly significant because PR proteins are a crucial part of innate immune responses in plants exposed to biotic or abiotic stress. The PR proteins accumulate locally in the infected and surrounding tissues, as well as in distant uninfected tissues, shielding the plants against further infection. Numerous plant species from various families have reported triggering PRs, indicating that these proteins play a general function in adaptation to biotic or abiotic stress situations (Jain and Khurana, 2018).

During incompatible interactions between a pathogen and a host plant, the plant's defense responses are triggered to limit the damage caused by the pathogen. These responses involve coordinated metabolic alterations and the synthesis of novel proteins collectively known as PRs (Antoniw et al., 1978). PRs are proteins encoded by the host plant but are only produced in pathological or closely related circumstances, including those that are not pathogenic. Numerous PRs have been found in various plant species, pointing to a broad protective function against biotic stress. These proteins are produced systemically, as well as locally at the site of infection, which helps to build SAR or the HR against subsequent infections by pathogenic bacteria, fungi, and viruses (Goodman and Novacky, 1994). Necrotic lesions caused by localized host cell death at the infection site define HR. Additionally, plants launch defensive mechanisms in areas of the plant that are not diseased, resulting in SAR that offers broad-based resistance to different diseases (Delaney, 1997). The mechanism of action of these proteins is still poorly understood, despite numerous publications on the function of several classes of PR proteins during biotic and abiotic stressors.

2.5.4 ROS REGULATION

Plants produce ROS as a byproduct of their aerobic metabolism in a variety of cellular compartments, including chloroplasts (Dietz et al., 2016), mitochondria (Huang et al., 2016), and peroxisomes (Sandalio et al., 2015). Even though ROS has the potential to cause irreparable harm and cell death, they also serve as essential signaling molecules that control plant growth and stress responses. The reactivity, manufacturing sites, and capacity to traverse biological membranes of ROS all play a part in their dual function in vivo (Miller et al., 2010). Since oxygen is the source of all ROS and may be transformed into high-energy ROS in a number of organelles, the birth of oxygen-releasing photosynthetic life has a tremendous effect on all living things. ROS oxidizes and modifies cellular components, interfering with their original functions. ROS oxidizes and alters biological components, preventing them from performing their intended tasks. In unfavorable environments, plants produce a variety of ROS species that control a number of functions, such as pathogen defense, programmed cell death, and stomatal behavior. These ROS species can have irreversible effects on tissue and organ development, resulting in abnormal plant growth or plant death. To regulate plant developmental processes and stress responses, ROS also interacts with hormones and epigenetic modifiers (Gill and Tuteja, 2010). Low ROS concentrations are essential for cellular proliferation and differentiation, but excessive ROS concentrations can be dangerous because they increase the risk of DNA damage and improper PCD timing (Xie et al., 2014).

2.5.5 OXIDATIVE BURST

The oxidative burst in plants refers to the rapid and transient production of ROS in response to various biotic and abiotic stressors. ROS, such as hydrogen peroxide, superoxide radicals, and singlet oxygen, are produced as byproducts of plant metabolism and can cause damage to cellular components, including DNA, proteins, and membranes. However, they also function as signaling molecules that regulate various physiological processes, such as PCD, plant growth, and stress responses. The oxidative burst is an important component of the plant defense response against pathogens. When a pathogen invades a plant, it triggers the production of ROS by various cellular compartments, such as chloroplasts, mitochondria, and peroxisomes. This ROS generation sets off a series of processes that result in the expression of defense-related genes and the use of additional defense mechanisms, like the creation of PRs.

ROS produced during the oxidative burst also play a crucial part in controlling the growth and development of plants. They control various processes, including cell differentiation, proliferation, and expansion.

However, excessive ROS production can lead to cell death and the inhibition of growth, which may result in stunted or abnormal plant development (Mendoza and Eui, 2011).

2.5.6 HYPERSENSITIVE RESPONSE (HR)

All higher plants have a hypersensitive defense response, which is characterized by fast cell death at the site of pathogen invasion. While it is typically linked to pathogen resistance, there are instances where it may lead to other outcomes, such as pathogen vulnerability, slowed growth, and even speciation over long periods of time. Given the high potential costs of an erroneous trigger, plants have developed numerous means of suppressing and regulating this response. HR is a widespread occurrence found in many, if not all, higher plants triggered by various classes of pathogens. The most frequently observed HRs are those induced by fungi, oomycetes, bacteria, and viruses, although sustained interactions with other organisms like nematodes and insects have also been reported to trigger HR (Ebert and Herre, 1996). In parasitic plant-host interactions, cell death has also been noticed, although it is unknown if this indicates an HR. HR is mostly associated with race-specific resistance in pathogen resistance to biotrophic diseases, which feed on living tissues. In contrast, it may benefit necrotrophs that need dead host tissue to complete their life cycle. However, HR's effect is more complicated in hemibiotrophic pathogens, such as *Phytophthora infestans* and *Colletotrichum graminicola*, which initially interact with the host as biotrophs and then switch to a necrotrophic lifestyle. HR may be helpful to the host in these situations early on but not later in the interaction (Jupe et al., 2013).

2.5.7 ROLE OF PHYTOHORMONES IN PLANT IMMUNITY

Plants have created complex mechanisms to sense external signals in order to prepare for optimal responses to biotic stresses. Small, low-molecular-weight endogenous chemicals referred to as phytohormones control the defense reaction to such stimuli. These hormones, including SA, JA, ET, ABA, brassinosteroids (BRs), auxins, cytokinins (CKs), and gibberellins (GAs), respond to stress through synergistic and antagonistic actions, which are referred to as signaling cross-talk (Mauch-Mani and Métraux, 2005). According to large-scale transcriptome analyses, various signaling networks have been shown to have cross-talk (Davletova et al., 2005). In response to developmental and environmental stimuli, phytohormones regulate growth, development, and physiological activities while preserving stability in a highly complex network. Phytohormones, like animal hormones, are also transported from one location to another to mediate various processes (Eyidogan et al., 2012). The production, distribution, or signal transduction of these hormones can be affected by stress circumstances, resulting in morphological, molecular, and physiological changes that equip plants to endure the stress. The lowest level transducers start the stress signal activation of the signaling cascade, the phytohormones, which also start the response process (Harrison and Bonner, 2012).

2.5.7.1 SALICYLIC ACID (SA)

SA is a plant hormone essential for enhancing plant defenses against biotic stressors such as pathogen infestations. In addition to promoting the production of PR proteins and turning on SAR in the plant, SA also controls the expression of genes related to defense. SAR is a robust, long-lasting defense mechanism that guards the plant against several diseases. SA also has a role in the plant's response to abiotic stresses such as drought, salt, and heavy metals. In these situations, SA levels increase in the plant, and the activation of defense-related genes can help the plant adapt to the stress and survive (Sticher et al., 1997).

2.5.7.2 JASMONIC ACID (JA)

JA is another essential plant hormone that plays a crucial part in plant immunity, particularly in the defense against herbivores and necrotrophic pathogens. JA signaling triggers a set of defense responses that are different from those activated by SA, which is primarily involved in resistance to biotrophic pathogens. When herbivores or necrotrophic pathogens attack a plant, JA levels in the plant tissue increase, which initiates a chain of events that leads to the expression of defense-related genes and the production of various secondary metabolites, such as alkaloids, terpenoids, and phenylpropanoids. These compounds can directly harm herbivores or pathogens or act as repellents to prevent further damage.

JA signaling also induces the expression of protease inhibitors, which can inhibit the digestive enzymes of herbivores and reduce their ability to digest plant tissue. This can limit the damage caused by herbivores and reduce their fitness. In addition to its role in defense against herbivores and necrotrophic pathogens, it has been demonstrated that JA influences how plants react to abiotic conditions such as drought and salt stress (Avanci et al., 2010).

2.5.7.3 ABSCISIC ACID (ABA)

ABA is a plant hormone that plays a multifaceted role in plant immunity. ABA plays a role in the control of numerous physiological systems, such as seed germination, stomatal closure, and responses to abiotic stress; it also has a role in plant defense against biotic stress. ABA has been demonstrated to act as a positive regulator of plant defense against some pathogens, such as bacteria and fungi. It can trigger several defensive mechanisms, including the creation of ROS, the accumulation of PR proteins, and the manufacturing of phytoalexins, as well as defense-related genes and their expression. On the other hand, ABA can act as a negative regulator of plant defense against other pathogens, such as viruses and insect herbivores. In these cases, ABA can block the creation of defense-related chemicals and the expression of genes relevant to the defense. "Recent studies have also shown that ABA can interact with other plant hormones, such as SA and JA, to modulate the plant's response to biotic stress. The interaction between ABA, SA, and JA can result in a complex signaling network that regulates the balance between defense against pathogens and tolerance to abiotic stress" (Ton et al., 2009).

2.5.7.4 ETHYLENE (ET)

ET is a plant hormone that plays a crucial part in plant immunity, particularly in defense against necrotrophic pathogens, which cause cell death and tissue decay. ET signaling triggers a set of defense responses that are different from those activated by SA and JA. The creation of several secondary metabolites, including ethylene and indole derivatives, as well as the activation of defense-related genes, are all triggered by an increase in ET levels in plant tissue that occurs when necrotrophic diseases attack a plant. These substances may directly poison pathogens or may function as signaling molecules that trigger the activation of further defense mechanisms. Additionally, ET signaling activates genes involved in creating ROS, lignin, and suberin, which can strengthen plant cell walls and increase their resistance to pathogen invasion (Takahashi et al., 2007).

2.5.8 PHYTOHORMONES AND THEIR ROLE IN PLANT DEFENSE

Primary metabolites and secondary metabolites are terms used to describe plant chemicals (Table 2.1). Plants' primary metabolites for growth, development, and reproduction include sugars, amino acids, and nucleic acids. Secondary metabolites, on the other hand, include proteins that are involved in defense or poisonous substances that serve as a deterrent to pathogen invasion (Freeman and Beattie, 2008). Byproducts of primary metabolism, such as enzymes participating in numerous metabolic pathways and important cell macromolecules, are used to make these secondary metabolites. They are employed to defend plants against microbial infections despite being poisonous by nature. Additionally, some of these metabolites help plants resist abiotic stressors like UV-B radiation and are employed in both friendly and antagonistic interactions with other organisms, such as luring pollinators or deterring diseases and herbivores (Schäfer et al., 2009).

Plants use the byproducts of primary metabolism, such as enzymes involved in metabolic pathways and important cell macromolecules, to make secondary metabolites. Due to their poisonous character, these metabolites are employed to defend plants against invasive diseases (Mes et al., 2000). They also protect plants from abiotic stress, communicate with other organisms, and restrain pathogens and herbivores. Terpenes/terpenoids, phenolics, and chemicals containing nitrogen and sulfur are the three primary categories of secondary metabolites, and they are primarily formed from common amino acids. Terpenes are poisonous and deter herbivores from eating plants, whereas phenolics, mostly produced by the shikimic acid pathway, shield plants against fungus and microbial diseases. According to studies, plants create about 200,000 secondary metabolites as a part of their biochemical defense mechanisms, which have been evolving for millions of years. Plant-defense substances can be divided into two categories: constitutive metabolites, which are always present in plants, and induced metabolites, which are produced after a pathogen is detected. Constitutive metabolites are also called "prohibitins" or "phytoanticipins," whereas induced metabolites, also called "phytoalexins," are created to shield the plant against disease. The main chemical barrier that a pathogen must overcome is phytoanticipins, while phytoalexin production takes two to three days because enzyme synthesis is needed to transform their precursors into the appropriate defensive protein. Plants from the same species or closely related species have the same secondary metabolites/toxic compounds (Hammond-Kosack et al., 1996).

TABLE 2.1 Common Antimicrobial Plant Chemicals (Hammond-Kosack et al., 1996)

S.No.	Chemicals
1	Terpenes/Terpenoids
2	Saponins
3	Lignin
4	Flavonoids
5	Tannins
6	Alkaloids
7	Glucosinolates (GSL)
8	Glutathione (GSH)
9	Cyanogenic Glycosides
10	Defensins

2.5.9 ROLE OF PLANT CELL WALL

The plant cell wall is the first line of defense against the invasion of pathogens and plays a crucial role in detecting external stresses and transmitting corresponding signals to initiate defense responses (Cosgrove, 2005). Although it was previously viewed as a passive barrier, recent studies have suggested that the cell wall actively participates in defense mechanisms.

Depending on the plant species, stage of growth, and response to stress, the composition and structure of the cell wall can change. The cell wall is dynamically remodeled as a monitoring system to adapt to external stresses, including pathogenic infections. Reinforcement of the cell wall is a typical physiological response to activated immune responses in plants (Underwood, 2012). It has been increasingly recognized that maintaining cell wall integrity (CWI) is essential for activating and monitoring defense responses (Vaahtera et al., 2019). Therefore, changes to the plant cell wall are thought to be connected to the two-layered innate immune system, which consists of ETI and PTI. Plant cell walls serve as barriers against biotic and abiotic stressors in addition to giving structure to the plant body. The cell wall, which is occasionally coated by a cuticle, is typically the first barrier that infections must overcome. To overcome this barrier, microorganisms have developed an arsenal of wall-degrading enzymes that are important virulence factors.

Most plant cell walls consist of a network of cellulose microfibrils cross-linked with hemicelluloses with pectic polysaccharides in the primary walls of growing parts and lignin in the secondary walls of mature tissues (Scheller and Ulvskov, 2010). Despite these fundamental elements, cell walls differ in their fine structures and architectures, as seen in the variety of methods pathogens have devised to breach them. Plants may create poisonous substances and deposit strengthening polymers like callose and phenolic complexes in response to an attack (Hückelhoven, 2007). However, the capacity of pathogens to break down cell wall components also comes at a cost; thus, these physical and chemical responses only make up a portion of the cell wall's defensive function. Plants monitor changes in cell wall status, which can activate defense mechanisms and any rupture of cell wall integrity, which can result in distortion of the plasma membrane and the release of degradation fragments (Hématy et al., 2009).

2.5.10 MAP KINASE CASCADE

In eukaryotes, mitogen-activated protein kinase (MAPK) cascades play a key role in signaling. They work in conjunction with sensors and receptors to coordinate cellular responses, enabling organisms to grow and develop normally and adapt to their constantly changing environment (Widmann et al., 1999). One MAPK (MPK), one MAPK kinase (MAPKK, MKK, or MEK), and one MAPKK kinase make up a normal MAPK cascade (MAPKKK, MKKK, or MEKK). Multiple members frequently exist at a single layer of the cascade. To diversify the signaling pathway, they have either redundant or partially overlapping functions. The top kinase(s) of a MAPK cascade, MAPKKK, are activated in response to a stimulus, which phosphorylates and activates the downstream MAPKK(s). The activated MAPKK(s) in turn phosphorylate and activate the MAPK(s), the bottom tier kinase(s) in a MAPK cascade. The phosphorylation of several downstream substrates, such as transcription factors, protein kinases, other enzymes, and structural proteins by the activated MAPKs, can then trigger cellular responses (Jonak et al., 2002).

The sequence closeness and distinctive TXY activation motif between kinase subdomains VII and VIII of plant MAPKs make them easy to recognize. Once the upstream MAPKKs dual-phosphorylate the Thr and Tyr residues in this motif, the MAPKs are activated. Based on the amino acid in the center of the TXY motif, plant MAPKs can be divided into two classes. The TDY subtype is unique to plants, and the TEY subtype is related to yeast and the animal ERK subfamily. Three further groupings are created from the TEY subtype. It is well recognized that group A MAPKs, including *Arabidopsis* MPK3 and MPK6, are essential for plant immunology, abiotic stress response, and growth and development. Plant immunology, reactivity to environmental changes, and growth/development are all impacted by group B MAPKs, which include *Arabidopsis* MPK4, MPK5, MPK11, MPK12, and MPK13. However, little information is available regarding group C MAPKs, including *Arabidopsis* MPK1, MPK2, MPK7, and MPK14 (Zhang et al., 2018).

On the other hand, *Arabidopsis* possesses eight members of the TDY subtype, also referred to as group D MAPKs, including MPK8, MPK9, MPK15, MPK16, MPK17, MPK18, MPK19, and MPK20 (Dóczi et al., 2012). These MAPKs lack the C-terminal common docking (CD) domain seen in the other plant MAPK groups, as well as in MAPKs from yeast and animals, but they do contain an expanded C-terminal area in comparison to group A, B, and C members. MAPKKs, MAPK phosphatases, and protein substrates all dock at the CD domain. It was discovered that Arabidopsis MPK9, a member of the TDY subtype, may auto-phosphorylate on the Thr and Tyr residues in the TDY motif without requiring upstream MAPKKs for its activation. This contrasts with canonical MAPKs. Members of the TDY subtype might not act as usual MAPKs and might not work in three-tier MAPK cascades.

2.6 PROGRAMMED CELL DEATH IN PLANTS

Inhibition of pathogen growth is caused by pathogen recognition via NLRs in animals and NB-LRRs in plants. In plants, the HR, a type of programmed cell death localized at the site of attempted pathogen invasion, is frequently but not always accompanied by pathogen recognition. In terms of morphology, HR represents a distinct and specific form of cell death, characterized by certain features that are common along with plant-specific traits like vacuolization and chloroplast disintegration. Several types of animal cell death, including cytoplasmic shrinkage and chromatin condensation, are seen in the later stages (Mur et al., 2008).

The defense response and HR in plants rely heavily on the chloroplast. First, defense signaling chemicals, including ROS, reactive nitrogen oxide intermediates (NOI), and the defense hormones SA and JA, are all crucially produced in the chloroplast. Secondly, light is often necessary for the development of HR. Thirdly, some pathogen effectors have chloroplast localization signals and have been shown to suppress immunity in some cases (Fu et al., 2007). During ETI, the molecular processes resulting from HR partially overlap with those involved in MTI, including SA, ROS, NOI accumulation, MAPK cascade activation, intracellular calcium level changes, transcriptional reprogramming, and antimicrobial compound synthesis (Mur et al., 2008). ETI frequently produces a faster and stronger response than MTI, indicating that quantitative rather than qualitative differences are what cause HR induction.

Increased levels of ROS are a characteristic feature of necroptosis and could be one of the key reasons for the occurrence of necroptotic cell death. The rise in ROS levels during necroptosis can be attributed to mitochondrial activity, which is stimulated by RIP3, leading to an increase in energy metabolism (Zhang et al., 2009). Furthermore, RIP1-mediated recruitment of the NADPH oxidase NOX1 to the plasma membrane might cause an increase in ROS generation. For the growth of HR and the activation of systemic immunity in plants, plasma membrane NADPH oxidases in the apoplast must produce ROS (superoxide) (Torres et al., 2005). According to this, there may be a mechanistic connection between these two types of cell death. The HR response is also influenced by ROS produced in other plant organelles, such as chloroplasts, mitochondria, and peroxisomes. In fact, compartmentalization might be essential for the defense-related roles of ROS signaling.

Necroptosis plays a crucial part in both inflammation and immunity. Similar to pyroptotic cells, necroptosis-inducing cells discharge a range of pro-inflammatory chemicals that activate PRRs (Zitvogel et al., 2010). This type of cell death has been observed in response to viral infections where the virus blocks apoptosis in the host cell to facilitate colonization. Necroptosis may serve as a substitute way for virus removal when apoptosis is inhibited, and its pro-inflammatory nature may also trigger an immune response, leading to a systemic reaction.

2.7 ADVANCED TECHNOLOGIES IN PLANT-PATHOGEN INTERACTION AND CLIMATE RESILIENCE CROPS

2.7.1 GENOME EDITING TOOLS

Genome editing is a promising technology that could revolutionize plant breeding and help ensure global food security. The first-generation gene-editing tools are meganucleases, which have a precise and broad recognition site and are often referred to as homing endonuclease enzymes. Studies have shown that double-stranded breaks caused by meganucleases are repaired through non-homologous end joining, resulting in gene knockout. However, because the DNA binding domains are attached to the catalytic region, meganucleases cannot be easily altered with other genome editing techniques. Second-generation gene-editing tools include ZFNs, which are more efficient and reliable due to their ability to generate double-stranded breaks (Durai et al., 2005). ZFNs have two primary domains: the DNA-binding domain containing 300 to 600 zinc-finger repeats and the DNA slice domain, which is regarded as a nonspecific domain of type 2 restriction endonucleases. The use of TALENs for genome editing is also very common. The DNA-binding domains of TALE proteins, which feature intricate duplications of 34 amino acids for efficient base pair modification, were combined with the FokI slicing domain to create TALENs. TALENs and ZFNs have shown potential for genome editing despite the fact that there hasn't been much research using them in plants. In general, TALENs are preferred over ZFNs for genome editing programs because they are more efficient and dependable (Gaj et al., 2012).

2.7.1.1 INTRODUCTION TO CRISPR

RNA and protein-based adaptive immune systems called clustered regularly interspaced short palindromic repeats (CRISPR)-Cas guard bacteria and archaea against invasive viruses and foreign nucleic acids (Bhaya et al., 2011). Bacterial CRISPR loci are comprised of a Cas operon and a repeat-spacer array. Archived sequences known as CRISPR spacers are 23–44 base pair long sequences that were copied from phage or plasmid DNA that had previously infected the bacteria (Mojica et al., 2005). These sequences are situated close to the CRISPR-associated (Cas) genes, which produce the Cas ribonucleases and are sandwiched between repetitive sequences. The bacterial genome's incorporation of spacers enables it to identify and destroy nucleic acids that match the stored sequence upon subsequent invasion.

Pre-CRISPR RNA (crRNA), which is produced by the CRISPR locus, is converted into mature crRNA by the Cas proteins and auxiliary factors (Deltcheva et al., 2011). In order to cut nucleic acid sequences that are surrounded by a brief protospacer adjacent motif (PAM) sequence, Cas proteins use crRNA (Makarova et al., 2006). There are three different CRISPR-Cas systems, and each one has a unique method of processing pre-crRNA. Since Type II systems only need one protein, Cas9, to find and cleave the target sequences, they are the most often utilized for eukaryotic genome engineering. The repetition portions of the crRNA transcript are paired with the trans-activating crRNA (tracrRNA). The tracrRNA and crRNA work in tandem to direct Cas9 to the target locations and activate its nuclease activity against DNA that is complementary to the crRNA (Deltcheva et al., 2011). TracrRNA and crRNA are frequently combined to create single-guide RNA (sgRNA) transcripts in order to engineer eukaryotic genomes. The complementary DNA strand to guide the Cas9 HNH domain cuts RNA, whereas the non-complementary strand is cut by the ruvc domain.

2.7.1.2 APPLICATION OF CRISPR/CAS9 FOR CROP IMPROVEMENT

The CRISPR/Cas9 technique has emerged as a highly popular and advantageous tool for genome editing. It is highly efficient, cost-effective, and adaptable, making it ideal for breeding polyploid crops that are challenging to improve through traditional methods. The technique enables the generation of monoallelic and biallelic mutants, along with a wide range of resulting allelic phenotypes in the first generation, which is typically not feasible through classical breeding methods. Compared to other genome editing techniques such as TALENs, ZFNs, and MNs, CRISPR/Cas9 is simpler, more efficient, and more versatile (Cong et al., 2013). It can edit multiple genomes simultaneously, providing an optimal solution to various issues faced in plant breeding. Many plant species, including monocots and dicots, have been improved using the CRISPR/Cas9 technique for various traits like disease resistance, yield, quality, and tolerance to climate change. Numerous crops' genomes, including those of wheat, maize, rice, cotton, tomatoes, potatoes, bananas, and apples, have been edited using this technique. The most typical use of CRISPR/Cas9 is to induce indels, which cause frameshift mutations in the target genes (Tan et al., 2020).

2.7.1.3 DEVELOPMENT OF DISEASE-RESISTANT VARIETIES USING CRISPR/CAS9

Recent developments in CRISPR technology have made it possible for researchers to create a wide variety of CRISPR variants with various uses. Plants with resistance to disease have been successfully engineered using CRISPR.

2.7.1.3.1 Gene Disruption By Indels In Coding Sequences

This is where the CRISPR-Cas9 technology is most frequently used. It benefits from the cellular NHEJ DNA repair system's erratic nature. An insertion or deletion (indel) of one or more nucleotides occurs as a result at the sgRNA-guided location, resulting in a frameshift mutation and preventing gene expression. This approach has been used to introduce desirable features into a variety of crops, including important grains like wheat and rice. Through the disruption of a plant susceptibility (S) gene, which modifies the interaction between plants and pathogens and reduces pathogen fitness on the host plant, this approach has been used to design disease resistance (Zaidi et al., 2018).

2.7.1.3.2 Gene Disruption by Indels in Promoter Regions

Instead of the coding area of a plant gene, the promoter region can be altered using a method similar to this one. A plant's susceptibility to pathogens can be disrupted by using CRISPR-mediated promoter editing in one of two ways: either to disrupt the promoter sequence, with the goal of completely blocking gene expression, or to disrupt an effector-binding site, with the goal of preventing a pathogen effector from binding to the promoter.

Targeting upstream open reading frame (ORF) regions and editing cis-regulatory elements have both been shown to be additional ways that CRISPR can change the regulation of genes in addition to promoter editing (Zhang et al., 2018).

2.7.1.3.3 Gene Deletion by Multiplex SgRNAs

Multiple sgRNAs can be utilized in CRISPR systems to introduce multiple DSBs at specific places in the target genome. For instance, two sgRNAs binding before and after the start and finish codons of the target gene will result in DSBs at the corresponding sites. Prior to the cellular repair of NHEJ machinery correcting the DSBs, these DSBs cause the loss of the DNA fragment carrying the gene of interest. This method can be used to delete huge chromosomal segments and specific genes because sgRNAs can be generated at any genomic region containing an adequate PAM trinucleotide sequence. This is made easier by creating SpCas9-NG, a Cas9 with a rational design with a softer recognition threshold than standard NGG PAMs. Gene clusters are especially advantageous in the context of engineered pathogen resistance. Deleting the chromosomal fragment is probably going to result in long-term resistance to the target pathogen in gene clusters, where numerous genes are found in close proximity to one another on the chromosome (Ashfield et al., 2012).

2.7.1.3.4 Gene Insertion by Homology-Directed Repair

The technique of homology-directed repair via gene insertion can be utilized with CRISPR to create disease resistance by modifying the S genes. However, since the proteins encoded by S genes are crucial and serve multiple functions, interfering with them can have negative impacts on plant health and productivity. There are instances where it is necessary to introduce resistance genes (R genes) to combat pathogens even though gene expression can be changed through strategies like cis-regulatory element and promoter editing. This is especially true when there is little understanding of the plant-pathogen interaction and the S genes involved. In such cases, R genes can be inserted into the plant genome using the CRISPR toolset (Butt et al., 2017).

After producing a DSB guided by sgRNA, CRISPR-mediated gene insertion utilizes cellular HDR machinery instead of NHEJ. This involves supplying a delivery fragment containing an *R* gene flanked by homologous sequences to the DSB ends along with Cas9 and sgRNA. The *R* gene is then precisely inserted between the two DSB sites through HDR-mediated repair, which has been successfully used to introduce one or more genes in specific genomic locations. However, the efficiency of HDR in plants is currently low, making gene insertion in plants a challenging process. Despite this, it is still possible to investigate the roles of *S* genes using CRISPR-mediated gene insertion. With the help of HDR, Wang et al., successfully inserted GFP fused with the glutathione S-transferase locus in rice, allowing for the investigation of protein localization and *S* gene expression control. Protein changes, however, may change expression and localization, highlighting the drawbacks of this strategy.

2.7.1.3.5 Biomimicking by the Promoter, Allele, or Gene Replacement

To avoid regulatory hurdles and consumer resistance to the insertion of foreign DNA into crops, CRISPR technology provides a biomimicking approach that offers an alternative solution. Specifically for resistance against biotic and abiotic challenges, this method makes use of the huge gene pool found in natural species and wild relatives of farmed crops. Recent studies have successfully transferred resistance traits by identifying functional R genes in the wild relatives of cultivated species and transferring them to the cultivated crop variety using multiplexed HDR methodology. The process of "biomimicking" entails introducing CRISPR-mediated mutations so that the target gene's sequence resembles that of a disease-resistant variation. Suppose the nucleotide variations between the target gene in cultivated and wild kinds do not significantly affect plant viability and productivity. In that case, just the specific mutations linked to disease resistance are introduced instead of the entire gene. A CRISPR system created for point mutagenesis, which has been utilized effectively in various species for gene modification, is employed to accomplish this targeted nucleotide change (Komor et al., 2016).

2.7.1.3.6 Development of Climate-Smart Crops Using CRISPR/Cas9

Climate change is causing significant alterations in the agricultural sector, leading to decreased crop yields and food security concerns. With the use of the groundbreaking genome editing tool CRISPR/Cas9, researchers may change crop genes, potentially improving their tolerance to climate change and other environmental stresses.

The development of climate-smart crops using CRISPR/Cas9 involves identifying genes that control various traits related to environmental stress tolerance, such as drought, heat, cold, and salinity. Scientists can then use CRISPR/Cas9 to edit those genes by inserting a desirable trait or silencing a detrimental one.

The Cas9 technique has been used to address various abiotic stresses in major crops, including rice, wheat, maize, cotton, and potatoes, leading to the development of climate-smart and stress-resistant crops (Razzaq et al., 2021). Using CRISPR/Cas9, several genes for abiotic stress resistance have been found and created in crops, such as the SlMAPK3 protein gene in tomatoes, TaDREB3 and TaDREB2 in wheat, and OsMKP2 and OsBADH2 in rice. Other genes, including OsAnn3, SAPK2, ARGOS8, Leafl,2, and OsRR22, have also been edited to confer resistance to cold, drought, and salinity stresses in rice. The creation of drought-tolerant variants of chick-peas was facilitated by the Cas9-mediated deletion of the 4CL and RVE7 genes. Herbicide resistance has been artificially evolved in crops using the CRISPR/Cas9 method, such as OsALS1 in rice. These studies demonstrate the potential of CRISPR/Cas9 in modernizing plant breeding systems and developing crops with enhanced stress tolerance.

A study examined the use of Cas-based mutation to modify two genetic factors, Drb2a and Drb2b, which controlled drought and salt tolerance in soybeans. Another study concentrated on the mitogen-activated protein kinase gene, which protects the plant's membrane from oxidative stress and regulates gene transcription to mitigate the effects of drought stress. The Cas9 technique was also used to produce knockout variants in the *SIMPAK3* gene in tomatoes to develop drought tolerance (Table 2.2).

TABLE 2.2 List of Climate-Smart Crops (Razzaq et al., 2021)

S.No.	Crop	Traits
1.	Rice	Cold tolerance
2.	Tomato	Drought tolerance
3.	Chickpea	Drought tolerance
4.	Wheat	Drought tolerance
5.	Maize	Tolerance to salinity stress
6.	Soyabean	Tolerance to drought and salinity stress

Several genes control crop yield and resilience to stress, and scientists have identified QTLs that regulate important traits in crop breeding systems. While these QTLs have been combined in elite lines, unintended transfer of non-target regions during this process can lead to negative outcomes. To address this issue, targeted mutagenesis using Cas9 can be employed. For example, the QTLs Gn1a and GS3 in rice cultivars were targeted using a Cas9-based QTL editing approach. This study demonstrates how CRISPR/Cas9 technology can be utilized to develop crops that are well-adapted to their environment and resistant to various abiotic stresses such as drought, high temperatures, heavy metals, and nutrient deficiencies (Shen et al., 2018).

2.8 ONCLUSION AND FUTURE OUTLOOK

In modern agriculture, disease-causing pathogens constantly threaten crop yield and quality, making it one of the most disruptive factors. Currently, the primary solution is the use of agrochemicals. However, locating and analyzing R genes may open up possibilities for creating plant kinds that are disease resistant. Future research in the field of plant pathology aims to determine the molecular basis of disease resistance against various phyto-pathogens and understand how R gene products recognize pathogen elicitors and block pathogen growth. This knowledge will facilitate the production of durable disease-resistant plants and reduce the reliance on environmentally harmful agrochemicals. Therefore, it is essential to understand the molecular basis of plant-pathogen interactions to equip plants with the best molecular defenses against pathogen attacks.

KEYWORDS

- **abscisic acid (ABA)**
- **climate-smart crops**
- **elicitors**
- **innate immunity**
- **interactions of host-pathogen**

REFERENCES

Antoniw, J. F., & Pierpoint, W. S. (1978). Purification of a tobacco leaf protein associated with resistance to virus infection. *Biochemical Society Transactions, 6*(1), 248–250.

Ashfield, T., Egan, A. N., Pfeil, B. E., Chen, N. W., Podicheti, R., Ratnaparkhe, M. B., & Innes, R. W. (2012). Evolution of a complex disease resistance gene cluster in diploid *Phaseolus* and tetraploid *Glycine*. *Plant Physiology, 159*(1), 336–354.

Attard, A., Gourgues, M., Galiana, E., Panabieres, F., Ponchet, M., & Keller, H. (2008). Strategies of attack and defense in plant–oomycete interactions, accentuated for *Phytophthora parasitica* Dastur (syn. *P. nicotianae* Breda de Haan). *Journal of Plant Physiology, 165*(1), 83–94.

Avanci, N. C., Luche, D. D., Goldman, G. H., & Goldman, M. H. S. (2010). Jasmonates are phytohormones with multiple functions, including plant defense and reproduction. *Genetics and Molecular Research, 9*(1), 484–505.

Bent, A. F., & Mackey, D. (2007). Elicitors, effectors, and R genes: The new paradigm and a lifetime supply of questions. *Annual Review of Phytopathology, 45*, 399–436.

Bhaya, D., Davison, M., & Barrangou, R. (2011). CRISPR-Cas systems in bacteria and archaea: Versatile small RNAs for adaptive defense and regulation. *Annual Review of Genetics, 45*, 273–297.

Bigeard, J., Colcombet, J., & Hirt, H. (2015). Signaling mechanisms in pattern-triggered immunity (PTI). *Molecular Plant, 8*(4), 521–539.

Butt, H., Eid, A., Ali, Z., Atia, M. A., Mokhtar, M. M., Hassan, N., & Mahfouz, M. M. (2017). Efficient CRISPR/Cas9-mediated genome editing using a chimeric single-guide RNA molecule. *Frontiers in Plant Science, 8*, 1441.

Chen, L. Q., Hou, B. H., Lalonde, S., Takanaga, H., Hartung, M. L., Qu, X. Q., & Frommer, W. B. (2010). Sugar transporters for intercellular exchange and nutrition of pathogens. *Nature, 468*(7323), 527–532.

Chisholm, S. T., Coaker, G., Day, B., & Staskawicz, B. J. (2006). Host-microbe interactions: Shaping the evolution of the plant immune response. *Cell, 124*(4), 803–814.

Choudhary, D. K., Prakash, A., & Johri, B. N. (2007). Induced systemic resistance (ISR) in plants: Mechanism of action. *Indian Journal of Microbiology, 47*, 289–297.

Cong, L., Ran, F. A., Cox, D., Lin, S., Barretto, R., Habib, N., & Zhang, F. (2013). Multiplex genome engineering using CRISPR/Cas systems. *Science, 339*(6121), 819–823.

Cosgrove, D. J. (2005). Growth of the plant cell wall. *Nature Reviews Molecular Cell Biology, 6*(11), 850–861.

Dangl, J. L., & Jones, J. D. (2001). Plant pathogens and integrated defense responses to infection. *Nature, 411*(6839), 826–833.

Davletova, S., Rizhsky, L., Liang, H., Shengqiang, Z., Oliver, D. J., Coutu, J., & Mittler, R. (2005). Cytosolic ascorbate peroxidase 1 is a central component of the reactive oxygen gene network of Arabidopsis. *The Plant Cell, 17*(1), 268–281.

Delaney, T. P. (1997). Genetic dissection of acquired resistance to disease. *Plant Physiology, 113*(1), 5.

Deltcheva, E., Chylinski, K., Sharma, C. M., Gonzales, K., Chao, Y., Pirzada, Z. A., & Charpentier, E. (2011). CRISPR RNA maturation by trans-encoded small RNA and host factor RNase III. *Nature, 471*(7340), 602–607.

Delventhal, R., Rajaraman, J., Stefanato, F. L., Rehman, S., Aghnoum, R., McGrann, G. R., & Schaffrath, U. (2017). A comparative analysis of non-host resistance across the two Triticeae crop species wheat and barley. *BMC Plant Biology, 17*, 1–22.

Dietz, K. J., Turkan, I., & Krieger-Liszkay, A. (2016). Redox-and reactive oxygen species-dependent signaling into and out of the photosynthesizing chloroplast. *Plant Physiology, 171*(3), 1541–1550.

Dóczi, R., Ökrész, L., Romero, A. E., Paccanaro, A., & Bögre, L. (2012). Exploring the evolutionary path of plant MAPK networks. *Trends in Plant Science, 17*(9), 518–525.

Durai, S., Mani, M., Kandavelou, K., Wu, J., Porteus, M. H., & Chandrasegaran, S. (2005). Zinc finger nucleases: Custom-designed molecular scissors for genome engineering of plant and mammalian cells. *Nucleic Acids Research, 33*(18), 5978–5990.

Ebert, D., & Herre, E. A. (1996). The evolution of parasitic diseases. *Parasitology Today, 12*(3), 96–101.

Eyidogan, F., Oz, M. T., Yucel, M., & Oktem, H. A. (2012). Signal transduction of phytohormones under abiotic stresses. In M. Ashraf & P. J. C. Harris (Eds.), *Phytohormones and Abiotic Stress Tolerance in Plants* (pp. 1–48). Springer Netherlands.

Freeman, B., & Beattie, G. (2008). An overview of plant defenses against pathogens and herbivores. *Plant Health Instructor, 149.*

Fu, Z. Q., Guo, M., Jeong, B. R., Tian, F., Elthon, T. E., Cerny, R. L., & Alfano, J. R. (2007). A type III effector ADP-ribosylates RNA-binding proteins and quells plant immunity. *Nature, 447*(7142), 284–288.

Gaj, T., Guo, J., Kato, Y., Sirk, S. J., & Barbas III, C. F. (2012). Targeted gene knockout by direct delivery of zinc-finger nuclease proteins. *Nature Methods, 9*(8), 805–807.

Gill, S. S., & Tuteja, N. (2010). Reactive oxygen species and antioxidant machinery in abiotic stress tolerance in crop plants. *Plant Physiology and Biochemistry, 48*(12), 909–930.

Goodman, R. N., & Novacky, A. J. (1994). The hypersensitive reaction in plants to pathogens: A resistance phenomenon. *American Phytopathological Society (APS), 4*, 671–683.

Hamann, T. (2012). Plant cell wall integrity maintenance as an essential component of biotic stress response mechanisms. *Frontiers in Plant Science, 3*, 77.

Hammond-Kosack, K. E., & Jones, J. D. G. (1996). Resistance gene-dependent plant defense responses. *The Plant Cell, 8*(10), 1773.

Harrison, M. A. (2012). Cross-talk between phytohormone signaling pathways under both optimal and stressful environmental conditions. In M. Ashraf & P. J. C. Harris (Eds.), *Phytohormones and Abiotic Stress Tolerance in Plants* (pp. 49–76). Springer Netherlands.

Hématy, K., Cherk, C., & Somerville, S. (2009). Host–Pathogen warfare at the plant cell wall. *Current Opinion in Plant Biology, 12*(4), 406–413.

Huang, S., Van Aken, O., Schwarzländer, M., Belt, K., & Millar, A. H. (2016). The roles of mitochondrial reactive oxygen species in cellular signaling and stress response in plants. *Plant Physiology, 171*(3), 1551–1559.

Hückelhoven, R. (2007). Cell wall–Associated mechanisms of disease resistance and susceptibility. *Annual Review of Phytopathology, 45*, 101–127.

Inohara, N., Chamaillard, M., McDonald, C., & Nunez, G. (2005). NOD-LRR proteins: Role in host-microbial interactions and inflammatory disease. *Annual Review of Biochemistry, 74*, 355–383.

Jain, D., & Khurana, J. P. (2018). Role of pathogenesis-related (PR) proteins in plant defense mechanism. In D. Jain & J. P. Khurana (Eds.), *Molecular Aspects of Plant-Pathogen Interaction* (pp. 265–281).

Jonak, C., Ökrész, L., Bögre, L., & Hirt, H. (2002). Complexity, cross talk and integration of plant MAP kinase signaling. *Current Opinion in Plant Biology, 5*(5), 415–424.

Jones, J. D., & Dangl, J. L. (2006). The plant immune system. *Nature, 444*(7117), 323–329.

Jupe, J., Stam, R., Howden, A. J., Morris, J. A., Zhang, R., Hedley, P. E., & Huitema, E. (2013). Phytophthora capsici-tomato interaction features dramatic shifts in gene expression associated with a hemi-biotrophic lifestyle. *Genome Biology, 14*, 1–18.

Komor, A. C., Kim, Y. B., Packer, M. S., Zuris, J. A., & Liu, D. R. (2016). Programmable editing of a target base in genomic DNA without double-stranded DNA cleavage. *Nature, 533*(7603), 420–424.

Kunz, M., Ahmed, N., Dandekar, T., & Naseem, M. (2013). Hormone signaling networks open multiple routes for immunity and disease in plants. *Biohelikon Immunity and Diseases, 1*(2).

Lotze, M. T., Zeh, H. J., Rubartelli, A., Sparvero, L. J., Amoscato, A. A., Washburn, N. R., & Billiar, T. (2007). The grateful dead: Damage-associated molecular pattern molecules and reduction/oxidation regulate immunity. *Immunological Reviews, 220*(1), 60–81.

Mahadevakumar, S., & Sridhar, K. R. (2020). Plant-microbe interaction: Current developments and future challenges. In S. Mahadevakumar & K. R. Sridhar (Eds.), *Advances in Plant Microbiome and Sustainable Agriculture: Diversity and Biotechnological Applications* (pp. 1–38).

Makarova, K. S., Grishin, N. V., Shabalina, S. A., Wolf, Y. I., & Koonin, E. V. (2006). A putative RNA-interference-based immune system in prokaryotes: Computational analysis of the predicted enzymatic machinery, functional analogies with eukaryotic RNAi, and hypothetical mechanisms of action. *Biology Direct, 1*(1), 1–26.

Mauch-Mani, B., & Mauch, F. (2005). The role of abscisic acid in plant–pathogen interactions. *Current Opinion in Plant Biology, 8*(4), 409–414.

Melotto, M., Underwood, W., & He, S. Y. (2008). Role of stomata in plant innate immunity and foliar bacterial diseases. *Annual Review of Phytopathology, 46*, 101–122.

Mendoza, M. (2011). Oxidative burst in plant-pathogen interaction. *Biotecnología Vegetal, 11*(2), 67–75.

Mengiste, T. (2012). Plant immunity to necrotrophs. *Annual Review of Phytopathology, 50*, 267–294.

Mes, J. J., Van Doorn, A. A., Wijbrandi, J., Simons, G., Cornelissen, B. J., & Haring, M. A. (2000). Expression of the Fusarium resistance gene I-2 colocalizes with the site of fungal containment. *The Plant Journal, 23*(2), 183–193.

Miller, G. A. D., Suzuki, N., Ciftci-Yilmaz, S. U. L. T. A. N., & Mittler, R. O. N. (2010). Reactive oxygen species homeostasis and signaling during drought and salinity stresses. *Plant, Cell & Environment, 33*(4), 453–467.

Mishra, R. (2018). Fungal and bacterial biotrophy and necrotrophy. In R. Mishra (Ed.), *Molecular Aspects of Plant-Pathogen Interaction* (pp. 21–42).

Mojica, F. J., Díez-Villaseñor, C. S., García-Martínez, J., & Soria, E. (2005). Intervening sequences of regularly spaced prokaryotic repeats derive from foreign genetic elements. *Journal of Molecular Evolution, 60*, 174–182.

Mur, L. A., Kenton, P., Lloyd, A. J., Ougham, H., & Prats, E. (2008). The hypersensitive response; the centenary is upon us but how much do we know? *Journal of Experimental Botany, 59*(3), 501–520.

Mur, L. A., Kenton, P., Lloyd, A. J., Ougham, H., & Prats, E. (2008). The hypersensitive response; the centenary is upon us but how much do we know? *Journal of Experimental Botany, 59*(3), 501–520.

Oliver, R. P., & Ipcho, S. V. (2004). Arabidopsis pathology breathes new life into the necrotrophs-vs.-biotrophs classification of fungal pathogens. *Molecular Plant Pathology, 5*(4), 347–352.

Osbourn, A. E. (1996). Preformed antimicrobial compounds and plant defense against fungal attack. *The Plant Cell, 8*(10), 1821.

Pan, Q., Wendel, J., & Fluhr, R. (2000). Divergent evolution of plant NBS-LRR resistance gene homologues in dicot and cereal genomes. *Journal of Molecular Evolution, 50*, 203–213.

Razzaq, M. K., Aleem, M., Mansoor, S., Khan, M. A., Rauf, S., Iqbal, S., & Siddique, K. H. (2021). Omics and CRISPR-Cas9 approaches for molecular insight, functional gene analysis, and stress tolerance development in crops. *International Journal of Molecular Sciences, 22*(3), 1292.

Sandalio, L. M., & Romero-Puertas, M. C. (2015). Peroxisomes sense and respond to environmental cues by regulating ROS and RNS signaling networks. *Annals of Botany, 116*(4), 475–485.

Schäfer, H., & Wink, M. (2009). Medicinally important secondary metabolites in recombinant microorganisms or plants: Progress in alkaloid biosynthesis. *Biotechnology Journal: Healthcare Nutrition Technology, 4*(12), 1684–1703.

Scheller, H. V., & Ulvskov, P. (2010). Hemicelluloses. *Annual Review of Plant Biology, 61*, 263–289.

Schulze-Lefert, P., & Panstruga, R. (2003). Establishment of biotrophy by parasitic fungi and reprogramming of host cells for disease resistance. *Annual Review of Phytopathology, 41*(1), 641–667.

Shen, L., Wang, C., Fu, Y., Wang, J., Liu, Q., Zhang, X., & Wang, K. (2018). QTL editing confers opposing yield performance in different rice varieties. *Journal of Integrative Plant Biology, 60*(2), 89–93.

Silverstein, K. A., Graham, M. A., Paape, T. D., & VandenBosch, K. A. (2005). Genome organization of more than 300 defensin-like genes in Arabidopsis. *Plant Physiology, 138*(2), 600–610.

Sticher, L., Mauch-Mani, B., & Métraux, A. J. (1997). Systemic acquired resistance. *Annual Review of Phytopathology, 35*(1), 235–270.

Swain, A., & Pan, A. (2023). Protein therapeutic target candidates against Acinetobacter baumannii, a pathogen of concern to planetary health: A network-based integrative omics drug discovery approach. *OMICS: A Journal of Integrative Biology, 27*(2), 62–74.

Takahashi, F., Yoshida, R., Ichimura, K., Mizoguchi, T., Seo, S., Yonezawa, M., & Shinozaki, K. (2007). The mitogen-activated protein kinase cascade MKK3–MPK6 is an important part of the jasmonate signal transduction pathway in Arabidopsis. *The Plant Cell, 19*(3), 805–818.

Tan, Y. Y., Du, H., Wu, X., Liu, Y. H., Jiang, M., Song, S. Y., & Shu, Q. Y. (2020). Gene editing: An instrument for practical application of gene biology to plant breeding. *Journal of Zhejiang University Science B, 21*(6), 460.

Thakur, M., & Sohal, B. S. (2013). Role of elicitors in inducing resistance in plants against pathogen infection: A review. *International Scholarly Research Notices, 2013*, 762412.

Ton, J., Flors, V., & Mauch-Mani, B. (2009). The multifaceted role of ABA in disease resistance. *Trends in Plant Science, 14*(6), 310–317.

Torres, M. A., & Dangl, J. L. (2005). Functions of the respiratory burst oxidase in biotic interactions, abiotic stress and development. *Current Opinion in Plant Biology, 8*(4), 397–403.

Torres, M. A., & Dangl, J. L. (2005). Functions of the respiratory burst oxidase in biotic interactions, abiotic stress and Development. *Current Opinion in Plant Biology, 8*(4), 397–403.

Underwood, W. (2012). The plant cell wall: A dynamic barrier against pathogen invasion. *Frontiers in Plant Science, 3*, 85.

Vaahtera, L., Schulz, J., & Hamann, T. (2019). Cell wall integrity maintenance during plant development and interaction with the environment. *Nature Plants, 5*(9), 924–932.

Vleeshouwers, V. G., & Oliver, R. P. (2014). Effectors as tools in disease resistance breeding against biotrophic, hemibiotrophic, and necrotrophic plant pathogens. *Molecular Plant-Microbe Interactions, 27*(3), 196–206.

Widmann, C., Gibson, S., Jarpe, M. B., & Johnson, G. L. (1999). Mitogen-activated protein kinase: Conservation of a three-kinase module from yeast to human. *Physiological Reviews, 79*(1), 143–180.

Wolpert, T. J., Dunkle, L. D., & Ciuffetti, L. M. (2002). Host-selective toxins and a virulence determinants: What's in a name? *Annual Review of Phytopathology, 40*(1), 251–285.

Xie, H. T., Wan, Z. Y., Li, S., & Zhang, Y. (2014). Spatiotemporal production of reactive oxygen species by NADPH oxidase is critical for tapetal programmed cell death and pollen development in Arabidopsis. *The Plant Cell, 26*(5), 2007–2023.

Yeats, T. H., & Rose, J. K. (2013). The formation and function of plant cuticles. *Plant Physiology, 163*(1), 5–20.

Zaidi, S. S. E. A., Mukhtar, M. S., & Mansoor, S. (2018). Genome editing: Targeting susceptibility genes for plant disease resistance. *Trends in Biotechnology, 36*(9), 898–906.

Zhang, D. W., Shao, J., Lin, J., Zhang, N., Lu, B. J., Lin, S. C., & Han, J. (2009). RIP3, an energy metabolism regulator that switches TNF-induced cell death from apoptosis to necrosis. *Science, 325*(5938), 332–336.

Zhang, H., Si, X., Ji, X., Fan, R., Liu, J., Chen, K., & Gao, C. (2018). Genome editing of upstream open reading frames enables translational control in plants. *Nature Biotechnology, 36*(9), 894–898.

Zhang, M., Su, J., Zhang, Y., Xu, J., & Zhang, S. (2018). Conveying endogenous and exogenous signals: MAPK cascades in plant growth and defense. *Current Opinion in Plant Biology, 45*, 1–10.

Zitvogel, L., Kepp, O., & Kroemer, G. (2010). Decoding cell death signals in inflammation and immunity. *Cell, 140*(6), 798–804.

CHAPTER 3

Insights into the Molecular Mechanism of Plant-Pathogen Interactions in Fungal Diseases of Crop Plants

VISHAL DINKAR,[1] JAGMOHAN SINGH,[2] CHARISHMA KRISHNAPPA,[3] SWAGATA THAKUR,[3] GOPI KISHAN,[4] YAMINI TAK,[5] ASHARANI PATEL,[6] KULESHWAR PRASAD SAHU,[6] SAJAD UN NABI,[1] and SAHIL MEHTA[7]

[1]*ICAR–Central Institute of Temperate Horticulture, Srinagar, Jammu and Kashmir, India*

[2]*Department of Vegetable Science, CCS Haryana Agricultural University, Hisar, Haryana, India*

[3]*Division of Plant Pathology, ICAR–Indian Agricultural Research Institute, New Delhi, India*

[4]*ICAR–Indian Institute of Seed Science, Kushmaur, Mau, Uttar Pradesh, India*

[5]*Agricultural Research Station, Agriculture University, Kota, Rajasthan, India*

[6]*School of Agricultural Sciences, GD Goenka University, Sohna, Gurugram, Haryana, India*

[7]*Department of Botany, Hansraj College, University of Delhi, New Delhi, India*

Climate Change and Biotic Factors: A Molecular Approach.
Ashutosh Singh, Saurabh Pandey, and Amarjeet Kumar, (Eds.)

ABSTRACT

The plant pathogen interacts at the interface through molecular signals that decide the fate of the relationship. The disease reaction compatibility is governed by the interaction of pathogen molecules such as Pathogen-Associated Molecular Patterns (PAMPs) and effectors with plant receptor molecules like Pattern Recognition Receptors (PRRs) & NBS-LRR proteins (R-proteins). Innate immunity (PTI), the first line of defense, is the result of the recognition of PAMP by PRRs, where the breakdown of PTI by effector molecules of pathogen leads to susceptibility called effector-triggered susceptibility (ETS). The second and strong line of defense is effector-triggered immunity (ETI) which is prompted by the recognition of pathogen-secreted effectors by NBS-LRR proteins. The recognition of pathogen molecules by plant receptors initiates the defense signaling in plants through components like MAPK cascades and regulatory phytohormones such as salicylic acid (SA), jasmonic acid (JA), and ethylene (ET). Further, the defense responses are activated through transcriptional regulations by TFs like WRKY, and defense-related proteins are synthesized, such as PR-proteins and antimicrobial secondary metabolites. This book chapter comprehensively covers the plant–pathogen interactions from recognition of the pathogen to biochemical alterations that provide tolerance or resistance to the plant against a pathogen.

3.1 INTRODUCTION

Erstwhile, the importance of pathogens was realized owing to their ability to cause yield losses. This was further enumerated to be the result of the breakdown of host resistance by the pathogen. The effect was considered significant if an enormous yield loss was observed. Pathogen stress is the most complex stress faced by a plant. Host resistance and susceptibility are two dynamic concepts in a single host. A relay of responses is often overlooked during host-pathogen interaction. Despite the evidence that there is a ~400 million years back history of coevolution between plants and pathogens, the mechanisms both at the cellular and molecular levels remained obscure (Upson et al., 2018). The understanding was limited to a few morphological changes observed during pathogen progression in plants. The realization that the interaction between plants and pathogens is more than just the morphological changes is not very long ago. The gene-for-gene interaction revolutionized the understanding of plant-pathogen interaction. This realized the genetic interdependence between plants and pathogens (Flor, 1971). Since this inception, several other concepts revealing the molecular mechanism of plant-pathogen interaction were further developed. The saga started with genes and went on to products, and now the understanding is at the molecular motifs or signature sequences. The interplay is found to be very fierce but silent (Zhang et al., 2022). It is after 50 years that the fungi represent the most diversified kingdom on Earth and are the major group of organisms causing huge agronomic as well as economic losses. They are responsible for both plant nutrition and disease outbreaks (Polme et al., 2017). Fungi were majorly known for their plundering action on dead organic materials for nutrient sources. Most known fungi species are saprobes, and a few are plant pathogens. It was in this group of pathogen and plant systems that the basis of molecular interaction started to evolve. Fungi have evolved mechanisms to overcome plant defense systems. The infection and colonization behavior is different for different groups of fungi. For example, a few penetrate directly, and some enter the host through natural openings or wounds. Some fungi colonize above-ground plant parts, and some strictly adhere to roots and vascular tissues. The drivers of this organ specificity or their infection behavior were poorly understood (Schafer, 1994). Further, the genes responsible for causing pathogenicity in plants were referred to as fungal pathogenicity genes. These genes are involved directly and inherently in pathogenicity. It took nearly 50 years before the first fungal avirulence (*Avr*) gene in support of the gene-for-gene hypothesis was cloned.

The interaction of plants and pathogens starts initially at contact, mostly plant membranes. Pathogens use a series of motifs referred to as PAMPs (Pathogen-Associated Molecular Patterns) to recognize plant motifs referred to as Pattern Recognition Receptors (PRRs), which are mainly receptor-like kinases (RLKs) and receptor-like proteins (RLPs). A few of the fungal PAMPs are chitin, cell wall derivative glucan, lipopolysaccharide, and ergosterol. PAMPs are often referred to as general elicitors, which help the plants differentiate self and non-self-motifs (Ingle et al., 2006). Plants always employ a phased approach to combat pathogens. One is at the membrane level through PRRs, and the other is inside the cell employing polymorphic NB-LRR protein products of R genes. Initially, the plant employs PRRs to recognize pathogen PAMPs, such as chitin and ergosterol. Once the recognition is confirmed, the plant exhibits a resistance show termed PAMP-triggered immunity (PTI); this is also considered basal resistance. If the pathogen is weak, it is stopped at this point, or this event can trigger the pathogen to release specific elicitors to knock out PTI. These elicitors are called effectors. This process of knocking down PTI is referred to as Effector-triggered susceptibility (ETS). This raises an arms race where the plant tries to evade ETS through R gene-mediated resistance, referred to as Effector-triggered immunity (ETI). This leads to rapid and enhanced defense responses in plants, including HR. Jones and Dangl have summarized this whole concept as a zigzag model of plant immunity, which was later referred to as the central dogma of plant pathology (de Wit et al., 2009; Jones and Dangl, 2006; Zhang et al., 2022).

Is it just the recognition that helps the plant to obliterate the pathogen? What drives the defense action after recognition of plant PRRs and pathogen PAMPs? Here comes the role of signaling cascades. Interactions between stress-derived stimuli such as PAMPs or effectors and a signal perception apparatus in plants result in the activation of a wide range of signaling molecules triggering local and global responses via multiple (frequently intercrossing) signal transduction pathways. Initial or early response of pattern recognition is usually associated with ion influx, mostly calcium, followed by callose deposition, the rigorous burst of reactive oxygen species (ROS). Delayed responses such as miRNA pathway activation, MAPK cascade activation, and induced expression of a large number of defense-related genes, such as disease-process-related protein (PR), were also reported to be activated by effector and receptor recognition (Boyko and Kovalchuk, 2011; De Jonge et al., 2011; Jones and Dangl, 2006). G protein-coupled reception and MAPK signaling are major signaling pathways found in fungi.

There is a network of signaling pathways to help fungal pathogens. In fungi, heterotrimeric G protein participates in the regulation of various activities like vegetative growth, pathogenicity, sporulation, and differentiation of infection structure by regulating the activities of adenylate cyclase, phospholipase, and ion channels. MAPK pathways are very important fungal systems involved in many physiological and developmental processes, including stress response, virulence, interaction with other organisms, etc. In many pathogenic fungi, the HOG pathway mainly plays a role in adapting to high osmotic pressure in the external environment. The MAPK pathway is known to participate in plant PTI reaction mainly through the following pathways. First, MAPK induces the expression of downstream plant immune-related genes. ETI represents the efficient interaction between the R protein and Avr protein. Up to now, three main modes of interaction between R protein and Avr have been demonstrated: direct interaction mode, indirect interaction mode, and transcriptional regulation model. For a long time, PTI and ETI have been considered as two independent systems. However, with the deepening of research, PTI and ETI have become interactive from relative independence. There are many intersections and similarities between them in early signal transduction and downstream immune response (Martinez-Soto & Ruiz-Herrera, 2017; Li et al., 2022).

The important pathogens that breach the basal resistance of plants are effectors. There is a lack of understanding of effectors from eukaryotic pathogenic fungi. There is continuous selection pressure on the pathogen to adapt to new resistance strategies employed by plants. The forces acting on the evolution of effectors are less understood. Universally, two pressures are presumed to cause effector evolution. The first scope of development is to adapt to host targets through evolution, and the second one is to evade/escape detection by producing an arsenal of effectors. These dynamic selection pressures create a continuously unsettled biotic environment for effectors, accelerating their evolution (Upson et al., 2018).

The mechanisms or physiological interactions between plants and different groups of fungi are different, but overall, they share a common mechanism for pathogenesis. Biotrophic pathogens require a living host to complete their life cycle, and, therefore, they take care to secrete effectors just sufficient enough to suppress host immunity while minimizing damage to the host. In other cases, necrotrophic pathogens secrete effectors to proliferate on plant tissues. Hemibiotrophic pathogens require a living host at the beginning and tend to induce necrosis at later stages. There is a selective integration of biotrophic and necrotrophic mechanisms at different spatial and temporal levels. Host physiology is reprogrammed to favor pathogen colonization by effectors (Zhang et al., 2022).

To bring down the pathogen invasion, the plant employs two mechanisms: resistance and tolerance. The traits governing these two mechanisms are under constant pressure, which might have led to modifications in the plant genome at certain loci. The modifications could be direct or indirect, and a few instances of epigenetic modification were also reported. This modification could get converted to a stable (epi)genetic trait. DNA methylation, histone modifications, chromatin remodeling, and RNA interference are the most common epigenetic interactions reported and studied (Boyko and Kovalchuk, 2011; Thakur et al., 2022). This chapter is a collective attempt to gather information about plant and pathogenic fungi interaction.

3.2 MECHANISM OF PATHOGEN-TRIGGERED IMMUNITY (PTI) AT THE MOLECULAR LEVEL

3.2.1 PATHOGEN-ASSOCIATED MOLECULAR PATTERNS AND THEIR PERCEPTION IN PLANTS

When someone comes and knocks on the door, that is the first signal to us that some person is trying to come inside. PAMP is that very 'knock on' to the plants from the pathogens' side. Pathogen-associated molecular patterns (PAMPs) are molecules with unique conserved motifs that are associated with pathogens serving as ligands for host pattern recognition molecules known as Pattern Recognition Receptors (PRRs) (Heise, 2014). PAMPs are also known as Microbe-associated molecular patterns (MAMPs), as most of the pathogens are microbes. As these give the host plant the first sense of foreign or non-self-trying to intrude inside, they contribute to the initiation of host immunity. The innate immune system of plants has two levels of perception or switches to get started (Jones & Dangl, 2006). Recognition of PAMPs by PRRs is the first level where the host initiates PTI. PRRs are membrane-localized proteins that may be generally categorized as either receptor-like kinase (RLK) or receptor-like protein (RLP). RLK and RLP have a multidomain architecture that includes cytoplasmic and transmembrane domains and extracellular ectodomains. RLK and RLP vary significantly in their cytoplasmic domains, which are crucial for subsequent signaling. RLKs have a lengthy cytoplasmic domain since their primary job as kinases is to carry out signaling. RLPs, on the other hand, have a limited cytoplasmic domain; therefore, they rely on other cytoplasmic proteins for signaling that occurs subsequently (Rathore & Ghosh, 2018). PTI is a basal or generalized immunity, the mechanism of which is more or less the same for diverse pathogens and provides incomplete but horizontal resistance to the host (Nürnberger & Kemmerling, 2009). To cross this first-line barrier, pathogens secrete effector proteins, and plants also try to recognize them directly or indirectly via R proteins or resistant proteins. Their continuous evolution extends this battle to eternity, and the winning baton keeps interchanging between these two competitors for survival. Recent research has proved that neither PTI nor ETI confers immunity alone; rather, the overall resistance output is greater than the sum of the individual mechanisms. This is due to numerous 'mutual potentiation's' between the two systems (Ngou et al., 2021; Yuan et al., 2021). Let us know about the PAMPs originating from different pathogens and their recognition by and interaction with PRRs.

3.2.2 FUNGAL PAMPS AND THEIR INTERACTION WITH PRRS

Fungal ethylene-inducing xylanase (EIX) is detected by the receptors, namely LeEIX1/2, in tomatoes. The primary amino acid sequence of LeEix1 and LeEix2 is 81.4% identical (Ron & Avni, 2004). Although both receptors have the ability to interact with EIX, only LeEix2 causes defensive reactions.

BAK1 binds to LeEix1 but not to LeEix2. LeEix1 is no longer able to attenuate plant responses to EIX in plants when BAK1 is silenced, demonstrating that BAK1 is necessary for attenuating EIX-induced internalization and signaling of the LeEix2 receptor. Here, LeEix1 serves as a decoy receptor for LeEix2, a role that necessitates the kinase activity of BAK1 (Bar et al., 2011). EIX1 (ß-1,4-endo xylanase) is a proteinaceous enzyme with a 22 kDa molecular weight. It has been identified from *Trichoderma viride* and triggers innate immunity in tobacco and tomato plants (Furman-Matarasso et al., 1999).

Additionally, oligosaccharides from fungi and oomycetes function as MAMP/PAMPs. Induced innate immunity occurs in both monocot and dicot plants as a result of fungal chitin and its derivative, N-acetylglucosamine. Fungal chitin is sensed by two LysM (Lysin Motif) chitin elicitor-binding proteins (CEBiP) and CERK1 in rice (Shimizu et al., 2010). Five predicted LysM-containing receptor-like kinases (LYKs) are encoded by the *Arabidopsis* genome; these are AtCERK1/LysMRLK1 (CERK1) and AtLYK2 to AtLYK5 (LYK2–5), of which only CERK1 and LYK3 are expected to have a functional kinase domain. Along with three LYPs (LysM-containing receptor proteins): AtLYP1–3 are likely to be expressed on the plasma membrane (Tanaka et al., 2013). Another novel PAMP, Cell Death Inducing 1 (RcCDI1), has recently been characterized from *Rhynchosporium commune* against the young stage of barley. RcCDI1 and its homologs from different fungal species, including Zymoseptoria tritici, *Magnaporthe oryzae*, and *Neurospora crassa*, exhibited PAMP activity, inducing cell death in Solanaceae but not in other families of dicots or monocots. RcCDI1-triggered cell death was shown to require *Nicotiana benthamiana* Brassinosteroid insensitive 1-Associated Kinase 1 (NbBAK1), *N. benthamiana* suppressor of BIR1–1 (NbSOBIR1), and *N. benthamiana* SGT1 (NbSGT1), but was not suppressed by PiAVR3a or PexRD2 (effectors released by *Phytophthora infestans*). RcCDI1 activates cell death through a pathway that is distinct from that triggered by the oomycete PAMP INF1 (Franco-Orozco et al., 2017). Fungal PAMPs bind to plant PRRs via homodimerization and hetero-multimerization (Rathore & Ghosh, 2018).

N. benthamiana NbSERK3 controls the immunological responses brought on by the Oomycetes PAMP protein INF1 and greatly aids in *Phytophthora infestans* resistance. However, it is unable to change resistance to *P. mirabilis,* a sibling species of *P. infestans*, that prefers a different host plant. Two homologs of NbSERK3 – NbSERK3A and NbSERK3B have been cloned (Chaparro-Garcia et al., 2011). Some other known Oomycetes PAMPs are: OPEL, CBEL, Pep-13, Nep1-like proteins, Arachidonic acids, etc., but most of their receptors on behalf of the host are yet not known (Nürnberger et al., 1994; Gaulin et al., 2006; Gijzen & Nürnberger, 2006; Chang et al., 2015). The PRRs vary in how they depend on BRI1-associated kinase 1/somatic embryo receptor kinase 3 (BAK1/SERK3), an LRR-RLK that was initially identified as a co-receptor for the brassinosteroid insensitive 1 (BRI1) but is now known to regulate a number of signaling cascades against bacteria as well as filamentous pathogens. On the other hand, certain PTI pathways, like the fungal chitin signaling pathway in *Arabidopsis*, do not depend on BAK1 activity. Instead, LysM-receptor kinase CERK1 initiates chitin signaling, which results in resistance against fungal pathogens (Chinchilla et al., 2009) (Figure 3.1).

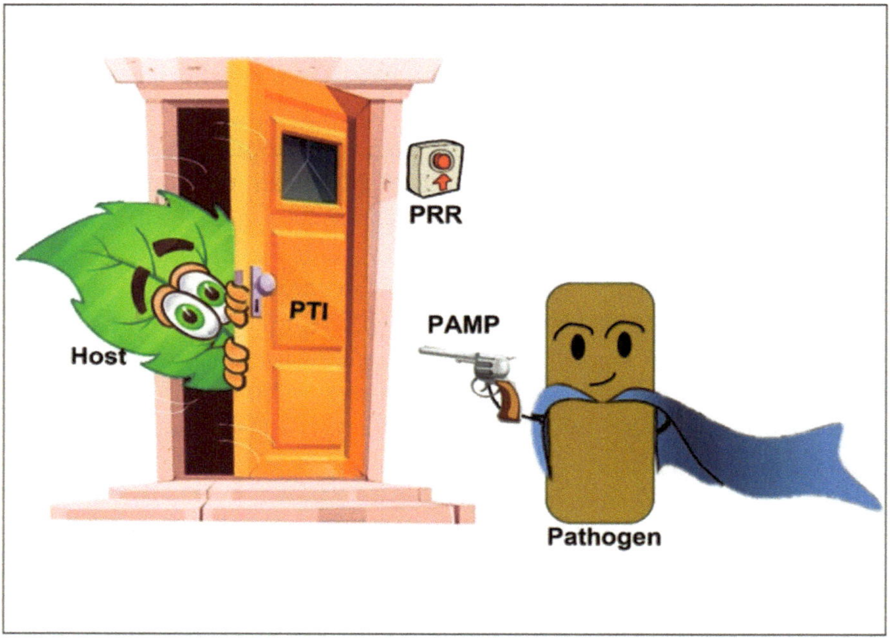

FIGURE 3.1 A cartoonified representation of the development of Pattern Triggered Immunity (PTI). When the pathogen with its Pathogen-Associated Molecular Pattern (PAMP, represented as a gun) presses the doorbell, the doorbell acts as a Pattern Recognition Receptor (PRR), and the host initiates PTI via cell wall strengthening mechanisms represented as the closing of the door.

3.2.3 DAMPS AND THEIR INTERACTION WITH PRRS

Plant-derived molecules, in certain cases, are responsible for inducing immune responses. Here, the immune system can detect danger through hydrophobic portions of innate biological molecules known as DAMPs (Damage-associated molecular patterns) (Seong & Matzinger, 2004) and can cause an inflammatory response. Systemin, a peptide of 18-amino-acid length, induces an immune response in tomato (McGurl et al., 1992). The cytoplasmic protein pro-systemin serves as the precursor of systemin (Narváez-Vásquez & Ryan, 2004). Systemin is secreted in response to cell injury and functions as a DAMP. Previous studies revealed that RLK SR160, an ortholog of BRI1 of tomato, functions as a systemin receptor (Scheer & Ryan, 1999).

In *Arabidopsis*, a putative cytoplasmic peptide called Pep1 turns on the defense genes and alkalinizes the cell culture. Pep1 is a 23-amino-acid-long peptide with seven homologs derived from PROPEP1–7 (Huffaker et al., 2006). PEPR1, a member of the LRR X1 subfamily, recognizes Pep1. On the basis of the primary sequence, a second receptor for Pep peptide was found and designated as PEPR2. Studies on the interaction between peptide and receptor revealed redundancy, as Pep1 and Pep2 bind to both PEPR1 and PEPR2, while PEPR1 binds to Pep3–6 (Yamaguchi et al., 2010). Plant cell walls release cutin and oligogalacturonides (OGs), which also function as DAMPs (Denoux et al., 2008). Wall-associated kinase1 (WAK1) acts as a receptor for OG (Brutus et al., 2010). The receptor for cutin is yet unknown, though. By activating MAPK, accumulating phytoalexins, glucanase and chitinase, ROS generation, callose deposition, increasing nitric oxide and calcium release and influx in the cytoplasm, and inducing expression of several defense genes, OG elicits immunological responses in plants (Navazio et al., 2002; Denoux et al., 2008; Ferrari et al., 2007, 2013; Galletti et al., 2011). Extracellular ATP (eATP) is another elicitor found in plants that belongs to the DAMP family. The surface-associated receptor for eATP perception 'does not respond to nucleotide 1 (DORN1),' the founding member of a purinoceptor subfamily, P2K (P2 receptor kinase) (Choi et al., 2014; Tanaka et al., 2014). A wound-inducible DORN1 mutant was shown to reduce the transcriptional response to wounding and upregulate genes after the injection of eATP. The ability of eATP to activate MAPK, Ca2+ influx, JA, and ethylene production, as well as to stimulate defense genes, results in the induction of innate immunological response (Tanaka et al., 2014).

AtHMGB3, a protein from the HMGB (High Mobility Group Box) family, is proven to function as DAMP in *Arabidopsis*. There are 15 genes in *Arabidopsis* that code for proteins with HMG-box domains (Merkle & Grasser, 2011; Vénéreau et al., 2015). When recombinant AtHMGB3 was injected into leaves, it caused the activation of MAPK, the formation of calloses, the induction of defense genes, and the development of resistance to *Botrytis cinerea*. Through the conserved Arg and Lys residues on AtHMGB3, Salicylic acid (SA) binds to the protein and stops its DAMP action (Choi et al., 2016). SA, on the other hand, promotes a healthy immune response in plants and confers resistance to pathogens that are both biotrophic and hemibiotrophic. Jasmonic acid (JA) is in charge of supplying resistance to insects and necrotrophic pathogens. The JA and SA signaling pathways are antagonistic (Huang et al., 2020). Infection and cellular damage brought about by the necrotrophic pathogen resulted in the release of AtHMGB3 into the extracellular environment, which in turn activated defense genes linked to ethylene and JA. However, the presence of a biotrophic pathogen elevated the SA level, which reduced AtHMGB3 DAMP-mediated immune protection and promoted the expression of the SA-associated defense genes (Choi et al., 2016). So, these studies should be critically studied so that in the future, they can open up a new way of protecting plants from necrotrophic pathogens via early induction of DAMPs.

3.3 MECHANISM OF EFFECTOR-TRIGGERED IMMUNITY (ETI) AT THE MOLECULAR LEVEL

3.3.1 EFFECTOR MOLECULES AND THEIR TYPES

Effectors are pathogen-originated molecules that bring structural and/or functional changes in host plants to favor host infection, colonization, and/or trigger host defense machinery (Kamoun 2006, Lo Presti, 2015). Effectors can act either as *virulence factors* – helping to overcome PAMP or MAMP-associated PTI and thereby establish effector-triggered susceptibility (ETS) or as *avirulence factors* (commonly termed as *elicitors*) – triggering effector-triggered immunity (ETI) in the host. Thus, the functional characteristics of virulence/avirulence are essential to consider a molecule as an effector. During evolution, plants have accumulated genomic regions that code for factors that identify effectors and bring resistance responses termed *resistance genes* (R-genes). Conversely, plants also possess genes that promote susceptibility to pathogens, termed *susceptibility genes* (S-genes).

3.3.1.1 PROTEINACEOUS AND NON-PROTEINACEOUS EFFECTORS

Effectors were initially thought to be only proteinaceous. In fact, most of the fungal effectors are small proteins. Proteinaceous effectors lack motifs to predict their function, and their functional analysis remains challenging, most likely due to their functional redundancy in pathogens and evolution by divergent selection (Giraldo and Valent 2013; Seong et al., 2023). Identification of conserved motifs of the fungal effectors has led to a novel approach termed 'effectoromics,' which uses these motifs as functional markers to characterize R-genes in the host, e.g., potato (Vleeshouwers et al., 2014). In addition to well-studied secreted effector proteins, non-proteinaceous fungal secondary metabolites, cytotoxins, and small RNAs (sRNA) are known to fit the definition of effectors, but their knowledge is in its infancy (reviewed in Collemare et al., 2019). Unlike bacterial pathogens, fungal pathogens lack a specialized secretion system; however, invasive hyphae such as haustoria can perform the function, e.g., rust fungi. Fungi secrete effector molecules that either attach to the cell wall, accumulate in the apoplast, or directly enter the cytoplasm (via haustoria), which can disrupt cell integrity, regulate programmed cell death (PCD), hormone signaling, Ca^{2+}-dependent pathways, or regulate transcription in host cells via sRNAs (Lorrain et al., 2019; Shao et al., 2021).

Fungal secondary metabolites (SMs) can be host-specific and nonhost-specific toxins (chemical effectors) and can function as elicitors, suppress plant defense, and harden fungal CW (Collemare et al., 2019). For example, trichothecenes and tenuazonic acid inhibit ribosomal protein synthesis, radicicol interferes with Hsp90 protein folding, ophiobolin, and herbarumin inhibit calmodulin signaling, HC-toxin, and depudecin interfere with histone acetylation, and some SMs interfere with plant hormone signaling (reviewed in Collemare et al., 2019). Necrotrophic fungi release host-specific toxins (HSTs), e.g., victorin by *Cochliobolus victoriae* (Victoria blight of oat), T-toxin by *Cochliobolus heterostrophus* (Southern corn leaf blight), and ToxA by *Pyrenophora tritici-repentis* (wheat tan spot), which decide their presence and virulence on the respective host (Friesen et al., 2008). HSTs produce toxic effects in a host species-dependent manner.

Fungal pathogens can trigger RNA interference (RNAi) in the host plant by sending sRNA effectors. *Botrytis cinerea*-derived small RNAs can function as virulence sRNA effectors and can selectively silence multiple targets involved in the immune response in *Arabidopsis* and tomato (Weiberg et al., 2013). The stripe rust fungus *Puccinia striiformis* f.sp. tritici (Pst) secretes microRNA-like RNA 1 (Pst-milR1) to trigger RNAi and suppress the wheat Pathogenesis-Related 2 (*PR2*) gene (Wang et al., 2017). Similar sRNA effector-induced virulence is also reported for *Magnaporthe oryzae* on rice (Raman et al., 2017), *Fusarium graminearum* on wheat (Zeng et al., 2018), and *Blumeria graminis* f. sp. hordei and *B. graminis* f.sp. tritici on barley and wheat, respectively (Kusch et al., 2018).

3.3.1.2 APOPLASTIC AND SYMPLASTIC EFFECTORS

Fungal effectors can either be present in the microbe and plant cell interphase, termed apoplastic effectors, or can move into the plant cell through the PM and target different cellular compartments, termed symplastic/translocated effectors. Translocated effectors can move between cells through plasmodesmata. The *Fusarium oxysporum* secreted effectors Avr2 move between host cells when interacting with another effector, Six5. The Avr2-Six5 effector causes virulence in susceptible plants (lacking R-gene) but is required for resistance induction in hosts with R-gene (Cao et al., 2018). The *Ustilago maydis* secretes PIT2, PEP1, RSP3, and FLY1 effectors that localize in maize apoplast, whereas CMU1, TIN2, SEE1, JSI1, and MER1 translocate into maize cells (Hoang et al., 2021).

3.3.1.3 SPECIFIC AND CORE EFFECTORS

Fungal species secrete effectors that can be unique to the species or isolate, i.e., race-specific or variable effectors. Core effectors are common to a broad species range, even across taxa (Giraldo and Valent, 2013). *Ustilago maydis* PEP1, STA1, CCE1, and RSP3 effectors are core effectors, and their null mutants can be rescued by transforming with orthologs from related smut fungi (Hoang et al., 2021). UmVp1 is functionally specialized in *U. maydis*. *U. hordei* effector UhVp1 showed distinct cellular localization compared to UmVp1 and could not complement the loss of UmVp1 (Hoang et al., 2021). Likewise, NIS1 is a core effector conserved in Ascomycota and Basidiomycota fungi, recognized by BAK1 and BIK1 receptor kinases in the host (Irieda et al., 2019).

Fungal genomics suggests one-, two-, and multi-compartment genome models for rapid evolution and diversification of fungal effectors. Most advocated two-compartment (or two-speed) model says that effector genes reside in specific genomic compartments richer in repetitive sequences and transposable elements, which help them to evolve relatively faster than the rest of the genome and confer adaptive advantage (Dong et al., 2015; Faino et al., 2016). However, gene duplications or deletions, copy-number variation (CNV), and heterozygosity of the effector loci explain the rapid evolution of virulence in one-compartment (one-speed) genomes (Frantzeskakis et al., 2019). For instance, *Blumeria graminis* f.sp. hordei (Frantzeskakis et al., 2018), *Ramularia collo-cygni* (Stam et al., 2018), and *Puccinia striiformis* f.sp. tritici (Schwessinger et al., 2018) fit into a two-speed genome model. Further, the acquisition or loss of accessory chromosomes carrying effector genes loaded with non-synonymous substitutions and epigenetic marks than the core genome may confer genomic plasticity and adaptive advantage to the pathogen, e.g., *Fusarium spp*. (Vlaardingerbroek et al., 2016; Fokkens et al., 2018; Frantzeskakis et al., 2019; Witte et al., 2021).

3.3.2 EFFECTOR-TRIGGERED SUSCEPTIBILITY (ETS)

Successful plant pathogens have evolved various molecular strategies to breach basal immunity, i.e., PTI. These molecular patterns (effectors) may be the modified version of the PAMPs/MAMPs or can be entirely novel players. Effectors have undergone diversifying selection which conferred a single effector can target multiple plant defense strategies or species-specific virulence functions (Giraldo and Valent, 2013). These novel specialized fungal effectors overcome PTI and result in compatible plant-pathogen interaction and disease establishment, i.e., ETS. Various fungal pathogen strategies are discussed below to suppress plant basal immunity and support disease progression.

3.3.2.1 EFFECTORS PERTURB PLANT CELL WALL-ASSOCIATED RESPONSE

Pathogen-secreted plant cell wall degrading enzymes (CWDEs) alter the physical barrier on plant cells by degrading cellulose, hemicellulose, lignin, pectin, and structural protein matrix. This is an essential virulence mechanism in nearly all phytopathogens (Kubicek et al., 2014). Genome-wide analysis has revealed numerous CWDE-encoding loci in fungal genomes. For example, the *Curvularia lunata (*leaf spot of maize) genome contains 13 cutinase-coding genes (Liu et al., 2016), and 32 CWDE-encoding genes were differentially expressed in the biotrophic fungus *Ustilago maydis* during maize infection (Lanver et al., 2018). *M. oryzae* secreted GH6 and GH7 cellulases promote virulence in rice (Van Vu et al., 2012). Likewise, two cutinase-coding genes (*AaCut3* and *AaCut7*) expression and host-specific toxin are needed for full virulence of *Alternaria alternata* in citrus (Ma et al., 2019).

Arginine-Glycine-Aspartic acid (RGD) motif-containing proteins alter CW and plasma membrane (PM) adhesion and integrity, which interfere with signal transduction and favor pathogen virulence (Wang and Wang,

2018). For example, the *Phytophthora* effector IPI-O contains RGD-motif targets LecRK-I.9 protein in *Arabidopsis* and disrupts CW-PM adhesion, compromising resistance against Phytophthora (Bouwmeester et al., 2011). Likewise, the necrotrophic fungus *Pyrenophora tritici-repentis* causing wheat tan spot secretes two selective host toxins (RGD-motif containing effectors), PtrToxA and PtrToxB, critically required for plant cell entry and disease establishment (Shao et al., 2021).

The necrotrophic fungus *Ascochyta rabiei* secretes *Ascochyta rabiei* PEXEL-like Effector Candidate 25 (ArPEC25), which interferes with lignin biosynthesis in chickpea (Singh et al., 2023). ArPEC25 moves to plant nuclei and interferes with the DNA-binding ability of host TF CaβLIM1a and, thus, negatively regulates the *CaPAL1* gene, which encodes phenylalanine ammonia-lyase (PAL) involved in lignin biosynthesis.

3.3.2.2 EFFECTORS BLOCK CHITIN-INDUCED RESPONSE

Chitin is perceived as MAMPs by plant basal defense machinery and triggers PTI. Fungal effectors can interfere with chitin-mediated PTI – protect mycelium from plant-secreted apoplastic chitinases, inhibition of LysM receptor recognition, block plant perception of chitin oligomers, degradation of chitinases and/or modification of fungal CW components (Zhang et al., 2022). For example, tomato leaf mold *C. fulvum* secretes Avr4 (chitin-binding lectin), binds, and masks fungal CW hydrolysis from chitinases (van den Burg et al., 2006). Likewise, the accumulation of α-1,3-glucan on *Magnaporthe oryzae* CW is critical for the successful infection of rice (Fujikawa et al., 2012). The α-1,3-glucan coating protects fungal CW from antimicrobials, including chitinases. LysM domains containing LysM effectors are chitin-binding effector proteins, identified in *C. fulvum* (e.g., Extracellular protein 6, Ecp6), *M. oryzae*, and *Zymoseptoria tritici* (e.g., Mg1LysM and Mg3LysM), outdo plant receptor proteins for chitin binding (Sanchez-Vallet et al., 2015).

3.3.2.3 EFFECTORS INHIBITION OF SECRETED PROTEASES

Plants secrete several kinds of proteases which confer basal immunity. Successful pathogens have evolved protease inhibitors (effectors) that block these plant proteases for enhanced virulence. Due to diversifying selection pressure on both plant proteases and protease inhibitors in pathogens, a single effector can target multiple plant proteases and vice versa and can differ in their binding affinities (Wang and Wang, 2018). *P. infestans* secretes the extracellular Kazal-like serine protease inhibitor EPI1, which targets the subtilisin-like protease P69B and thus blocks RCR3 activation in tomato required for basal immunity (Tian et al., 2004; Paulus et al., 2020). Also, *P. infestans* secretes two cystatin-like proteases – EPIC1 inhibits the papain-like extracellular cysteine protease RCR3, and EPIC2B inhibits RCR3 and PIP1 (PHYTOPHTHORA INHIBITED PROTEASE 1, closely related to RCR3) in potato. Thus, both promote *P. infestans* virulence (Tian et al., 2007; Song et al., 2009). Likewise, *Cladosporium fulvum* effector Avr2 physically interacts with and inhibits the extracellularly secreted RCR3 in tomato (van Esse et al., 2008). Heterologous expression of Avr2 in *Arabidopsis* enhances susceptibility against *Botrytis cinerea* and *Verticillium dahliae*. Based on current knowledge, cystatin-like protease RCR3 seems to be the central plant protease blocked by different pathogen effectors (Song et al., 2009; Paulus et al., 2020). RXLR-type effector AVRblb2 significantly enhances the virulence of *P. infestans* by targeting the host papain-like cysteine protease C14 secreted for basal defense (Bozkurt et al., 2011). Enzyme inhibitors secreted by fungal pathogens also suppress plant immunity. For example, *Fusarium verticillioides* releases a toxin (Fumonisin B1) to inhibit the activity of maize β-1,3-glucanase (Sánchez-Rangel et al., 2012).

3.3.2.4 EFFECTORS INTERFERE ROS ACCUMULATION

Rapid oxidative burst is a typical feature of PTI. Fungal pathogens have evolved effectors that interfere with the function of the reactive oxygen species (ROS) production machinery of plants. For example, smut fungi (*Ustilago maydis* and *U. hordei*) secrete apoplastic effector PEP1 (Protein essential during penetration 1), which targets a key ROS-producing enzyme PEROXIDASE 12 (POX12) of maize and barley (Hemetsberger et al., 2015). Peroxidase inhibition by PEP1 was also observed in *Blumeria graminis* f. sp. hordei and barley pathosystem. Likewise, biotrophic fungus *Puccinia striiformis* f. sp. *tritici* (*Pst*) secretes a glycine-serine-rich effector PstGSRE4, inhibits wheat enzyme copper-zinc superoxide dismutase TaCZSOD2, causing reduced H_2O_2 accumulation and HR, and promotes *Pst* infection (Liu et al., 2022).

3.3.2.5 PSEUDO-EFFECTOR DECOYS

Fungal pathogens may secrete decoy effectors mimicking the main virulence-causing effector to deceive the host plant. For instance, *Phytophthora* spp. Effector XEG1 has xyloglucanase activity to produce reduced sugars in the soybean and *Nicotiana benthamiana* apoplast, which promote disease progression. Host-secreted apoplastic glucanase inhibitors (e.g., GmGIP1 in soybean) can directly interact and inhibit XEG1, but the pathogen deploys a decoy effector XLP1 (paralogous to XEG1) to competitively block the glucanase inhibitor, thus interfering with the interaction between XEG1 and the glucanase inhibitor and favoring pathogen virulence (Ma et al., 2017).

3.3.2.6 DETOXIFICATION OF ANTIMICROBIAL COMPOUNDS

Detoxification of host antimicrobials by biotrophic fungi is rare, whereas necrotrophic and hemibiotrophic fungi induce cell death and plant tissue damage, and thus, detoxification is indispensable. Plant-origin compounds with antimicrobial properties can be *phytoalexins*, which prevent successful pathogen infection, and *phytoanticipins*, whose activity depends on the damage to host tissues (Westrick et al., 2021). Antimicrobial compounds, including pterocarpans, isoflavones, flavonoids, stilbenoids, indoles, terpenoids, cyanogenic glucosides, saponins, benzoxazinoids, glucosinolates, and phenylpropanoid derivatives such as ferulic acid and cinnamic acid, are enzymatically transformed into non-toxic or less toxic compounds by pathogens for their proliferation (Westrick et al., 2021). For example, *Phaseollidin hydratase* detoxifies phaseollidin, and kievitone accumulates in common bean during *Phaseolus vulgaris* infection (Turbek et al., 1992). Pea pisatin detoxification by *Nectria haematococca* is carried out by microsomal cytochrome P450s (called Pisatin Demethylases, George and Van Etten, 2001). Grapevine-secreted stilbenoid resveratrol protects it from bunch rot fungus *Botrytis cinerea*, which detoxifies resveratrol by secreted laccase Bclcc2 (Schouten et al., 2002). Brassinin hydrolases of *Alternaria brassicicola* and brassinin hydrolases and oxidase in *Leptosphaeria maculans* detoxify brassinin (indole phytoalexins) of crucifers (Pedras et al., 2008, 2009). *Gaeumannomyces graminis* produce avenacinase, *Stagonospora avenae*, and *Septoria avenae* produce multiple saponin hydrolases (avenacosides) to detoxify oat saponins (Morrissey et al., 2000; Westrick et al., 2021). *Septoria lycopersici* secretes the tomatinase enzyme (homologous to avenacinase) to degrade tomato saponin α-tomatine (Sandrock and Van Etten, 2001). Glutathione S-transferases of *Alternaria brassicicola* and isothiocyanate hydrolases of *Sclerotinia sclerotiorum* detoxify toxic isothiocyanates in crucifers.

3.3.3 EFFECTOR AND R-PROTEIN INTERACTION

Plant R-proteins recognize specific molecular patterns in pathogen effectors, which could be understood via different models. Several theories have been proposed based on credible experiments regarding the interaction/perception of pathogen effectors and plant R-proteins. Both direct and indirect interaction models have evolved over time largely to explain the specific and non-specific interactions between pathogen races and plant species. However, nine mechanisms are broadly suggested for plant R-protein defense function (see Kourelis and Van Der Hoorn, 2018). Huge diversity exists for receptors and effectors, and these hypotheses are only based on the current understanding of such interactions; thus, variations/exceptions for these models shall prevail in the future. As per direct interaction models, R-proteins physically interact with effectors, which trigger resistance response in plants. Genome sequencing and protein-protein interaction experiments have revealed that R-proteins and effector molecules show different recognition/binding affinities, which indicate strong diversifying selection and antagonistic co-evolution of host and pathogen (Dodds and Rathjen, 2010). As per the direct recognition model, effectors and R-proteins can diversify due to sequence polymorphism. Effectors evolve to decrease binding affinity with R-proteins, while R-proteins strive to enhance recognition. During indirect interaction, R-proteins do not physically interact with effectors. R-proteins function as guards and sense the effector-driven conformational changes in pathogen target hubs (termed as guardee) or target-related proteins acting as '*decoys*.'

3.3.3.1 ELICITOR-RECEPTOR MODEL

R-protein is assumed to be a receptor and has an affinity to directly bind its corresponding Avr-protein/elicitor, which triggers downstream defense responses in diverse pathogens, including fungi (Jia et al., 2000; Deslandes et al., 2003; Dodds et al., 2006; Catanzariti et al., 2009; Gassmann and Bhattacharjee, 2012). The model supports the gene-for-gene specificity and diversification of both *R* and *Avr* genes. Jia et al. (2000) first showed that the rice blast pathogen *Magnaporthe grisea*-derived Avr-Pita effector binds to rice cytoplasmic Pi-ta protein in yeast two-hybrid (Y2H) and *in vitro* assays. Loss of physical interaction and, consequently, loss of resistance was observed due to amino acid substitutions in the Pi-ta leucine-rich domain

(LRD) or in the Avr-Pita$_{176}$ (variant of Avr-Pita) protease motif. Likewise, the rice blast fungus also secretes an avirulence Avr-Pik protein which can physically interact with rice Pik (R-protein) in Y2H and *in planta* co-immunoprecipitation assays (Kanzaki et al., 2012). Flax rust fungus (*Melampsora lini*) AvrL567 variants bind to L5, L6, and L7 variants of R-proteins (Dodds et al., 2006), and R-protein M and effector AvrM directly and specifically interact in a Y2H assay (Catanzariti et al., 2009).

3.3.3.2 GUARD HYPOTHESIS

According to the Guard hypothesis, cellular hubs (guardee) targeted by the pathogen effector, which are indispensable for virulence, are monitored by R-proteins (guard). The conformational changes in guardee induced by pathogen effectors are perceived by R-protein, leading to binding with guardee protein and activation of downstream resistance cascade (Dangl and Jones, 2001). The theory argues that indirect R-protein and effector interact to solve the mystery of recognition of the broad range of pathogen effectors by a small R-protein repertoire (van der Hoorn and Kamoun, 2008). Obviously, guarded effector targets may undergo selection for better interaction with pathogen effectors. However, the theory assumes that guardee are the virulence targets and thus fails to explain the recognition of Avr-proteins, causing no interference in cellular processes such as viral coat protein (Collier and Moffett, 2009). Several variants of the Guard hypothesis have been proposed to better explain pathogen effector recognition, including decoy, bait and switch, and integrated decoy models.

3.3.3.3 DECOY MODEL

Decoy/trap proteins are closely related to the cellular targets of effector proteins. These decoys evolve by gene duplication or simply mimic the effector targets which are required for R-protein function. These decoys are monitored by R-proteins, and upon effector perception, conformational changes in decoys trigger the R-protein-mediated resistance signaling cascade (van der Hoorn and Kamoun, 2008). However, in the absence of the corresponding R-protein in the host, decoys have nearly no function related to defense or susceptibility. The defense mechanism against leaf mold pathogen *Cladosporium fulvum* in tomato is a classic example of such effector and R-protein interaction. *C. fulvum* effector Avr2 acts by inhibiting several defense-related proteases, including PIP1 and RCR3 proteins required for basal host defense (van Esse et al., 2008). Tomato profusely secretes RCR3, an apoplastic PR-protein (cysteine protease) that acts as a decoy for the Avr2 effector. RCR3 binds and traps Avr2 to inhibit binding with PIP1, and the Avr2-RCR3 complex activates Cf-2-mediated HR and disease resistance (Rooney et al., 2005). Moreover, RCR3 is conserved across the Solanaceae family and has a variable binding affinity with the Avr2 effector; however, only tomato and potato RCR3 can trigger Cf-2-mediated resistance (Kourelis et al., 2020).

3.3.3.4 BAIT AND SWITCH MODEL

Effector recognition and resistance signaling by R-protein is completed in two sequential steps, involving cooperative action of different domains within R-protein (Collier and Moffett, 2009). First, an effector interacts and either causes structural changes such as modification, removal, or conformational change in the bait protein, or forms a complex with the bait protein. Secondly, the effector protein then binds to the cognate R-protein via the bait protein and triggers the signaling switch. In contrast to the guard or decoy model, here the R-protein directly recognizes the pathogen effector facilitated by the effector target, i.e., the bait protein.

3.3.3.5 INTEGRATED DECOY

The model explains the effector perception and defense signaling by auto-repressed R-protein pair. A variable non-canonical integrated domain in one of the NLRs (termed as sensor NLR, sNLR) resembles the target protein of the pathogen effector and acts like an integrated decoy. These integrated domains sense effectors either directly by physical binding or indirectly by sensing the effector activity in the host cell (Xi et al., 2022). Subsequently, the second member, helper NLR (hNLR), in the R-protein duplex is auto-active but repressed by sNLR in the absence of an effector, gets activated, and triggers defense signaling upon effector sensing by sNLR (Cesari et al., 2014; Xi et al., 2022). The *Arabidopsis* TIR-NB-LRR proteins – resistance to *Pseudomonas syringae* 4 (RPS4) and resistance to *Ralstonia solanacearum* 1 (RRS1) interact and pair to trigger ETI is a well-studied example to explain the integrated decoy model. The RPS4/RRS1-R complex recognizes both bacterial effectors AvrRps4 and PopP2 via interactions with the WRKY domain (DNA-binding domain, as an integrated decoy) in RRS1-R, and RPS4 triggers a resistance response upon effector perception (Le Roux et al., 2015). The RPS4/RRS1 pair also triggers resistance against the fungal pathogen *Colletotrichum higginsianum* in a similar fashion (Xi et al., 2022). Effector sensing and signaling against rice blast, wheat, and barley rust pathogens support the integrated decoy model.

3.4 ROLE OF R-PROTEIN IN PLANT DEFENSE

The specificity of the plant immune response is determined by the disease resistance genes (R). The host R gene and the pathogen avirulence gene (avr) interact together in an allele-specific manner to form this simple but sophisticated immune system (Flor, 1971). A defense reaction is initiated when this genetic interaction occurs. This leads to hypersensitive response (HR), which is characterized by accelerated calcium and ion fluxes, an extra-cellular oxidative burst, transcriptional reprogramming at and near infection sites, and, most frequently, localized programmed cell death (Greenberg and Vinatzer, 2003; Nimchuk et al., 2003). It is believed that the culmination of these actions stops pathogen development. A basal defense response also takes place in the absence of particular identification, and it appears that PAMPs, like flagellin and lipopolysaccharides (LPS), trigger it (Gomez, 2002). Although it is slower in time and has a smaller amplitude than R-protein-mediated defense, the basal defense response largely overlaps with it. Although it does not prevent pathogen colonization, the basal defense does restrict the spread of pathogens (Glazebrook et al., 1997). Therefore, it appears that R-protein action speeds up and enhances inherent basal defense responses.

3.4.1 STRUCTURAL DOMAINS OF NBS-LRR PROTEINS

For ATP binding and hydrolysis, each NBS-LRR protein (Figure 3.2) encodes a conserved NBS (Tameling et al., 2002). R proteins and some eukaryotic cell-death effectors, such as Apoptotic protease activating factor-1 (Apaf-1) and cell death protein-3 (Ced3), have similarities, which can be used to expand a domain that includes the NBS (Inohara et al., 2002; Inohara et al., 2001). The varied number of carboxy-terminal LRRs in their structure is what stands out the most. The architecture of the mammalian CARD/Nod proteins frequently arranges the activation of downstream effectors as follows: the amino-terminus is necessary for protein-protein interactions with an adaptor protein, whilst the NBS domain is in charge of ATP hydrolysis and signal release. The LRRs' carboxy- and amino-terminal domains appear to serve distinct purposes. Specific residues in the LRR's carboxy-terminal domain appear to be responsible for bacterial identification, whereas the amino-terminal domain appears to influence activation (Inohara et al., 2002; Tanabe et al., 2004; Chen et al., 2004). This defines the carboxy-terminal domain of the LRR as an interaction platform for upstream activators. These investigations led to the development of the "jack-knife" model for Apaf-1 activation (Moffett et al., 2002).

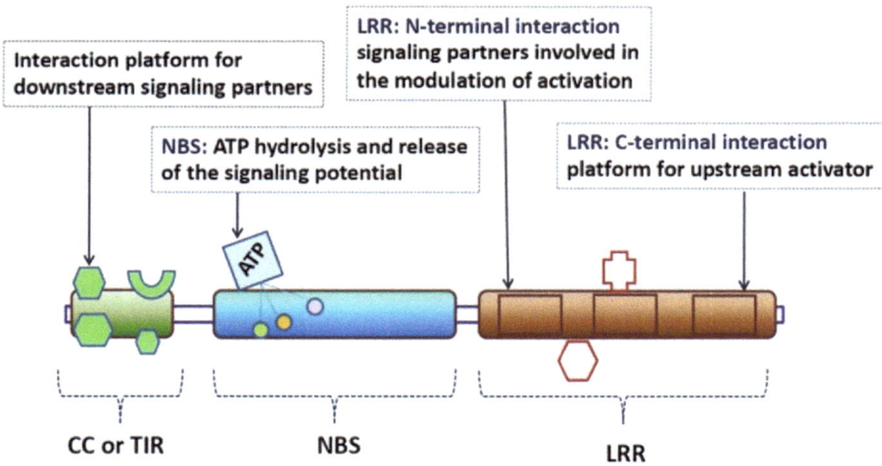

FIGURE 3.2 Structural domains of NBS-LRR proteins (i.e., CC, TIR, NBS, and amino-terminal [N-terminal] LRR regions) are essential for defense signaling.

3.4.2 MAJOR CLASSES OF R PROTEINS

Before, the bulk of R proteins that are activated upon effector recognition can be divided into five types mostly based on the patterns in which they combine specific structural motifs. Pto from tomato, the unique member of class 1, has a myristoylation motif at its N-terminus and a serine/threonine kinase catalytic domain (Loh et al., 1998; Martin et al., 1993). Leucine-rich repeats (LRRs), a putative nucleotide-binding site (NBS), and a putative leucine-zipper (LZ) or other coiled-coil (CC) sequence are all characteristics of a large number of proteins that are classified as class 2. Class 3 is identical to class 2, but instead of the CC sequence, these proteins have what is known as the TIR region because of their resemblance to the N-terminus of the Toll and Interleukin 1 receptor (IL-1R) proteins. This has implications, which are discussed below, for how pathogen effector proteins are delivered to the plant cell and where R proteins act to intercept these proteins. The R proteins belonging to the first three classes of R proteins do not contain transmembrane (TM) domains and are all believed to be localized intracellularly. The tomato's Cf proteins belong to class 4. They lack an NBS and instead possess a TM, an extra-cellular LRR, a tiny putative cytoplasmic tail, and no recognizable motifs (Dixon et al., 1996; Jones et al., 1994). Lastly, class 5 primarily includes the rice protein Xa21, which has a cytoplasmic serine/threonine kinase region in addition to an extracellular LRR and a TM (Song et al., 1995). Certain R proteins do not fall into any of these five categories. A toxin reductase called Hm1 provides resistance to a fungal disease that affects maize (Johal and Briggs, 1992). Recessive mutant alleles for Mlo in barley confer resistance to powdery mildew, and it may be a negative regulator of defense responses (Buschges et al., 1997). In a non-race-specific mechanism, RPW8 provides resistance to powdery mildew in *Arabidopsis* (Xiao et al., 2001). Although they act in particular recognition, other R proteins have new architectures. For example, the sugar beet nematode resistance protein Hs1pro1 lacks any recognizable protein interaction domains in its structure (Cai et al., 1997). The tomato Ve proteins for resistance to Verticillium are likely cell-surface glycoproteins with signals that are similar to receptor-mediated endocytosis (Kawchuk et al., 2001). In resistant *Arabidopsis* ecotypes, RTM1 and RTM2 prevent the tobacco etch potyvirus from spreading throughout the plant system and, respectively, have sequences comparable to jacalin and a small heat shock protein (Chisholm et al., 2000; Whitham et al., 2000). RRS1-R is a TIR-NBS-LRR protein, but it differs from other TIR-NBS-LRR proteins in that it confers *Ralstonia solanacearum* resistance across strains, is recessive genetically, and contains a putative nuclear localization signal and a 60-amino acid motif typical of the WRKY family of plant transcriptional activator proteins in its C terminus (Deslandes et al., 2002). Two tandem protein kinase domains and a predicted weak transmembrane domain were recently discovered to be present in the Rpg1 protein, which is utilized to combat barley stem rust (Brueggeman et al., 2002).

According to a recent study, nine groupings can be used to categorize plant resistance genes. Based on the arrangement of their amino acid motif and membrane-spanning domains, this classification is made (Figure 3.3). The LRR domain is found in the majority of R proteins and is crucial for the specificity of recognition (Gururani et al., 2012). The first significant class of R genes is represented by the CC-NBS-LRR genes. Cytoplasm contains the proteins that NBS-LRR genes encode. The first NBS-LRR protein, which the *NBS-LRR* gene encodes, was shown to be a cytoplasmic protein. It has an NBS, a C-terminal LRR, and a putative CC at the N-terminus. The first major class of R genes is represented by the *Arabidopsis* RPM1 resistance gene (I-2). Class-II: The cytoplasmic proteins of the second class of resistance genes have an N-terminal TIR domain, an NBS motif, and a C-terminal LRR domain. Examples of the second class of R genes include the *flax L6* gene, the *RPP5* gene, and the tobacco *N* gene (Gururani et al., 2012; Lawrence et al., 1995). Class-III: The third class of R genes is made up of additional cytoplasmic leucine-rich repeats and does not contain the NBS motif (eLRR). There is a transmembrane domain connected to this eLRR (TrD). Although eLRRs are thought to be the primary component of several defense proteins, such as polygalacturonase-inhibiting proteins (PGIPs) (Gururani et al., 2012; Jones and Jones, 1997), no clear evidence of an eLRR function in pathogen identification or the activation of defense genes has been found (Gururani et al., 2012; Jones, 2001). Examples of this class include the *Cladosporium fulvum* resistance genes (Cf-9, Cf-4, and Cf-2). Class-IV: The fourth major class of resistance genes has a transmembrane domain (TrD), an intracellular KIN (serine-threonine kinase) domain, and an LRR extracellular domain (Gururani et al., 2012; Song et al., 1995). This class includes the rice *Xa21* gene, which exhibits Xanthomonas resistance. Class-V: The extracellular LRRs, the Pro-Glu-Ser-Thr (PEST) domain, and the ECS are all present in the fifth class of resistance genes (short protein motifs). The PEST domain, which plays a role in protein breakdown, is unique to tomato Ve2 and not tomato Ve1. Proteins are believed to be targeted for endocytosis by receptors in the ECS. *Arabidopsis*, Class-VI: A representative of the important seventh class of R genes is the *RRS1-R* gene. *Ralstonia solanacearum* is a bacterial phyto-pathogen that is protected by the *RRS1-R* gene. The TIR-NBS-LRR class just gained a new member with the addition of the *RRS1-R* gene. The R-protein class includes TIR-NBS-LRR. RRS1-R has a putative nuclear localization signal (NLS), a WRKY domain, and a C-terminal extension (Deslandes et al., 2002, 2003). The WRKY domain comprises a region of 60 amino acids. Class-VII: The ninth class of *R*-genes has NBS, LRR, and RPW8 domains at the N-terminus. Examples of this class of genes are the *Arabidopsis* RPW8.1 and RPW8.2 genes (Xiao et al., 2005). Class VIII: The N-terminus of the genes in this class of NBS-LRRs has a BED finger domain (Kohler et al., 2008). Around 50-60 amino acid residues make up this domain, which has a common pattern of cysteine and histidine residues that together form a zinc-finger DNA-binding domain. Both the rice *Xa1* gene and the poplar's poptr 1:787192 have this BED domain (Bai et al., 2002; Kohler et al., 2008). A genome-wide analysis of NBS-LRR in Populus trichocarpa has suggested the existence of a further class of NBS-LRR genes (Kohler et al., 2008). This class, known as TCNL, has TIR and CC domains at the N-terminus (TIR-CC-NBS-LRR). Given that they include both TIR and CC domains, this might have developed as a result of gene recombination events.

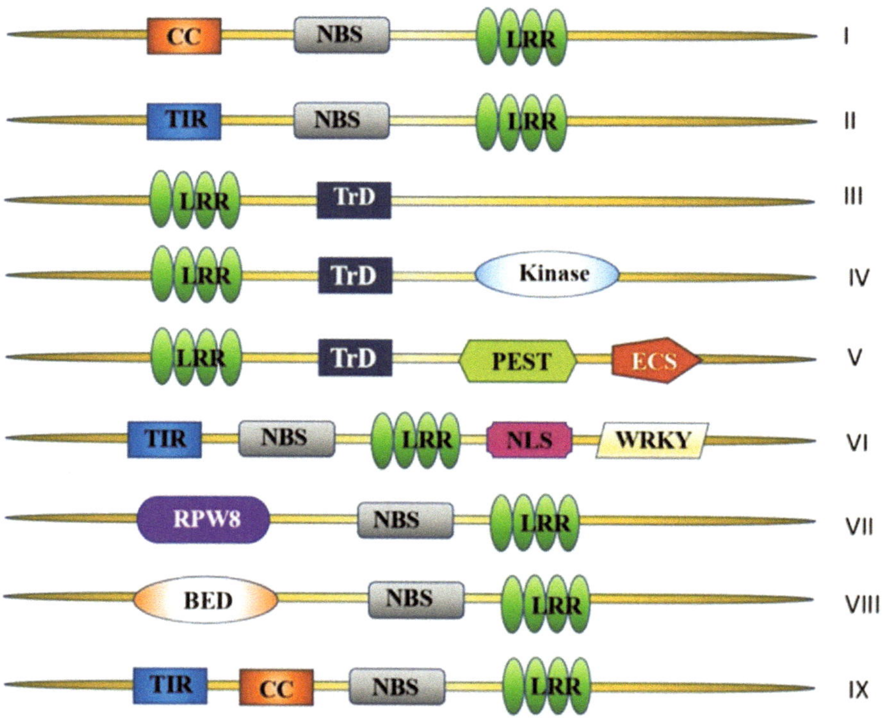

FIGURE 3.3 Classes of plant NBS-LRR genes.

Source: Reprinted with permission from Dubey and Singh, 2018. Copyright © 2018, Springer Nature Singapore Pte Ltd.

3.4.3 INTERACTION OF AN EFFECTOR

Specificity to pathogen detection. The LRR's amino-terminal portion is thought to function in intramolecular signal transduction, whereas the carboxy-terminal portion aids in pathogen identification. The tertiary structure of the LRR domain of the wild emmer wheat (*Triticum dicoccoides*) *Lr10* gene provided structural evidence for the LRR domain's function (Sela et al., 2012). LRR domains play a significant role in pathogen recognition in a variety of NBS-LRR proteins, including L, Mla, and Rx. These proteins all have a high number of hypervariable amino acid residues (Dodds et al., 2006; Seeholzer et al., 2010). Pathogen recognition specificity is partly located in the C-terminal of the LRR domain, according to studies on domain swapping and individual mutations (Bendahmane et al., 2000; Sela et al., 2012). The LRR detects the presence of pathogens, and the intramolecular signal spreads to the rest of the protein as a result. The NB-ARC domain interacts with the LRR and TIR-CC domains of the NBS-LRR protein while it is in a relaxed state.

(Note: LRR leucine-rich repeats, NBS nucleotide-binding site, ECS endocytosis cell signaling domain, TIR toll/interleukin-1-receptors, CC coiled-coil, TrD transmembrane domain, PEST amino acid domain, NLS nuclear localization signal, WRKY amino acid domain.)

A closed conformation is formed by all of these domains. It has been demonstrated through biochemical research on tomato I-2, flax M, Mi-1, and L6, as well as barley MLA27. ADP binds to the NB-ARC domain in the auto-inhibited state, or the auto-inhibited state is the ADP-bound form on the NB-ARC domain (Tameling et al., 2002; Williams et al., 2014). According to studies on CED4's crystal structure, the NB-ARC domain binds to ATP when it is in an open configuration. A conformational shift in the NB-ARC domain is thought to be caused by LRR in the activated state. This conformational change facilitates the conversion of ADP to ATP and promotes the adoption of a more open shape. When ATP hydrolysis activity is disturbed, NBS-LRR genes have been seen to express themselves constitutively. It has been demonstrated through studies regarding I-2 mutations. Studies on the NBS-LRR protein M in flax rust provide more support. An auto-active mutant (D555V) co-purified with ATP more frequently than the wild-type M protein, which also co-purified with ADP (Williams et al., 2014). A portion of the tobacco N protein was present in the NBS-LRR's resting state in ATP-bound form, but when it came into contact with the effector molecules, ATP was exchanged for ADP (Ueda et al., 2006). Both occurrences suggest that changes in the nucleotide-binding state are responsible for NBS-LRR protein activation. The NB-ARC domain of three NBS-LRR proteins, PSiP from corn, R1 from rice (NB), and RPM1 from *Arabidopsis*, has recently been found to have an unexpected and extra enzymatic activity. According to studies conducted by Mayerhofer et al. (2016), this domain hydrolyzes ATP in anticipation of the nucleoside rather than the diphosphate, demonstrating that it exhibits phosphatase activity. After activation of the NBS-LRR protein, the phosphatase activity for NB is reliable and returns to an adenosine-bound state as opposed to an ADP-bound conformation. The structure of the NBS-LRR protein undergoes conformational changes once it is activated. For interactions with other proteins, conformational variations in the NBS-LRR protein may reveal various binding sites. When an NB-LRR protein is activated, the protein's conformation may change, exposing many possible binding sites that permit various interactions with other proteins.

3.5 MECHANISM OF MAP KINASE SIGNALING CASCADE

3.5.1 FUNCTION OF MAPK ACTIVATION IN PLANT DEFENSE RESPONSES

Throughout their lives, plants are frequently attacked by bacteria, fungi, viruses, nematodes, and various insects. To counteract pathogen infection, plants have developed various defense mechanisms, including programmed cell death, thickening of the cell wall, Reactive Oxygen Species accumulation, and biosynthesis of pathogenesis-related proteins (Mithöfer and Maffei, 2016). The MAPK cascade is one of the first pathways to be activated during the defense response to pathogenic infection, and it is involved in plant defense/stress hormone signaling (ethylene, jasmonic acid, and salicylic acid), generation of ROS, phytoalexin biosynthesis, the closing of stomata, strengthening of the cell wall, defense gene activation, and hypersensitive response (HR) cell death (Taj et al., 2010; Meng and Zhang, 2013; Jagodzik et al., 2018). MAPK cascades in fungi and host plants do not operate independently but rather contribute to a complex molecular dialogue (He et al., 2020). Fungal MAPKs aid in mechanical or enzymatic penetration of host tissues, whereas plant MAPKs are required for plant immunity activation. When compared to other conserved genes, genome-scale transcriptome analysis reveals that MAPK genes are comparatively highly expressed in free-living mycelium, fruiting bodies, and ectomycorrhiza (Hamel et al., 2012). There is currently very little functional information on MAPK signaling in symbiotic fungi. MAPK signaling may be imperative during the presymbiotic colonization phase, probably by modulating the expression of target genes required for host infection and the establishment of functional ectomycorrhiza (Chow et al., 2018). Meng et al. (2013) discovered that MPK3/MPK6 and ETHYLENE RESPONSE FACTOR6 (ERF6) play critical roles in *Arabidopsis thaliana* resistance to the necrotrophic fungal pathogen *Botrytis cinerea*. MPK3/MPK6 plays a crucial role in the biosynthesis of pathogen-responsive camalexin, which leads to fighting against *B. cinerea* in *Arabidopsis* mpk3/mpk6 mutants (Ren et al., 2008). MPK3/MPK6 promotes ethylene (ET) biosynthesis at the post-translational level by phosphorylating rate-limiting isoforms of ACS (ACS2 and ACS6) to avert their degradation via the ubiquitin-mediated pathway (Wang et al., 2022). Through the WRKY33 transcription factor, the MPK3/MPK6 cascade controls the expression of the ACS2 and ACS6 genes. When plants interact with necrotrophic fungal pathogens, MAPK signaling cascades are triggered, regulating ACS and increasing ET production in plants (Li et al., 2016). *P. syringae* infection or treatment with flg22 activates MPK4, which phosphorylates MKS1 and causes the release of WRKY33, which directly expresses the camalexin biosynthetic genes in plants (Qiu et al., 2008).

3.5.2 MAPKS IN PAMPS-TRIGGERED IMMUNITY

In the natural habitat of plants, most pathogenic fungi produce molecules such as pathogen-associated molecular patterns (PAMPs) and effectors during contamination. Elicitors are molecules that plant-specific receptors can recognize to activate plant immunity, such as PTI (PAMP-triggered immunity) and ETI (effector-triggered immunity) (Abdul Malik et al., 2020). Recognition of fungal chitin in *Arabidopsis* can trigger a number of receptor-like cytoplasmic kinases (RLCKs) that induce PTI signals such as ROS generation, the closing of stomata, deposition of callose, and activation of MAPK cascade. PAMPs are typically recognized by plasma membrane-localized pattern recognition receptors (PRRs); however, due to their low abundance, identifying the PM-localized PRR proteins is problematic (Tang et al., 2017; Huang et al., 2020). Hatsugai et al. (2017) discovered that PTI inhibits an ETI signaling sector, implying that PTI regulates immune signal pathways to fine-tune plant defenses and avoid unnecessary fitness costs. Recently, *M. oryzae* snodprot1 homolog protein (MSP1), which functions as PAMP and triggers PTI responses in rice, has been identified. Rice OsPT8 protein, a member of the phosphate transporters (PTs) Pht1 family, is suppressed after infection with rice pathogens and treatment with pathogen-associated molecular patterns (Gupta et al., 2019). OsPT8 overexpression reduces rice disease confrontation with the pathogens *Magnaportheoryzae* and *Xanthomonasoryzae*. OsPT8 interacts with the rice mitogen-activated protein kinase BWMK1, which is a disease-resistance regulator (Dong et al., 2019). Two MAPK cascades have received a lot of attention in the context of immunity. Both the MAPKKK3/5-MKK4/5-MPK3/6 and the

MEKK1-MKK1/2-MPK4 cascades are thus rapidly activated upon PAMP perception, and these rapid but transient activations are critical for the transcriptional reprogramming seen in PTI (Asai et al., 2002; Sun et al., 2018). BWMK1 phosphorylates OsEREBP1, which triggers OsEREBP1 binding to the GCC-box element of pathogenesis-related (*PR*) gene promoters. BWMK1 overexpression in tobacco increases *PR* gene expression and thus resistance to *Pseudomonas syringae* and *Phytophthora parasitica* (Cheong et al., 2003). The first cloned *MAPK* gene in rice, WMK1 (for blast- and wound-induced MAP kinase), is induced by *M. oryzae* infection. Plant defense responses are regulated by BWMK1, OsMAPK5, OsMAPK6, and OsBIMK2. In transgenic rice, RNAi silencing of OsMAPK5 results in constitutive expression of PR genes such as PR1 and PR10, as well as enhanced resistance to *M. oryzae* and the bacterial pathogen *Burkholderia glumae* (Xiong and Yang, 2003).

3.5.3 MAPKS IN EFFECTOR-TRIGGERED IMMUNITY

ETI comprises host plants that detect pathogen effectors directly or indirectly through resistance proteins and mount a more vigorous and fast defense response. Nucleotide-binding and leucine-rich-repeat-containing receptors are the most common types of resistance proteins (NLRs). Dimerization or oligomerization of NLRs upon effector recognition signals to downstream components activates "helper" NLRs and concludes in the ETI response (Nguyen et al., 2021). Two non-exclusive hypotheses may be there to explain the activities of MPK3/6 sustained during ETI. According to the first theory, the pathogen may directly trigger MAPKs, or they may be indirectly stimulated by a positive feedback regulation that occurs downstream of the MAPK cascade. The second theory is that ETI blocks the mechanisms that typically reduce MAPK signaling, such as phosphatase deactivation (Thulasi Devendrakumar et al., 2018). The Toll Interleukin-1 Receptor NLR (TNL) and the Coiled-Coil NLR (CNL) are the two main classes of plant NLRs that vary in their N-terminal domains. When the effectors AvrRpt2, AvrB, and AvrRpm1 are recognized by the CNLs RPS2 and RPM1 in the plasma membrane, MPK3/6 is activated. The nuclear translocation of MPK3/6 may be required for this activation. When MPK3/6 is activated, it stimulates the expression of WRKY33 in a way that might be reliant on EDS1, as well as reactive oxygen species, salicylic acid, and ethylene signaling (Lang et al., 2020). Gain-of-function mutation activates the TNL receptor SSI4 in conditions of moderate humidity, which activates MPK3/6 constitutively. Plant ETI-specific scaffold proteins may be in charge of the sustained activations of MPK3/6 by promoting and/or sustaining the association between active MAP2K and MPK3/6 (Saijo and Loo, 2020). The ongoing functions of SIPK, WIPK, and Ntf4 strengthen plant defensive responses by actively regulating an HR-like cell death, whose execution is defined by a disruption of chloroplast metabolism and light-dependent production of ROS (Huang et al., 2019). Transcriptional regulator Both MPK3/6 and the Calcium-Dependent Protein Kinases (CDPK) have the ability to phosphorylate WRKY33, with each having a different impact on the protein's capacity for transactivation and DNA binding. Both the temporal pattern of phosphorylation and the actions of other kinases may affect how ETI-specific MAPK (Asai et al., 2002; Zhou et al., 2020). Two rice MAPKs (OsMPK3 and OsMPK6) and one MAPK kinase are activated by the fungal MAMP that acts as the chitin elicitor in rice (OsMKK4). OsMPK6 was necessary for the chitin elicitor-induced diterpenoid phytoalexins' biosynthesis. It also caused dynamic shifts in the metabolic flux from glycolysis to the formation of secondary metabolites and suppressed fundamental cellular processes, including translation and cell division. Plant metabolism is reprogrammed during MAMP-triggered defense responses through the OsMKK4-OsMPK6 cascade (Kishi-Kaboshi et al., 2010). Elicitors PAMPs from *Phytophthora* spp. are known to be recognized by plant receptors and to induce defense responses during a PTI process. Additionally, it has been discovered that certain *Phytophthora* species include inhibitors of several host proteases that are released, including subtilisin and papain-like cysteine proteases (Naveed et al., 2020).

Plants have evolved more complicated mechanisms to deal with biotic stress conditions, allowing them to perceive signals from the environment and prepare for an appropriate response. The response of the organism to biotic stresses is regulated by phytohormones, which are small, low-molecular-weight endogenous compounds. Auxins, cytokinins (CKs), gibberellins (GAs), jasmonic acid (JA), ethylene (ET), abscisic acid (ABA), salicylic acid (SA), and brassinosteroids (BRs) constitute only a few phytohormones that respond to stress (Mauch-Mani and Mauch, 2005). The existence of cross-talk between several signaling networks over time has been supported by a study using large-scale transcriptome data (Davletova et al., 2005). Along with signaling for defense, these phytohormones also control physiological processes such as growth and development. By preserving homeostasis in a very intricate network, these phytohormones respond to environmental and developmental signals.

3.6 PHYTOHORMONES SIGNALING AND CROSS TALK FOR PLANT DEFENSE

3.6.1 PHYTOHORMONE AND THEIR ROLE IN BIOTIC STRESS VIA DIFFERENT SIGNALING PATHWAYS

3.6.1.1 SALICYLIC ACID (SA)

According to the shikimate-phenylpropanoid pathway, plants produce SA, a phenolic molecule, to control defense mechanisms against biotrophic and hemibiotrophic diseases (Sticher et al., 1997). A defense response is triggered in remote areas of the plant when the SA pathway is activated at the infection site. By diverting resources and energy away from the growth and reproductive processes, constitutive SA buildup frequently results in a loss in plant fitness (Ishihara et al., 2008). As a result, the cell has tight control over its biosynthesis and signaling. The chloroplast produces SA, which is then transferred to the cytosol to trigger immunological reactions. The level of physiologically active SA in the cytoplasm is controlled by these modifications, such as the glucosylation of SA at its hydroxyl group, which results in the production of a glucoside (SAG), and the glucosylation of SA at its carboxyl group, which results in the production of an ester (SGE). MeSA is a mobile phloem signal that moves from the infected leaf to the systemic tissues, where it activates defense mechanisms before being transformed back to SA by the methylation of SA (Park et al., 2007). SA has an impact on a variety of physiological processes. For the onset of various stress symptoms in plants and cell death resembling the hypersensitive response (HR), SA is necessary. The PR genes are an extensive class of genes that produce antimicrobial proteins and increase resistance to a wide range of infections. With the exogenous application of SA, the expression of numerous genes encoding pathogenesis-related (PR) proteins has also been noticed (Malamy et al., 1990). Additionally, SA is necessary for plants under stress to develop systemic acquired resistance (SAR), which builds up in distant tissues (Vernooij et al., 1994). It is hypothesized that SA plays a part in biotic stress circumstances because several analyses showed that it is involved in numerous processes. According to Yalpani et al. (1994), exposure to ozone and ultraviolet light causes SA to build up over time. Ozone increases levels of oxidative stress; thus, SA increases antioxidant defenses in plants, indicating that it may play a part in other abiotic stress conditions. Additionally, research indicates that mechanisms for pathogen- and ozone-induced resistance overlap, supporting the significance of SA in plants (Sharma et al.,

1996). Furthermore, it has been proposed that SA controls calcium-mediated signaling pathways directly or indirectly. This is clear from research in which it was shown that SA caused several calcium-dependent protein kinases (CDPKs) to express themselves in plants (Leclercq et al., 2005). It is believed that SA is an essential component of the intricate signaling transduction networks due to its role in triggering numerous essential genes in signaling pathways. The suggested SA receptors are NON-EXPRESSOR OF PR GENE (NPR) family members. The SA marker gene PR-1, whose activation necessitates binding the SA-dependent transcriptional enhanceosome to its promoter, is used to monitor SA function (Rochon et al., 2006).

3.6.1.2 ETHYLENE (ET)

This is because ethylene is necessary for plants to respond to stress and to prepare for the tolerance mechanism necessary for their survival. But persistent stress-related conditions that cause excessive ethylene synthesis also limit plant growth and development, which, if prolonged, results in plant mortality. For the plant to survive under stressful conditions and to recover so that it might grow later, it is believed that strict management of ethylene homeostasis is maintained. Ethylene influences the signaling response by interacting with other phytohormones like jasmonic acid and salicylic acid. Depending on the various lifestyles of the pathogens, it may have either a beneficial or detrimental impact on SA-mediated defense responses (Derksen et al., 2013). The susceptibility to infections in plants is influenced by the gain and loss of function of ethylene-related genes. Greater levels of ethylene and greater resistance to *Magnaporthe oryzae* and *R. solani* were produced by rice when ACS2 (1-aminocyclopropane-1-carboxylic acid synthase) was overexpressed (Helliwell et al., 2013). The susceptibility of rice plants to *Magnaporthe oryzae* infection was increased by OsEIN2b RNAi silencing (Seo et al., 2011). Ethylene is perceived and transmitted as a signal by plants and is important for plant growth and survival. The genes that code for the proteins ethylene response sensors 1 and 2 (ERS1, ERS2) and ethylene resistant 2 (ETR2) are stimulated by the presence of ethylene. Ethylene response factors are the principal controlling variables in ethylene signaling (ERFs). However, it has been proposed that the ethylene insensitive 3 (*EIN3*) gene stimulates the expression of ERF 1 to activate the defense response. Therefore, ERF1 functions after EIN3 (Solano et al., 1998). A number of ERF subfamily members have been linked to the ability to withstand biotic stress. For instance, it has been demonstrated that ERF1 binds to the GCC box (AGCCGCC) in the promoters of genes that are induced by biotic stress in both rice and *Arabidopsis* (Xu et al., 2007; Cheng et al., 2013). *Magnaporthe oryzae*, which causes rice blast fungus, is demonstrated to be negatively regulated by ERF 922 in rice (Liu et al., 2012). It was found that rice with overexpression of ERF922 had higher amounts of ABA, which may be the reason for the grain's increased vulnerability to fungus (Koga et al., 2004). ERFs have been linked to long-distance leaf-to-leaf and leaf-to-root signaling in a number of cases to prepare the plant for future pathogen attacks (Dey et al., 2014). In contrast to untreated plants, treatment with Pseudomonas strain EA105 stimulates the expression of ERF1 in distal uninfected leaves and increases resistance to *Magnaporthe oryzae* (Spence et al., 2014). Additionally, the constitutive triple response inhibits EIN2 when ethylene is not present. ETR1 releases EIN2 from its inhibition in response to ethylene perception. EIN2's cytosolic C-terminal domain interacts with CTR1, which phosphorylates it and prevents EIN2 from signaling in the absence of ethylene (Ju et al., 2012). It has been discovered that biotic stress increases the rate of ethylene biosynthesis (Van Loon 1984; Abeles et al., 1992). A novel negative feedback system is demonstrated by the ethylene signaling mechanism in plants, which is interesting.

3.6.1.3 JASMONATES

Methyl jasmonate (MeJA) has previously been proven to be a signaling molecule that mediates intra- and interplant communication, changing the plant defense responses (Seo et al., 2011; Wasternack, 2007). This is because MeJA is volatile and has the capacity to diffuse through membranes. In-depth research has been done on tomato and *Arabidopsis* to better understand the biosynthesis and signaling pathways of jasmonates (Kazan and Manners, 2008). Additionally, jasmonates play a role in the control of the biosynthesis of numerous secondary metabolites, including phenylpropanoids, alkaloids, terpenoids, and antioxidants. Jasmonates cause the production of defense-related genes during pathogen attacks, particularly necrotrophic fungal and bacterial infections (Mei et al., 2006). Ethylene-responsive factors also operate as mediators of JA-regulated stress responses. ERF1, ERF2, and ERF5 regulate the ethylene-responsive factors PDF1.2, a JA-responsive marker gene.

3.6.1.4 GIBBERELLIN (GA)

Gibberellins, a type of phytohormone found in plants, belong to the diverse family of diterpenoid chemicals. In addition to higher plants, fungi and bacteria also synthesize it (MacMillan, 2001). According to a number of studies, GA, a signaling component, is crucial for plant disease susceptibility and resistance. The work of Zhu et al. (2005) provided the initial report on the function of GA. They established that a rice dwarf virus capsid protein interacts with plant ent-kaurene oxidases to cause a drop in GA levels in the infected plants. A group of nuclear growth-repressing proteins known as the DELLA proteins has been discovered to control the immunological response in plants that are controlled by JA and SA. They are the negative regulators of GAs. It can be hypothesized that indirectly GAs contribute to the oxidative stress caused by salinity and osmosis. By changing the expression of xyloglucan endotransglucosylase/endohydrolases (XTHs) and expansins, GA signaling inhibits cell wall relaxing and is crucial for the formation of cell walls. *Magnaporthe oryzae* and *Xanthomonas oryzae* pv. oryzae, two hemi-biotrophic rice diseases, are less resistant to GA when it is applied exogenously to rice (Qin et al., 2013). Salicylic acid and gibberellic acid levels in rice were reduced as a result of overexpressing a GA-deactivating enzyme, but the resistance to *Magnaporthe oryzae* and *Xanthomonas oryzae* pv. *oryzae* was increased. The opposite effect was obtained by loss of function, and plants became more susceptible (Yang et al., 2008).

3.6.1.5 CYTOKININS

Cytokinins have been discovered in large amounts in growing tissues such as cambium, shoot apex, and root tips. In *Arabidopsis* infected with *Plasmodiophora brassicae*, which is known to cause clubroot disease in plants, whole genome expression analysis revealed downregulation of genes involved in cytokinin homeostasis, specifically cytokinin synthases and cytokinin oxidases/dehydrogenases, demonstrating its role in the clubroot disease in *Arabidopsis* (Siemens et al., 2006). The phytohormone cytokinins have also been found to control nitrate reductase activity, which in turn controls how nitrogen is metabolized in plants (Sykorová et al., 2008). According to research by Sakakibara et al. (2006), root-derived cytokinin signals have been demonstrated to increase the activity of photosynthesis-related enzymes. Senescence has been associated with a drop in cytokinin concentration in Glycine max (Noodén et al., 1990). High-throughput study of *Arabidopsis* produced results that were comparable (Breeze et al., 2011). These findings imply that cytokinins slow plant senescence. It has been suggested that cytokinins can stimulate the production and storage of phytoalexins without the need for SA.

3.6.1.6 AUXINS

Low-molecular-weight chemical substances called auxins are among the most significant and varied phytohormones typically present in all plants. Auxins have been demonstrated to regulate senescence, react to different infections, and trigger an abiotic stress response in plants (Fu and Wang, 2011; Wang et al., 2010). One of the most prevalent endogenous auxins in plants, indole acetic acid (IAA), is produced by both tryptophan (Trp)-dependent and -independent mechanisms (Zhao, 2010). Auxin promotes the expression of auxin response genes in *Arabidopsis*, rice, and soybean (Javid et al., 2011). Auxins have a significant role in the biotic and abiotic stress response through talks and cross-talk with other phytohormones (Fahad et al., 2015). Auxins have been proven to reduce the defense response in plants in addition to their beneficial impact. According to studies (Navarro et al., 2006; Chen et al., 2007), its exogenous application promotes diseases caused by *Pseudomonas savastanoi* Pst DC3000 and *Agrobacterium tumefaciens*. The phytohormone auxins can travel great distances through the phloem from young, highly active tissues to the roots (Marchant et al., 2002). In addition, it travels over short distances between cells and is controlled by certain influx and efflux carrier proteins (Muday and Rahman, 2008). Although inverse interactions between ROS and auxins have been noted under diverse environmental stress situations, the auxin-responsive gene is still unknown. Due to the oxidation of IAA by peroxidase activity, the effects of oxidative stress have been seen in auxin stability (Kawano, 2003). During plant defense, auxin and SA act synergistically. According to recent data (Valls et al., 2006; Kazan and Manners, 2009; Bielach et al., 2017), many pathogens either produce auxin on their own upon infection or increase auxin production as a component of the host defense and developmental mechanism.

3.6.1.7 BRASSINOSTEROIDS (BRS)

Brassinosteroids (BR) are endogenous plant growth-promoting hormones. Brassinosteroids (BRs) are localized in almost all parts of the plants, namely, seeds, fruits, young vegetative tissues, and pollens. BRs are known to affect cellular proliferation and expansion (Clouse and Sasse, 1998). Analyses suggest the role of BRs in providing resistance against various biotic and abiotic stresses (Bajguz and Hayat, 2009; Gomes, 2011). Treatment with BR has been found to boost disease resistance in tobacco and rice (Nakashita et al., 2003). Therefore, it has been suggested that BRs, through causing an accumulation of apoplastic H_2O_2, also elevate the antioxidant system, leading to the development of plant stress tolerance (Jiang et al., 2012). Analysis indicates that BRs could play a role in adapting to biotic and abiotic stress (Bajguz and Hayat, 2009; Gomes, 2011). Brassinosteroid-insensitive 1 (BRI1) is a cell surface receptor kinase that has been linked to the perception of BRs in plants. Plants control BRs, which result in an array of responses based on the environment. Recent studies have demonstrated that exogenously applied BRs improved the resistance of different crops to a wide spectrum of diseases (Khripach, 2000). It has also been demonstrated that exogenous BR treatment can give barley tolerance to a number of *Fusarium* infections (Ali et al., 2013). The study found that in the infection of *Puccinia graminicola*, the pathogen grasps the host BR machinery as virulence factors to enhance infection (De Vleesschauwer et al., 2012). ROS generation can also be impacted by the limited range of BR concentrations (Baxter et al., 2014). Stress tolerance is significantly influenced by BR-induced ROS generation. BRs can scavenge ROS and promote the synthesis of antioxidants (Bajguz and Hayat, 2009; Fariduddin et al., 2013). By controlling the generation of secondary metabolites, they can also have an impact on disease resistance. Because of this, BR has a variety of functions in plant-pathogen interactions, including regulating PAMP perception, activating the stress-responsive gene, regulating oxidative metabolism, and producing secondary metabolites. The BRs are found to control the GA metabolism in rice (Tong et al., 2014). Following the perception of stress, BRs can communicate across defense signaling pathways with a variety of hormones, including SA, JA, ABA, auxins, and GA.

3.6.1.8 ABSCISIC ACID (ABA)

Abscisic acid (ABA), a phytohormone, is important for the germination, dormancy, and development of seeds, among other aspects of plant growth. Depending on the surrounding conditions, ABA accumulates after infection and can have either a good or harmful impact on the host's defense mechanisms. The plant becomes more vulnerable to disease and herbivore attack due to its main antagonistic actions with SA/JA/ethylene. To respond physiologically to a variety of external stimuli, ABA encourages stomatal closure. By preventing pathogens from entering through these passive ports, ABA produces preinvasive defense (Ton et al., 2009). The examination of rice plant susceptibility to *Magnaporthe grisea* as a result of ABA application demonstrates this (Koga et al., 2004). It has been discovered that viral infection in plants causes tobacco to have higher ABA concentrations (Whenham et al., 1986). When first exposed to stress, SA, JA, and ethylene may not be activated, but ABA operates as a critical endogenous component, playing a significant role by causing stomatal closure. After infection, the ABA signaling cascade sets in motion processes that increase the resistance phenotype through callose formation and prevent pathogen invasion in the attacked plants. In addition, the use of ABA reduced the transcript accumulation of defense genes like PDF1.2 (plant defensin 1.2), CHI (basic chitinase), HEL (hevein-like protein), and LEC (lectin-like protein), making *Arabidopsis* plants more vulnerable to infection by the fungus *Fusarium oxysporum* and the bacteria *Erwinia chrysanthemi* (Asselbergh et al., 2008).

3.6.2 CROSS-TALK IN VARIOUS PHYTOHORMONES

The interaction of numerous genes and gene families that are appropriately organized in the network results in all plant activities and reactions. Growth, development, and adaptation to diverse environmental stimuli essentially go hand in hand with plants. Through the modification of genes, several phytohormones are recognized to have a significant part in practically every activity. Additionally, plants maintain homeostasis and adjust to environmental changes through the balance of phytohormones. Only effective and systematic cross-talk between different phytohormones makes this possible. Together with BR, auxin, cytokinin, GA, SA, JA, and ET harmoniously mediate the response of plants to biotic stresses. The ability of a plant to defend itself does not only depend on one hormone; rather, all phytohormones interact with one another to control, either favorably or negatively, the defense response. Evidence suggests that JA and SA signaling interact negatively with herbivores and necrotrophic diseases (Glazebrook, 2005). In the conflicted conversations between SA and JA, NPR1 plays a crucial overlapping role. The JA-responsive genes lipoxygenase 2 (LOX2), plant defensin 1.2, and vegetative storage protein (VSP) were downregulated with the help of SA. In npr-1 mutants, this inhibition was completely removed. In the conflict between SA and JA, the WRKY transcription factor 70 family also has a significant influence. The constitutive responsive expression of PR genes and the suppression of JA-responsive PDF1.2 were disrupted by the recombinant WRKY-70 overexpressing line (Li et al., 2004). Similar research on *Arabidopsis* mpk4 (Map Kinase 4) mutants established the antagonistic connections between SA and JA (Petersen et al., 2000). A few reports further demonstrate their beneficial relationships. Together, JA and ET control the post-infection production of defense genes in a synergistic way. The development of root hairs by both phytohormones protects the plant from necrotrophs and helps to induce or stabilize EIN3 (Zhu et al., 2011). In tomato plants, where genes expressing proteinase inhibitors were activated in response to wounding, the synergy of JA interaction was seen (O'Donnell et al., 1996). Even the stimulation of ERF1-induced PR gene expression is dependent on their collaborative activity (Lorenzo et al., 2003). The JA-activated MYC2 inhibits its JA-regulated genes in a manner similar to how it represses the downstream activities of EIN3. Although auxins are well-known for playing an important role in controlling plant growth and development, research suggests that auxins also increase disease susceptibility and that resistance to biotic stresses necessitates suppressing auxin signaling. Auxin-responsive genes decrease due to SAR. Auxin-responsive genes were repressed when *Arabidopsis* plants were treated with the SA analog benzothiadiazole S-methyl ester (BTH) (Wang et al., 2007). Additionally, SA signaling inhibits the expression of the auxin repressor protein (AUX/IAA), which inhibits future auxin activity, as well as the auxin signaling F-box gene (AFB) and transport inhibitor resistant 1 (TIR1) genes. A number of studies have shown the important role of cytokinins in response to biotic stress (Reusche et al., 2013; Wang et al., 2007). *Verticillium longisporum*-resistant recombinant *Arabidopsis* plants showed increased resistance once their CK levels were stabilized (Reusche et al., 2013). As it has been observed that SA defense responses are increased by cytokinin-activated transcription factor (Wang et al., 2007), CK and SA signaling pathways interact to regulate plant defense. As a result of improved resistance to the rice blast fungus that was seen in a way that was dependent on OsNPR1 and WRKY 45, the cooperativity between SA and CK was also detected in rice plants (Jiang et al., 2013). As DELLA proteins are known to interact with important repressors of JA signaling JAZ1, GA, and JA also interact to inhibit JA-mediated transcriptional repression. The Repressor of GA1-3 (RGL3), which competes with MYC2 to favorably regulate JA-mediated response against necrotrophs, is known to be expressed in response to JA. BRs can interact with a variety of hormones, including SA, JA, ABA, auxins, and GA, to cross-talk between defense signaling pathways. It is well known that BR can promote pathogen resistance both dependently and independently of SA. Additionally, it counteracts rice plant resistance brought on by JA. Auxins and BRs can interact bilaterally, and this interaction can affect how the BR signaling cascade responds to disease and resistance. It activates GA repressor genes while suppressing GA biosynthesis genes. Additionally, facilitating cross-talk between BR and GA are known connections between the transcription factor BZR1 and the DELLA proteins. The cross-talk hypothesis of plant phytohormones has been supported by a number of studies, and it appears to be accurate in the sense that plants maintain a delicate balance between growth and environmental reaction, in which all phytohormones play a significant part. In conclusion, the complicated phytohormone signaling cascades are still being uncovered. Understanding the roles each hormone plays antagonistically and synergistically will be easier through the deconstruction of the plant immunity web. Knowledge of the mechanisms underlying plant hormone interactions and dissecting the plant immune network would therefore contribute to a fundamental knowledge of how plants coordinated immune systems work. Crop enhancement strategies may focus on the complex web of phytohormones. By using transgenic methods, plants with disease resistance may produce more. The strategies will either enhance signaling pathway resistance through an antagonistic pathway or modulation of the synergistic pathway.

3.7 ROLE OF WRKY SUPERFAMILY OF THE PROTEIN TRANSCRIPTION FACTOR IN DEFENSE RESPONSE

WRKY [Trp(W)-Arg(R)-Lys(K)-Try(Y)] is the largest group of transcription factors (TFs) of plants playing a pivotal role in the regulation of different defense-related genes and signaling pathways (Bakshi & Oelmüller, 2014). These TFs express themselves by up-regulation or down-regulation of various downstream target genes leading to maintaining homeostasis in plant systems against biotic and abiotic stresses. Since the cloning of sweet potato *WRKY* gene SPF1 (SWEET POTATO FACTORS1) (Ishiguro & Nakamura, 1994), an array of WRKY genes have been identified and isolated from various plants. A large number of WRKY superfamily genes have been identified in different crop plants, including *Glycine max* (197), *Oryza sativa* (105), *Arabidopsis thaliana* (75), *Hordeum vulgare* (45), *Sorghum bicolor* (68),

Carica papaya (52), *Vitis vinifera* (80), *Zea mays* (119), *Vigna angularis* (84), *Vigna radiata* (85), *Morus notabilis* (54), *Musa acuminata* (147), and *Musa balbisiana* (132), *Capsicum annuum* (73), and Pennisetum glaucum (97) (Baranwal et al., 2016; Chanwala et al., 2020; Goel et al., 2016; Hussain et al., 2018; Mangelsen et al., 2008; Pan & Jiang, 2014; SONG et al., 2010; Song & Gao, 2014; Srivastava et al., 2018; Wei et al., 2012; Zhang & Feng, 2014). WRKY TFs are a crucial part of the signaling web in plant systems responsible for various biotic and abiotic defense responses besides playing an important role in the growth and developmental processes (Bakshi & Oelmüller, 2014; Phukan et al., 2016). WRKY TFs are characterized by a highly conserved sequence of amino acids WRKYGQK along with zinc-finger-like motifs Cys(2)-His(2) or Cys(2)-HisCys, and bind with the (T) TGAC(C/T), W-box *cis-element* in the target gene's promoter resulting in downstream gene expression (Bakshi & Oelmüller, 2014; Wani et al., 2021).

3.7.1 DNA-BINDING DOMAIN OF WRKY PROTEINS

WRKY DNA-binding domain (DBD) is characterized by an almost 60 amino acids long 4-stranded β-sheet and an invariant heptapeptide motif WRKYGQK at their N-terminus and a zinc-finger motif at their C-terminus (Jiang et al., 2017). Variations in binding have been observed in the WRKY genes across plant species despite having a conserved W-box. This variability could be attributed to different DBDs and zinc-finger motifs. Eulgem et al. (2000) classified the WRKY superfamily into three different groups based on the number of DBDs and the pattern of the zinc finger motif. Group I contains WRKYs having two DBDs; group II has a single DBD with a different C2–H2 zinc finger, while group III contains a single DBD with a C2–HC zinc finger. Group II is further divided into different subgroups like IIa, IIb, IIc, IId, and IIe using sequences of primary amino acids and additional domains like nuclear localization domain, kinase domain, leucine zippers, serine-threonine-rich region, a proline-rich region, glutamine-rich region, and TIR-NBS-LRR domain (Rushton et al., 2010).

Substitutions in highly conserved WRKYGQK amino acid sequences have also been reported across different crop plants. Among them, WRKYGKK (maize, banana, mulberry, and soybean), WRKYGRK (banana), WRKYGEK (rice and banana), WRKYGNK, and WRKYGHK (banana) are the important ones (Baranwal et al., 2016; Goel et al., 2016; Wani et al., 2021). The WRKY family members in rice have 19 WRKY domain variants where WRKYGEK and WRKYGKK are the common variants, along with other variants like WRICGQK, WRMCGQK, WKKYGQK, WIKYGQK, WKRYGQK, WSKYEQK, and WRKYSEK (Zhang & Wang, 2005). WRKY TFs act as activators or repressors by binding specifically to W-box cis-elements in target gene promoters (Kim et al., 2008; Phukan et al., 2016). Functional W-box elements are also present in most WRKY TFs that bind with their own promoters or other WRKY TFs for auto- or cross-regulation (S. Liu et al., 2015). Exceptionally, NtWRKY12 containing WRKYGKK sequence binds with WK-box, which is required for the PR-1a expression by salicylic acid (SA) and bacterial elicitors (Van Verk et al., 2008). Mutation from GKK to GQK or GEK is responsible for variation in the DNA-binding activity of NtWRKY12 (Van Verk et al., 2008). OsWRKY13 from rice binds with W-box & PRE4, and AtWRKY50 binds with TGA2 and TGA5 TFs and PR1 promoter (Hussain et al., 2018; Qiu et al., 2007).

3.7.2 REGULATION OF WRKY TFS

WEKY TFs are tightly regulated at transcriptional and post-transcriptional levels to balance plants' homeostasis. ETI (effector-triggered immunity) mediated regulation is required for the activation of certain WRKYs like HvWRKY1 and HvWRKY2. These TFs are activated by recognition of avirulence AVR10 effector by resistance protein MLA in the cytoplasm and subsequent binding of MLA with WRKY in the nucleus (Shen et al., 2007). A key regulator (PAD4) of SA signaling regulates AtWRKY33 for *B. cinerea* resistance and genes involved in SA signaling, ET-JA-mediated cross-communication, redox homeostasis, and camalexin biosynthesis (Qiu et al., 2008). MAP Kinase MPK4 plays an important role in the activation of this WRKY to regulate downstream responses by forming the MPK4–MKS1–WRKY33 complex in the nucleus. MAMP or PAMP perception triggers MPK, MKK, and MEKK for activation, which leads to nuclear dissociation of the MPK4–MKS1–WRKY33 complex releasing AtWRKY33 and MKS1. This leads to the activation of PAD3 (phytoalexin deficient 3) by AtWRKY33 (Qiu et al., 2008). MAPK4 binding with MKS1 results in interaction with AtWRKY25 and AtWRKY33 to act as a negative regulator of SA-mediated defense responses to P. syringae. PKS2 (SOS2-like protein kinase 5) expression induced by *P. syringae* pv. Tomato DC3000 infection and binds with the AKR (Ankyrin Repeats) motif leading to phosphorylation of the *NPR1* gene. Due to this interaction, up-regulation of AtWRKY38 and

AtWRKY62 takes place, ultimately leading to plant defense responses (Xie et al., 2010). OsWRKY45 expression is regulated by MAPK at the time of *M. oryzae* or *Xanthomonas oryzae* pv *oryzae* attacks on rice, which further regulates expression of downstream SA/benzothiadiazole (BTH)-mediated defense (Matsushita et al., 2013). AtWRKY38 and AtWRKY62 are repressed by Histone Deacetylase 19 (HDA19) through the removal of acetyl groups from histone tails, resulting in negative regulation of basal defense (PTI) (Kim et al., 2008). Expressions of WRKYs are also regulated by proteasome-mediated degradation under normal conditions to keep them in check and maintain homeostasis. OsWRKY45 (role in SA/BTH induced defense) is regulated by the nuclear UPS (ubiquitin proteasome system) and rapidly degraded under normal conditions to suppress defense responses. These proteasomes are inhibited by pathogen attacks, leading to the accumulation of polyubiquitinated OsWRKY45 (Matsushita et al., 2013). Inter-organelle retrograde signaling also regulates WRKY expression (Phukan et al., 2016).

3.7.3 INTERACTION OF WRKY TFS WITH SA AND JA SIGNALING PATHWAY

Salicylic acid (SAs) and jasmonic acid (JAs) are two crucial players in regulating defense responses against pathogens. SA-induced defense responses are involved in fending off biotrophic and hemibiotrophic plant pathogens, whereas jasmonates-induced defense reactions help the plants to fight off necrotrophic pathogens and herbivorous insects (Koornneef & Pieterse, 2008). Signaling pathways of these two hormones act in a synergistic or antagonistic way and make the plant immune system more complex and flexible (Koornneef & Pieterse, 2008). NPR1 is the master transcriptional co-regulator of SA-dependent genes, regulated by several other SA-dependent modifications and involved in SA/JA crosstalk (Fu & Dong, 2013), while F-box protein COI1 and the JAZ repressor proteins are the master co-regulators of JA-signaling pathways (Caarls et al., 2015). WRKY transcription factors play an important role in the regulation of SA-dependent defense responses, and some of them are involved in SA/JA cross-talk (Phukan et al., 2016). Leon-Reyes et al. (2010) identified AtWRKY70 as a node of convergence between SA and JA signaling. Overexpression of WRKY70 caused enhanced expression of SA-responsive PR genes and concomitantly suppressed methyl jasmonate (MeJA)-induced expression of the JA-responsive gene PLANT DEFENSIN1.2 (PDF1.2). Hence, WRKY70

TABLE 3.1 A List of WRKY Tfs Reported Playing a Role in the Regulation of Different Signaling Pathways and Defense-Related Genes During Plant-Pathogen Interactions

WRKY TF	Host Plant	Pathogen	Target	Regulatory Role	Reference
WRKY3, WRKY4	*Arabidopsis*	*Botrytis cinerea*	JA-mediated pathway	Positive	(Lai et al., 2008)
WRKY4	*Arabidopsis*	P. *syringae*	SA-mediated pathway	Negative	
WRKY27	*Arabidopsis*	*Ralstonia solanacearum*	nitrogen metabolism and nitric oxide (NO)	Negative	(Mukhtar et al., 2008)
WRKY33	*Arabidopsis*	*Botrytis cinerea and Alternaria brassicicola*	JA-mediated pathway	Positive	(Zheng et al., 2006)
		Botrytis cinerea	ABA-signaling pathway	Negative (Resistance to *B. cinerea*)	(S. Liu et al., 2015)
		P. *syringae*	SA-mediated pathway	Negative	(Zheng et al., 2006)
WRKY38, WRKY62	*Arabidopsis*	*Pseudomonas syringae*	Basal defense (*HDA19* gene)	Negative	(Kim et al., 2008)
WRKY46, WRKY70, WRKY53	*Arabidopsis*	*P. syringae*	SA-mediated pathway	Positive (Synergistic)	(Hu et al., 2012)
OsWRKY51	Rice	*Xanthomonas oryzae pv. Oryzae (*Xoo*)*	Defense gene OsPR10a	Positive	(Hwang et al., 2016)
OsWRKY13	Rice	Xoo	SA and JA-mediated pathway	Positive- SA	(Qiu et al., 2007)
		Magnaporthe grisea		Negative- JA (Resistance against biotrophs)	
OsWRKY45	Rice	*Magnaporthe grisea*	SA-mediated pathway	Positive	(Shimono et al., 2007)
OsWRKY11	Rice	Xoo	*CHITINASE 2*	Positive	(Lee et al., 2018)
OsWRKY53	Rice	*Magnaporthe grisea*	elicitor-induced defense signaling pathways	Positive	(Chujo et al., 2007)
OsWRKY62	Rice	Xoo	Basal and Xa21-mediated defense response	Negative	(Peng et al., 2008)

TABLE 3.1 *(Continued)*

WRKY TF	Host Plant	Pathogen	Target	Regulatory Role	Reference
OsWRKY67	Rice	Xoo	Basal and Xa21-mediated defense response	Positive	(Vo et al., 2018)
OsWRKY71	Rice	Xoo	OsNPR1 and OsPR1b genes	Positive	(X. Liu et al., 2007)
OsWRKY89	Rice	*Magnaporthe grisea*	Several pathways	Positive	(H. Wang et al., 2007)
CaWRKY40b	Capsicum	*Ralstonia solanacearum*	Several defense genes	Negative	(Khan et al., 2018)
CaWRKY22	Capsicum	*R solanacearum*	Several defense genes	Positive	(Hussain et al., 2018)
CaWRKY6	Capsicum	*R solanacearum*	CaWRKY40	Positive	
CaWRKY40, CaWRKY27	Capsicum	*R solanacearum*	SA/JA/ ET signaling	Positive	(Dang et al., 2013)
SpWRKY1	Tomato	*Phytophthora infestans*	Diverse signaling pathways	Positive	(Li et al., 2015)
SlWRKY8	Tomato	*P. syringae* pv. tomato DC3000	Several defense genes	Positive	(Y. F. Gao et al., 2020)
VlWRKY3	Grape	*Golovinomyces cichoracearum*	Diverse signaling pathways	Positive	(Guo et al., 2018)
ScWRKY5	Sugarcane	*Sporisorium scitamineum*	Diverse signaling pathways	Positive	(D. Wang et al., 2020)

acts as a positive regulator of the SA-mediated defenses while repressing the JA response. Several WRKY transcription factors, such as WRKY62 (Mao et al., 2007), WRKY41 (Higashi et al., 2008), WRKY50, and WRKY51 (Q. M. Gao et al., 2011), were also involved in the suppression of JA-signaling pathways through SA. Several researchers have reported the differential role of WRKY TFs in various plant-pathogen systems by up- or down-regulation of defense-related genes involved in SA- or JA-mediated signaling pathways (Table 3.1).

3.8 FUTURE PROSPECTS AND CONCLUSION

Annual crop yield losses and threats to global food security and agricultural sustainability are significant consequences of plant diseases. To ensure sustainable crop production for the growing world population, enhancing crop resistance against various diseases is of utmost importance. Uncovering the molecular mechanisms that govern crop-phytopathogen interactions is crucial for improving crop resistance and disease management. Plants and pathogens have engaged in a long-term coevolutionary relationship, leading to the development of advanced defense mechanisms in crops against phyto-pathogens. As a result, invading pathogens have developed various tactics to enhance their ability to invade and infect. Understanding the complex interactions between these two entities can shed light on the ever-changing and dynamic coevolutionary "arms race" between them.

KEYWORDS

- **biotic stress responses**
- **effectors**
- **ETI**
- **NBS-LRRs**
- **PAMPs/DAMPs**
- **plant immunity**
- **PR protiens**
- **PRRs**
- **PTI**

REFERENCES

Abdul Malik, N. A., Kumar, I. S., & Nadarajah, K. (2020). Elicitor and receptor molecules: Orchestrators of plant defense and immunity. *International Journal of Molecular Sciences, 21*(3), 963. https://doi.org/10.3390/ijms21030963.

Abeles, F. B., Morgan, P. W. and Saltveit Jr, M. E. (1992). Ethylene. *Plant Biology, 2nd ed.*, Academic Press, New York.

Ali, S. S., Kumar, G. S., Khan, M. & Doohan, F. M. (2013). Brassinosteroid enhances resistance to fusarium diseases of barley. *Phytopathology, 103*(12), 1260–1267. https://doi.org/10.1094/PHYTO-01-13-0011-R.

Asai, T., Tena, G., Plotnikova, J., Willmann, M. R., Chiu, W. L., Gomez-Gomez, L., Boller, T., Ausubel, F. M., & Sheen, J. (2002). MAP kinase signaling cascade in Arabidopsis innate immunity. *Nature, 415*(6875), 977–983. https://doi.org/10.1038/415977a.

Asselbergh, B., De Vleesschauwer, D. & Höfte, M. (2008). Global switches and fine-tuning—ABA modulates plant pathogen defense. *Molecular Plant-Microbe Interactions, 21*(6), 709–719. https://doi.org/10.1094/MPMI-21-6-0709.

Bai, J., Pennill, L. A., Ning, J., Lee, S. W., Ramalingam, J., Webb, C. A., Zhao, B., Sun, Q., Nelson, J. C., Leach, J. E. & Hulbert, S. H. (2002). Diversity in nucleotide binding site–leucine-rich repeat genes in cereals. *Genome Research, 12*(12), 1871–1884. https://doi.org/10.1101/gr.454902.

Bajguz, A. & Hayat, S. (2009). Effects of brassinosteroids on the plant responses to environmental stresses. *Plant Physiology and Biochemistry, 47*(1), 1–8. https://doi.org/10.1016/j.plaphy.2008.10.002.

Bakshi, M., & Oelmüller, R. (2014). WRKY transcription factors: Jack of many trades in plants. *Plant Signaling & Behavior, 9*(2), e27700. https://doi.org/10.4161/psb.27700.

Bar, M., Sharfman, M., & Avni, A. (2011). LeEix1 functions as a decoy receptor to attenuate LeEix2 signaling. *Plant Signaling & Behavior, 6*(3), 455–457. https://doi.org/10.4161/psb.6.3.14714.

Baranwal, V. K., Negi, N., & Khurana, P. (2016). Genome-wide identification and structural, functional and evolutionary analysis of WRKY components of mulberry. *Scientific Reports, 6*, 30794. https://doi.org/10.1038/srep30794.

Baxter, A., Mittler, R., & Suzuki, N. (2014). ROS as key players in plant stress signaling. *Journal of Experimental Botany, 65*(5), 1229–1240. https://doi.org/10.1093/jxb/ert375.

Bendahmane, A., Querci, M., Kanyuka, K., & Baulcombe, D. C. (2000). Agrobacterium transient expression system as a tool for the isolation of disease resistance genes: Application to the Rx2 locus in potato. *The Plant Journal, 21*(1), 73–81. https://doi.org/10.1046/j.1365-313x.2000.00653.x.

Bielach, A., Hrtyan, M., & Tognetti, V. B. (2017). Plants under stress: Involvement of auxin and cytokinin. *International Journal of Molecular Sciences, 18*(7), 1427. https://doi.org/10.3390/ijms18071427.

Bouwmeester, K., de Sain, M., Weide, R., Gouget, A., Klamer, S., Canut, H., & Govers, F. (2011). The lectin receptor kinase LecRK-I.9 is a novel Phytophthora resistance component and a potential host target for an RXLR effector. *PLoS Pathogens, 7*(3), e1001327. https://doi.org/10.1371/journal.ppat.1001327.

Boyko, A., & Kovalchuk, I. (2011). Genetic and epigenetic effects of plant–pathogen interactions: An evolutionary perspective. *Molecular Plant, 4*(6), 1014–1023. https://doi.org/10.1093/mp/ssr024.

Bozkurt, T. O., Schornack, S., Win, J., Shindo, T., Ilyas, M., Oliva, R., Cano, L. M., Jones, A. M., Huitema, E., van der Hoorn, R. A., & Kamoun, S. (2011). Phytophthora infestans effector AVRblb2 prevents secretion of a plant immune protease at the haustorial interface. *Proceedings of the National Academy of Sciences, 108*(51), 20832–20837. https://doi.org/10.1073/pnas.1112708108.

Breeze, E., Harrison, E., McHattie, S., Hughes, L., Hickman, R., Hill, C., Kiddle, S., Kim, Y. S., Penfold, C. A., Jenkins, D., & Zhang, C. (2011). High-resolution temporal profiling of transcripts during Arabidopsis leaf senescence reveals a distinct chronology of processes and regulation. *The Plant Cell, 23*(3), 873–894. https://doi.org/10.1105/tpc.111.083345.

Brueggeman, R., Rostoks, N., Kudrna, D., Kilian, A., Han, F., Chen, J., Druka, A., Steffenson, B., & Kleinhofs, A. (2002). The barley stem rust-resistance gene Rpg1 is a novel disease-resistance gene with homology to receptor kinases. *Proceedings of the National Academy of Sciences, 99*(14), 9328–9333. https://doi.org/10.1073/pnas.142284699.

Brutus, A., Sicilia, F., Macone, A., Cervone, F., & De Lorenzo, G. (2010). A domain swap approach reveals a role of the plant wall-associated kinase 1 (WAK1) as a receptor of oligogalacturonides. *Proceedings of the National Academy of Sciences of the United States of America, 107*(20), 9452–9457. https://doi.org/10.1073/pnas.1000675107.

Buschges, R., Hollricher, K., Panstruga, R., Simons, G., Wolter, M., Frijters, A., Van Daelen, R., van der Lee, T., Diergaarde, P., Groenendijk, J., & Töpsch, S. (1997). The barley Mlo gene: A novel control element of plant pathogen resistance. *Cell, 88*(5), 695–705. https://doi.org/10.1016/S0092-8674(00)81912-1.

Caarls, L., Pieterse, C. M. J., & Van Wees, S. C. M. (2015). How salicylic acid takes transcriptional control over jasmonic acid signaling. *Frontiers in Plant Science, 6*, 170. https://doi.org/10.3389/fpls.2015.00170.

Cai, D., Kleine, M., Kifle, S., Harloff, H. J., Sandal, N. N., Marcker, K. A., Klein-Lankhorst, R. M., Salentijn, E. M., Lange, W., Stiekema, W. J., & Wyss, U. (1997). Positional cloning of a gene for nematode resistance in sugar beet. *Science, 275*(5301), 832–834. https://doi.org/10.1126/science.275.5301.832.

Cao, L., Blekemolen, M. C., Tintor, N., Cornelissen, B. J. C., & Takken, F. L. W. (2018). The Fusarium oxysporum Avr2-Six5 effector pair alters plasmodesmatal exclusion selectivity to facilitate cell-to-cell movement of Avr2. *Molecular Plant, 11*(5), 691–705. https://doi.org/10.1016/j.molp.2018.02.003.

Catanzariti, A. M., Dodds, P. N., Ve, T., Kobe, B., Ellis, J. G., & Staskawicz, B. J. (2010). The AvrM effector from flax rust has a structured C-terminal domain and interacts directly with the M resistance protein. *Molecular Plant-Microbe Interactions, 23*(1), 49–57. https://doi.org/10.1094/MPMI-23-1-0049.

Cesari, S., Bernoux, M., Moncuquet, P., Kroj, T., & Dodds, P. N. (2014). A novel conserved mechanism for plant NLR protein pairs: The "integrated decoy" hypothesis. *Frontiers in Plant Science, 5*, 606. https://doi.org/10.3389/fpls.2014.00606.

Chang, Y. H., Yan, H. Z., & Liou, R. F. (2015). A novel elicitor protein from Phytophthora parasitica induces plant basal immunity and systemic acquired resistance. *Molecular Plant Pathology, 16*(2), 123–136. https://doi.org/10.1111/mpp.12166.

Chanwala, J., Satpati, S., Dixit, A., Parida, A., Giri, M. K., & Dey, N. (2020). Genome-wide identification and expression analysis of WRKY transcription factors in pearl millet (*Pennisetum glaucum*) under dehydration and salinity stress. *BMC Genomics, 21*(1), 1–16. https://doi.org/10.1186/s12864-020-6622-0.

Chaparro-Garcia, A., Wilkinson, R. C., Gimenez-Ibanez, S., Findlay, K., Coffey, M. D., Zipfel, C., Rathjen, J. P., Kamoun, S., & Schornack, S. (2011). The receptor-like kinase serk3/bak1 is required for basal resistance against the late blight pathogen *Phytophthora infestans* in *Nicotiana benthamiana*. *PLoS ONE, 6*(1), e16608. https://doi.org/10.1371/journal.pone.0016608.

Chen, C. M., Gong, Y., Zhang, M., & Chen, J. J. (2004). Reciprocal cross-talk between Nod2 and TAK1 signaling pathways. *Journal of Biological Chemistry, 279*(24), 25876–25882.

Chen, Z., Agnew, J. L., Cohen, J. D., He, P., Shan, L., Sheen, J., & Kunkel, B. N. (2007). Pseudomonas syringae type III effector AvrRpt2 alters Arabidopsis thaliana auxin physiology. *Proceedings of the National Academy of Sciences, 104*(50), 20131–20136.

Cheng, M. C., Liao, P. M., Kuo, W. W., & Lin, T. P. (2013). The Arabidopsis ethylene response factor1 regulates abiotic stress-responsive gene expression by binding to different cis-acting elements in response to different stress signals. *Plant Physiology, 162*(3), 1566–1582.

Cheong, Y. H., Moon, B. C., Kim, J. K., Kim, C. Y., Kim, M. C., Kim, I. H., Park, C. Y., Kim, J. C., Park, B. O., & Koo, S. C. (2003). BWMK1, a rice mitogen-activated protein kinase, locates in the nucleus and mediates pathogenesis-related gene expression by activation of a transcription factor. *Plant Physiology, 132*, 1961–1972.

Chinchilla, D., Shan, L., He, P., de Vries, S., & Kemmerling, B. (2009). One for all: The receptor-associated kinase BAK1. *Trends in Plant Science, 14*(10), 535–541. https://doi.org/10.1016/j.tplants.2009.08.002.

Chisholm, S. T., Mahajan, S. K., Whitham, S. A., Yamamoto, M. L., & Carrington, J. C. (2000). Cloning of the Arabidopsis RTM1 gene, which controls restriction of long-distance movement of tobacco etch virus. *Proceedings of the National Academy of Sciences, 97*(1), 489–494.

Choi, H. W., Manohar, M., Manosalva, P., Tian, M., Moreau, M., & Klessig, D. F. (2016). Activation of plant innate immunity by extracellular high mobility group box 3 and its inhibition by salicylic acid. *PLoS Pathogens, 12*(3), e1005518. https://doi.org/10.1371/journal.ppat.1005518.

Choi, J., Tanaka, K., Cao, Y., Qi, Y., Qiu, J., Liang, Y., Lee, S. Y., & Stacey, G. (2014). Identification of a plant receptor for extracellular ATP. *Science, 343*(6168), 290–294. https://doi.org/10.1126/science.343.6168.290.

Chow, J., Notaro, M., Prabhakar, A., Free, S., & Cullen, P. (2018). Impact of fungal mapk pathway targets on the cell wall. *Journal of Fungi, 4*(3), 93. https://doi.org/10.3390/jof4030093.

Chujo, T., Takai, R., Akimoto-Tomiyama, C., Ando, S., Minami, E., Nagamura, Y., Kaku, H., Shibuya, N., Yasuda, M., Nakashita, H., Umemura, K., Okada, A., Okada, K., Nojiri, H., & Yamane, H. (2007). Involvement of the elicitor-induced gene OsWRKY53 in the expression of defense-related genes in rice. *Biochimica et Biophysica Acta (BBA)–Gene Structure and Expression, 1769*(7–8), 497–505. https://doi.org/10.1016/j.bbaexp.2007.04.006.

Clouse, S. D., & Sasse, J. M. (1998). Brassinosteroids: Essential regulators of plant growth and development. *Annual Review of Plant Biology, 49*, 427–451.

Collemare, J., O'Connell, R., & Lebrun, M. H. (2019). Nonproteinaceous effectors: The terra incognita of plant–fungal interactions. *New Phytologist, 223*(2), 590–596. https://doi.org/10.1111/nph.15708.

Collier, S. M., & Moffett, P. (2009). NB-LRRs work a "bait and switch" on pathogens. *Trends in Plant Science, 14*(10), 521–529. https://doi.org/10.1016/j.tplants.2009.07.005.

Dang, F. F., Wang, Y. N., Yu, L., Eulgem, T., Lai, Y., Liu, Z. Q., Wang, X., Qiu, A. L., Zhang, T. X., Lin, J., Chen, Y. S., Guan, D. Y., Cai, H. Y., Mou, S. L., & He, S. L. (2013). CaWRKY40, a WRKY protein of pepper, plays an important role in the regulation of tolerance to heat stress and resistance to *Ralstonia solanacearum* infection. *Plant, Cell & Environment, 36*(4), 757–774. https://doi.org/10.1111/pce.12011.

Dangl, J. L., & Jones, J. D. (2001). Plant pathogens and integrated defense responses to infection. *Nature, 411*, 826–833.

Davletova, S., Rizhsky, L., Liang, H., Shengqiang, Z., Oliver, D. J., Coutu, J., Shulaev, V., Schlauch, K., & Mittler, R. (2005). Cytosolic ascorbate peroxidase 1 is a central component of the reactive oxygen gene network of Arabidopsis. *The Plant Cell, 17*(1), 268–281. https://doi.org/10.1105/tpc.104.026971.

De Jonge, R., Bolton, M. D., & Thomma, B. P. (2011). How filamentous pathogens co-opt plants: The ins and outs of fungal effectors. *Current Opinion in Plant Biology, 14*(4), 400–406. https://doi.org/10.1016/j.pbi.2011.03.005.

De Vleesschauwer, D., Van Buyten, E., Satoh, K., Balidion, J., Mauleon, R., Choi, I. R., Vera-Cruz, C., Kikuchi, S., & Höfte, M. (2012). Brassinosteroids antagonize gibberellin- and salicylate-mediated root immunity in rice. *Plant Physiology, 158*(4), 1833–1846. https://doi.org/10.1104/pp.111.192104.

De Wit, P. J., Mehrabi, R., Van den Burg, H. A., & Stergiopoulos, I. (2009). Fungal effector proteins: Past, present and future. *Molecular Plant Pathology, 10*(6), 735–747. https://doi.org/10.1111/j.1364-3703.2009.00591.x.

Denoux, C., Galletti, R., Mammarella, N., Gopalan, S., Werck, D., De Lorenzo, G., Ferrari, S., Ausubel, F. M., & Dewdney, J. (2008). Activation of defense response pathways by OGs and Flg22 elicitors in Arabidopsis seedlings. *Molecular Plant, 1*(3), 423–445. https://doi.org/10.1093/mp/ssn019.

Derksen, H., Rampitsch, C., & Daayf, F. (2013). Signaling cross-talk in plant disease resistance. *Plant Science, 207*, 79–87. https://doi.org/10.1016/j.plantsci.2013.03.003.

Deslandes, L., Olivier, J., Peeters, N., Feng, D. X., Khounlotham, M., Boucher, C., Somssich, I., Genin, S., & Marco, Y. (2003). Physical interaction between RRS1-R, a protein conferring resistance to bacterial wilt, and PopP2, a type III effector targeted to the plant nucleus. *Proceedings of the National Academy of Sciences, 100*(13), 8024–8029. https://doi.org/10.1073/pnas.1331622100.

Deslandes, L., Olivier, J., Theulières, F., Hirsch, J., Feng, D. X., Bittner-Eddy, P., Beynon, J., & Marco, Y. (2002). Resistance to *Ralston solanacearum* in Arabidopsis thaliana is conferred by the recessive RRS1-R gene, a member of a novel family of resistance genes. *Proceedings of the National Academy of Sciences, 99*(4), 2404–2409. https://doi.org/10.1073/pnas.032485099.

Dey, S., Wenig, M., Langen, G., Sharma, S., Kugler, K. G., Knappe, C., Hause, B., Bichlmeier, M., Babaeizad, V., Imani, J., & Janzik, I. (2014). Bacteria-triggered systemic immunity in barley is associated with WRKY and ethylene responsive factors but not with salicylic acid. *Plant Physiology, 166*(4), 2133–2151. https://doi.org/10.1104/pp.114.244699.

Dixon, M. S., Jones, D. A., Keddie, J. S., Thomas, C. M., Harrison, K., & Jones, J. D. (1996). The tomato Cf-2 disease resistance locus comprises two functional genes encoding leucine-rich repeat proteins. *Cell, 84*(3), 451–459.

Dodds, P. N., & Rathjen, J. P. (2010). Plant immunity: Towards an integrated view of plant–Pathogen interactions. *Nature Reviews Genetics, 11*(8), 539–548. https://doi.org/10.1038/nrg2812.

Dodds, P. N., Lawrence, G. J., Catanzariti, A. M., Teh, T., Wang, C. I., Ayliffe, M. A., Kobe, B., & Ellis, J. G. (2006). Direct protein interaction underlies gene-for-gene specificity and coevolution of the flax resistance genes and flax rust avirulence genes. *Proceedings of the National Academy of Sciences, 103*(23), 8888–8893. https://doi.org/10.1073/pnas.0602577103.

Dong, S., Raffaele, S., & Kamoun, S. (2015). The two-speed genomes of filamentous pathogens: Waltz with plants. *Current Opinion in Genetics & Development, 35*, 57–65. https://doi.org/10.1016/j.gde.2015.09.001.

Dong, Z., Li, W., Liu, J., Li, L., Pan, S., Liu, S., Gao, J., Liu, L., Liu, X., Wang, G. L., & Dai, L. (2019). The rice phosphate transporter protein ospt8 regulates disease resistance and plant growth. *Scientific Reports, 9*(1), 5408. https://doi.org/10.1038/s41598-019-41872-8.

Eulgem, T., Rushton, P. J., Robatzek, S., & Somssich, I. E. (2000). The WRKY superfamily of plant transcription factors. *Trends in Plant Science, 5*(5), 199–206. https://doi.org/10.1016/S1360-1385(00)01600-9.

Fahad, S., Hussain, S., Matloob, A., Khan, F. A., Khaliq, A., Saud, S., Hassan, S., Shan, D., Khan, F., Ullah, N., & Faiq, M. (2015). Phytohormones and plant responses to salinity stress: A review. *Plant Growth Regulation, 75*, 391–404. https://doi.org/10.1007/s10725-014-0016-4.

Faino, L., Seidl, M. F., Shi-Kunne, X., Pauper, M., van den Berg, G. C., Wittenberg, A. H., & Thomma, B. P. (2016). Transposons passively and actively contribute to evolution of the two-speed genome of a fungal pathogen. *Genome Research, 26*(8), 1091–1100. https://doi.org/10.1101/gr.204974.116.

Fariduddin, Q., Khalil, R. R., Mir, B. A., Yusuf, M., & Ahmad, A. (2013). 24-Epibrassinolide regulates photosynthesis, antioxidant enzyme activities and proline content of *Cucumis sativus* under salt stress. *Environmental and Experimental Botany, 86*, 124–131. https://doi.org/10.1016/j.envexpbot.2011.06.002.

Fariduddin, Q., Khalil, R. R., Mir, B. A., Yusuf, M., & Ahmad, A. (2013). 24-Epibrassinolide regulates photosynthesis, antioxidant enzyme activities and proline content of *Cucumis sativus* under salt and/or copper stress. *Environmental Monitoring and Assessment, 185*, 7845–7856.

Ferrari, S., Galletti, R., Denoux, C., De Lorenzo, G., Ausubel, F. M., & Dewdney, J. (2007). Resistance to *Botrytis cinerea* induced in *Arabidopsis* by elicitors is independent of salicylic acid, ethylene, or jasmonate signaling but requires phytoalexin deficient3. *Plant Physiology, 144*(1), 367–379. https://doi.org/10.1104/pp.107.095596.

Ferrari, S., Savatin, D. V., Sicilia, F., Gramegna, G., Cervone, F., & De Lorenzo, G. (2013). Oligogalacturonides: Plant damage-associated molecular patterns and regulators of growth and development. *Frontiers in Plant Science, 4*(MAR), 1–9. https://doi.org/10.3389/fpls.2013.00049.

Flor, H. H. (1971). Current status of the gene-for-gene concept. *Annual Review of Phytopathology, 9*(1), 275–296.

Fokkens, L., Shahi, S., Connolly, L. R., Stam, R., Schmidt, S. M., Smith, K. M., Freitag, M., & Rep, M. (2018). The multi-speed genome of *Fusarium oxysporum* reveals association of histone modifications with sequence divergence and footprints of past horizontal chromosome transfer events. *BioRxiv, 465070*. https://doi.org/10.1101/465070.

Franco-Orozco, B., Berepiki, A., Ruiz, O., Gamble, L., Griffe, L. L., Wang, S., Birch, P. R. J., Kanyuka, K., & Avrova, A. (2017). A new proteinaceous pathogen-associated molecular pattern (PAMP) identified in Ascomycete fungi induces cell death in Solanaceae. *New Phytologist, 214*(4), 1657–1672. https://doi.org/10.1111/NPH.14542.

Frantzeskakis, L., Kracher, B., Kusch, S., Yoshikawa-Maekawa, M., Bauer, S., Pedersen, C., Spanu, P. D., Maekawa, T., Schulze-Lefert, P., & Panstruga, R. (2018). Signatures of host specialization and a recent transposable element burst in the dynamic one-speed genome of the fungal barley powdery mildew pathogen. *BMC Genomics, 19*(1), 1–23.

Frantzeskakis, L., Kusch, S., & Panstruga, R. (2019). The need for speed: Compartmentalized genome evolution in filamentous phytopathogens. *Molecular Plant Pathology, 20*(1), 3–7.

Friesen, T. L., Faris, J. D., Solomon, P. S., & Oliver, R. P. (2008). Host-specific toxins: Effectors of necrotrophic pathogenicity. *Cellular Microbiology, 10*(7), 1421–1428.

Fu, J., & Wang, S. (2011). Insights into auxin signaling in plant–pathogen interactions. *Frontiers in Plant Science, 2*, 74.

Fu, Z. Q., & Dong, X. (2013). Systemic acquired resistance: Turning local infection into global defense. *Annual Review of Plant Biology, 64*, 839–863. https://doi.org/10.1146/ANNUREV-ARPLANT-042811-105606.

Fujikawa, T., Sakaguchi, A., Nishizawa, Y., Kouzai, Y., Minami, E., Yano, S., Koga, H., Meshi, T., & Nishimura, M. (2012). Surface α-1,3-glucan facilitates fungal stealth infection by interfering with innate immunity in plants. *PLoS Pathogens, 8*(8), e1002882. https://doi.org/10.1371/journal.ppat.1002882.

Furman-Matarasso, N., Cohen, E., Du, Q., Chejanovsky, N., Hanania, U., & Avni, A. (1999). A point mutation in the ethylene-inducing xylanase elicitor inhibits the β-1-4-endoxylanase activity but not the elicitation activity. *Plant Physiology, 121*(2), 345–350. https://doi.org/10.1104/PP.121.2.345.

Galletti, R., Ferrari, S., & de Lorenzo, G. (2011). Arabidopsis MPK3 and MPK6 play different roles in basal and oligogalacturonide- or flagellin-induced resistance against *Botrytis cinerea*. *Plant Physiology, 157*(2), 804–814. https://doi.org/10.1104/PP.111.174003.

Gao, Q. M., Venugopal, S., Navarre, D., & Kachroo, A. (2011). Low oleic acid-derived repression of jasmonic acid-inducible defense responses requires the WRKY50 and WRKY51 proteins. *Plant Physiology, 155*(1), 464–476. https://doi.org/10.1104/PP.110.166876.

Gao, Y. F., Liu, J. K., Yang, F. M., Zhang, G. Y., Wang, D., Zhang, L., Ou, Y. Bin, & Yao, Y. A. (2020). The WRKY transcription factor WRKY8 promotes resistance to pathogen infection and mediates drought and salt stress tolerance in *Solanum lycopersicum*. *Physiologia Plantarum, 168*(1), 98–117. https://doi.org/10.1111/PPL.12978.

Gassmann, W., & Bhattacharjee, S. (2012). Effector-triggered immunity signaling: From gene-for-gene pathways to protein-protein interaction networks. *Molecular Plant-Microbe Interactions, 25*(7), 862–868.

Gaulin, E., Dramé, N., Lafitte, C., Torto-Alalibo, T., Martinez, Y., Ameline-Torregrosa, C., Khatib, M., Mazarguil, H., Villalba-Mateos, F., Kamoun, S., Mazars, C., Dumas, B., Bottin, A., Esquerré-Tugayé, M. T., & Rickauer, M. (2006). Cellulose binding domains of a *Phytophthora* cell wall protein are novel pathogen-associated molecular patterns. *Plant Cell, 18*(7), 1766–1777. https://doi.org/10.1105/tpc.105.038687.

George, H. L., & Van Etten, H. D. (2001). Characterization of pisatin-inducible cytochrome P450s in fungal pathogens of pea that detoxify the pea phytoalexin pisatin. *Fungal Genetics and Biology, 33*, 37–48.

Gijzen, M., & Nürnberger, T. (2006). Nep1-like proteins from plant pathogens: Recruitment and diversification of the NPP1 domain across taxa. *Phytochemistry, 67*(16), 1800–1807. https://doi.org/10.1016/J.PHYTOCHEM.2005.12.008.

Giraldo, M. C., & Valent, B. (2013). Filamentous plant pathogen effectors in action. *Nature Reviews Microbiology, 11*(11), 800–814.

Glazebrook, J. (2005). Contrasting mechanisms of defense against biotrophic and necrotrophic pathogens. *Annual Review of Phytopathology, 43*, 205–227.

Glazebrook, J., Rogers, E. E., & Ausubel, F. M. (1997). Use of *Arabidopsis* for genetic dissection of plant defense responses. *Annual Review of Genetics, 31*(1), 547–-569.

Goel, R., Pandey, A., Trivedi, P. K., & Asif, M. H. (2016). Genome-wide analysis of the musa WRKY gene family: Evolution and differential expression during development and stress. *Frontiers in Plant Science, 7*(MAR2016), Article 299. https://doi.org/10.3389/FPLS.2016.00299/BIBTEX.

Gomes, M. M. A. (2011). Physiological effects related to brassinosteroid application in plants. In S. Hayat & A. Ahmad (Eds.), *Brassinosteroids: A Class of Plant Hormone* (pp. 193–242). Springer.

Gómez-Gómez, L., & Boller, T. (2002). Flagellin perception: A paradigm for innate immunity. *Trends in Plant Science, 7*(6), 251–256.

Greenberg, J. T., & Vinatzer, B. A. (2003). Identifying type III effectors of plant pathogens and analyzing their interaction with plant cells. *Current Opinion in Microbiology, 6*(1), 20–28.

Guo, R., Qiao, H., Zhao, J., Wang, X., Tu, M., Guo, C., Wan, R., Li, Z., & Wang, X. (2018). The grape VLWRKY3 gene promotes abiotic and biotic stress tolerance in transgenic *Arabidopsis thaliana. Frontiers in Plant Science, 9*, 545. https://doi.org/10.3389/fpls.2018.00545.

Gupta, R., Min, C. W., Kim, Y. J., & Kim, S. T. (2019). Identification of Msp1-induced signaling components in rice leaves by integrated proteomic and phosphoproteomic analysis. *International Journal of Molecular Sciences, 20*(17), 4135.

Gururani, M. A., Venkatesh, J., Upadhyaya, C. P., Nookaraju, A., Pandey, S. K., & Park, S. W. (2012). Plant disease resistance genes: Current status and future directions. *Physiological and Molecular Plant Pathology, 78*, 51–65.

Hamel, L. P., Nicole, M. C., Duplessis, S., & Ellis, B. E. (2012). Mitogen-activated protein kinase signaling in plant-interacting fungi: Distinct messages from conserved messengers. *The Plant Cell, 24*(4), 1327–1351.

Hatsugai, N., Igarashi, D., Mase, K., Lu, Y., Tsuda, Y., Chakravarthy, S., Wei, H. L., Foley, J. W., Collmer, A., & Glazebrook, J. (2017). A plant effector-triggered immunity signaling sector is inhibited by pattern-triggered immunity. *EMBO Journal, 36*, 2758–2769.

He, X., Wang, C., Wang, H., Li, L., & Wang, C. (2020). The function of MAPK cascades in response to various stresses in horticultural plants. *Frontiers in Plant Science, 11*, 952.

Heise, M. T. (2014). Viral Pathogenesis. In Reference Module in Biomedical Sciences. https://doi.org/10.1016/B978-0-12-801238-3.00079-9.

Helliwell, E. E., Wang, Q., & Yang, Y. (2013). Transgenic rice with inducible ethylene production exhibits broad-spectrum disease resistance to the fungal pathogens *Magnaporthe oryzae* and *Rhizoctonia solani. Plant Biotechnology Journal, 11*(1), 33–42.

Hemetsberger, C., Mueller, A. N., Matei, A., Herrberger, C., Hensel, G., Kumlehn, J., Mishra, B., Sharma, R., Thines, M., Hückelhoven, R., & Doehlemann, G. (2015). The fungal core effector Pep1 is conserved across smuts of dicots and monocots. *New Phytologist, 206*(3), 1116–1126.

Higashi, K., Ishiga, Y., Inagaki, Y., Toyoda, K., Shiraishi, T., & Ichinose, Y. (2008). Modulation of defense signal transduction by flagellin-induced WRKY41 transcription factor in *Arabidopsis thaliana. Molecular Genetics and Genomics, 279*(3), 303–312. https://doi.org/10.1007/s00438-007-0315-0.

Hoang, C. V., Bhaskar, C. K., & Ma, L. S. (2021). A novel core effector Vp1 promotes fungal colonization and virulence of *Ustilago maydis. Journal of Fungi, 7*(8), 589. https://doi.org/10.3390/jof7080589.

Hu, Y., Dong, Q., & Yu, D. (2012). Arabidopsis WRKY46 coordinates with WRKY70 and WRKY53 in basal resistance against pathogen *Pseudomonas syringae. Plant Science, 185–186*, 288–297. https://doi.org/10.1016/j.plantsci.2011.12.003.

Huang, C., Yan, Y., Zhao, H., Ye, Y., & Cao, Y. (2020). Arabidopsis CPK5 phosphorylates the chitin receptor LYK5 to regulate plant innate immunity. *Frontiers in Plant Science, 11*, 702.

Huang, H., Ullah, F., Zhou, D. X., Yi, M., & Zhao, Y. (2019). Mechanisms of ROS regulation of plant development and stress responses. *Frontiers in Plant Science, 10*, 800.

Huang, S., Zhang, X., & Fernando, W. G. D. (2020). Directing trophic divergence in plant-pathogen interactions: Antagonistic phytohormones with no doubt? *Frontiers in Plant Science, 11*, 1–9. https://doi.org/10.3389/fpls.2020.600063.

Huffaker, A., Pearce, G., & Ryan, C. A. (2006). An endogenous peptide signal in *Arabidopsis* activates components of the innate immune response. *Proceedings of the National Academy of Sciences of the United States of America, 103*(26), 10098–10103. https://doi.org/10.1073/pnas.0603727103.

Hussain, A., Li, X., Weng, Y., Liu, Z., Ashraf, M. F., Noman, A., Yang, S., Ifnan, M., Qiu, S., Yang, Y., Guan, D., & He, S. (2018). CaWRKY22 acts as a positive regulator in pepper response to *Ralstonia solanacearum* by constituting networks with CaWRKY6, CaWRKY27, CaWRKY40, and CaWRKY58. *International Journal of Molecular Sciences, 19*(5), 1426. https://doi.org/10.3390/ijms19051426.

Hwang, S. H., Kwon, S. I., Jang, J. Y., Fang, I. L., Lee, H., Choi, C., Park, S., Ahn, I., Bae, S. C., & Hwang, D. J. (2016). OsWRKY51, a rice transcription factor, functions as a positive regulator in defense response against *Xanthomonas oryzae* pv. *oryzae. Plant Cell Reports, 35*(9), 1975–1985. https://doi.org/10.1007/s00299-016-2012-0.

Ingle, R. A., Carstens, M., & Denby, K. J. (2006). PAMP recognition and the plant–Pathogen arms race. *Bioessays, 28*(9), 880–889.

Inohara, N., & Nuñez, G. (2001). The NOD: A signaling module that regulates apoptosis and host defense against pathogens. *Oncogene, 20*(44), 6473–6481.

Inohara, N., Ogura, Y., & Nuñez, G. (2002). Nods: A family of cytosolic proteins that regulate the host response to pathogens. *Current Opinion in Microbiology, 5*(1), 76–80.

Irieda, H., Inoue, Y., Mori, M., Yamada, K., Oshikawa, Y., Saitoh, H., Uemura, A., Terauchi, R., Kitakura, S., Kosaka, A., & Singkaravanit-Ogawa, S. (2019). Conserved fungal effector suppresses PAMP-triggered immunity by targeting plant immune kinases. *Proceedings of the National Academy of Sciences, 116*(2), 496–505.

Ishiguro, S., & Nakamura, K. (1994). Characterization of a cDNA encoding a novel DNA-binding protein, SPF1, that recognizes SP8 sequences in the 5′ upstream regions of genes coding for sporamin and β-amylase from sweet potato. *Molecular & General Genetics, 244*(6), 563–571. https://doi.org/10.1007/BF00282746.

Ishihara, T., Sekine, K. T., Hase, S., Kanayama, Y., Seo, S., Ohashi, Y., Kusano, T., Shibata, D., Shah, J., & Takahashi, H. (2008). Overexpression of the *Arabidopsis thaliana* EDS5 gene enhances resistance to viruses. *Plant Biology, 10*(4), 451–461.

Jagodzik, P., Tajdel-Zielinska, M., Ciesla, A., Marczak, M., & Ludwikow, A. (2018). Mitogen-activated protein kinase cascades in plant hormone signaling. *Frontiers in Plant Science, 9*, 1387.

Javid, M. G., Sorooshzadeh, A., Moradi, F., Modarres Sanavy, S. A. M., & Allahdadi, I. (2011). The role of phytohormones in alleviating salt stress in crop plants. *Australian Journal of Crop Science, 5*(6), 726–734.

Jia, Y., McAdams, S. A., Bryan, G. T., Hershey, H. P., & Valent, B. (2000). Direct interaction of resistance gene and avirulence gene products confers rice blast resistance. *EMBO Journal, 19*(15), 4004–4014.

Jiang, C. J., Shimono, M., Sugano, S., Kojima, M., Liu, X., Inoue, H., Sakakibara, H., & Takatsuji, H. (2013). Cytokinins act synergistically with salicylic acid to activate defense gene expression in rice. *Molecular Plant-Microbe Interactions, 26*(3), 287–296.

Jiang, J., Ma, S., Ye, N., Jiang, M., Cao, J., & Zhang, J. (2017). WRKY transcription factors in plant responses to stresses. *Journal of Integrative Plant Biology, 59*(2), 86–101. https://doi.org/10.1111/JIPB.12513.

Johal, G. S., & Briggs, S. P. (1992). Reductase activity encoded by the HM1 disease resistance gene in maize. *Science, 258*(5084), 985–987.

Jones, D. A., & Jones, J. D. (1997). The role of leucine-rich repeat proteins in plant defenses. In *Advances in Botanical Research* (Vol. 24, pp. 89–167). Academic Press.

Jones, D. A., Thomas, C. M., Hammond-Kosack, K. E., Balint-Kurti, P. J., & Jones, J. D. (1994). Isolation of the tomato Cf-9 gene for resistance to *Cladosporium fulvum* by transposon tagging. *Science, 266*(5186), 789–793.

Jones, J. D. G., & Dangl, J. L. (2006). The plant immune system. *Nature, 444*(7117), 323–329.

Ju, C., Yoon, G. M., Shemansky, J. M., Lin, D. Y., Ying, Z. I., Chang, J., Garrett, W. M., Kessenbrock, M., Groth, G., Tucker, M. L., & Cooper, B. (2012). CTR1 phosphorylates the central regulator EIN2 to control ethylene hormone signaling from the ER membrane to the nucleus in *Arabidopsis*. *Proceedings of the National Academy of Sciences, 109*(47), 19486–19491.

Kamoun, S. (2006). A catalogue of the effector secretome of plant pathogenic oomycetes. *Annual Review of Phytopathology, 44*, 41–60.

Kanzaki, H., Yoshida, K., Saitoh, H., Fujisaki, K., Hirabuchi, A., Alaux, L., Fournier, E., Tharreau, D., & Terauchi, R. (2012). Arms race co-evolution of *Magnaporthe oryzae* AVR-Pik and rice Pik genes driven by their physical interactions. *The Plant Journal, 72*(6), 894–907.

Kawchuk, L. M., Hachey, J., Lynch, D. R., Kulcsar, F., Van Rooijen, G., Waterer, D. R., Robertson, A., Kokko, E., Byers, R., Howard, R. J., & Fischer, R. (2001). Tomato Ve disease resistance genes encode cell surface-like receptors. *Proceedings of the National Academy of Sciences, 98*(11), 6511–6515.

Kazan, K., & Manners, J. M. (2008). Jasmonate signaling: Toward an integrated view. *Plant Physiology, 146*(4), 1459–1468.

Kazan, K., & Manners, J. M. (2009). Linking development to defense: Auxin in plant–Pathogen interactions. *Trends in Plant Science, 14*(7), 373–382. https://doi.org/10.1016/j.tplants.2009.04.005.

Khan, M. I., Zhang, Y., Liu, Z., Hu, J., Liu, C., Yang, S., Hussain, A., Ashraf, M. F., Noman, A., Shen, L., Xia, X., Yang, F., Guan, D., & He, S. (2018). CaWRKY40b in pepper acts as a negative regulator in response to *Ralston solanacearum* by directly modulating defense genes including CaWRKY40. *International Journal of Molecular Sciences, 19*(5), 1403. https://doi.org/10.3390/ijms19051403.

Khripach, V., Zhabinskii, V., & de Groot, A. (2000). Twenty years of brassinosteroids: Steroidal plant hormones warrant better crops for the XXI century. *Annals of Botany, 86*(3), 441–447. https://doi.org/10.1006/anbo.2000.1227.

Kim, K. C., Lai, Z., Fan, B., & Chen, Z. (2008). Arabidopsis WRKY38 and WRKY62 transcription factors interact with histone deacetylase 19 in basal defense. *The Plant Cell, 20*(9), 2357–2371. https://doi.org/10.1105/tpc.107.055566.

Kishi-Kaboshi, M., Okada, K., Kurimoto, L., Murakami, S., Umezawa, T., Shibuya, N., Yamane, H., Miyao, A., Takatsuji, H., Takahashi, A., & Hirochika, H. (2010). A rice fungal MAMP-responsive MAPK cascade regulates metabolic flow to antimicrobial metabolite synthesis. *Plant Journal, 63*(4), 599–612. https://doi.org/10.1111/j.1365-313X.2010.04267.x.

Koga, H., Dohi, K., & Mori, M. (2004). Abscisic acid and low temperatures suppress the whole plant-specific resistance reaction of rice plants to the infection of *Magnaporthe grisea*. *Physiological and Molecular Plant Pathology, 65*(1), 3–9. https://doi.org/10.1016/j.pmpp.2004.10.002.

Kohler, A., Rinaldi, C., Duplessis, S., Baucher, M., Geelen, D., Duchaussoy, F., Meyers, B. C., Boerjan, W., & Martin, F. (2008). Genome-wide identification of NBS resistance genes in Populus trichocarpa. *Plant Molecular Biology, 66*, 619–636. https://doi.org/10.1007/s11103-007-9293-9.

Koornneef, A., & Pieterse, C. M. J. (2008). Cross talk in defense signaling. *Plant Physiology, 146*(3), 839–844. https://doi.org/10.1104/pp.107.112029.

Kourelis, J., & Van Der Hoorn, R. A. (2018). Defended to the nines: 25 years of resistance gene cloning identifies nine mechanisms for R protein function. *Plant Cell, 30*(2), 285–299. https://doi.org/10.1105/tpc.17.00562.

Kourelis, J., Malik, S., Mattinson, O., Krauter, S., Kahlon, P. S., Paulus, J. K., & van der Hoorn, R. A. (2020). Evolution of a guarded decoy protease and its receptor in solanaceous plants. *Nature Communications, 11*(1), 4393. https://doi.org/10.1038/s41467-020-18207-6.

Kubicek, C. P., Starr, T. L., & Glass, N. L. (2014). Plant cell wall–Degrading enzymes and their secretion in plant-pathogenic fungi. *Annual Review of Phytopathology, 52*, 427–451. https://doi.org/10.1146/annurev-phyto-102313-045831.

Kusch, S., Frantzeskakis, L., Thieron, H., & Panstruga, R. (2018). Small RNAs from cereal powdery mildew pathogens may target host plant genes. *Fungal Biology, 122*(11), 1050–1063. https://doi.org/10.1016/j.funbio.2018.09.002.

Lai, Z., Vinod, K., Zheng, Z., Fan, B., & Chen, Z. (2008). Roles of Arabidopsis WRKY3 and WRKY4 transcription factors in plant responses to pathogens. *BMC Plant Biology, 8*, 68. https://doi.org/10.1186/1471-2229-8-68.

Lang, J., & Colcombet, J. (2020). Sustained incompatibility between MAPK signaling and pathogen effectors. *International Journal of Molecular Sciences, 21*(21), 7954. https://doi.org/10.3390/ijms21217954.

Lanver, D., Müller, A. N., Happel, P., Schweizer, G., Haas, F. B., Franitza, M., Pellegrin, C., Reissmann, S., Altmüller, J., Rensing, S. A., & Kahmann, R. (2018). The biotrophic development of *Ustilago maydis* studied by RNA-seq analysis. *Plant Cell, 30*(2), 300–323. https://doi.org/10.1105/tpc.17.00857.

Lawrence, G. J., Finnegan, E. J., Ayliffe, M. A., & Ellis, J. G. (1995). The L6 gene for flax rust resistance is related to the Arabidopsis bacterial resistance gene RPS2 and the tobacco viral resistance gene N. *The Plant Cell, 7*(8), 1195–1206. https://doi.org/10.2307/3869908.

Le Roux, C., Huet, G., Jauneau, A., Camborde, L., Trémousaygue, D., Kraut, A., Zhou, B., Levaillant, M., Adachi, H., Yoshioka, H., & Raffaele, S. (2015). A receptor pair with an integrated decoy converts pathogen disabling of transcription factors to immunity. *Cell, 161*(5), 1074–1088. https://doi.org/10.1016/j.cell.2015.04.025.

Leclercq, J., Ranty, B., Sanchez-Ballesta, M. T., Li, Z., Jones, B., Jauneau, A., Pech, J. C., Latché, A., Ranjeva, R., & Bouzayen, M. (2005). Molecular and biochemical characterization of LeCRK1, a ripening-associated tomato CDPK-related kinase. *Journal of Experimental Botany, 56*(409), 25–35. https://doi.org/10.1093/jxb/eri011.

Lee, H., Cha, J., Choi, C., Choi, N., Ji, H. S., Park, S. R., Lee, S., & Hwang, D. J. (2018). Rice WRKY11 plays a role in pathogen defense and drought tolerance. *Rice, 11*(1), 1–12. https://doi.org/10.1186/s12284-018-0199-0.

Leon-Reyes, A., Van der Does, D., De Lange, E. S., Delker, C., Wasternack, C., Van Wees, S. C. M., Ritsema, T., & Pieterse, C. M. J. (2010). Salicylate-mediated suppression of jasmonate-responsive gene expression in Arabidopsis is targeted downstream of the jasmonate biosynthesis pathway. *Planta, 232*(6), 1423–1432. https://doi.org/10.1007/s00425-010-1265-z.

Li, B., Meng, X., Shan, L., & He, P. (2016). Transcriptional regulation of pattern-triggered immunity in plants. *Cell Host & Microbe, 19*(5), 641–650. https://doi.org/10.1016/j.chom.2016.04.008.

Li, J. bin, Luan, Y. shi, & Liu, Z. (2015). Overexpression of SpWRKY1 promotes resistance to *Phytophthora nicotianae* and tolerance to salt and drought stress in transgenic tobacco. *Physiologia Plantarum, 155*(3), 248–266. https://doi.org/10.1111/ppl.12315.

Li, J., Brader, G., & Palva, E. T. (2004). The WRKY70 transcription factor: A node of convergence for jasmonate-mediated and salicylate-mediated signals in plant defense. *The Plant Cell, 16*(2), 319–331. https://doi.org/10.1105/tpc.016980.

Li, L., Zhu, X. M., Zhang, Y. R., Cai, Y. Y., Wang, J. Y., Liu, M. Y., & Lin, F. C. (2022). Research on the molecular interaction mechanism between plants and pathogenic fungi. *International Journal of Molecular Sciences, 23*(9), 4658. https://doi.org/10.3390/ijms23094658.

Liu, C., Wang, Y., Wang, Y., Du, Y., Song, C., Song, P., Yang, Q., He, F., Bai, X., Huang, L., & Guo, J. (2022). Glycine-serine-rich effector PstGSRE4 in *Puccinia striiformis* f.sp. tritici inhibits the activity of copper zinc superoxide dismutase to modulate immunity in wheat. *PLoS Pathogens, 18*(7), e1010702. https://doi.org/10.1371/journal.ppat.1010702.

Liu, S., Kracher, B., Ziegler, J., Birkenbihl, R. P., & Somssich, I. E. (2015). Negative regulation of ABA signaling by WRKY33 is critical for Arabidopsis immunity towards Botrytis cinerea 2100. *eLife*. https://doi.org/10.7554/eLife.07295.

Liu, T., Hou, J., Wang, Y., Jin, Y., Borth, W., Zhao, F., Liu, Z., Hu, J., & Zuo, Y. (2016). Genome-wide identification, classification and expression analysis in fungal–plant interactions of cutinase gene family and functional analysis of a putative ClCUT7 in *Curvularia lunata. Molecular Genetics and Genomics, 291*, 1105–1115. https://doi.org/10.1007/s00438-016-1185-y.

Liu, X., Bai, X., Wang, X., & Chu, C. (2007). OsWRKY71, a rice transcription factor, is involved in rice defense response. *Journal of Plant Physiology, 164*(8), 969–979. https://doi.org/10.1016/j.jplph.2006.07.006.

Lo Presti, L., Lanver, D., Schweizer, G., Tanaka, S., Liang, L., Tollot, M., Zuccaro, A., Reissmann, S., & Kahmann, R. (2015). Fungal effectors and plant susceptibility. *Annual Review of Plant Biology, 66*, 513–545. https://doi.org/10.1146/annurev-arplant-043014-114623.

Loh, Y. T., Zhou, J., & Martin, G. B. (1998). The myristylation motif of Pto is not required for disease resistance. *Molecular Plant-Microbe Interactions, 11*(6), 572–576. https://doi.org/10.1094/MPMI.1998.11.6.572.

Lorenzo, O., Piqueras, R., Sánchez-Serrano, J. J., & Solano, R. (2003). Ethylene response factor1 integrates signals from ethylene and jasmonate pathways in plant defense. *The Plant Cell, 15*(1), 165–178. https://doi.org/10.1105/tpc.006130.

Lorrain, C., Gonçalves dos Santos, K. C., Germain, H., Hecker, A., & Duplessis, S. (2019). Advances in understanding obligate biotrophy in rust fungi. *New Phytologist, 222*(3), 1190–1206. https://doi.org/10.1111/nph.15606.

Ma, H., Zhang, B., Gai, Y., Sun, X., Chung, K. R., & Li, H. (2019). Cell-wall-degrading enzymes required for virulence in the host selective toxin-producing necrotroph *Alternaria alternata* of citrus. *Frontiers in Microbiology, 10*, 2514. https://doi.org/10.3389/fmicb.2019.02514.

Ma, Z., Zhu, L., Song, T., Wang, Y., Zhang, Q., Xia, Y., Qiu, M., Lin, Y., Li, H., Kong, L., & Fang, Y. (2017). A paralogous decoy protects *Phytophthora sojae* apoplastic effector PsXEG1 from a host inhibitor. *Science, 355*(6326), 710–714. https://doi.org/10.1126/science.aai7919.

MacMillan, J. (2001). Occurrence of gibberellins in vascular plants, fungi, and bacteria. *Journal of Plant Growth Regulation, 20*(4), 387–442. https://doi.org/10.1007/s003440010041.

Malamy, J., Carr, J. P., Klessig, D. F., & Raskin, I. (1990). Salicylic acid: A likely endogenous signal in the resistance response of tobacco to viral infection. *Science, 250*(4983), 1002–1004. https://doi.org/10.1126/science.250.4983.1002.

Mangelsen, E., Kilian, J., Berendzen, K. W., Kolukisaoglu, Ü. H., Harter, K., Jansson, C., & Wanke, D. (2008). Phylogenetic and comparative gene expression analysis of barley (*Hordeum vulgare*) WRKY transcription factor family reveals putatively retained functions between monocots and dicots. *BMC Genomics, 9*, 194. https://doi.org/10.1186/1471-2164-9-194.

Mao, P., Duan, M., Wei, C., & Li, Y. (2007). WRKY62 transcription factor acts downstream of cytosolic NPR1 and negatively regulates jasmonate-responsive gene expression. *Plant and Cell Physiology, 48*(6), 833–842. https://doi.org/10.1093/pcp/pcm058.

Marchant, A., Bhalerao, R., Casimiro, I., Eklof, J., Casero, P. J., Bennett, M., & Sandberg, G. (2002). AUX1 promotes lateral root formation by facilitating indole-3-acetic acid distribution between sink and source tissues in the *Arabidopsis* seedling. *The Plant Cell, 14*(3), 589–597. https://doi.org/10.1105/tpc.001414.

Martin, G. B., Brommonschenkel, S. H., Chunwongse, J., Frary, A., Ganal, M. W., Spivey, R., Wu, T., Earle, E. D. & Tanksley, S. D. (1993). Map-based cloning of a protein kinase gene conferring disease resistance in tomato. *Science, 262*(5138), 1432–1436. https://doi.org/10.1126/science.262.5138.1432.

Martinez-Soto, D., & Ruiz-Herrera, J. (2017). Functional analysis of fungal MAPK pathways. *Ibero-American Journal of Mycology, 34*(4), 192–202.

Matsushita, A., Inoue, H., Goto, S., Nakayama, A., Sugano, S., Hayashi, N., & Takatsuji, H. (2013). Nuclear ubiquitin proteasome degradation affects WRKY45 function in the rice defense program. *The Plant Journal, 73*(2), 302–313. https://doi.org/10.1111/tpj.12035.

Mauch-Mani, B., & Mauch, F. (2005). The role of abscisic acid in plant–Pathogen interactions. *Current Opinion in Plant Biology, 8*(4), 409–414. https://doi.org/10.1016/j.pbi.2005.05.015.
Mayerhofer, H., Sautron, E., Rolland, N., Catty, P., Seigneurin-Berny, D., Pebay-Peyroula, E., & Ravaud, S. (2016). Structural insights into the nucleotide-binding domains of the P1B-type ATPases HMA6 and HMA8 from Arabidopsis thaliana. *PLOS ONE, 11*(11), e0165666. https://doi.org/10.1371/journal.pone.0165666.
McGurl, B., Pearce, G., Orozco-Cardenas, M., & Ryan, C. A. (1992). Structure, expression, and antisense inhibition of the systemin precursor gene. *Science, 255*(5051), 1570–1573. https://doi.org/10.1126/science.1549783.
Mei, C., Qi, M., Sheng, G., & Yang, Y. (2006). Inducible overexpression of a rice allene oxide synthase gene increases the endogenous jasmonic acid level, PR gene expression, and host resistance to fungal infection. *Molecular Plant-Microbe Interactions, 19*(10), 1127–1137. https://doi.org/10.1094/MPMI-19-1127.
Meng, X., & Zhang, S. (2013). MAPK cascades in plant disease resistance signaling. *Annual Review of Phytopathology, 51*, 245–266. https://doi.org/10.1146/annurev-phyto-082712-102314.
Mithöfer, A., & Maffei, M. E. (2016). General mechanisms of plant defense and plant toxins. *Plant Toxin*, 1–22.
Moffett, P., Farnham, G., Peart, J., & Baulcombe, D. C. (2002). Interaction between domains of a plant NBS-LRR protein in disease resistance-related cell death. *The EMBO Journal, 21*(17), 4511–4519. https://doi.org/10.1093/emboj/cdf453.
Morrissey, J. P., Wubben, J. P., & Osbourn, A. E. (2000). *Stagonospora avenae* secretes multiple enzymes that hydrolyze oat leaf saponins. *Molecular Plant-Microbe Interactions, 13*, 1041–1052.
Muday, G. K., & Rahman, A. (2008). Auxin transport and the integration of gravitropic growth. *Plant Tropisms, 3*, 47–78.
Mukhtar, M. S., Deslandes, L., Auriac, M. C., Marco, Y., & Somssich, I. E. (2008). The Arabidopsis transcription factor WRKY27 influences wilt disease symptom development caused by *Ralstonia solanacearum*. *The Plant Journal, 56*(6), 935–947. https://doi.org/10.1111/j.1365-313X.2008.03651.x.
Narváez-Vásquez, J., & Ryan, C. A. (2004). The cellular localization of prosystemin: A functional role for phloem parenchyma in systemic wound signaling. *Planta, 218*(3), 360–369. https://doi.org/10.1007/s00425-003-1115-3.
Navarro, L., Dunoyer, P., Jay, F., Arnold, B., Dharmasiri, N., Estelle, M., Voinnet, O., & Jones, J. D. (2006). A plant miRNA contributes to antibacterial resistance by repressing auxin signaling. *Science, 312*(5772), 436–439. https://doi.org/10.1126/science.1126088.
Navazio, L., Moscatiello, R., Bellincampi, D., Baldan, B., Meggio, F., Brini, M., Bowler, C., & Mariani, P. (2002). The role of calcium in oligogalacturonide-activated signaling in soybean cells. *Planta, 215*(4), 596–605. https://doi.org/10.1007/s00425-002-0776-7.
Naveed, Z. A., Wei, X., Chen, J., Mubeen, H., & Ali, G. S. (2020). The PTI to ETI continuum in *Phytophthora*-plant interactions. *Frontiers in Plant Science, 11*, 593905. https://doi.org/10.3389/fpls.2020.593905.
Ngou, B. P. M., Ahn, H. K., Ding, P., & Jones, J. D. G. (2021). Mutual potentiation of plant immunity by cell-surface and intracellular receptors. *Nature, 592*(7852), 110–115. https://doi.org/10.1038/s41586-021-03315-7.
Nguyen, Q. M., Iswanto, A. B. B., Son, G. H., & Kim, S. H. (2021). Recent advances in effector-triggered immunity in plants: New pieces in the puzzle create a different paradigm.

International Journal of Molecular Sciences, 22(9), 4709. https://doi.org/10.3390/ijms22094709.

Nimchuk, Z., Eulgem, T., Holt III, B. F., & Dangl, J. L. (2003). Recognition and response in the plant immune system. *Annual Review of Genetics, 37*, 579–609. https://doi.org/10.1146/annurev.genet.37.110801.142628.

Noodén, L. D., Singh, S., & Letham, D. S. (1990). Correlation of xylem sap cytokinin levels with monocarpic senescence in soybean. *Plant Physiology, 93*(1), 33–39. https://doi.org/10.1104/pp.93.1.33.

Nürnberger, T., & Kemmerling, B. (2009). PAMP-triggered basal immunity in plants. In T. Nürnberger & B. Kemmerling (Eds.), *Advances in Botanical Research, 51*, 1–38. https://doi.org/10.1016/S0065-2296(09)51001-4.

Nürnberger, T., Nennstiel, D., Jabs, T., Sacks, W. R., Hahlbrock, K., & Scheel, D. (1994). High-affinity binding of a fungal oligopeptide elicitor to parsley plasma membranes triggers multiple defense responses. *Cell, 78*(3), 449–460. https://doi.org/10.1016/0092-8674(94)90423-5.

O'donnell, P. J., Calvert, C., Atzorn, R., Wasternack, C. H. M. O., Leyser, H. M. O., & Bowles, D. J. (1996). Ethylene as a signal mediating the wound response of tomato plants. *Science, 274*(5294), 1914–1917. https://doi.org/10.1126/science.274.5294.1914.

Pan, L. J., & Jiang, L. (2014). Identification and expression of the WRKY transcription factors of *Carica papaya* in response to abiotic and biotic stresses. *Molecular Biology Reports, 41*(3), 1215–1225. https://doi.org/10.1007/s11033-013-2966-8.

Park, S. W., Kaimoyo, E., Kumar, D., Mosher, S., & Klessig, D. F. (2007). Methyl salicylate is a critical mobile signal for plant systemic acquired resistance. *Science, 318*(5847), 113–116. https://doi.org/10.1126/science.1147113.

Paulus, J. K., Kourelis, J., Ramasubramanian, S., Homma, F., Godson, A., Hörger, A. C., & Win, J. (2020). Extracellular proteolytic cascade in tomato activates immune protease Rcr3. *Proceedings of the National Academy of Sciences, 117*(29), 17409–17417. https://doi.org/10.1073/pnas.2001892117.

Pedras, M. S. C., Minic, Z., & Jha, M. (2008). Brassinin oxidase, a fungal detoxifying enzyme to overcome a plant defense–Purification, characterization and inhibition. *FEBS Journal, 275*, 3691–3705. https://doi.org/10.1111/j.1742-4658.2008.06506.x.

Pedras, M. S. C., Minic, Z., & Sarma-mamillapalle, V. K. (2009). Substrate specificity and inhibition of brassinin hydrolases, detoxifying enzymes from the plant pathogens *Leptosphaeria maculans* and *Alternaria brassicicola*. *FEBS Journal, 276*, 7412–7428. https://doi.org/10.1111/j.1742-4658.2009.07357.x.

Peng, Y., Bartley, L. E., Chen, X., Dardick, C., Chern, M., Ruan, R., & Ronald, P. C. (2008). OsWRKY62 is a negative regulator of basal and Xa21-mediated defense against *Xanthomonas oryzae* pv. *oryzae* in rice. *Molecular Plant, 1*(3), 446–458. https://doi.org/10.1093/mp/ssn024.

Petersen, M., Brodersen, P., Naested, H., Andreasson, E., Lindhart, U., Johansen, B., & Sharma, S. B. (2000). Arabidopsis MAP kinase 4 negatively regulates systemic acquired resistance. *Cell, 103*(7), 1111–1120. https://doi.org/10.1016/S0092-8674(00)00213-0.

Phukan, U. J., Jeena, G. S., & Shukla, R. K. (2016). WRKY transcription factors: Molecular regulation and stress responses in plants. *Frontiers in Plant Science, 7*, 760. https://doi.org/10.3389/fpls.2016.00760.

Põlme, S., Bahram, M., Jacquemyn, H., Kennedy, P., Kohout, P., Moora, M., & Tedersoo, L. (2018). Host preference and network properties in biotrophic plant–Fungal associations. *New Phytologist, 217*(3), 1230–1239. https://doi.org/10.1111/nph.14859.

Qin, X., Liu, J. H., Zhao, W. S., Chen, X. J., Guo, Z. J., & Peng, Y. L. (2013). Gibberellin 20-oxidase gene OsGA20ox3 regulates plant stature and disease development in rice. *Molecular Plant-Microbe Interactions, 26*(2), 227–239. https://doi.org/10.1094/MPMI-06-12-0156-R.

Qiu, D., Xiao, J., Ding, X., Xiong, M., Cai, M., Cao, Y., & Wang, S. (2007). OsWRKY13 mediates rice disease resistance by regulating defense-related genes in salicylate- and jasmonate-dependent signaling. *Molecular Plant-Microbe Interactions, 20*(5), 492–499. https://doi.org/10.1094/MPMI-20-5-0492.

Qiu, J. L., Fiil, B. K., Petersen, K., Nielsen, H. B., Botanga, C. J., Thorgrimsen, S., & Petersen, M. (2008). Arabidopsis MAP kinase 4 regulates gene expression through transcription factor release in the nucleus. *The EMBO Journal, 27*(16), 2214–2221. https://doi.org/10.1038/emboj.2008.147.

Raman, V., Simon, S. A., Demirci, F., Nakano, M., Meyers, B. C., & Donofrio, N. M. (2017). Small RNA functions are required for growth and development of *Magnaporthe oryzae. Molecular Plant-Microbe Interactions, 30*(7), 517–530. https://doi.org/10.1094/MPMI-02-17-0039-R.

Rathore, J. S., & Ghosh, C. (2018). Pathogen-associated molecular patterns and their perception in plants. In C. Ghosh (Ed.), *Molecular Aspects of Plant-Pathogen Interaction* (pp. 69–88). Springer. https://doi.org/10.1007/978-981-10-7371-7_4.

Ren, D., Liu, Y., Yang, K. Y., Han, L., Mao, G., Glazebrook, J., & Zhang, S. (2008). A fungal-responsive MAPK cascade regulates phytoalexin biosynthesis in Arabidopsis. *Proceedings of the National Academy of Sciences USA, 105*(15), 5638–5643. https://doi.org/10.1073/pnas.0709463105.

Reusche, M., Klásková, J., Thole, K., Truskina, J., Novák, O., Janz, D., Strnad, M., Spíchal, L., Lipka, V., & Teichmann, T. (2013). Stabilization of cytokinin levels enhances Arabidopsis resistance against *Verticillium longisporum. Molecular Plant-Microbe Interactions, 26*(8), 850—860. https://doi.org/10.1094/MPMI-01-13-0014-R.

Rochon, A., Boyle, P., Wignes, T., Fobert, P. R., & Després, C. (2006). The coactivator function of Arabidopsis NPR1 requires the core of its BTB/POZ domain and the oxidation of C-terminal cysteines. *The Plant Cell, 18*(12), 3670–3685. https://doi.org/10.1105/tpc.106.042762.

Ron, M., & Avni, A. (2004). The receptor for the fungal elicitor ethylene-inducing xylanase is a member of a resistance-like gene family in tomato. *The Plant Cell, 16*(6), 1604–1615. https://doi.org/10.1105/tpc.022475.

Rooney, H. C., Van't Klooster, J. W., van der Hoorn, R. A., Joosten, M. H., Jones, J. D., & de Wit, P. J. (2005). *Cladosporium* Avr2 inhibits tomato Rcr3 protease required for Cf-2-dependent disease resistance. *Science, 308*(5729), 1783–1786. https://doi.org/10.1126/science.1111404.

Rushton, P. J., Somssich, I. E., Ringler, P., & Shen, Q. J. (2010). WRKY transcription factors. *Trends in Plant Science, 15*(5), 247–258. https://doi.org/10.1016/j.tplants.2010.02.006.

Saijo, Y., & Loo, E. P. (2020). Plant immunity in signal integration between biotic and abiotic stress responses. *New Phytologist, 225*(1), 87–104. https://doi.org/10.1111/nph.16168.

Sakakibara, H., Takei, K., & Hirose, N. (2006). Interactions between nitrogen and cytokinin in the regulation of metabolism and development. *Trends in Plant Science, 11*(9), 440–448. https://doi.org/10.1016/j.tplants.2006.07.004.

Sánchez-Rangel, D., Sánchez-Nieto, S., & Plasencia, J. (2012). Fumonisin B1, a toxin produced by *Fusarium verticillioides*, modulates maize β-1,3-glucanase activities involved in defense response. *Planta, 235*(5), 965–978. https://doi.org/10.1007/s00425-011-1566-0.

Sanchez-Vallet, A., Mesters, J. R., & Thomma, B. P. (2015). The battle for chitin recognition in plant-microbe interactions. *FEMS Microbiology Reviews, 39*(2), 171–183. https://doi.org/10.1093/femsre/fuu003.

Sandrock, R. W., & Van Etten, H. D. (2001). The relevance of tomatinase activity in pathogens of tomato? Disruption of the B2 -Tomatinase gene in *Colletotrichum coccodes* and *Septoria lycopersici* and heterologous expression of the *Septoria lycopersici* B2 -Tomatinase in *Nectria haematococca*, a pathogen of tomato fruit. *Physiological and Molecular Plant Pathology, 58*(4), 159–171. https://doi.org/10.1006/pmpp.2001.0341.

Schäfer, W. (1994). Molecular mechanisms of fungal pathogenicity to plants. *Annual Review of Phytopathology, 32*(1), 461–477. https://doi.org/10.1146/annurev.py.32.090194.002333.

Scheer, J. M., & Ryan, C. A. (1999). A 160-kD systemin receptor on the surface of *Lycopersicon peruvianum* suspension-cultured cells. *The Plant Cell, 11*(8), 1525–1536. https://doi.org/10.1105/tpc.11.8.1525.

Schouten, A., Wagemakers, L., Stefanato, F. L., Van Der Kaaij, R. M., & Van Kan, J. A. L. (2002). Resveratrol acts as a natural profungicide and induces self-intoxication by a specific laccase. *Molecular Microbiology, 43*(4), 883–894. https://doi.org/10.1046/j.1365-2958.2002.02799.x.

Schwessinger, B., Sperschneider, J., Cuddy, W. S., Garnica, D. P., Miller, M. E., Taylor, J. M., Dodds, P. N., Figueroa, M., Park, R. F., & Rathjen, J. P. (2018). A near-complete haplotype-phased genome of the dikaryotic wheat stripe rust fungus *Puccinia striiformis* f.sp. *tritici* reveals high inter-haplotype diversity. *mBio, 9*(1), e02275–17. https://doi.org/10.1128/mBio.02275-17.

Seeholzer, S., Tsuchimatsu, T., Jordan, T., Bieri, S., Pajonk, S., Yang, W., Jahoor, A., Shimizu, K. K., Keller, B., & Schulze-Lefert, P. (2010). Diversity at the *Mla* powdery mildew resistance locus from cultivated barley reveals sites of positive selection. *Molecular Plant-Microbe Interactions, 23*(4), 497–509. https://doi.org/10.1094/MPMI-23-4-0497.

Sela, H., Spiridon, L. N., Petrescu, A. J., Akerman, M., Mandel-Gutfreund, Y. A. E. L., Nevo, E., Loutre, C., Keller, B., Schulman, A. H., & Fahima, T. (2012). Ancient diversity of splicing motifs and protein surfaces in the wild emmer wheat (*Triticum dicoccoides*) LR10 coiled coil (CC) and leucine-rich repeat (LRR) domains. *Molecular Plant Pathology, 13*(3), 276–287. https://doi.org/10.1111/j.1364-3703.2011.00751.x.

Seo, S., Mitsuhara, I., Feng, J., Iwai, T., Hasegawa, M., & Ohashi, Y. (2011). Cyanide, a coproduct of plant hormone ethylene biosynthesis, contributes to the resistance of rice to blast fungus. *Plant Physiology, 155*(1), 502–514. https://doi.org/10.1104/pp.110.166456.

Seong, K., & Krasileva, K. V. (2023). Prediction of effector protein structures from fungal phytopathogens enables evolutionary analyses. *Nature Microbiology, 8*(1), 174–187. https://doi.org/10.1038/s41564-022-01067-2.

Seong, S. Y., & Matzinger, P. (2004). Hydrophobicity: An ancient damage-associated molecular pattern that initiates innate immune responses. *Nature Reviews Immunology, 4*(6), 469–478. https://doi.org/10.1038/nri1372.

Shao, D., Smith, D. L., Kabbage, M., & Roth, M. G. (2021). Effectors of plant necrotrophic fungi. *Frontiers in Plant Science, 12*, 687713. https://doi.org/10.3389/fpls.2021.687713.

Sharma, Y. K., León, J., Raskin, I., & Davis, K. R. (1996). Ozone-induced responses in *Arabidopsis thaliana*: The role of salicylic acid in the accumulation of defense-related transcripts and induced resistance. *Proceedings of the National Academy of Sciences USA, 93*(10), 5099–5104. https://doi.org/10.1073/pnas.93.10.5099.

Shen, Q. H., Saijo, Y., Mauch, S., Biskup, C., Bieri, S., Keller, B., Seki, H., Ülker, B., Somssich, I. E., & Schulze-Lefert, P. (2007). Nuclear activity of MLA immune receptors links isolate-specific and basal disease-resistance responses. *Science, 315*(5815), 1098–1103. https://doi.org/10.1126/science.1136372.

Shimizu, T., Nakano, T., Takamizawa, D., Desaki, Y., Ishii-Minami, N., Nishizawa, Y., Minami, E., Okada, K., Yamane, H., Kaku, H., & Shibuya, N. (2010). Two LysM receptor molecules, CEBiP and OsCERK1, cooperatively regulate chitin elicitor signaling in rice. *The Plant Journal, 64*(2), 204–214. https://doi.org/10.1111/j.1365-313X.2010.04324.x.

Shimono, M., Sugano, S., Nakayama, A., Jiang, C. J., Ono, K., Toki, S., & Takatsuji, H. (2007). Rice WRKY45 plays a crucial role in benzothiadiazole-inducible blast resistance. *The Plant Cell, 19*(6), 2064–2076. https://doi.org/10.1105/TPC.106.046250.

Siemens, J., Keller, I., Sarx, J., Kunz, S., Schuller, A., Nagel, W., Schmülling, T., Parniske, M., & Ludwig-Müller, J. (2006). Transcriptome analysis of Arabidopsis clubroots indicate a key role for cytokinins in disease development. *Molecular Plant-Microbe Interactions, 19*(5), 480–494.

Singh, S. K., Shree, A., Verma, S., Singh, K., Kumar, K., Srivastava, V., Singh, R., Saxena, S., Singh, A. P., Pandey, A., & Verma, P. K. (2023). The nuclear effector ArPEC25 from the necrotrophic fungus *Ascochyta rabiei* targets the chickpea transcription factor CaβLIM1a and negatively modulates lignin biosynthesis, increasing host susceptibility. *Plant Cell, 35*(3), 1134–1159.

Solano, R., Stepanova, A., Chao, Q., & Ecker, J. R. (1998). Nuclear events in ethylene signaling: A transcriptional cascade mediated by ethylene-insensitive3 and ethylene-response-factor1. *Genes & Development, 12*(23), 3703–3714.

Song, J., Win, J., Tian, M., Schornack, S., Kaschani, F., Ilyas, M., van der Hoorn, R. A., & Kamoun, S. (2009). Apoplastic effectors secreted by two unrelated eukaryotic plant pathogens target the tomato defense protease Rcr3. *Proceedings of the National Academy of Sciences, 106*(5), 1654–1659.

Song, W. Y., Wang, G. L., Chen, L. L., Kim, H. S., Pi, L. Y., Holsten, T., Gardner, J., Wang, B., Zhai, W. X., Zhu, L. H., & Fauquet, C. (1995). A receptor kinase-like protein encoded by the rice disease resistance gene, Xa21. *Science, 270*(5243), 1804–1806.

Song, Y., & Gao, J. (2014). Genome-wide analysis of WRKY gene family in *Arabidopsis lyrata* and comparison with *Arabidopsis thaliana* and *Populus trichocarpa*. *Chinese Science Bulletin, 8*(59), 754–765. https://doi.org/10.1007/S11434-013-0057-9.

SONG, Y., AI, C. rui, JING, S. juan, & YU, D. qiu. (2010). Research progress on functional analysis of rice wrky genes. *Rice Science, 17*(1), 60–72. https://doi.org/10.1016/S1672-6308(08)60105-5.

Spence, C., Alff, E., Johnson, C., Ramos, C., Donofrio, N., Sundaresan, V., & Bais, H. (2014). Natural rice rhizosphere microbes suppress rice blast infections. *BMC Plant Biology, 14*, 1–17.

Srivastava, R., Kumar, S., Kobayashi, Y., Kusunoki, K., Tripathi, P., Kobayashi, Y., Koyama, H., & Sahoo, L. (2018). Comparative genome-wide analysis of WRKY transcription factors in two Asian legume crops: Adzuki bean and Mung bean. *Scientific Reports, 8*(1). https://doi.org/10.1038/S41598-018-34920-8.

Stam, R., Münsterkötter, M., Pophaly, S. D., Fokkens, L., Sghyer, H., Güldener, U., Hückelhoven, R., & Hess, M. (2018). A new reference genome shows the one-speed genome structure of the barley pathogen *Ramularia collo-cygni. Genome Biology and Evolution, 10*(12), 3243–3249.

Sticher, L., Mauch-Mani, B., & Métraux, A. J. (1997). Systemic acquired resistance. *Annual Review of Phytopathology, 35*(1), 235–270.

Sun, T., Nitta, Y., Zhang, Q., Wu, D., Tian, H., Lee, J. S., & Zhang, Y. (2018). Antagonistic interactions between two MAP kinase cascades in plant development and immune signaling. *EMBO Reports, 19*, e45324.

Sýkorová, B., Kurešová, G., Daskalova, S., Trčková, M., Hoyerová, K., Raimanová, I., Motyka, V., Trávníčková, A., Elliott, M. C., & Kamínek, M. (2008). Senescence-induced ectopic expression of the *A. tumefaciens* ipt gene in wheat delays leaf senescence, increases cytokinin content, nitrate influx, and nitrate reductase activity, but does not affect grain yield. *Journal of Experimental Botany, 59*(2), 377–387.

Taj, G., Agarwal, P., Grant, M., & Kumar, A. (2010). MAPK machinery in plants: Recognition and response to different stresses through multiple signal transduction pathways. *Plant Signaling & Behavior, 5*(11), 1370–1378.

Tameling, W. I., Elzinga, S. D., Darmin, P. S., Vossen, J. H., Takken, F. L., Haring, M. A., & Cornelissen, B. J. (2002). The tomato R gene products I-2 and MI-1 are functional ATP binding proteins with ATPase activity. *The Plant Cell, 14*(11), 2929–2939.

Tanabe, T., Chamaillard, M., Ogura, Y., Zhu, L., Qiu, S., Masumoto, J., Ghosh, P., Moran, A., Predergast, M. M., Tromp, G., & Williams, C. J. (2004). Regulatory regions and critical residues of NOD2 involved in muramyl dipeptide recognition. *The EMBO Journal, 23*(7), 1587–1597.

Tanaka, K., Choi, J., Cao, Y., & Stacey, G. (2014). Extracellular ATP acts as a damage-associated molecular pattern (DAMP) signal in plants. *Frontiers in Plant Science, 5*, 1–9. https://doi.org/10.3389/fpls.2014.00446.

Tanaka, K., Nguyen, C. T., Liang, Y., Cao, Y., & Stacey, G. (2013). Role of LysM receptors in chitin-triggered plant innate immunity. *Plant Signaling & Behavior, 8*(1), 147–153. https://doi.org/10.4161/psb.22598.

Tang, D., Wang, G., & Zhou, J. M. (2017). Receptor kinases in plant-pathogen interactions: More than pattern recognition. *The Plant Cell, 29*(4), 618–637.

Thakur, R. K., Prasad, P., Bhardwaj, S. C., Gangwar, O. P., & Kumar, S. (2022). Epigenetics of wheat–rust interaction: An update. *Planta, 255*(2), 50.

Thulasi Devendrakumar, K., Li, X., & Zhang, Y. (2018). MAP kinase signaling: Interplays between plant PAMP- and effector-triggered immunity. *Cellular and Molecular Life Sciences, 75*(16), 2981–2989.

Tian, M., Huitema, E., Da Cunha, L., Torto-Alalibo, T., & Kamoun, S. (2004). A Kazal-like extracellular serine protease inhibitor from *Phytophthora infestans* targets the tomato pathogenesis-related protease P69B. *Journal of Biological Chemistry, 279*(25), 26370–26377.

Ton, J., Flors, V., & Mauch-Mani, B. (2009). The multifaceted role of ABA in disease resistance. *Trends in Plant Science, 14*(6), 310–317.

Tong, H., Xiao, Y., Liu, D., Gao, S., Liu, L., Yin, Y., Jin, Y., Qian, Q., & Chu, C. (2014). Brassinosteroid regulates cell elongation by modulating gibberellin metabolism in rice. *The Plant Cell, 26*(11), 4376–4393.

Turbek, C. S., Smith, D. A., & Schardl, C. L. (1992). An extracellular enzyme from *Fusarium solani* f.sp. *phaseoli* which catalyses hydration of the isoflavonoid phytoalexin, phaseollidin. *FEMS Microbiology Letters, 94*, 187–190.

Upson, J. L., Zess, E. K., Białas, A., Wu, C. H., & Kamoun, S. (2018). The coming of age of EvoMPMI: Evolutionary molecular plant–microbe interactions across multiple timescales. *Current Opinion in Plant Biology, 44*, 108–16.

Valls, M., Genin, S., & Boucher, C. (2006). Integrated regulation of the type III secretion system and other virulence determinants in *Ralstonia solanacearum*. *PLoS Pathogens, 2*(8), e82.

van den Burg, H. A., Harrison, S. J., Joosten, M. H., Vervoort, J., & de Wit, P. J. (2006). *Cladosporium fulvum* Avr4 protects fungal cell walls against hydrolysis by plant chitinases accumulating during infection. *Molecular Plant-Microbe Interactions, 19*(12), 1420–1430.

van der Hoorn, R. A., & Kamoun, S. (2008). From guard to decoy: A new model for perception of plant pathogen effectors. *The Plant Cell, 20*(8), 2009–2017.

van Esse, H. P., Van't Klooster, J. W., Bolton, M. D., Yadeta, K. A., van Baarlen, P., Boeren, S., Vervoort, J., de Wit, P. J., & Thomma, B. P. (2008). The *Cladosporium fulvum* virulence protein Avr2 inhibits host proteases required for basal defense. *The Plant Cell, 20*(7), 1948–1963.

Van Loon, L. C. (1984). Regulation of pathogenesis and symptom expression in diseased plants by ethylene. In L. T. Evans (Ed.), *Ethylene: Biochemical, Physiological and Applied Aspects* (pp. 171–180). Springer Netherlands.

Van Verk, M. C., Pappaioannou, D., Neeleman, L., Bol, J. F., & Linthorst, H. J. M. (2008). A novel WRKY transcription factor is required for induction of PR-1a gene expression by salicylic acid and bacterial elicitors. *Plant Physiology, 146*(4), 1983–1995. https://doi.org/10.1104/PP.107.112789.

Van Vu, B., Itoh, K., Nguyen, Q. B., Tosa, Y., & Nakayashiki, H. (2012). Cellulases belonging to glycoside hydrolase families 6 and 7 contribute to the virulence of *Magnaporthe oryzae*. *Molecular Plant-Microbe Interactions, 25*(9), 1135–1141.

Vénéreau, E., Ceriotti, C., & Bianchi, M. E. (2015). DAMPs from cell death to new life. *Frontiers in Immunology, 6*, 1–11. https://doi.org/10.3389/fimmu.2015.00422.

Vernooij, B., Friedrich, L., Morse, A., Reist, R., Kolditz-Jawhar, R., Ward, E., Uknes, S., Kessmann, H., & Ryals, J. (1994). Salicylic acid is not the translocated signal responsible for inducing systemic acquired resistance but is required in signal transduction. *The Plant Cell, 6*(7), 959–965.

Vlaardingerbroek, I., Beerens, B., Schmidt, S. M., Cornelissen, B. J., & Rep, M. (2016). Dispensable chromosomes in *Fusarium oxysporum* f.sp. *lycopersici*. *Molecular Plant Pathology, 17*(9), 1455–1466.

Vleeshouwers, V. G., & Oliver, R. P. (2014). Effectors as tools in disease resistance breeding against biotrophic, hemibiotrophic, and necrotrophic plant pathogens. *Molecular Plant Microbe Interactions, 27*(3), 196–206.

Vo, K. T. X., Kim, C. Y., Hoang, T. V., Lee, S. K., Shirsekar, G., Seo, Y. S., & Jeon, J. S. (2018). *OsWRKY67* plays a positive role in basal and *XA21*-mediated resistance in rice. *Frontiers in Plant Science, 8*, Article 2220. https://doi.org/10.3389/fpls.2017.02220.

Wang, B., Sun, Y., Song, N., Zhao, M., Liu, R., Feng, H., & Kang, Z. (2017). *Puccinia striiformis* f.sp. *tritici* microRNA-like RNA 1 (*Pst-milR1*), an important pathogenicity factor of *Pst*, impairs wheat resistance to *Pst* by suppressing the wheat pathogenesis-related 2 gene. *New Phytologist, 215*(1), 338–350.

Wang, D., Pajerowska-Mukhtar, K., Culler, A. H., & Dong, X. (2007). Salicylic acid inhibits pathogen growth in plants through repression of the auxin signaling pathway. *Current Biology, 17*(20), 1784–1790.

Wang, D., Wang, L., Su, W., Ren, Y., You, C., Zhang, C., & Su, Y. (2020). A class III WRKY transcription factor in sugarcane was involved in biotic and abiotic stress responses. *Scientific Reports, 10*(1), Article 1–15. https://doi.org/10.1038/s41598-020-78007-9.

Wang, H., Hao, J., Chen, X., Hao, Z., Wang, X., Lou, Y., & Guo, Z. (2007). Overexpression of rice *WRKY89* enhances ultraviolet B tolerance and disease resistance in rice plants. *Plant Molecular Biology, 65*(6), 799–815. https://doi.org/10.1007/S11103-007-9244-X.

Wang, S., Bai, Y., Shen, C., Wu, Y., Zhang, S., Jiang, D., & Qi, Y. (2010). Auxin-related gene families in abiotic stress response in *Sorghum bicolor*. *Functional & Integrative Genomics, 10*, 533–546.

Wang, X., Meng, H., Tang, Y., Zhang, Y., He, Y., Zhou, J., & Meng, X. (2022). Phosphorylation of an ethylene response factor by MPK3/MPK6 mediates negative feedback regulation of pathogen-induced ethylene biosynthesis in *Arabidopsis*. *Journal of Genetics and Genomics, 49*(8), 810–822.

Wang, Y., & Wang, Y. (2018). Trick or treat: Microbial pathogens evolved apoplastic effectors modulating plant susceptibility to infection. *Molecular Plant Microbe Interactions, 31*(1), 6–12.

Wani, S. H., Anand, S., Singh, B., Bohra, A., & Joshi, R. (2021). WRKY transcription factors and plant defense responses: Latest discoveries and future prospects. *Plant Cell Reports, 40*(7), 1071–1085. https://doi.org/10.1007/s00299-021-02691-8.

Wasternack, C. (2007). Jasmonates: An update on biosynthesis, signal transduction and action in plant stress response, growth and development. *Annals of Botany, 100*(4), 681–697.

Wei, K. F., Chen, J., Chen, Y. F., Wu, L. J., & Xie, D. X. (2012). Molecular phylogenetic and expression analysis of the complete WRKY transcription factor family in maize. *DNA Research: An International Journal for Rapid Publication of Reports on Genes and Genomes, 19*(2), 153–164. https://doi.org/10.1093/dnares/dsr048.

Weiberg, A., Wang, M., Lin, F. M., Zhao, H., Zhang, Z., Kaloshian, I., & Jin, H. (2013). Fungal small RNAs suppress plant immunity by hijacking host RNA interference pathways. *Science, 342*(6154), 118–123.

Westrick, N. M., Smith, D. L., & Kabbage, M. (2021). Disarming the host: Detoxification of plant defense compounds during fungal necrotrophy. *Frontiers in Plant Science, 12*, Article 651716.

Whenham, R. J., Fraser, R. S. S., Brown, L. P., & Payne, J. A. (1986). Tobacco-mosaic-virus-induced increase in abscisic-acid concentration in tobacco leaves: Intracellular location in light and dark-green areas, and relationship to symptom development. *Planta, 168*(4), 592–598.

Whitham, S. A., Anderberg, R. J., Chisholm, S. T., & Carrington, J. C. (2000). *Arabidopsis RTM2* gene is necessary for specific restriction of tobacco etch virus and encodes an unusual small heat shock–like protein. *The Plant Cell, 12*(4), 569–582.

Williams, S. J., Sohn, K. H., Wan, L., Bernoux, M., Sarris, P. F., Segonzac, C., & Casey, L. W. (2014). Structural basis for assembly and function of a heterodimeric plant immune receptor. *Science, 344*(6181), 299–303.

Witte, T. E., Villeneuve, N., Boddy, C. N., & Overy, D. P. (2021). Accessory chromosome-acquired secondary metabolism in plant pathogenic fungi: The evolution of biotrophs into host-specific pathogens. *Frontiers in Microbiology, 12*, Article 664276.

Xi, Y., Cesari, S., & Kroj, T. (2022). Insight into the structure and molecular mode of action of plant paired NLR immune receptors. *Essays in Biochemistry, 66*(5), 513–526.

Xiao, S., Calis, O., Patrick, E., Zhang, G., Charoenwattana, P., Muskett, P., & Turner, J. G. (2005). The atypical resistance gene, *RPW8*, recruits' components of basal defense for powdery mildew resistance in *Arabidopsis*. *The Plant Journal, 42*(1), 95–110.

Xiao, S., Ellwood, S., Calis, O., Patrick, E., Li, T., Coleman, M., & Turner, J. G. (2001). Broad-spectrum mildew resistance in *Arabidopsis thaliana* mediated by *RPW8*. *Science, 291*(5501), 118–120.

Xie, C., Zhou, X., Deng, X., & Guo, Y. (2010). *PKS5*, a SNF1-related kinase, interacts with and phosphorylates *NPR1*, and modulates expression of *WRKY38* and *WRKY62*. *Journal of Genetics and Genomics, 37*(6), 359–369. https://doi.org/10.1016/S1673-8527(09)60054-0.

Xiong, L., & Yang, Y. (2003). Disease resistance and abiotic stress tolerance in rice are inversely modulated by an abscisic acid-inducible mitogen-activated protein kinase. *Plant Cell, 15*, 745–759.

Xu, Z. S., Xia, L. Q., Chen, M., Cheng, X. G., Zhang, R. Y., Li, L. C., & Qiu, Z. G. (2007). Isolation and molecular characterization of the *Triticum aestivum* L. ethylene-responsive factor 1 (*TaERF1*) that increases multiple stress tolerance. *Plant Molecular Biology, 65*, 719–732.

Yalpani, N., Enyedi, A. J., León, J., & Raskin, I. (1994). Ultraviolet light and ozone stimulate accumulation of salicylic acid, pathogenesis-related proteins and virus resistance in tobacco. *Planta, 193*(3), 372–376.

Yamaguchi, Y., Huffaker, A., Bryan, A. C., Tax, F. E., & Ryan, C. A. (2010). *PEPR2* is a second receptor for the *Pep1* and *Pep2* peptides and contributes to defense responses in *Arabidopsis*. *The Plant Cell, 22*(2), 508–522. https://doi.org/10.1105/tpc.109.068874.

Yang, D. L., Li, Q., Deng, Y. W., Lou, Y. G., Wang, M. Y., Zhou, G. X., & He, Z. H. (2008). Altered disease development in the *eui* mutants and *Eui* overexpressors indicates that gibberellins negatively regulate rice basal disease resistance. *Molecular Plant, 1*(3), 528–537.

Yuan, M., Ngou, B. P. M., Ding, P., & Xin, X. F. (2021). PTI-ETI crosstalk: An integrative view of plant immunity. *Current Opinion in Plant Biology, 62*, 102030. https://doi.org/10.1016/j.pbi.2021.102030.

Zeng, W., Wang, J., Wang, Y., Lin, J., Fu, Y., Xie, J., & Cheng, J. (2018). Dicer-like proteins regulate sexual development via the biogenesis of perithecium-specific microRNAs in a plant pathogenic fungus *Fusarium graminearum*. *Frontiers in Microbiology, 9*, Article 818. https://doi.org/10.3389/fmicb.2018.00818.

Zhang, S., Li, C., Si, J., Han, Z., & Chen, D. (2022). Action mechanisms of effectors in plant-pathogen interaction. *International Journal of Molecular Sciences, 23*(12), Article 6758.

Zhang, Y., & Feng, J. C. (2014). Identification and Characterization of the Grape WRKY Family. *BioMed Research International, 2014*. https://doi.org/10.1155/2014/787680.

Zhang, Y., & Wang, L. (2005). The WRKY transcription factor superfamily: Its origin in eukaryotes and expansion in plants. *BMC Evolutionary Biology, 5*(1), Article 1–12. https://doi.org/10.1186/1471-2148-5-1/FIGURES/3.

Zhao, Y. (2010). Auxin biosynthesis and its role in plant development. *Annual Review of Plant Biology, 61*, 49–64.

Zheng, Z., Qamar, S. A., Chen, Z., & Mengiste, T. (2006). *Arabidopsis WRKY33* transcription factor is required for resistance to necrotrophic fungal pathogens. *The Plant Journal: For Cell and Molecular Biology, 48*(4), 592–605. https://doi.org/10.1111/j.1365-313X.2006.02901.x.

Zhou, J., Wang, X., He, Y., Sang, T., Wang, P., Dai, S., & Meng, X. (2020). Differential phosphorylation of the transcription factor *WRKY33* by the protein kinases *CPK5/CPK6* and *MPK3/MPK6* cooperatively regulates camalexin biosynthesis in *Arabidopsis*. *Plant Cell, 32*(8), 2621–2638.

Zhu, J., Verslues, P. E., Zheng, X., Lee, B. H., Zhan, X., Manabe, Y., & Hasegawa, P. M. (2005). *HOS10* encodes an *R2R3*-type MYB transcription factor essential for cold acclimation in plants. *Proceedings of the National Academy of Sciences, 102*(28), 9966–9971.

Zhu, Z., An, F., Feng, Y., Li, P., Xue, L., Jiang, Z., & Yu, Q. (2011). Derepression of ethylene-stabilized transcription factors (*EIN3/EIL1*) mediates jasmonate and ethylene signaling synergy in *Arabidopsis*. *Proceedings of the National Academy of Sciences, 108*(30), 12539–12544.

CHAPTER 4

Molecular Basis of the Evolution of Pathogens Under Changing Climate Conditions

CHANDER SINGH,[1] SUPRIYA PANDEY,[2] UMARA RAHMANI,[3] MANISHA CHAUDHARY,[4] AMAN CHAUHAN,[1] ANSHUMAN SEMWAL,[1] SUMIT RAI,[2] RAMKUMAR,[5] VARSHA MISHRA,[6] VISHAL DINKAR,[7] and ASHISH KUMAR SINGH[8]

[1]*Dr. Yashwant Singh Parmar University of Horticulture and Forestry, Nauni, Solan, Himachal Pradesh, India*

[2]*Govind Ballabh Pant National Institute of Himalayan Environment, Kosi-Katarmal, Almora, Uttarakhand, India*

[3]*College of Agriculture, Sardar Vallabhbhai Patel University of Agriculture and Technology, Modipuram, Meerut, Uttar Pradesh, India*

[4]*College of Agriculture, Chaudhary Sarwan Kumar Himachal Pradesh Krishi Vishwavidyalaya, Palampur, Kangra, Himachal Pradesh, India*

[5]*Department of Entomology and Agriculture Zoology, Institute of Agriculture Sciences, Banaras Hindu University, Varanasi, Uttar Pradesh, India*

[6]*Sir J.C. Bose Technical Campus, Department of Biotechnology, Kumaun University, Bhimtal, Nainital, Uttarakhand, India*

[7]*ICAR–Central Institute of Temperate Horticulture, Rangreth, Srinagar, Jammu and Kashmir, India*

[8]*Crop Protection Division, ICAR-VPKAS, Almora, Uttarakhand, India*

Climate Change and Biotic Factors: A Molecular Approach.
Ashutosh Singh, Saurabh Pandey, and Amarjeet Kumar, (Eds.)

ABSTRACT

In the last 200 years, environmental changes brought on by both natural and human activity have increased globally. Long-term changes in weather patterns and extreme weather event frequency are referred to as climate change. It is widely understood that infectious diseases can be impacted by climate change. The majority of research on the connections between disease and climate change has been on particular infections, modes of transmission, or the results of a single form of extreme weather. The environmental effects of climate change are becoming increasingly clear. All life forms are affected by the range and survivability of extreme weather events, rising average global temperatures, changing precipitation patterns, and increasing frequency of such events. Each disease may react differently to changes in CO_2 concentrations, temperature, and water availability, which can have positive, neutral, or adverse effects on disease development. Temperature and humidity have an impact on the virulence mechanisms of pathogens, including the generation of toxins and virulence proteins, as well as pathogen reproduction and survival. Plant diseases can adapt to climate change by taking advantage of the current phenotypic plasticity, migrating to regions with favorable climates, or evolving new traits. Research is underway to develop indicators and predictive methods for locating disease outbreaks in the future. The populations of pathogens exhibit a wide range of genetic characteristics that have been a topic of interest among researchers in the origins and evolution of infectious disease. It is impossible to forecast how climate change will affect plant pathogens, but given that they have a greater range of adaptive mechanisms than their hosts and faster generation rates, they will likely have more opportunities to adapt and evolve. Due to the complex and interacting nature of these events, it can be challenging to identify the key point(s) that triggered the onset of a disease, which may have occurred millions of years ago during the coevolution of the host with its pathogen. Therefore, it is not unexpected that the molecular processes leading to the genesis of any disease have largely remained undiscovered. Therefore, in this manuscript, we are attempting to summarize recent research on the evolution of pathogens, the link between pathogens and climate change, the molecular basis of evolution, and the mechanisms underlying pathogenicity.

4.1 INTRODUCTION

The plant disease is caused by a harmful organism, a sensitive plant, and environmental factors. The presence of microbial antagonists to the pathogen as well as human activities (such as agronomic methods, fungicide applications, transportation of plant material in the global market, etc.) may also contribute to the development of a disease. Changes in environmental conditions are strongly correlated with differences in the level of losses incurred by a disease, and climatic changes are frequently linked to the development of new diseases because the environment has a significant, direct or indirect, impact on plants, pathogens, and their antagonists (Anderson et al., 2004). Because of these factors, global warming-induced changes (e.g., higher temperatures, changes in precipitation rates and precipitation patterns, higher CO_2 and ozone concentrations, droughts, etc.) could affect the frequency and intensity of plant diseases, as well as plant-pathogen coevolution (Chakraborty, 2005; Burdon et al., 2006; Garrett et al., 2006; Crowl et al., 2008; Eastburn et al., 2011; Elad and Pertot, 2014).

Global crop output is severely hampered by plant diseases and pests. The burden can be quantified indirectly by the social, environmental, and economic costs of control as well as directly through the reduction in yield (Fones et al., 2020). Crop diseases and pests have specific tolerances for or need for specific environmental conditions, much like all other species. These limitations specify their ecological niche, which establishes the geographic areas and seasons during which pests and diseases can spread and destroy crops (Chaloner et al., 2020). As the climate changes, favorable circumstances for pest outbreaks move in time and geography, changing the risks farmers confront and the management strategies required to control them (Bebber, 2015). Infection and disease are governed by complicated and species-specific interactions among diverse biotic and abiotic variables (Fones et al., 2020). Temperature is a crucial regulator of disease risk (Chaloner et al., 2020; Delgado-Baquerizo et al., 2020), and plant pathogen ranges have varied in response to past global warming (Bebber et al., 2013; Chaloner et al., 2021).

A disease develops only when a virulent pathogen infects a vulnerable host under the right environmental conditions at the right moment (Agrios, 2005). Climatic change has the potential to modify host resistance to infections by changing host physiology and pathogen virulence. Rises in global temperature will cause a shift in agroclimatic zones towards the poles, altering the geographic distribution of existing diseases by bringing infections into new places. New infections and diseases may evolve concurrently in existing agroecosystems. Existing diseases may evolve into more aggressive, climate-resilient variants. Each pathogen group may have different impacts depending on its biology, host niche, survival and distribution strategies, and climate change vulnerability.

Fungi are the most common plant pathogens, having many types of parasitic associations with their host plants. The current climatic conditions, particularly temperature and moisture, significantly impact fungal pathogenesis. Temperature is an important element during the many life phases of a fungal pathogen. Any temperature variation substantially impacts fungal reproduction, rate of infection, number of disease cycles, long and short-range dispersal, and off-season survival. Temperature changes may activate dormant races of diseases, resulting in an unexpected emergence of an epidemic. The combination of rising temperatures and excessive soil moisture generates a warm and humid atmosphere that is ideal for soilborne pathogens and foliar diseases. Every pathogen has an ideal temperature range, such as *Xanthomonas oryzae* pv. oryzae (*Xoo*), which fails to colonize the rice xylem effectively when the daytime temperature surpasses 35°C (Horino et al., 1982).

More than 40 bacterial genera have been documented to be pathogenic to plants. The appearance of several bacterial species as a severe concern worldwide might be attributed to global warming. *Acidovorax avenae* subsp. *avenae* infecting highland rice in southern Europe, *Burkholderia andropogonis* infecting jojoba in eastern Australia, *B. glumae* affecting rice in the southern USA, and *Dickeya zeae* infecting rice in north India, particularly at high temperatures (Hunjan and Lore, 2020). Temperature increases disrupt bacterial physiology, genetics, and bacterial-plant interactions. Temperature increases the fitness of plant pathogenic bacteria on their respective hosts. During their contact with the host plant, plant pathogenic bacteria, particularly xanthomonads and pseudomonads, create enormous quantities of extracellular polysaccharides (EPS). Typically, biotic and abiotic factors trigger this production. Under natural settings, bacteria employ EPS to aggregate and form biofilms to protect themselves from adverse environmental conditions, antimicrobial chemicals, and so forth. Bacteria often use EPS for communication, adhesion, energy storage, and other purposes (Wingender et al., 1999; Vardharajula and Ali, 2015). During pathogenesis, temperature influences the expression of effector genes in plant pathogenic bacteria as well as their recognition events in host plants.

Viruses have a close relationship with vectors; viral pathogens infect their plant hosts, providing a new dimension to the disease triangle. The environmental needs of virus vectors also limit the spread of a virus disease pandemic (Malmstrom et al., 2011). Climatic change may modify both host plants and the kinds of vectors associated with them, dictating the degree of the virus's impact on that host (Jones, 2009). It may also have an impact on vector range, migration, and biology (Canto et al., 2009). High temperatures and carbon dioxide (CO_2) levels can drastically affect plant biochemistry, causing plants to behave differently to insect vectors of viruses. This has the potential to influence insect fertility, feeding rates, off-season survival, and dispersion (Trebicki et al., 2017).

4.2 CURRENT UNDERSTANDING OF PATHOGENS

Any organism that causes plant illnesses, including nematodes, viruses, bacteria, protists, and fungi, is referred to as a plant pathogen (Alberts et al., 2002). Plant diseases are significant for a variety of reasons, including concerns over fragile ecosystems and the need to protect the food supply. Plant viruses can infect all types of plants and limit a grower's capacity to produce crops. The plant's vascular system serves as a delivery route for invaders and a rich supply of nutrients (Lucas et al., 2013). Phloem and xylem tissues make up this structure, providing two distinct host habitats for plant diseases. Macro-organisms are also primarily nutrient-reliant on the plant vascular system, in addition to viral and bacterial microorganisms (Cavicchioli et al., 2019). Different pathogen groups attack plants separately, or occasionally more than one pathogen-producing complex, and a more severe illness will attack the plant simultaneously. In the majority of cases, the appearance of these external signs will tell us what type of infection is causing the symptoms.

Plant pathogens have a variety of strategies to attack. Some establish themselves on the plant's surface, while others may target certain regions like the roots, stems, and leaves. Some colonize the plant's tissue. Pathogens frequently result in issues including tissue death, browning, a reduction in fruiting, difficulties setting flowers, etc. Extreme cases can result in the host plant's death. Signs and symptoms of fungal, bacterial, and viral plant diseases are as follows: Fungal diseases: birds-eye spots on berries Damping of seedlings: Phytophthora Leaf spot: Septoria brown spot Chlorosis: Chlorosis Bacterial diseases: Leaf spot with yellow halo Fruit spot: Canker Crown Gall: Shepherd's Crook stem ends Woody plant diseases: Mosaic Yellowing: Crinkling Stunting:

4.2.1 ABIOTIC OR NON-INFECTIOUS DISEASES

Abiotic or non-infectious diseases are brought on by environmental factors outside the plant, not by live organisms. Although they cannot transfer from one plant to another, they are quite prevalent and need to be considered when determining any plant's health (Nazarov et al., 2020). Nutritional inadequacies, soil compaction, salt damage, ice, and sun scorch are a few examples of abiotic illnesses. Rarely do certain plant diseases, known as non-infectious diseases, have abiotic origins, such as environmental influences. Extreme abiotic environmental factors, such as the presence of significant toxic concentrations in the soil or the air, a lack of water, oxygen, or minerals, and unfavorable temperature, humidity, oxygen, CO_2, or light conditions, can all have detrimental effects on plants (Kumar et al., 2010). These factors can also cause them to die.

4.3 BIOTIC OR INFECTIOUS DISEASES

Biotic or infectious diseases are caused by living things. When they infect plants, they are referred to as plant pathogens. All plant tissues (leaf, shoot, stem, crown, root, tuber, fruit, seed, and vascular tissues) are at risk of being infected by pathogens that can pass from plant to plant. Many phytopathogens are responsible for infectious plant illnesses, including fungi, bacteria, viruses, viroids, mollicutes, parasitic higher plants, and protozoa. A phytopathogen that may spread from an infected plant to a healthy one, causing the same disease and exhibiting the same symptoms, is said to create an infectious disease (Abdulkhair and Alghuthaymi, 2016). Most phytopathogens can survive inside plants, but some others, such as fungi, bacteria, and parasitic higher plants, can live on the plant's surface.

4.4 TYPES OF PLANT PATHOGENS

4.4.1 FUNGI AND FUNGAL-LIKE ORGANISMS (FLOS)

Fungi and FLOs together cause more plant diseases than any other group of plant pathogens. They cannot produce their own food. They lack chlorophyll. They grow in filaments. They can or cannot reproduce through spores (Messel, 2014). FLOs and fungi can survive the winter on plant detritus or in the soil. Surprisingly, saprophytic and pathogenic fungi are the two main fungi affecting plants. Whereas saprophytic fungi grow in or on dead tissues, pathogenic fungi thrive in or on plant tissues and seriously impair the physiological processes of plants (Köhl et al., 2019). As a result, a precise diagnosis of plant disease must be made. Microscopic inspection is used to precisely diagnose and classify fungi by identifying the mycelial morphological traits and any fruiting structures and spores. Following thorough identification of the fungus and plant disease symptoms, the latter should be contrasted with information from the reference. This investigation will precisely establish whether the fungus is a saprophyte or a pathogen. Although microscopic examination is a crucial and efficient tool for identifying fungi, it occasionally falls short of precise identification because infected plant tissue lacks fungal spores and fruiting structures (Senanayake et al., 2020). As a result, a different approach must be taken, such as the use of a selective medium for sporulation promotion, identification, or isolation. However, in order for some fungi to develop spores, they must be incubated under specific temperatures, aeration, or light conditions (Narayanasamy and Narayanasamy, 2008).

4.4.2 BACTERIA

Bacteria are small, unicellular organisms with cell walls. They reproduce through binary fission (a single cell divides into two). Introduction to the plant must take place through the plant's natural wounds or apertures. Bacteria primarily overwinter in the soil and on non-decomposing plant material; however, some also persist inside insect vectors. Due to the presence of saprophytes, the appearance of bacterial growth in or on plant tissues indicates the possibility of bacterial plant disease. Hence, microscopy analysis and physiological parameter assessment are required for reliable bacterial identification (Senanayake et al., 2020). In some instances, selective media are essentially employed in bacterial identification to identify the genus and species of bacteria. Moreover, a single pure bacterial colony inoculated into a healthy plant can be used to reproduce the identical symptoms that the infected plant displayed as the confirmatory test for bacterial pathogenicity. Furthermore, serological assays or immunodiagnostic methods such as agglutination and precipitation, fluorescent antibody labeling, and enzyme-linked immunosorbent assay (ELISA) can be used (Atmar, 2014). These approaches have a number of benefits, including reasonable sensitivity, reasonable specificity, speed, and ease of use. It is anticipated that standardized, trustworthy antisera will soon be accessible.

4.4.3 VIRUSES AND VIROIDS

Viruses are protein-coated, intracellular nucleic acid particles that replicate in their hosts, which are other living things (Gelderblom, 1996). Viruses can infect and replicate in other living things. Viroids are particles that resemble viruses but lack a protein covering. Most viruses and viroids are transmitted by food-borne vectors like nematodes, fungus, and insects. Moreover, seed, vegetative propagation, and pruning all have the potential to spread phytopathogenic viruses and viroids (Hadidi et al., 2022).

4.4.4 *NEMATODES*

Nematodes are tiny, worm-like organisms. Most nematodes are soil-dwelling organisms that move within the soil (Basyoni and Rizk, 2016). Yet, certain nematodes can infect sections of above-ground plants and spread via insects. One of the most prevalent phytopathogens with distinguishable symptoms is the nematode. These symptoms only manifest in the diseased area. The spread of nematode infections in or on plants is widespread, particularly in favorable conditions like moderate temperature and high humidity (Bahadur, 2021).

4.5 EVOLUTION IN PATHOGENS AND THEIR PATHOGENICITY

The term 'virulence' has a long and complicated history with plant pathologists. Virulence can be roughly defined as the degree of damage a parasite infection inflicts on a host, which is believed to be inversely related to host health (Casadevall and Pirofski, 2001), and pathogenicity as the specific ability of a parasite to infect and harm the host (Casadevall and Pirofski, 2001). Both virulence and pathogenicity may be affected by selection, and changes in parasite populations may promote both host-parasitic coevolution and parasite evolution (Day et al., 2007). Over the past 30 years, many theoretical papers have been published on the factors that influence pathogenicity evolution and virulence. It is widely accepted that virulent parasites are capable of infecting and killing a host. As virulence is the primary characteristic of pathogens, it has long been a significant objective of pathology, especially plant pathology, to comprehend the evolution of virulence. Virulence can affect the effectiveness of control measures for infectious diseases that affect humans, pets, and plants. The development of virulence can influence important events such as the appearance and reappearance of pathogens, the switching and expansion of host ranges, and the development of host resistance. Also, the crucial function that diseases play in determining ecosystem dynamics and composition may be modified by virulence evolution (Sacristán and García-Arenal, 2008).

Despite the significance of virulence for the health and welfare of people, animals, and plants, there hasn't been much consensus among scientists from different fields on the term's definition beyond an intuitive one. The damage parasite infection inflicts on the host is often associated with virulence in animal pathology and in evolutionary biology; to be more precise, it is often referred to as the 'host fitness' effect of parasite infection (Day et al., 2007). It has been proposed that evolution contributes to the escape from natural foes. Introduced plant species have the potential to both leave behind co-evolved diseases and pick up new ones. Both plants and pathogens are prone to fast evolutionary changes in traits crucial to infection and pathogenicity. Although plant pathologists have given the evolution of pathogenicity a great deal of attention, animal pathologists and evolutionary biologists have often focused on the evolution of virulence. In order to gain an understanding of the life cycle of parasites, it is necessary to provide an explanation of their virulence.

It is crucial for effective agricultural systems and a rapidly developing area of research, the examination of plant-pathogen interactions. Every time a pathogen or plant species is introduced to an area outside of its native range, novel plant-pathogen combinations occur. The ability of a pathogen to survive in a new host is determined by the genetic compatibility of the pathogen with its host, which is determined by either preadaptation by the pathogen or later evolutionary changes by the host. The environmental impact of a novel interaction depends on the pathogen's developmental history, the speed of virus or resistance development, and the environment. For instance, the propagation of a disease pandemic or the extinction of an impending plant invasion.

4.5.1 THE EVOLUTION OF VIRULENCE

While plant pathologists study pathogenicity evolution, animal pathologists and evolutionary biologists focus on virulence evolution. Understanding parasite life history involves explaining virulence. It is unclear why parasites hurt their hosts because virulence does not clearly benefit parasites dependent on their hosts for survival and fitness. One of the most popular theories is that virulence is a by-product of the proliferation of parasites within the host (Lenski and May, 1994). If this were the case, selection would be based on virulence, which would be related to within-host reproduction, a key factor in parasite fitness. Yet, increased virulence will severely impact between-host transmission, a crucial aspect of parasite fitness, by increasing host death and morbidity. The trade-offs hypothesis is derived from the concept that virulence results in a trade-off between intra-host proliferation and inter-host transmission (Alizon et al., 2009). Since its inception more than ten years ago, questions have been raised about the utility of the compromise hypothesis and its central assumption that there is an additive relationship between parasite growth and virulence. To gain a better understanding of virulence, there are a number of theories that consider the diversity of parasite life cycles and the unique characteristics of host-pathogen interactions (Bull

1994; Ebert and Bull, 2003). The baseline reproduction ratio, often known as R_0 (Anderson and May, 1979), is represented as,

$$R_0 = \frac{\beta . S}{\mu + a + \gamma}$$

where β is the rate at which the parasite is transmitted; S is the number of susceptible hosts present in the population; μ is the natural mortality rate of the host; and β is the rate of mortality of the host due to infection (virulence) and γ the recovery rate (Figure 4.1). However, formal estimates of virulence evolution are based on the compromise hypothesis, which was first developed by Anderson and May in 1982, and has led to a substantial body of theoretical work aimed at explaining many aspects of pathogen evolution and host-pathogen coevolution. Importantly, the evolution of virulence has been extensively modeled in relation to epidemiological and ecological aspects. For instance, the role of epidemiology and ecology in the development of virulence has only been studied in a few experimental studies, which have yielded mixed results. The majority of the evidence for plant-parasite systems comes from trials with fungi and oomycetes. Reports of research explicitly created to investigate the effect of various factors on virulence evolution are uncommon.

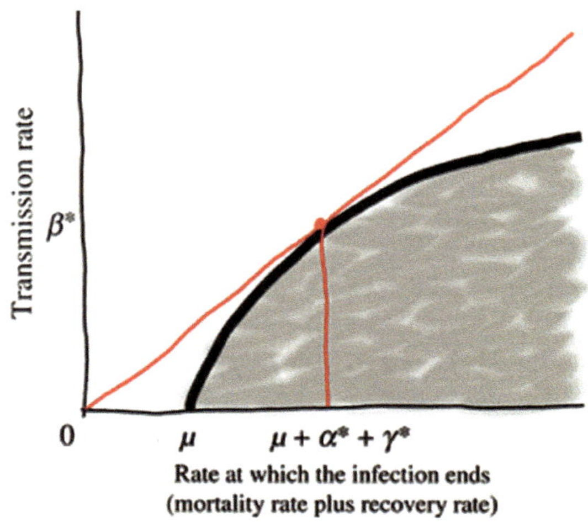

FIGURE 4.1 The trade-off curve.

Source: Modified from Alizon et al., 2009

4.5.2 THE EVOLUTION OF PATHOGENICITY

Pathogenicity in parasites and host resistance are targets for selection because parasites must infect hosts to survive, and parasite infection reduces host fitness. According to De Meaux and Mitchell-Olds (2003) and Nurnberger et al. (2004), plants have a number of naturally occurring and artificial barriers to infection that help them fight against the disease. Pathogens, however, use virulence factors to get past these barriers and infect plants.

Plant immunity operates on various tiers. Basal defense responses are initiated by the identification of conserved pathogen-associated molecular patterns in one layer (PAMP-triggered immunity, PTI). Knowing the parasite's pathogenicity inhibitors (which may have been designed to lower PTI) adds an extra layer of protection (Chisholm et al., 2006; Jones and Dangl, 2006). The coevolution of an antagonist host with a coevolved pathogen may be due to the selection of resistance proteins that target both the pathogenic effectors and pathogenic factors (Dawkins and Krebs, 1979, Stahl and Bishop, 2000). The gene-for-gene and matching-allele models, with contrasting underlying premises and hypotheses, have been offered as the two main models of host-parasite interaction affecting the success of infection. The gene-for-gene and matching-allele models may be useful for comprehending the evolution of plant-pathogen interactions, even though they have typically been used in the context of plant and animal systems.

4.6 MAJOR PATHOGENS AND THEIR CLASSIFICATIONS

Plant species can be affected by one or more pathogens all over the globe. Each kind of crop may be affected by one or more pathogens, causing losses, i.e., diseases of different magnitudes. Pathogens may belong to various groups of organisms, predominantly fungi, bacteria, viruses, and viroids, which can be classified based on morphological characters and physiological characteristics.

4.6.1 FUNGI

Fungi are eukaryotic, achlorophyllous, thallophytic organisms that reproduce sexually or asexually. These organisms have no chlorophyll, reproduce by sexual and asexual spores, and generally have a mass of intertwined thread-like structures called hyphae or mycelium with distinct nuclei. According to Hawksworth (1991), about 4,300 reported genera and 70,000 species live on other organisms as parasites or on their residues as saprophytes. More than 8,000 species of fungi are pathogens to plants. Kirk et al. (2008) classify fungi into three kingdoms and eleven phyla as follows:

Kingdom of Fungi:

- Protists are divided into four phyla:
- Endoparasitic Slime Molds
- Dictyosteliid Slime Molds
- Acrasid Slime Molds
- Myxomycota: True Slime Molds

Kingdom of Stramenopila:

- Grouped into three phyla:
- Oomycota
- Hyphochytridiomycota
- Labyrinthulomycota (Slime molds)

Kingdom of Fungi:

- Grouped into four phyla as follows:
- Chytridiomycota: Chytridiaceae
- Zygomycota: Zygomycaceae
- Ascomycota: Ascomycaceae
- Basidiomycota: Basidiomycaceae

Thalli: Naked, pleomorphic without cell walls: Plasmodia, Pseudoplasmodia. Pseudopodia are produced, and they are motile. Plasma flow is mediated by multinucleated aseptate mycelium. Mycelium of the following orders are also motile: mycota (a.k.a. multinucleated mycelium), mycelium of the zygomycotic family (mycota), mycelium (zygomycotic mycelium), and coenocytic mycelium. Asexual reproduction takes place in sporangia (spore-bearing sacs). Sexual spores (zygospores) are produced by the fusion of sex cells. Sex cells are morphologically related to sex cells. Sexual spores are produced by zygospores. Sex spores are created by the fusion of morphologically related sex cells. Sexual spores are produced by oomycetes through gametangial (sexual) contact between antheridium (the body's male sex organ) and the body's female sex organ, oogonium. Sexual spores are not produced by the Chytridiomycetes. The Ascomycetes and Basidiomycetes are septate and produce sexual spores (ascospores in asci and basidiospores on basidia, respectively).

4.6.2 BACTERIA

Bacteria are prokaryotic organisms, lack a well-ordered nucleus (no nuclear membrane), and are surrounded by a cell membrane and a cell wall. According to a study published in 1878 by University of Illinois Professor T. J. Burrill, the cause of the so-called apple and pear blight was the presence of a bacterium in the affected tissues (Agrios, 2005). In 1882, he named that bacterium as *Micrococcus amylovorus.* There are about 80 bacterial species that cause plant diseases. Some of them are made up of several pathovars (pathogens). Bacteria are classified according to Gram reaction (Gram +ve, Gram –ve), morphology (Bacillus rod-shaped, coccus, spirilla, spiral, vibrio, comma-shaped, spirochaete, corkscrew-shaped), monocystis (mono), or in a cluster (Di cluster, Tri cluster, Tetra cluster, Sarcina cluster, cuboid cluster, four bacterial cells), respiration (aerobic, anaerobic, or aerobic), and flagella (monotrichous, monophagous, amphitrichous, lophotrichous, or lophophagous flagella as tuft). Flagella are distributed throughout the cell (except at the poles) (Mondal, 2011; Verma and Jayashree, 2002) (Figure 4.2).

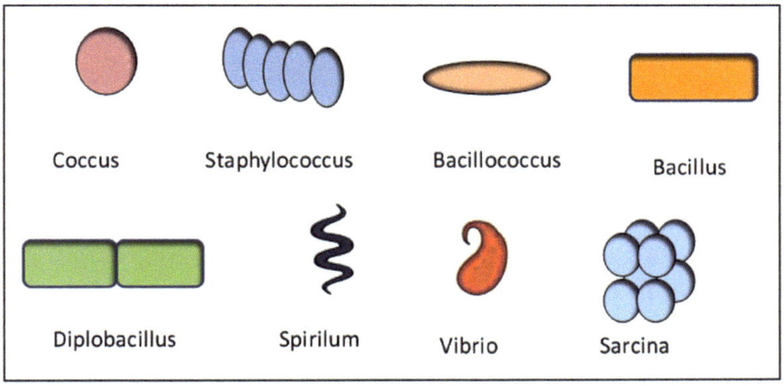

FIGURE 4.2 Morphological classification of bacteria.

A general classification of plant pathogenic bacteria is as follows:

Part I: Bacteria which are Gram-negative aerobic rods and cocci consist of Family: Pseudomonadaceae having Genus: *Pseudomonas, Xanthomonas, Xylella, Burkholderia, Acidovorax, Rhizomonas, Rhizobacter, Ralstonia* and *Xylophilus* and Family: Rhizobiaceae having Genus: *Agrobacterium.*

Part II: Bacteria that are Gram-negative facultative anaerobic rods consist of Family: Enterobacteriaceae having two Genus: *Erwinia* & *Pantoea.*

Part III: Irregular Gram-positive nonsporinating rod-forming bacteria fall into the following families: mycobacterium with genera: clavibacterium, curribacterium, rhodococcus, anthrobacterium, rathoyibacter.

Part IV: Bacteria which are Actinomycetes, bacteria forming branching filaments consist of Family: Streptomycetaceae having Genus: *Streptomyces*

Part V: Contains Mollicutes, prokaryotes that have a cell membrane but no cell wall grouped in Family: Mycoplasmataceae, the plant mycoplasma-like organisms phytoplasmas and Family: Spiroplasmataceae with Genus: *Spiroplasma*, helical, motile but lacking flagella.

4.6.3 VIRUSES

The word 'virus' means 'poison.' Viruses are nucleoproteins that can replicate only in living cells, i.e., obligate parasites. Plant viruses are classified as "sub-cellular infective entities which are capable of reproducing only within the cell and may be pathogenic" according to Bawden (1943). In the 16th century, the disease tulipomania of tulip causing floral discoloration was described as caused by a virus and was the first report of a disease caused by a virus. There are about 850 viral species described to cause plant diseases. The virus contains genetic material, either RNA or DNA, not both. Based on morphological studies with the help of X-ray diffraction and electron microscopy, viruses are either rods, filiform, isometric, or polyhedral in shape. They apparently do not act as organisms and affect the host like other pathogens but disturb the host metabolism of nucleic acid. Some viruses have great economic significance and target a variety of plants, whereas others are restricted to a single host. Although there are various viral symptoms, chlorosis is a prevalent one. Other symptoms include mottling, mosaic, ring spot, curling, rolling, enation, yellows, mosaic, stunting, rosettes, witches' brooms, etc. Sometimes symptoms are "masked," not manifesting in specific situations, such as in hot temperatures, or latent, not showing up until another virus is also present. Necrotic signs can result in death. They are transmitted from infected plants to healthy ones by mechanical methods (rubbing or abrasion), grafting, or biotic means (insects, mites, fungi, and nematodes). Murphy et al. (1995) and Brunt et al. (1996) classified the virus as follows:

- **(+) ssRNA viruses:** Rigid rod-shaped viruses are Tobamovirus, Tobravirus, Hordeivirus, Furovirus, Pecluvirus, Pomovirus, and Benyvirus; Filamentous viruses are assigned to two families, namely Potyviridae and Closteroviridae, while some genera are not assigned to any family, such as PotatotubervirusX, Carlavirus, Foveavirus, Capillovirus, Trichovirus, Vitivirus, and Allexvirus; Isometric viruses are assigned to four families – Sequiviridae, Comoviridae, Bromoviridae, Tombusviridae, Leutoviridae. However, some genera are not assigned to any family, such as Sobemovirus, Marafivirus, Umbravirus, Tymovirus, and Idaeovirus, and Bacillus virus (Ourmiavirus).
- **(-)ssRNA viruses:** Viruses of Family Rhabdoviridae and Phenuiviridae are bacilliform, Bunyaviridae are membranous, while viruses of Tenuiviridae and Ophioviridae are thin flexuous.

- **dsRNA viruses:** These viruses are assigned to two families, namely Reoviridae & Partitiviridae.
- **dsDNA viruses:** These viruses are assigned to the family Caulimoviridae.
- **(+)ssDNA viruses:** These viruses are assigned to two families, namely Geminiviridae & Circoviridae.

4.6.4 VIROIDS

Viroids are small, circular ssRNAs that are naked and parasitize the host machinery for their replication. Some examples of viroids include Potatoes Spindle Tuber Viroid, Apple Dimple Fruit Viroid, Australian Grapevine Viroid, Apple Scar Skin Viroid, Avocado Sunblotch Viroid, Citrus Exocortis Viroid, Citrus Stunt Viroid, Chrysanthemum Chlorotic Mottle Viroid, Citrus Bent Leaf Viroid, Coleus Blumei Viroid, Columnea latent Viroid, Coconut Cadang-Cadang Viroid, Tomato planta macho Viroid, Grapevine Yellow Speckle Viroid, Coconut Tinangaja Viroid, Hop latent Iresine Viroid, Peach latent mosaic Viroid, Mexican Papita Viroid, Pear Blister Canker Viroid, and Tomato Apical Stunt Viroid.

4.7 PHYSIOLOGICAL CLASSIFICATION OF PATHOGENS

In any organism, there are certain characteristics that are common, either morphological, biochemical, or pathogenic. If the organism is pathogenic, this common group of characteristics may constitute a group within the living world. Within the living world, there are kingdoms (Eight kingdoms in the living world). The hierarchical order within the kingdoms follows as Phylum, Class, Order, Family, Genus, and Species. For example, *Puccinia graminis* indicates a group of individuals from the genus Puccinia which can infect plants of the family Gramineae. Within the individuals of *Puccinia graminis*, certain individuals can infect wheat (*Triticum aestivum*), certain can infect barley (*Hordeum vulgare*), and certain can infect oat (*Avena sativa*). Accordingly, they have been given the names *Puccinia graminis tritici* (Capable of infecting wheat), *Puccinia graminis hordei* (Capable of infecting barley), and *Puccinia graminis* avanae (Capable of infecting oat). Thus, the special groups within the species are called variety, special form, or forma specialis, and the correct way of writing them is "*Puccinia graminis* f.sp. tritici," "*Puccinia graminis* f.sp. hordei," or "*Puccinia graminis* f.sp.

avanae." Thus, as a conclusion, it can be said that within the species of a pathogen, certain individuals may exist which are morphologically similar, but they differ with respect to their physiology, biochemistry, and pathogenic capacity. These individuals can be differentiated on the basis of their symptoms or on certain host cultivars. They are also called *Forma specialis* (f.sp.). Individuals within the species of a pathogen are morphologically similar, but they show variation with respect to their pathogenic ability on a particular set of host varieties. Genetically and geographically, they are also distinct and known as the Physiological race. When the progeny of an organism shows variation in the traits (character) from their parents, such progeny will be called a variant. A progeny developed by a variant having similar heredity is called a biotype. It is the subgroup of individuals within a species or race, usually characterized by the common possession of a single or a few new characters.

4.8 THE LINK BETWEEN PATHOGENICITY AND CLIMATE CHANGE

According to a report by FAO in 2021, agricultural communities are very concerned about climate change. The three basic components of the agricultural process are pathogen, host, and environmental conditions. The interaction between these components is what determines whether an infection will develop or not, and climate change has a significant impact on all of these aspects. Crop output, pest susceptibility, and insect pest lifetime are all significantly impacted by changes in climatic factors (Pareek et al., 2017). Crop health is affected by crop pests and disease susceptibility, which in turn impacts crop health. As a result of these changes, farming practices have to be altered in order to deal with their consequences and avoid a drop in productivity.

In order to produce the food and fiber required to sustain human life, agriculture is an economic activity that is heavily dependent on climate and weather. However, agriculture is a sector that is quite susceptible to climate change. Different kinds of uncertainty characterize how climate change will affect agriculture. First, there are questions about the speed and scope of climate change itself. Second, there are questions regarding how agricultural-based outputs, including CO_2 fertilization, would react. Thirdly, there are questions about how society will react to these anticipated effects or even about its capacity to do so. These uncertainties place restrictions on several parts of climate change research. We have a limited grasp of how future climate change will manifest itself because the majority of these uncertainties are not quantifiable (Anita et al., 2010).

At the global, regional, and local levels, food security may be impacted by climate change. Food supply and quality can be affected by climate change, which can also diminish availability. Increasing temperatures, changing extreme weather patterns, altering precipitation patterns, and decreasing water availability, for instance, might all have a negative impact on agricultural output. Due to the likelihood of harsh weather, which is anticipated to occur more frequently in the future, deliveries of food may be halted, and prices may rise as a result of a lack of supplies. Additionally, rising temperatures can hasten contamination and food degradation. Floods caused by increasing sea levels, which can vanish in low-lying regions and result in significant crop losses, are a prime illustration of the disastrous effects of climate change (Mimura, 2013). Drought is another illustration of when inadequate soil water levels result in plants losing their biological processes and even becoming more vulnerable to pests and disease (Seleiman et al., 2021).

4.8.1 TEMPERATURE

The impact of high temperatures on any kind of crop is greatly influenced by the temperature at which the crop can grow and reproduce at its best. Increased temperatures in some areas may be advantageous to the kinds of crops that thrive there or may allow farmers to switch to planting crops that thrive in warmer climates. That is not always the case; if the higher temperature is higher than the crop's optimal temperature, yields will fall, or even worse, disease infection and appearance may happen. Many viruses are expected to expand into new regions as temperatures rise, where they will come into contact with new potential hosts (Amari et al., 2021). A pathogen's biology can be directly impacted by environmental conditions in a number of ways.

Temperature, relative humidity (R.H.), and, in the case of foliar diseases, frequent leaf wetness all significantly impact propagule production, germination, and pathogen growth rates (Wu et al., 2016). The ideal temperature for infection varies greatly between pathogen species. Generally speaking, extended periods of environmental factors (such as temperature, precipitation, and humidity) that are nearly ideal for developing a pathogen result in more destructive epidemics (Wu et al., 2016). Temperature and R.H. can have an impact on a pathogen's ability to survive in the absence of a host (such as overwintering and oversummering). The strongest effects of global warming are anticipated to be seen in the tropics, even though the increase in temperatures there is predicted to be less than that in temperate climates. This is because tropical species have a limited temperature range for growth, making them more susceptible to temperature changes. Additionally, many species are currently residing quite close to their ideal temperature range. Higher latitude pathogens might be more tolerant to a wider range of temperatures (Bita et al., 2013). Since these viruses typically exist in climates that are below their physiologically ideal range, warming is anticipated to increase both their fitness and the danger of outbreaks of the diseases they are linked to. Climate change factors can have an impact on both the host and the pathogen. For instance, a certain range of temperatures can encourage the growth of the pathogen, while another range of temperatures can increase the host's resistance to pathogenic diseases. Wheat and oats are a good illustration of these occurrences since they are more prone to rust infections as the temperature rises, whilst some forage species grow more resistant (Petzoldt et al., 2006). Furthermore, many pests may experience 1 to 5 additional life cycles per season as a result of temperature variations as slight as CO_2 fluctuations, which improves their capacity to overcome plant resistance.

4.8.2 MOISTURE

Temperature and moisture cannot be separated from one another. Reduced rainfall may result in a decline in the incidence of grape downy mildew infections. However, in a warmer scenario, the temperature rises more than make up for the shorter duration of leaf wetness, in part because earlier infections in the growing season give epidemics more time to spread (Gautum et al., 2013). In some circumstances, less precipitation translates into a decreased risk of disease, even though rising temperatures are often linked to an increased risk of disease development for most pathosystems (Velásquez et al., 2018). By modifying plant architecture and, consequently, the microenvironment, environmental changes can also indirectly impact the biology of disease (Garin et al., 2014). For instance, canopy shape and density can influence the temperature, moisture, and UV light availability at an infection site. More plants and leaves tend to be wetter, which encourages the growth of infections that love humid environments (Velásquez et al., 2018).

4.8.3 CO_2 LEVEL

The physiology of a host species may vary as a result of conditions that affect plant development, such as high CO_2 levels, higher warmth, or drought, which would significantly modify how biotrophic infections colonize host tissues

(Eland et al., 2014). High levels of CO_2 have been demonstrated to boost pathogen fecundity, hastening the evolution in response to climatic change (Chakraborty, 2005). High quantities of CO_2 can affect crop productivity. Several lab tests revealed that elevated CO_2 levels could have a favorable impact on growth. However, several elements like fluctuating temperatures, water, ozone, and poor nutrient levels might work against these potential production increases. For instance, if the temperature is higher than what a crop prefers, if there is a lack of water or nutrients, the gain in yield may be modest. Lower nitrogen and protein content in soybean and alfalfa plants are associated with higher CO_2 levels, which significantly negatively impacts quality. The capacity of rangeland and pasture to maintain cattle that rely on grazing can be reduced by reduced forage and grain quality (Capstaff and Miller, 2018). While increased CO_2 might promote plant growth, it also reduces the nutritional value of most food crops. Reduced protein and critical mineral content in a range of plant species, such as rice, soybeans, and wheat, directly result from rising atmospheric CO_2 levels (Taub, 2010). Consequently, the impact of increased CO_2 on the nutritional value of crops is seen as a potential and indirect threat to human health. Additionally, pesticides and their residual toxicity in people pose a concern to human health due to the increased use and decreased efficacy of pesticides because of the emergence of insect resistance.

4.8.4 DROUGHT AND OTHER FACTORS

Drought spells are a critical issue in areas with warmer summer temperatures, resulting in soil dryness. Although higher irrigation may be feasible in certain areas, water resources may also be reduced in other areas, making less water available for irrigation when more is required (Malhi et al., 2021). Warmer, wetter, and higher CO_2 levels are favorable to the growth of many weeds, pests, and fungi (Zayan, 2019). Climate change can affect both the host and the pathogen. Specific temperature ranges can promote pathogen growth, while other ranges can enhance the host's resistance to diseases.

Due to high humidity and warmth during harvest, several mycotoxins, such as *Fusarium* spp., are secreted. Mycotoxins have higher concentrations (Perincherry et al., 2019). Weed growth is also accelerated by humidity, and weed biomass rises as temperature rises (Ramesh et al., 2017). Depending on the physiology of the plant, a given parameter may have a variety of consequences. For instance, higher CO_2 concentrations may result in slower plant breakdown rates, which raises the fungal inoculum levels and may encourage the creation of more fungal spores. On the other hand, elevated CO_2 levels may alter the physiological makeup of plants and increase their resistance to some diseases. Low water levels and soil erosion, which reduce soil fertility and hence plant health, are two other extreme situations that may exist. Another important determining factor is fungicide activity, and climate change may significantly impact fungicide effectiveness. Highly frequent rainfalls have a significant negative impact on the effectiveness of contact fungicides because rain has the power to whisk contact fungicides off the surface of the hosts, rendering them useless. However, this characteristic doesn't significantly impact plants with high metabolic rates because they consume more fungicides.

Research published in Smithsonian magazine in 2021 predicted that crops grown at higher latitudes would face more challenges and increased rates of illness. Over a given amount of time, the number of generations of a polycyclic pathogen significantly impacts epidemics. Numerous diseases' rates of reproduction are influenced by temperature and moisture (Caffarra et al., 2012). The lengthened growing seasons brought on by climate change will increase the amount of time available for pathogen reproduction and spread (Figure. 4.3). The sexual reproduction of diseases may potentially be affected by climate change (Legler et al., 2012), enhancing the potential for particular populations to evolve. Given their shorter generation times and, frequently, their capacity to spread swiftly via the wind, most diseases will have an edge over plants when it comes to migration and climate change adaptation (Davis et al., 2005).

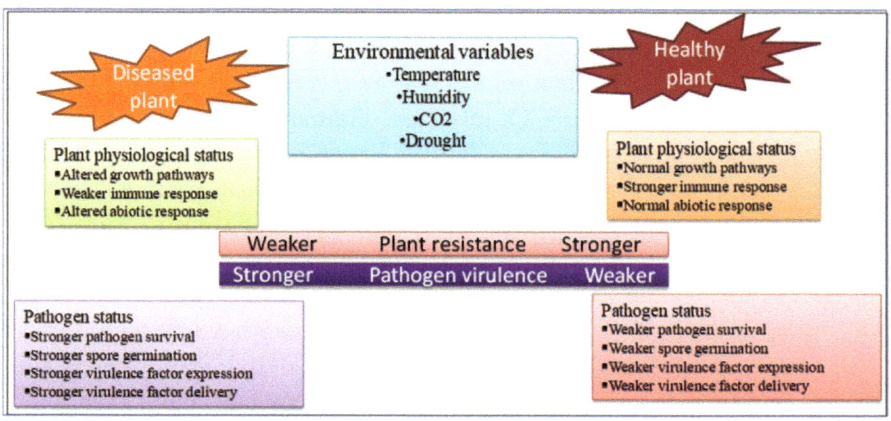

FIGURE. 4.3 Impact of environmental conditions on phytopathogens.

Source: Modified from Velásquez et al., 2018

4.9 IMPACT OF CLIMATE CHANGE ON THE INCREASING THREAT OF PATHOGENS

Many environmental factors influence the molecular evolutionary processes of disease-causing organisms. Climate change has notable effects on host plant resistance mechanisms and the virulence capacity of pathogens. Long-term drought and certain other persistent climate changes may cause hosts, pathogens, or both to evolve, changing their susceptibility and virulence (O'Hara et al., 2021). To understand whether environmental factors influence variations in DNA, several researchers conducted parallel experiments in controlled and field conditions distinct from those in host plants. Stressful situations were replicated to determine whether these situations are the primary factors affecting cell competence and mutations (Bartoli et al., 2016). Natural ecosystems are more prone to infectious plant diseases than agricultural ecosystems due to climate change (Burdon and Zhan, 2020). Natural plants exhibit higher genetic variation compared to those in agricultural ecosystems.

In agricultural plants, the adaptation to climatic change can be greatly improved by continuous and timely integration of adaptive traits into crop varieties through breeding, making natural plants more vulnerable to climate change and pathogen infection due to the lack of artificial evolution in these plants (Rasheed et al., 2017; Yang et al., 2022).

The wheat pathogen – *Zymoseptoria tritici*, and potato pathogen – *Phytophthora infestans* adapted quite well to changing climate conditions. This adaptation is caused by genomic structure changes and gene expression changes (Yang et al., 2022). *Z. tritici* produces chlamydospore structures that are highly resistant to cold, heat, and drought compared to other cell types. Based on transcriptomic analysis, the morphogenesis of this pathogen is co-regulated with the expression of the virulence factors. As the environmental conditions changed, the cellular morphology of the pathogen changed, resulting in new morphotypes. Each morphotype has the potential to affect both the survival of the pathogen and disease epidemiology. When pathogens developed, new, virulent races/types emerged on different crops. These new races will rapidly spread by long-distance dispersal (Singh et al., 2011).

In recent years, the incidence of late blight has increased in Fennoscandia (Europe). On average, it occurs 30 days earlier today than it did in the 1980s (Lehsten et al., 2017). Plant pathogens adapt genetic changes and strive to overcome the resistant mechanisms of the host, thus complicating disease management tactics. In areas where interannual variations in climate are high, it becomes challenging to implement control measures as diseases spread more rapidly. When large groups of pathogens are influenced by evolutionary pressures, there is a high likelihood that new races will evolve over the years due to increased temperatures and CO_2 levels, leading to increased fecundation and infection cycles in a favorable microclimate under an expanded canopy. Changing climatic conditions alter the host plant resistance, affecting the pathogen biology (Lake and Wade, 2009). Elevated levels of CO_2 affect the disease development capacity, production, and saprophytic survival of the pathogen inoculum. The increased CO_2 levels enhance the fecundity of pathogens. Rising temperatures and CO_2 levels over the years extend the plant growing season and alter canopy structure, microclimate, and the amount of susceptible tissue (Pangga et al., 2013).

The constant increase in temperature has caused the replacement of pathogens like *Fusarium spp.* in Europe by the more thermogenic and toxic species of *Fusarium* pseudograminearum. This has increased the risk of contamination with mycotoxins (Paterson, Lima, 2010). These adaptations also led to the replacement of the existing population of *Fusarium spp.* in Canada by highly toxic forms of *Fusarium* pseudograminearum. Race-9 Xoo was able to overcome the R gene Xa7 found in rice. The genetic adaptation of Race-9 Xoo played a critical role in increasing its virulence and aggression over 11 consecutive years of increasing temperatures to become the dominant pathogen population in Europe (Webb, 2011). At the same time, the *Xa7* gene gained effectiveness at higher temperatures to control the pathogen race that had evolved due to adaptive radiations in climatic conditions.

The enhancement in infection cycles due to elevated CO_2 boosts the fecundity of *Colletotrichum gloeosporioides*, which increases its population; also, this strain evolves with increasing aggressiveness after completing about 25 successive infection cycles. A favorable microclimate within an enlarged canopy and warming created by elevated CO_2, in turn, enhances infection cycles (Siebold and Tiedemann, 2012) and the number of asexual generations (Pangga et al., 2013).

The great genotypic diversity of pathogens is reflected in their evolutionary history. While an increase in diversity would enhance evolutionary potential, the fitness of recently developed genotypes is also essential (Webb et al., 2011). Future virulence trends can be predicted using the link between pathogen fitness and virulence (Parker and Gilbert, 2004). The saprophytic fitness does not inhibit when the population of necrophyte *F. pseudograminearum* increases due to elevated levels of CO_2 (Melloy et al.,

2010). Phenotypic plasticity favors the genotypes to mold themselves to adjust to changing climatic conditions and produce different phenotypes. These studies were made in fungal genera *Phytophthora* spp. (Schrey and Richards, 2012). Gene expression and enzymatic activity regulate plasticity, which is assumed to be more suitable for adapting to random fluctuations in climate events (Kelly, 2019; Ashapkin et al., 2020; McGuigan et al., 2021).

During the long time span of climate change, data on the impact of short-term weather on pathogens used in forecasting (Seem, 2004; Webb et al., 2011) are often not applicable. A twelve-year research revealed that more pathogens attacked warmer plots and caused more damage (Roy et al., 2004). In the past, it was considered that *P. striiformis* f. sp. *tritici* was active at somewhat lower temperature ranges. *P. striiformis f. sp. tritici* populations from more recent times, however, have acclimated to warmer temperature ranges (Milus et al., 2006). Markell and Milus (2008) examined isolates from the 1960s to 2004 with genetic markers and morphological comparisons and differentiated the isolates obtained as pre and post-2000 (better adapted to a warmer temperature range). All post-2000 isolates examined were able to cause disease on wheat plants with resistance genes Yr8 and Yr9, while these resistance genes were effective at preventing disease for pre-2000 isolates. The main cause of the current epidemic outbreaks and the dramatic shift in *P. striiformis f. sp. tritici* populations was probably the new isolates' capacity to overcome these resistance genes.

So, it's clear that weather plays a role in plant diseases, and changes in climate-induced weather patterns are likely to impact the frequency and intensity of epidemics in the future (see Garrett et al., 2021). At climate-relevant scales, rapidly evolving pathogens and changing geographic distribution will be the primary drivers of change, and will therefore drive future disease control efforts (Chakraborty 2013, Lehsten et al., 2017).

4.10 MOLECULAR BASIS OF EVOLUTION IN PATHOGENS

Plant diseases are a long-term threat to the sustainability of crop production around the world. Plants provide many of the defense mechanisms that pathogens need to break down in order to infect and proliferate. Plants recognize the pathogen-related molecular patterns (PAMPs) and induce an immune response that pathogens must overcome using secreted proteins and effectors (Jones and Dangl, 2008).

TABLE 4.1 Some Examples of Pathogen Races, Their Host, and Mechanism of Evolution

Pathogen	Race/Gene Involved	**Host**	**Mechanism of Evolution**	**Effect of Change**	**Reference**
Puccinia gramanis f. sp. *tritici*	Ug99	Wheat	Mutation	Enhanced virulence	Hawkins 2020
Melampsora lini	*AvrL567* gene	Australian flax	Mutation	Interruption in R-Avr recognition	Dodds 2006
Fusarium oxysporum	Lineage-specific (L.S.) region	Multiple host range	Horizontal gene transfer	Conversion from non-pathogenic to a pathogenic strain	Ma 2010
Melampsora medusae-populina	-	Populus	Hybridization between *M. medusa* and *M. laricipopulina*	Wider host range	Spiers and Hopcroft 1994
Phytophthora fragariae	-	*Alnus* spp.	Hybridization between *P. cambivora* and *P. fragariae*	Give rise to heteroploid hybrids (Enhanced host range)	Brasier 2004
Pseudomonas syringae pv. Tomato	Race-1	Tomato	Mutation	Enhanced virulence	Kunkeaw 2010
Banana Bunchy Top Virus	DNA-U3 sequence	Banana	Intragenomic recombination	Enhanced virulence	Hyder 2011
Puccinia gramanis f. sp. *Tritici*	-	Wheat	Insertion or deletion of transposable elements	Novel pathotypes differ from pre-existing ones by one or two avirulence genes.	Anderson et al., 2010
Pyrenophora tritici-repentis	*Tox A* gene	Wheat	Interspecific virulent gene transfer	Enhanced virulence	Friesen et al., 2006

On the other hand, plant R-genes recognize the effector protein (Avr) and induce local cell death, thus preventing the pathogen's further spread to healthy plant cells. The host plays a major role in the evolution of virulence in the pathogen, as pathogen fitness depends on the ability to replicate in a host and spread to another host. In nature, host-pathogen interactions are typically co-evolved, and pathogen virulence may evolve in tandem with host resistance. It is generally accepted that only pathogenic pathogens are capable of infecting and damaging the host. Pathogens access the plant interior directly by penetrating the leaf or root surface and indirectly by entering through wounds or natural openings.

During the invasion process, the pathogen secretes cell wall-degrading enzymes, which facilitate the delivery of pathogen effectors into host cells. Since the host and pathogen are in a continuous evolutionary battle against each other, plants also have to evolve defense mechanisms to prevent pathogen invasion. Plant defense mechanisms work either by blocking pathogen entrance or activating a range of defense responses.

Pathogen evolution is a complex process driven by the interaction of multiple ecological and genetic factors. This results in the emergence of new pathogen-evolved strains, which can effectively break the plant defense barriers. The critical factors that make the pathogen escape from plant defense are population size and generation time, which tend to be much larger and shorter for pathogens. Due to these factors, microbial populations are more likely to experience new mutations, resulting in the evolution of more virulent races/strains. Due to this, pathogens are able to re-emerge and expand their host range (McDonald and Stukenbrock, 2016). However, the evolutionary rate does not simply depend on population size but varies with the virulence gene. For instance, the wheat stem rust pathogen has a mutation rate of 8.3×10^{-6} to 6.9×10^{-11} asexual generation^{-1}. The R-gene *Sr31* in wheat remained effective for over 30 years until the appearance of the 'Ug99' race (Singh et al., 2011).

4.11 PATHOGENS EVOLUTION AND THEIR VIRULENCE MECHANISMS

4.11.1 DNA MUTATIONS

Genetic variation is ultimately caused by a DNA mutation that creates new codon sequences or changes gene regulation. Mutations are any heritable alteration that occurs in an organism's genome. A simple nucleotide mutation may lead to a more pathogenic race. While mutation rate is usually considered an innate characteristic of an organism, there is growing evidence that mutation bias is adaptive (Yoshida et al., 2009). Some parts of the genome are more susceptible to mutation than others. In the case of the bacterium *Magnaporta grisea* (*Magnaporta* spp.), the loss or gain of effector genes is often related to the instability of the telomeric region of the chromosome. Genomes can be divided into gene-rich, slowly evolving regions and repeat-rich, gene-sparse regions that show high variability. A prime example of this is the two-speed genome of the bacterium Pneumocystis aeruginosa (Dun et al., 2015; Monroe et al., 2020).

4.11.2 TRANSPOSABLE ELEMENTS

Transposable elements are the major mutators in the prokaryotic and eukaryotic genome, and are responsible for deletions, silencing, duplications, and rearrangements (Moller et al., 2017; Gijzen et al., 2014; Seidl et al., 2017). The gain or loss of genes associated with transposable elements has been linked to the ability to infect various hosts. For instance, in the microorganism Magnaporthe grisea, the inclusion of transposable elements in avirulent genes led to an increase in virulence because the host was unaware of the presence of the transposable elements (Fudal, 2005). In certain fungi, the close association of virulence factors with transposable elements can lead to an increase in the local mutation rate (RIP). For example, in Leptosphaeria maculans, effector alleles containing premature stop codons (RIP) or non-synonymous RIP mutations evolved in a few years to evade plant resistance mediated by immune receptors (Rouxel, 2011).

4.11.3 GENETIC EXCHANGE

Genetic exchange between individuals, either sexually or through a parasexual relationship, can lead to the generation of new pathogens that are more virulent. As a result, recombination can generate new combinations of susceptibility factors and, can generate novel pathogens more quickly than mutation. For example, the highly pathogenic blue13 lineage is derived from a sexually reproducing population of *P. infestans* in northern Europe. In addition to sexual reproduction, there are other mechanisms of asexual nuclei exchange and asexual recombination. Parasexuality is a mechanism of asexual transmission of nuclei between two organisms. The virus population can be composed of the same or a variety of virus species. This results in increased virulence as well as a wide host range (Cooke et al., 2012; Ruiz et al., 2018). In RNA viruses, the mechanism of recombination is the switching of copy choice template during RNA replication. In DNA viruses, the mechanism is less well understood and may involve two-step break repair events on DsDNA intermediates via host cell factors. Genetic exchange in viral populations may also be caused by reassortment (also called pseudo-recombinant) of genomic segments in virus populations with segmented (or multipartite) genomes. Multipartite (or segmented) genomes are common in plant viruses, although the cost of transmission is high as most segments must be transmissible between hosts to cause novel infections (Zwart, Elena, 2020). Reassortment is closely related to virus development and speciation and is considered an advantage of multipartite viruses as it allows for the generation of fitter variants (Rossinck, 2005).

4.12 BALANCING PATHOGEN AND CLIMATE CHANGE

About 15% of crop yield is lost to bacterial and fungal pathogens, while 3% is lost to viruses (Oerke and Dehne, 2004). The loss brought on by microbial infection is reported to be as high as 30% for some crops, such as potatoes. Instead of applying chemical agents, researchers are experimenting with other approaches to enhancing the resistance of plants to microbial infections through their genetic composition. Crop improvement relies heavily on conventional breeding, which involves growing and evaluating large populations of crops over many generations, a labor-intensive and lengthy process. However, biotechnology has increased the pace of crop breeding. In the case of genetic engineering, an organism's genetic material is directly altered by biotechnology (Christou, 2013). This has several advantages over conventional breeding: First, it allows for the addition, removal, alteration, or fine-tuning of specific targeted genes with very few unintended changes to the entire crop genome. Second, conventional breeding requires more generations to create crops with desirable agronomic properties. Third, it facilitates the exchange of genetic data between different species. In other words, the raw genetic material that can be used for this purpose is not limited to the genes that are present within the species itself. Fourth, the plant transformation during the process of genetic engineering allows new genes to be introduced into vegetatively proliferated crops such as bananas, cassava, or potatoes in order to increase their yields. Fifth, genetic engineering has several characteristics that make it an effective tool for improving plant resistance to disease.

4.13 DISEASE RESISTANCE IN PLANTS: BREAKTHROUGHS IN GENETIC ENGINEERING

4.13.1 PATHOGEN-DERIVED RESISTANCE AND RNA INTERFERENCE

Transgenic plants expressing virus pathogen genes are known to be immunocompromised against the virus and related strains of the pathogen (Lomooseff, 1995). Based on these findings, it was proposed that ectopic transgenic expression of genes coding for wild type or mutational viral proteins could affect viral replication (Swanford and Johnston, 1985). The genetic code for most plant viruses is a single-strand RNA molecule.

When the viral genome is replicated by RNA-dependent RNA polymerase, double-stranded RNA (dsRNA) replicative intermediates frequently develop, inducing RNAi in the host. Transgenic viral RNA overexpression frequently results in the creation of dsRNA, which also activates RNAi. This process is commonly referred to as co-suppression.

4.13.2 RNAI-BASED ANTIVIRAL RESISTANCE MANAGEMENT TECHNIQUES

RNAi-based techniques can be of three types: co-suppression, antisense RNA, and miRNA-based. In co-suppression, viral RNAs are silenced by RNAi when their genes are overexpressed in transgenic organisms. When nucleotide sequences are expressed along with their reverse complements (antisense RNA), self-complementary transcripts are produced. A double-stranded RNA molecule is formed by processing these transcripts, which triggers RNAi and silences corresponding viral RNA molecules. Artificial microRNA (amiRNA) is an engineered RNA that confers resistance to viruses. A plant-infecting virus cannot infect these molecules since their nucleotide sequences are replaced with virus-specific sequences.

4.13.3 TARGETING VIRUS BY CRISPR/CAS PLATFORM

In recent years, CRISPR-Cas technology, derived from bacteria, has been used to engineer plant virus resistance. CRISPR/Cas is used by many bacterial species to fight viral infections. In this procedure, an RNA-guided nuclease (typically a Cas protein) cleaves at specified target locations on the viral DNA or RNA source, causing their destruction. DNA or RNA molecules are cleaved with specificity when CRISPR RNA and their complementary bases are present. There are a number of Cas proteins that have been shown to have sequence-specific nucleoside cleavage activity (see Wu et al., 2018). For instance, the Cas9 RNA-guided exonuclease (Cas9) from *Streptococcus pyogenes* (SpCas9) induces double-strand breaks in DNA in vivo. Cas13a, a RNA-guided nucleoside cleave (Cas13a, ShShCas13a, Cas13a), targets RNA. CRISPR/Cas acts like molecular scissors, enabling the cleavage of DNA or RNA at specific sequences (see Price et al., 2015). CRISPR/Cas has been shown to be a powerful tool for engineering plant antiviral defense through the development of plant resistance to DNA or RNA viruses. The Tomato yellow leaf. SpCas9 overexploitation and the use of artificially generated guide RNAs focused on different regions of the TYLCV virus have induced resistance to TYLCV in the *Nicotiana Benthamiana* and the *Solanum Lycopersicum* (Mahfouz et al., 2017).

4.14 CONCLUSION

In this chapter, we discussed the different phytopathogens, the classification and evolution of virulent types, and their molecular mechanisms with respect to changing climatic factors. We have also discussed how climate change can worsen the impact of plant diseases and modern breeding interventions that have been employed or have the potential to assist in disease resistance breeding, with examples such as RNAi and CRISPR/Cas modules. Obviously, continuous changes in climate patterns have the potential to endanger the world's already vulnerable food security in a variety of ways, including the promotion of major plant diseases in crop fields and conferring a conducive environment for the resurgence of virulent races/biotypes and evolution of new deadly diseases in crucial food-producing regions. We expect some diseases will be a serious worry for global food security in this century based on current climate trends and plant and pathogen responses to important climatic influences. Based on the limited (and often inconsistent) literature available in this critical area of study, there is an urgent need to scale up global research efforts to gain a better understanding of how individual and combinations of environmental factors affect plant immune responses, pathogen emergence, and disease progression. The impact of environmental factors such as temperature, humidity, and CO2 on disease development may vary depending on the plant and the pathogen species. As a result, forecasting the overall fate of all diseases may be challenging. Nonetheless, one recurring theme is that there are optimum environmental conditions for pathogen virulence and plant immune systems to work concomitantly and in a mutually opposite direction. Environmental variations from these optimum levels necessarily impair plant immunity or pathogen virulence. The combined environmental influences on plant immunity and pathogen virulence result in the ultimate impact on disease. Phytopathogen adaptability does not augur well for an agricultural system that will almost certainly be tested by climate change since the diseases are likely to follow their hosts throughout the world and develop to escape accompanying pressures. Moreover, pathogen mobility, whether through trade commodities or as a result of climate-induced migration, is likely to provide possibilities for infections to arise in new regions or on new hosts, as shown throughout history. Our best weapon is a thorough understanding of these infections. What are the specifics of their on- and off-plant life cycles? What selection pressures do they experience from us, their hosts, and the environment, and how do these forces affect their evolutionary potential? Next, we must create models that account for all of these factors operating on a specific pathogen while being basic enough to effectively predict illness. So, fundamental research efforts should be directed towards such problems, with the goal of having an impact by seeking to use the ensuing knowledge to influence the creation of novel disease management measures.

KEYWORDS

- **climate change**
- **disease**
- **evolution**
- **pathogen**
- **pathogenicity**
- **virulence**

REFERENCES

Abdulkhair, W. M., & Alghuthaymi, M. A. (2016). Plant pathogens. In E. Rigobelo (Ed.), *Plant Growth* (pp. 49–59). IntechOpen. London, UK.

Agrios, G. N. (2005). *Plant Pathology* (5th ed.). Elsevier Academic Press.

Alberts, B., Johnson, A., Lewis, J., Raff, M., Roberts, K., & Walter, P. (2002). *Molecular Biology of the Cell* (4th ed.). Garland Science.

Alizon, S., Hurford, A., Mideo, N., & Van Baalen, M. (2009). Virulence evolution and the trade-off hypothesis: History, current state of affairs and the future. *Journal of Evolutionary Biology, 22*(2), 245–259.

Amari, K., Huang, C., & Heinlein, M. (2021). Potential impact of global warming on virus propagation in infected plants and agricultural productivity. *Frontiers in Plant Science, 12*, 649768.

Anderson, J. P., Gleason, C. A., Foley, R. C., Thrall, P. H., Burdon, J. B., & Singh, K. B. (2010). Plants versus pathogens: An evolutionary arms race. *Functional Plant Biology, 37*(6), 499–512.

Anderson, P. K., Cunningham, A. A., Patel, N. G., Morales, F. J., Epstein, P. R., & Daszak, P. (2004). Emerging infectious diseases of plants: Pathogen pollution, climate change and agrotechnology drivers. *Trends in Ecology & Evolution, 19*, 535–544.

Anderson, R. M., & May, R. M. (1979). Population biology of infectious diseases: Part I. *Nature, 280*(5721), 361–367.

Anderson, R. M., & May, R. M. (1982). Coevolution of hosts and parasites. *Parasitology, 85*(2), 411–426.

Anita, W., Dominic, M., & Neil, A. (2010). *Climate Change and Agriculture Impacts, Adaptation and Mitigation*. OECD Publishing.

Ashapkin, V. V., Kutueva, L. I., Aleksandrushkina, N. I., & Vanyushin, B. F. (2020). Epigenetic mechanisms of plant adaptation to biotic and abiotic stresses. *International Journal of Molecular Sciences, 21*(20), 7457.

Atmar, R. L. (2014). Immunological detection and characterization. In R. Kaslow, L. Stanberry, & J. Le Duc (Eds.), *Viral Infections of Humans: Epidemiology and Control* (pp. 47–62). Springer.

Bahadur, A. (2021). Nematodes diseases of fruits and vegetables crops in India. In C. Belle (Ed.), *Nematodes-Recent Advances, Management and New Perspectives* (pp. 1–18). IntechOpen. London, UK.

Bartoli, C., Roux, F., & Lamichhane, J. R. (2016). Molecular mechanisms underlying the emergence of bacterial pathogens: An ecological perspective. *Molecular Plant Pathology, 17*(2), 303–310.

Basyoni, M. M., & Rizk, E. M. (2016). Nematodes ultrastructure: Complex systems and processes. *Journal of Parasitic Diseases, 40*, 1130–1140.

Bawden, F. C. (1943). *Plant Viruses and Virus Diseases* (2nd ed.). Chronica Botanica Company.

Bebber, D. P. (2015). Range-expanding pests and pathogens in a warming world. *Annual Review of Phytopathology, 53*, 335–356.

Bebber, D. P., Ramotowski, M. A. T., & Gurr, S. J. (2013). Crop pests and pathogens move polewards in a warming world. *Nature Climate Change, 3*, 985–988.

Bita, C. E., & Gerats, T. (2013). Plant tolerance to high temperature in a changing environment: Scientific fundamentals and production of heat stress-tolerant crops. *Frontiers in Plant Science, 4*, 273.

Brasier, C. M., Kirk, S. A., Delcan, J., Cooke, D. E. L., Jung, T., & Man In't Veld, W. A. (2004). *Phytophthora alni* sp. nov. and its variants: Designation of emerging heteroploid hybrid pathogens spreading on *Alnus* trees. *Mycological Research, 108*, 1172–1184.

Brunt, A. A., Crabtree, K., Dallwitz, M. J., Gibbs, A. J., & Watson, L. (Eds.). (1996). *Viruses of plants.* Descriptions and lists from the VIDE database (pp. 1484). ISBN: 9780851987941(CABI Publisher). Viruses of Plants | CABI Books (cabidigitallibrary.org).
Bull, J. J. (1994). Virulence. *Evolution, 48*(5), 1423–1437.
Burdon, J. J., & Zhan, J. (2020). Climate change and disease in plant communities. *PLoS Biology, 18*(11), e3000949.
Burdon, J. J., Thrall, P. H., & Ericson, A. L. (2006). The current and future dynamics of disease in plant communities. *Annual Review of Phytopathology, 44*, 19–39.
Caffarra, A., Rinaldi, M., Eccel, E., Rossi, V., & Pertot, I. (2012). Modelling the impact of climate change on the interaction between grapevine and its pests and pathogens: European grapevine moth and powdery mildew. *Agriculture, Ecosystems & Environment, 14*, 889–101.
Canto, T., Aranda, M. A., & Fereres, A. (2009). Climate change effects on physiology and population processes of hosts and vectors that influence the spread of hemipteran-borne plant viruses. *Global Change Biology, 15*, 1884–1894.
Capstaff, N. M., & Miller, A. J. (2018). Improving the yield and nutritional quality of forage crops. *Frontiers in Plant Science, 9*, 535.
Casadevall, A., & Pirofski, L. A. (2001). Host-pathogen interactions: The attributes of virulence. *Journal of Infectious Diseases, 184*(3), 337–344.
Cavicchioli, R., Ripple, W. J., Timmis, K. N., Azam, F., Bakken, L. R., Baylis, M., & Crowther, T. W. (2019). Scientists' warning to humanity: Microorganisms and climate change. *Nature Reviews Microbiology, 17*(9), 569–586.
Chakraborty, S. (2005). Potential impact of climate change on plant-pathogen interactions. *Australasian Plant Pathology, 34*(4), 443–448.
Chakraborty, S. (2013). Migrate or evolve: Options for plant pathogens under climate change. *Global Change Biology, 19*, 1985–2000.
Chaloner, T. M., Gurr, S. J., & Bebber, D. P. (2020). Geometry and evolution of the ecological niche in plant-associated microbes. *Nature Communications, 11*, 2955.
Chaloner, T. M., Gurr, S. J., & Bebber, D. P. (2021). Plant pathogen infection risk tracks global crop yields under climate change. *Nature Climate Change, 11*, 710–715.
Chisholm, S. T., Coaker, G., Day, B., & Staskawicz, B. J. (2006). Host-microbe interactions: Shaping the evolution of the plant immune response. *Cell, 124*(4), 803–814.
Christou, P. (2013). Plant genetic engineering and agricultural biotechnology 1983–2013. *Trends in Biotechnology, 31*, 125–127.
Cooke, D. E. L., Cano, L. M., Raffaele, S., Bain, R. A., Cooke, L. R., Etherington, G. J., & Kamoun, S. (2012). Genome analyses of an aggressive and invasive lineage of the Irish potato famine pathogen. *PLoS Pathogens, 8*(10), e1002940.
Crowl, T. A., Crist, T. O., Parmenter, R. R., Belovsky, G., & Lugo, A. E. (2008). The spread of invasive species and infectious disease as drivers of ecosystem change. *Frontiers in Ecology and the Environment, 6*(5), 238–246.
Davis, M. B., Shaw, R. G., & Etterson, J. R. (2005). Evolutionary responses to changing climate. *Ecology, 86*(7), 1704–1714.
Dawkins, R., & Krebs, J. R. (1979). Arms races between and within species. *Proceedings of the Royal Society of London. Series B. Biological Sciences, 205*(1161), 489–511.
Day, T., Graham, A. L., & Read, A. F. (2007). Evolution of parasite virulence when host responses cause disease. *Proceedings of the Royal Society of London. Series B. Biological Sciences, 274*(1626), 2685–2692.

De Meaux, J., & Mitchell-Olds, T. (2003). Evolution of plant resistance at the molecular level: Ecological context of species interactions. *Heredity, 91*(4), 345–352.
Delgado-Baquerizo, M., Guerra, C. A., Cano-Diaz, C., Egidi, E., Wang, J. T., Eisenhauer, N., Singh, B. K., & Maestre, F. T. (2020). The proportion of soilborne pathogens increases with warming at the global scale. *Nature Climate Change, 10*(6), 550–554.
Dodds, P. N., Lawrence, G. J., Catanzariti, A. M., Teh, T., Wang, C. I. A., Ayliffe, M. A., Kobe, B., & Ellis, J. G. (2006). Direct protein interaction underlies gene-for-gene specificity and coevolution of the flax resistance genes and flax rust avirulence genes. *Proceedings of the National Academy of Sciences, 103*(23), 8888–8893.
Dong, S., Raffaele, S., & Kamoun, S. (2015). The two-speed genomes of filamentous pathogens: Waltz with plants. *Current Opinion in Genetics & Development, 35*, 57–65.
Eastburn, D. M., McElrone, A. J., & Bilgin, D. D. (2011). Influence of atmospheric and climatic change on plant–pathogen interactions. *Plant Pathology, 60*(1), 54–69.
Ebert, D., & Bull, J. J. (2003). Challenging the trade-off model for the evolution of virulence: Is virulence management feasible? *Trends in Microbiology, 11*(1), 15–20.
Eland, Y., & Pertot, I. (2014). Climate change impacts on plant pathogens and plant diseases. *Journal of Crop Improvement, 28*(1), 99–139.
Fones, H. N., Bebber, D. P., Chaloner, T. M., Kay, W. T., Steinberg, G., & Gurr, S. J. (2020). Threats to global food security from emerging fungal and oomycete crop pathogens. *Nature Food, 1*(6), 332–342.
Francisco, C. S., Ma, X., Zwyssig, M. M., McDonald, B. A., & Palma-Guerrero, J. (2019). Morphological changes in response to environmental stresses in the fungal plant pathogen *Zymoseptoria tritici*. *Scientific Reports, 9*(1), 1–18.
Friesen, T. L., Stukenbrock, E. H., Liu, Z., Meinhardt, S., Ling, H., Faris, J. D., Rasmussen, J. B., Solomon, P. S., McDonald, B. A., & Oliver, R. P. (2006). Emergence of a new disease as a result of interspecies virulence gene transfer. *Nature Genetics, 38*(8), 953–956.
Fudal, I., Bohnert, H. U., Tharreau, D., & Lebrun, M. H. (2005). Transposition of MINE, a composite retrotransposon, in the avirulence gene ACE1 of the rice blast fungus *Magnaporthe grisea*. *Fungal Genetics and Biology, 42*(9), 761–772.
Garin, G., Fournier, C., Andrieu, B., Houlès, V., Robert, C., & Pradal, C. (2014). A modelling framework to simulate foliar fungal epidemics using functional–Structural plant models. *Annals of Botany, 114*(4), 795–812.
Garrett, K. A., Dendy, S. P., Frank, E. E., Rouse, M. N., & Travers, S. E. (2006). Climate change effects on plant disease: Genomes to ecosystems. *Annual Review of Phytopathology, 44*, 489–509.
Garrett, K. A., Nita, M., De Wolf, E. D., Esker, P. D., Gomez-Montano, L., & Sparks, A. H. (2021). Plant pathogens as indicators of climate change. In T. M. Letcher (Ed.), *Climate Change: Observed Impacts on Planet Earth* (pp. 499–513). Elsevier.
Gautam, H. R., Bhardwaj, M. L., & Kumar, R. (2013). Climate change and its impact on plant diseases. *Current Science, 105*(12), 1685–1691.
Gelderblom, H. R. (1996). Structure and classification of viruses. In S. Baron (Ed.), *Medical Microbiology* (4th ed.). University of Texas.
Gijzen, M., Ishmael, C., & Shrestha, S. D. (2014). Epigenetic control of effectors in plant pathogens. *Frontiers in Plant Science, 5*, 638.
Glass, N. L., & Dementhon, K. (2006). Non-self-recognition and programmed cell death in filamentous fungi. *Current Opinion in Microbiology, 9*(6), 553–558.

Gonsalves, D. (1998). Control of papaya ringspot virus in papaya: A case study. *Annual Review of Phytopathology, 36*, 415–437.

Gonsalves, D. (2006). Transgenic papaya: Development, release, impact, and challenges. *Advances in Virus Research, 67*, 317–354.

Hadidi, A., Sun, L., & Randles, J. W. (2022). Modes of viroid transmission. *Cells, 11*(4), 719.

Haque, S., Park, R. F., Keiper, F. J., Bariana, H. S., & Wellings, C. R. (2008). Pathogenic and molecular variation support the presence of genetically distinct clonal lineages in Australian populations of *Puccinia graminis* f.sp. avenae. *Mycological Research, 112*(6), 663–673.

Hawksworth, D. L. (1991). The fungal dimension of biodiversity: Magnitude, significance, and conservation. *Mycological Research, 95*(6), 641–655.

Horino, O., Mew, T. W., & Yamada, T. (1982). The effect of temperature on the development of bacterial leaf blight on rice. *Annals of the Phytopathological Society of Japan, 48*(1), 72–75.

Hunjan, M. S., & Lore, J. S. (2020). Climate change: Impact on plant pathogens, diseases, and their management. In K. Jabran, S. Florentine, & B. Chauhan (Eds.), *Crop Protection Under Changing Climate* (pp. 85–100). Springer.

Hyder, M. Z., Shah, S. H., Hameed, S., & Naqvi, S. M. S. (2011). Evidence of recombination in the banana bunchy top virus genome. *Infection, Genetics and Evolution, 11*(6), 1293–1300.

Isebaert, S., De Saeger, S., Devreese, R., Verhoeven, R., Maene, P., Heremans, B., & Haesaert, G. (2009). Mycotoxin-producing Fusarium species occurring in winter wheat in Belgium (Flanders) during 2002–2005. *Journal of Phytopathology, 157*(2), 108–116.

Jackson, R. W., Johnson, L. J., Clarke, S. R., & Arnold, D. L. (2011). Bacterial pathogen evolution: Breaking news. *Trends in Genetics, 27*(1), 32–40.

Jones, J. D., & Dangl, J. L. (2006). The plant immune system. *Nature, 444*(7117), 323–329.

Jones, R. A. C. (2009). Plant virus emergence and evolution: Origins, new encounter scenarios, factors driving emergence, effects of changing world conditions, and prospects for control. *Virus Research, 141*(2), 113–130.

Karki, S. J., Karki, B. S., & Baral, A. (2016). Climate change and its effects on agricultural crops in Nepal. *Nepal Journal of Science and Technology, 17*(1), 41–50.

Keesing, F., Holt, R. D., & Ostfeld, R. S. (2006). Effects of species diversity on disease risk. *Ecology Letters, 9*(4), 485–498.

Kelly, M. (2019). Adaptation to climate change through genetic accommodation and assimilation of plastic phenotypes. *Philosophical Transactions of the Royal Society B, 374*(1768), 1–10.

Kirk, P. M., Cannon, P. F., Minter, D. W., & Stalpers, J. A. (2008). *Ainsworth & Bisby's Dictionary of the Fungi* (10th ed.). CAB International.

Köhl, J., Kolnaar, R., & Ravensberg, W. J. (2019). Mode of action of microbial biological control agents against plant diseases: Relevance beyond efficacy. *Frontiers in Plant Science, 10*, 845.

Kumar, K. S. A., & Karthika, K. S. (2020). Abiotic and biotic factors influencing soil health and/or soil degradation. In B. Giri & A. Varma (Eds.), *Soil Health* (pp. 145–161). Springer.

Kunkeaw, S., Tan, S., & Coaker, G. (2010). Molecular and evolutionary analyses of *Pseudomonas syringae* pv. tomato race 1. *Molecular Plant-Microbe Interactions, 23*(4), 415–424.

Lake, J. A., & Wade, R. N. (2009). Plant–pathogen interactions and elevated CO_2: Morphological changes in favor of pathogens. *Journal of Experimental Botany, 60*(11), 3123–3131.

Legler, S. E., Caffi, T., & Rossi, V. (2012). A nonlinear model for temperature-dependent development of *Erysiphe necator* chasmothecia on grapevine leaves. *Plant Pathology, 61*(1), 96–105.
Lehsten, V., Wiik, L., Hannukkala, A., Andreasson, E., Chen, D., Ou, T., Liljeroth, E., Lankinen, A., & Briggs, L. G. (2017). Earlier occurrence and increased explanatory power of climate for the first incidence of potato late blight caused by *Phytophthora infestans* in Fennoscandia. *PLoS One, 12*(5), 1–21.
Lenski, R. E., & May, R. M. (1994). The evolution of virulence in parasites and pathogens: Reconciliation between two competing hypotheses. *Journal of Theoretical Biology, 169*(3), 253–265.
Lomonossoff, G. P. (1995). Pathogen-derived resistance to plant viruses. *Annual Review of Phytopathology, 33*, 323–343.
Lucas, W. J., Groover, A., Lichtenberger, R., Furuta, K., Yadav, S. R., Helariutta, Y., He, X. Q., Fukuda, H., Kang, J., Brady, S. M., & Patrick, J. W. (2013). The plant vascular system: Evolution, development and functions. *Journal of Integrative Plant Biology, 55*(4), 294–388.
Ma, L. J., Van Der Does, H. C., Borkovich, K. A., Coleman, J. J., Daboussi, M. J., Di Pietro, A., Rep, M., et al. (2010). Comparative genomics reveals mobile pathogenicity chromosomes in Fusarium. *Nature, 464*(7287), 367–373.
Mahfouz, M., Tashkandi, M., Ali, Z., Aljedaani, F., & Shami, A. (2017). Engineering resistance against Tomato yellow leaf curl virus via the CRISPR/Cas9 system in tomato. *bioRxiv, 237735*. https://doi.org/10.1101/237735.
Malhi, G. S., Kaur, M., & Kaushik, P. (2021). Impact of climate change on agriculture and its mitigation strategies: A review. *Sustainability, 13*(3), 1318.
Malmstrom, C. M., Melcher, U., & Bosque-Pérez, N. A. (2011). The expanding field of plant virus ecology: Historical foundations, knowledge gaps, and research directions. *Virus Research, 159*(2), 84–94.
Markell, S. G., & Milus, E. A. (2008). Emergence of a novel population of *Puccinia striiformis* f. sp. tritici in eastern United States. *Phytopathology, 98*(6), 632–639.
McDonald, B. A., & Stukenbrock, E. H. (2016). Rapid emergence of pathogens in agro-ecosystems: Global threats to agricultural sustainability and food security. *Philosophical Transactions of the Royal Society B, 371*(1709), 20160026.
McGuigan, K., Hoffmann, A. A., & Sgro, C. M. (2021). How is epigenetics predicted to contribute to climate change adaptation? What evidence do we need? *Philosophical Transactions of the Royal Society B, 376*(1826), 1–10.
Melloy, P., Hollaway, G., Luck, J. E., Norton, R. M., Aitken, E. A. B., & Chakraborty, S. (2010). Production and fitness of *Fusarium pseudograminearum* inoculum at elevated carbon dioxide in FACE. *Global Change Biology, 16*(12), 3363–3373.
Messel, H. (2014). *Abridged Science for High School Students: The Nuclear Research Foundation School Certificate Integrated*. Elsevier.
Milus, E. A., Seyran, E., & McNew, R. (2006). Aggressiveness of *Puccinia striiformis* f.sp. tritici isolates in the South-Central United States. *Plant Disease, 90*(7), 847–852.
Mimura, N. (2013). Sea-level rise caused by climate change and its implications for society. *Proceedings of the Japan Academy, Series B, 89*(7), 281–301.
Moller, M., & Stukenbrock, E. H. (2017). Evolution and genome architecture in fungal plant pathogens. *Nature Reviews Microbiology, 15*(12), 771–783.

Murphy, F. A., Fauquet, C. M., Bishop, D. H. L., Ghabrial, S. A., Jarvis, A. W., Martelli, G. P., Mayo, M. A., & Summers, M. D. (1995). *Virus Taxonomy: Classification and Nomenclature of Viruses*. Springer, Amsterdam.

Narayanasamy, P. (2008). Molecular variability of microbial plant pathogens. In P. Narayanasamy (Ed.), *Molecular Biology in Plant Pathogenesis and Disease Management: Microbial Plant Pathogens* (Vol. 1, pp. 159–225). Springer, Dordrecht.

Nazarov, P. A., Baleev, D. N., Ivanova, M. I., Sokolova, L. M., & Karakozova, M. V. (2020). Infectious plant diseases: Etiology, current status, problems and prospects in plant protection. *Acta Naturae, 12*(3), 46–59.

Nürnberger, T., Brunner, F., Kemmerling, B., & Piater, L. (2004). Innate immunity in plants and animals: Striking similarities and obvious differences. *Immunological Reviews, 198*(1), 249–266.

Oerke, E. C., & Dehne, H. W. (2004). Safeguarding production: Losses in major crops and the role of crop protection. *Crop Protection, 23*(4), 275–285.

O'Hara, N. B., Franks, S. J., Kane, N. C., Tittes, S., & Rest, J. S. (2021). Evolution of pathogen response genes associated with increased disease susceptibility during adaptation to an extreme drought in a *Brassica rapa* plant population. *BMC Ecology and Evolution, 21*(1), 1–11.

Pangga, I. B., Hanan, J., & Chakraborty, S. (2013). Climate change impacts on plant canopy architecture: Implications for pest and pathogen management. *European Journal of Plant Pathology, 135*(3), 595–610.

Pareek, A., Meena, B. M., Sharma, S., Tetarwal, M. L., Kalyan, R. K., & Meena, B. L. (2017). Impact of climate change on insect pests and their management strategies. In P. S. Kumar, M. Kanwat, P. D. Meena, V. Kumar, & R. A. Alone (Eds.), *Climate Change and Sustainable Agriculture* (pp. 388). New India Publishing Agency, New Delhi, India.

Park, R. F., & Wellings, C. R. (2012). Somatic hybridization in the Uredinales. *Annual Review of Phytopathology, 50*, 219–239.

Parker, I. M., & Gilbert, G. S. (2004). The evolutionary ecology of novel plant-pathogen interactions. *Annual Review of Ecology, Evolution, and Systematics, 35*, 675–700.

Paterson, R. R. M., & Lima, N. (2010). How will climate change affect mycotoxins in food? *Food Research International, 43*(7), 1902–1914.

Perincherry, L., Lalak-Kańczugowska, J., & Stępień, Ł. (2019). Fusarium-produced mycotoxins in plant-pathogen interactions. *Toxins, 11*(11), 664.

Petzoldt, C., & Seaman, A. (2006). Climate change effects on insects and pathogens. *Climate Change and Agriculture: Promoting Practical and Profitable Responses* (pp. 6–16). Retrieved from Pesticide Action Network North America

Raffaele, S., Win, J., Cano, L. M., & Kamoun, S. (2010). Analyses of genome architecture and gene expression reveal novel candidate virulence factors in the secretome of *Phytophthora infestans*. *BMC Genomics, 11*(1), 637.

Ramesh, K., Matloob, A., Aslam, F., Florentine, S. K., & Chauhan, B. S. (2017). Weeds in a changing climate: Vulnerabilities, consequences, and implications for future weed management. *Frontiers in Plant Science, 8*, 95.

Rasheed, A., Hao, Y., Xia, X., Khan, A., Xu, Y., Varshney, R. K., & He, Z. (2017). Crop breeding chips and genotyping platforms: Progress, challenges, and perspectives. *Molecular Plant, 10*(8), 1047–1064.

Roossinck, M. J. (2005). Symbiosis versus competition in plant virus evolution. *Nature Reviews Microbiology, 3*(12), 917–924.

Rouxel, T., Grandaubert, J., Hane, J. K., Hoede, C., van de Wouw, A. P., Couloux, A., & Howlett, B. J. (2011). Effector diversification within compartments of the *Leptosphaeria maculans* genome affected by repeat induced point mutations. *Nature Communications, 2*, 202.

Roy, B. A., Gusewell, S., & Harte, J. (2004). Response of plant pathogens and herbivores to a warming experiment. *Ecology, 85*(9), 2570–2581.

Ruiz, L., Simon, A., Garcia, C., Velasco, L., & Janssen, D. (2018). First natural crossover recombination between two distinct species of the family Closteroviridae leads to the emergence of a new disease. *PLoS One, 13*(5), e0198228.

Sacristán, S., & García-Arenal, F. (2008). The evolution of virulence and pathogenicity in plant pathogen populations. *Molecular Plant Pathology, 9*(3), 369–384.

Sanford, J. C., & Johnston, S. A. (1985). The concept of parasite-derived resistance—Deriving resistance genes from the parasite's own genome. *Journal of Theoretical Biology, 113*(2), 395–405.

Schrey, A. W., & Richards, C. L. (2012). Within-genotype epigenetic variation enables broad niche width in a flower-living yeast. *Molecular Ecology, 21*(11), 2559–2561.

Seem, R. C. (2004). Forecasting plant disease in a changing climate: A question of scale. *Canadian Journal of Plant Pathology, 26*(3), 274–283.

Seidl, M. F., & Thomma, B. P. H. J. (2017). Transposable elements direct the coevolution between plants and microbes. *Trends in Genetics, 33*(11), 842–851.

Seleiman, M. F., Al-Suhaibani, N., Ali, N., Akmal, M., Alotaibi, M., Refay, Y., Dindaroglu, T., Abdul-Wajid, H. H., & Battaglia, M. L. (2021). Drought stress impacts on plants and different approaches to alleviate its adverse effects. *Plants, 10*(2), 259.

Senanayake, I. C., Rathnayake, A. R., Marasinghe, D. S., Calabon, M. S., Gentekaki, E., Hurdeal, V. G., Pem, D., Dissanayake, L. S., Wijesinghe, S. N., & Bundhun, D. (2020). Morphological approaches in studying fungi: Collection, examination, isolation, sporulation and preservation. *Mycosphere, 11*(1), 2678–2754.

Siebold, M., & Von Tiedemann, A. (2012). Potential effects of global warming on oilseed rape pathogens in Northern Germany. *Fungal Ecology, 5*(1), 62–72.

Singh, R. P., Hodson, D. P., Huerta-Espino, J., Jin, Y., Bhavani, S., Njau, P., Herrera-Foessel, S., Singh, P. K., Singh, S., & Govindan, V. (2011). The emergence of Ug99 races of the stem rust fungus is a threat to world wheat production. *Annual Review of Phytopathology, 49*(1), 465–481.

Spiers, A. G., & Hopcroft, D. H. (1994). Comparative studies of the poplar rusts *Melampsora medusae*, *M. larici-populina* and their interspecific hybrid *M. medusae-populina*. *Mycological Research, 98*(8), 889–903.

Stahl, E. A., & Bishop, J. G. (2000). Plant–Pathogen arms races at the molecular level. *Current Opinion in Plant Biology, 3*(4), 299–304.

Taub, D. (2010). Effects of rising atmospheric concentrations of carbon dioxide on plants. *Nature Education Knowledge, 3*(10), 21–32.

Trębicki, P., Dáder, B., Vassiliadis, S., & Fereres, A. (2017). Insect–plant–pathogen interactions as shaped by future climate: Effects on biology, distribution, and implications for agriculture. *Insect Science, 24*(6), 975–989.

Tricoli, D. M., Carney, K. J., Russell, P. F., McMaster, J. R., Groff, D. W., Hadden, K. C., Himmel, P. T., Hubbard, J. P., Boeshore, M. L., & Quemada, H. D. (1995). Field evaluation of transgenic squash containing single or multiple virus coat protein gene constructs for resistance to cucumber mosaic virus, watermelon mosaic virus 2, and zucchini yellow mosaic virus. *Nature Biotechnology, 13*(13), 1458–1465.

Vardharajula, S., & Ali, S. Z. (2015). The production of exopolysaccharide by Pseudomonas putida GAP-P45 under various abiotic stress conditions and its role in soil aggregation. *Microbiology, 84*(4), 512–519.

Varsani, A., Lefeuvre, P., Roumagnac, P., & Martin, D. (2018). Notes on recombination and reassortment in multipartite/segmented viruses. *Current Opinion in Virology, 33*, 156–166.

Velásquez, A. C., Castroverde, C. D. M., & He, S. Y. (2018). Plant–Pathogen warfare under changing climate conditions. *Current Biology, 28*(10), R619–R634.

Verma, J. P., & Jayashree, J. (2002). *Fundamentals of Plant Bacteriology*. Kalyani Publishers, New Delhi.

Webb, K. M., Ona, I., Bai, J., Garrett, K. A., Mew, T., Cruz, C. M. V., & Leach, J. E. (2011). A benefit of high temperature: Increased effectiveness of a rice bacterial blight disease resistance gene. *New Phytologist, 185*(2), 568–576.

Wingender, J., Neu, T. R., & Flemming, H. C. (1999). What are bacterial extracellular polymeric substances? In J. Wingender, T. R. Neu, & H. C. Flemming (Eds.), *Microbial Extracellular Polymeric Substances: Characterization, Structure and Function* (pp. 1–19). Springer, Berlin.

Wu, W. Y., Lebbink, J. H. G., Kanaar, R., Geijsen, N., & van der Oost, J. (2018). Genome editing by natural and engineered CRISPR-associated nucleases. *Nature Chemical Biology, 14*(7), 642–651.

Wu, X., Beecher, G. R., Holden, J. M., Haytowitz, D. B., Gebhardt, S. E., & Prior, R. L. (2006). Concentrations of anthocyanins in common foods in the United States and estimation of normal consumption. *Journal of Agricultural and Food Chemistry, 54*(11), 4069–4075.

Wu, X., Lu, Y., Zhou, S., Chen, L., & Xu, B. (2016). Impact of climate change on human infectious diseases: Empirical evidence and human adaptation. *Environment International, 86*, 14–23.

Yang, L., Ren, M., & Zhan, J. (2022). Modeling plant diseases under climate change: Evolutionary perspectives. *Trends in Plant Science, 20*(1), 1–8.

Yoshida, K., Saunders, D. G. O., Mitsuoka, C., Natsume, S., Kosugi, S., Saitoh, H., Inoue, Y., Chuma, I., Tosa, Y., Cano, L. M., Kamoun, S., & Terauchi, R. (2016). Host specialization of the blast fungus *Magnaporthe oryzae* is associated with dynamic gain and loss of genes linked to transposable elements. *BMC Genomics, 17*(1), 370.

Zayan, S. A. (2019). Impact of climate change on plant diseases and IPM strategies. In S. Topolovec-Pintaric (Ed.), *Plant Diseases: Current Threats and Management Trends* (pp. 1–11). IntechOpen, London, U.K.

Zhu, Y. X., Ou-Yang, W. J., Zhang, Y. F., & Chen, Z. L. (1996). Transgenic sweet pepper plants from Agrobacterium-mediated transformation. *Plant Cell Reports, 16*(1–2), 71–75.

Zwart, M. P., & Elena, S. F. (2020). Modeling multipartite virus evolution: The genome formula facilitates rapid adaptation to heterogeneous environments. *Virus Evolution, 6*(1), veaa022.

CHAPTER 5

Recent Advances on the Impact of Climate Change on Host Plant Resistance and Its Molecular Mechanisms

R. M. NAGALAKSHMI,[1] R. SANGEETHA VISHNU PRABHA,[2] K. SHAMINI,[3] and R. PONMANI[4]

[1]Sethu Bhaskara Agricultural College and Research Foundation, Karaikudi, Tamil Nadu, India

[2]Vanavarayar Institute of Agriculture, Pollachi, Tamil Nadu, India

[3]Center for Plant Breeding and Genetics, Tamil Nadu Agricultural University, Coimbatore, Tamil Nadu, India

[4]Mother Teresa College of Agriculture, Pudukkottai, Tamil Nadu, India

ABSTRACT

Global climate change due to anthropogenic activities has an intense effect on all biological systems. The increased levels of carbon dioxide, ozone, methane, nitrous oxide, hot extremes, erratic rainfall, and continued drought have affected the physiological processes of both pests and hosts. Climate change has altered the survival rate, mating behavior, fertility rate, fecundity, and adaptive mechanisms of pests. Similarly, the host plants also show varied defense mechanisms against pests. The altered behavior of insects affects crop yield through increased or decreased survival mechanisms, pollination, and seed set in entomophily crops. Many pests and host plants have coevolved, leading to a balance between pest activity and host resistance mechanisms. However, climate change can disrupt this balance and may result in sudden pest outbreaks. Understanding the molecular mechanisms behind this is

Climate Change and Biotic Factors: A Molecular Approach.
Ashutosh Singh, Saurabh Pandey, and Amarjeet Kumar, (Eds.)

essential to overcome the adverse effects of climate change on yield due to biotic stress in food crops. Among all the environmental factors affected by climate change, increased temperature has an intense impact on host plant resistance due to its significant role in the production of phytochemicals in plants. Molecular and yield expression studies of genes involved will help combat the adverse effects of climate change on crop production. The yield of crops is largely influenced by pest and disease management practices which need to be improved with integrated mechanisms considering climate change and its impact on hosts and pests. The changing environment has increased the diversity, geographical range, new biotypes, altered survival rate and mechanisms, life span, number of generations, interspecific interactions, migration and invasion to non-native areas, the spread of insect-borne diseases, and population of natural enemies. Insects are more adaptable species and evolve rapidly than cultivated crops and pose serious trouble in food production. Climate change has proposed a serious threat to food insecurity in the near future. Proper monitoring of insect population, behavior, migration, and attack on crops precisely will help to predict and overcome the adverse effects of climate change on crop production.

5.1 INTRODUCTION

Climate change is one of the major threats to the wellness of mankind (Crowley T J. 2000). From the day of the formation of Earth, it has undergone continuous changes in its nature and composition. It was on fire when thrown out from the sun and its composition and nature varied over millions of years and reached a form that can support life. Even after the origin of life on Earth, several catastrophic events resulted in mass extinctions of living organisms and the Earth rejuvenated every time. Several living organisms went extinct when they couldn't manage the changing environment. The earlier changes in the environment were not influenced by any of the living organisms but now, the most intelligent living organism on Earth, humans, are responsible for climate change. Extensive use of fossil fuels, industrial and livelihood waste disposal into natural bodies, use of hazardous chemicals in agriculture, exploitation of natural resources, and decreasing forest cover are the major reasons behind climate change. The immediate effects of climate change are an increase in temperature, an increase in carbon dioxide and ozone levels in the atmosphere, erratic rainfall, loss of vegetation, sudden outbreaks of pests and diseases, loss of diversity in flora and fauna, and a higher rate of extinction of living organisms, frequent drought, unstable ecology, and an increase in biotic and abiotic stresses to agricultural crops (Chandra et al., 2022). Though these changes are a part of evolution, the rate of these changes is accelerated several times by human activities, particularly post-industrial revolution.

Climate change directly affects the production and productivity of crops through abiotic stress and indirectly through increased incidence of pests and diseases (Skendzic et al., 2021). Climate change influences the life cycle of insect pests and pathogens in several ways. The coevolution process between agricultural crops and pests and pathogens is an ongoing phenomenon. Climate change interferes with this coevolution and is mostly detrimental to the crops, resulting in the fall of agricultural production. During the early stages of agricultural research, the main objective of the research is yield, which then shifted to biotic and abiotic stress resistance and now specifically to the stresses created by climate change. Interdisciplinary approaches are needed to combat and overcome the effects of changing climate on agriculture and humankind. Basic research on climate change is essential to finding solutions to various hindrances in agricultural production due to adverse climatic factors. Changes in climate need to be studied along with the changes in insect pest behavior. The risk assessment due to climate change on agriculture is a very big challenge. The expected impact of climate change on crop yield is predicted by various modeling mechanisms, whereas the after-effects of climate change on insect and disease incidence are less studied and need to be explored.

5.2 ROLE OF GLOBAL WARMING ON INSECT PEST AND DISEASE INCIDENCE

Temperature is one of the most important environmental factors determining the distribution, survival, and reproduction of any living organism. Different stages of plant growth require different temperature regimes. Similarly, insects, pests, and pathogens also require specific temperatures in their developmental stages (Abarca and Spahn, 2021). Extreme temperatures favor only a few organisms. Long sunny days and optimum temperature are widely preferred. Higher temperatures affect crop growth both directly and indirectly. Temperature affects crop growth indirectly through additional moisture requirements for the plants, poor water availability due to increased evaporation leading to soil moisture loss, increased transpirational water loss, and all related defects in crop physiology. The rise in temperature rapidly increases the metabolic rate of insects. A 10°C increase in temperature doubles the metabolic rate of most insects. Accelerated temperature affects the insect's food consumption, growth, development, mobility, metamorphosis, mating behavior, fecundity, life cycle, population size, population dynamics, survival rate, and geographic range (Abarca and Spahn, 2021).

Global warming in general will increase the geographical range, which may turn some minor pests into major pests (Skendzic et al., 2021). The increase in temperature increases herbivory. The heightened temperature doesn't have a uniform impact on all insects and environments. Tropical insects are expected to experience a decline in survival, while temperate insects will show increased survival rates. Many insects may even become extinct within 100 years. Based on the geographical distribution of crops and insects, a mixed response in the population dynamics of insects is expected. The global temperature is projected to increase by 1.8–4.0°C by the end of the century. Although there may be some hindrances in immediate survival rates, the overall temperature rise creates an optimal environment for insects. A slight temperature increase will boost growth and reproduction, leading to an increase in population sizes of various species and overall dynamic equilibrium. An increased reproduction rate and shorter life cycle result in a higher number of generations per year. More generations throughout the year make plants more susceptible to insect damage during their vegetative and reproductive phases.

An increase in the growing degree days (GDD) affects the phenology of plants and insects. Increased GDD significantly changes the phenology of univoltine and multivoltine insects. Both univoltine and multivoltine insects experience shorter larval and pupal stages and longer adult phases (Skendzic et al., 2021). The longer adult phase increases the extent of feeding, flight duration, and reproduction rates. Multivoltine insects specifically get more generations due to additional flight duration. A 2°C increase in temperature may bring one to five additional generations. The early onset of the adult flight period is the main reason behind the increased number of generations. Though the insects are multivoltine, their lifecycle needs a specific temperature. Global warming lengthens the duration of ample temperature for insect growth and thus favors more generations within a specific time period. The aphid population is a well-known example of an increased number of generations due to increased temperature. Some insects strive hard to survive in the higher temperature, but most of the insects have increased growth and reproduction. Some pests follow the "stop and go" method of development, i.e., they will accelerate their growth during favorable environments and halt during unfavorable conditions. These pests also have more generations. The shorter the life cycle, the larger the population size. Spring flights of the potato aphid started two weeks earlier for every 1°C increase in temperature during the months of January and February.

Pests of the temperate world have a major period of their life cycle during warmer days. An increase in temperature provides more days with optimum temperature and increases the population size. Also, the winter mortality of insects is reduced due to higher winter temperatures. The duration of winter inactivity is also reduced. Insects follow a diapause strategy to overcome adverse environmental conditions. Aestivation and hibernation are the two diapause mechanisms of insects. Aestivation is adapted to higher temperatures and hibernation to lower temperatures. Higher temperatures shorten the hibernation period, making the active phase of the insect life cycle longer (Abarca and Spahn, 2021). Reduction in frost due to higher average temperature has reduced the limiting factors of survival in *Fusarium circinatum* (Quesada et al., 2019). Reduced frost decreases frost damage in plants, which is the site of entry for many pathogens. In such cases, disease incidence is reduced. The gender ratio also varies in some pests like thrips. The interaction of pests with their predators and parasitoids also varies with climate. If the pest completes its life cycle before the onset of feeding by the predators, the predators will have no role in pest control. The diversity in the insect population is a great indicator of climate change. Apart from the damage to agricultural crops, insects play a key role in the food chain. Insects dwelling in soil are less affected than insects having an entire life cycle on plants. Soil acts as an insulator maintaining a near-optimum temperature. Insect population decreases with an increase in altitude and latitude. As climate change increases temperature, colder regions of higher altitudes and latitudes get warmer and support new insect populations, expanding their geographical range. It is estimated that a 1°C increase in temperature will expand the geographical range up to 200 km towards the north or south and 40 m in altitude (Saroj S, 2019).

Potato aphid, European grape vine moth, and many lepidopteran pests have shown a clear increase in the number of generations due to the hike in temperature (Du Plessis et al., 2020). Assessment of the effect of increased temperature is difficult due to the wide variation in plant and insect phenology, ecosystems, and other climate-changing factors. Global trading has made alien invasion a serious threat to agricultural production. Even after strict quarantine screening, new pests continue to emerge. The damage due to the increased activity of local pests is less compared to the outbreak of an alien species. *Tuta absoluta* was introduced in Nepal in 2016 and is now the major pest of tomatoes in the country (Bajracharya et al., 2016).

Insects play a pioneering role in the spread of viral and phytoplasma diseases. As viruses need a live vector for their spread and establishment, the population size and phenology of insect vectors play a main role in the spread of viral diseases (Pan et al., 2021). Viruses and vectors may be generalists or specialists. An increase in temperature has a clear impact on the phenology and population of insects, and hence the spread and establishment of viruses are also impacted by the rise in temperature. The transmission efficiency of vectors varies. Hence, the occurrence of a disease is determined by the nature of viruses, the efficiency of vectors, host susceptibility, and the prevalent environment (Pandey et al., 2018). Therefore, a changing environment alters the epidemiology of diseases. The efficiency of transmission may be increased, decreased, or remain neutral due to an increase in temperature. It depends on various characteristics of viruses, vectors, host plants, and the altered temperature. Most of the insect vectors belong to the order Hemiptera. Aphids and whiteflies belonging to this order spread many disastrous diseases. Aphids are widely found in temperate regions and whiteflies are in the tropic zone. As mentioned earlier, aphids had a significant increase in the number of generations and reproduction rate with an increase in temperature, hence the increased transmission of viruses. Aphids are known for their long-distance travels; an increase in the activity of aphids results in severe disease epidemics. The disease incidence is determined by the time of infection and the viral inoculum transmitted. Aphids have increased inoculum during milder winters and hotter summers. The rise in temperature and duration of exposure determine the efficiency of vectors in transmitting the diseases. The effect of increased temperature on the bird cherry-oat aphid, greenhouse whitefly, and American grapevine leafhopper, and the corresponding disease spread has obviously proved the role of temperature fluctuations in the spread of viral diseases. In tropical regions, tomato yellow leaf curl virus and chlorosis, cucurbit virus diseases, cassava mosaic, and torrado-like disease of tomato are well-known examples of increased disease incidence due to increased temperature (Pandey et al., 2019). The role of temperature is more than the available vegetation in determining the geographical range of pests. An increase in temperature will increase the geographical range of insects from tropical to subtropical and temperate regions. Loss of agricultural production due to insect pests is higher in tropical regions than subtropical and temperate regions. Continuous increases in temperature will induce frequent pest outbreaks in temperate regions as well (Eastburn et al., 2011). Cases where an increase in temperature resulted in an increase in resistance were also reported.

Apart from the insect vectors, the pathogens themselves need a specific environment for their entry, survival, development, reproduction, and spread (Pandey et al., 2020). Thus, elevated temperature will have a direct influence on disease severity. Many host plants that are resistant to rust have become susceptible to an increase in temperature. Also, many inactive races of pathogens have become virulent and resulted in disease epidemics. Additionally, temperature stress and moisture stress weaken the plant and make them susceptible to diseases. The developmental stages, pathogenicity, and reproduction rate are altered by temperature instability, specifically fungal spores proliferate more under high temperature and moisture (Eastburn et al., 2011). Increased temperature will favor the survival and spread of pathogens up to the upper threshold level. Warm winters and high night temperatures favor the pathogens, increasing their sporulation, aerial infection, aggressiveness, and their survival between seasons. *Puccinia striiformis* has shown increased aggressiveness during higher temperatures (War et al., 2016). High temperature with sufficient moisture increases evapotranspiration, which in turn makes the microclimate more humid. A humid microclimate is favored by many fungal pathogens. Global warming increases the soil temperature, and when it increases beyond the buffering capacity of the soil, the diversity of microbes, soil-dwelling insect pests, and plants will be significantly altered. A higher incidence of *Fusarium* wilt in chickpea is observed when the temperature is above 33°C. Few resistant genes-specific diseases have become susceptible to increased temperature. High rates of mutations due to shortened life cycle make the pathogens more adaptable and alter the volatile components of plants, making them more attractive to insect vectors (Sood et al., 2021).

The overall effect expected in global warming is an increased occurrence and outbreak of pests due to the expansion of geographical range, shortened life cycle, the spread of insect-borne diseases, and increased chances of invasive pests, affecting agricultural production.

5.3 IMPACT OF ELEVATED CO_2 LEVELS IN THE ATMOSPHERE ON INSECT PEST AND DISEASE INCIDENCES

CO_2 is a highly accelerated environmental gas due to the industrial revolution. Increased CO_2 levels affect plants, insects, and pathogens both directly and indirectly (Tonnang et al., 2022). CO_2 concentration in the atmosphere now is above 400 µmol/mol, whereas before the industrial revolution it was around 280 µmol/mol. Increased CO_2 levels lie behind many of the global environmental changes, including a rise in temperature, erratic rainfall, and altered weather patterns. CO_2 levels have a direct and significant role in the biochemistry, physiology, and biological mechanisms of plants, insects, and pathogens. Increased CO_2 levels change the nutritional status of the host plant, which in turn affects the insect pests feeding on the plants. Changes in the stomatal closure mechanisms and increased photosynthesis are favorable to the plants, resulting in increased biomass production and yield. Increased CO_2 levels decrease the seed protein content, accumulate more non-structural carbohydrates, and alter the C:N ratio, which results in increased feeding behavior of insects to meet their nutritional requirements (Tonnang et al., 2022). Insect pests and pathogens rely on the water and nutritional content of plants for their survival and reproduction. Reduced stomatal conductance leads to reduced micronutrient composition due to variations in water uptake.

Increased CO_2 levels alter the disease triangle in several ways (Juroszek et al., 2020). It is difficult to give a generalized prediction on the interaction of host plants with pathogens and the environment. The effect varies with the crop, insect, pathogen, and environment. The direct effect of enhanced CO_2 levels on insects and pathogens is less compared to its effects on plant chemistry, but the chemical composition of the diet taken by the insects plays a significant role in their feeding and behavior. The indirect effects of higher CO_2 concentrations overshadow the direct effects. Plants grown in artificially higher CO_2 conditions can be fed to the insects, and their behavioral changes can be studied. *Rhopalosiphum maidis,* the corn aphid, reported a very high population increase under conditions of high temperature and CO_2 concentrations (Csorba et al., 2022). CO_2 is the substrate for photosynthesis in plants. Increased CO_2 levels, along with other climate change factors, affect host and pest interaction. The aphid population had a significant increase under high temperatures and CO_2 levels. The feeding behavior of aphids was studied using an electrical penetration graph (EPG). EPG observations showed reduced salivation, higher phloem sap intake, and a shorter non-pathway phase in *Aphis pisum* (Jhou et al., 2021). Late instar larvae of *Operophtera brumata* fed voraciously on the leaves of *Betula pendula* grown under higher CO_2 levels due to inadequate nutrition in the plants. To meet their nutrient requirements, insects consume more vegetation with poor nutrition. Increased CO_2 levels decrease the nitrogen accumulation in leaves. Reduced nitrogen increases the quantity of consumption by up to 40%.

Increased biomass of host plants due to increased CO_2 levels keeps the pathogen reservoir viable, available, and attractive for longer periods. The onset of the viral symptoms starts earlier, making the hosts attractive to insect vectors. Increased photosynthesis results in increased carbon accumulation in plants. The excess carbon is allotted to pectin, cellulose, phenolics, tannins, and terpenoids. These biochemical changes in host plants affect the physiology and behavior of the insect pests feeding on them. When pests attack the plants, defensive secondary metabolites are produced. These secondary metabolites are signaled by phytohormones like jasmonic acid, salicylic acid, and ethylene (El Sabagh et al., 2022). Increased CO_2 levels affect the synthesis and transport of these phytohormones, which in turn affects the resistance mechanisms. Down-regulation of the jasmonic acid pathway is observed in soybean under higher CO_2 levels, which increases the susceptibility to Japanese beetle and western corn root worms. Japanese beetles, when fed with soybean leaves, showed increased fecundity possibly due to the increased sugar identified in the leaves. In some cases, increased CO_2 levels increase the resistance to certain pests. Chickpea plants, when grown under 750 μmol/mol concentration of CO_2, had higher amounts of phenols, tannins, peroxidase, polyphenol oxidase, phenylalanine, and tyrosine ammonia lyase. The chemicals had toxic and anti-feeding properties on *Helicoverpa armigera* (Mallikarjuna et al., 2020). Higher CO_2 levels and soil moisture increase the production of DIMBOA. Aphids prefer plants with lower DIMBOA concentrations in the leaves. Hence, plants with higher DIMBOA levels are resistant to barley yellow virus, which is transmitted by aphids. The viruses present in the plants change the amino acid composition of plants and trigger them to produce volatile compounds, making them attractive to aphids. Increased CO_2 levels change the level of oxalic and folic acid in chickpea plants, which results in altered pest resistance. The biochemical changes in the plants may be due to the insect feeding or the pathogen. It is difficult to distinguish the effects of both in plants. When tomato plants infected with tomato yellow leaf curl virus and tobacco mosaic virus are grown under higher CO_2 concentration (700 μmol/mol), the production of salicylic acid increased several folds and reduced the disease severity. The effect of CO_2 concentration on susceptibility and resistance to diseases is less predictable. Higher CO_2 levels reduced the severity of *Pseudomonas syringae* and increased the severity of *Botrytis cinerea* in *Arabidopsis* and tomato (Zhou et al., 2017). The predators and parasitoids of the pest become less efficient in attacking the pests as they are new to the modified insect phenology. Under certain conditions, the elevated CO_2 levels increase the efficiency of predators. The development time of the parasitoids *Cotesia plutella* of aphids and *Lysiphlebia japonica* of lepidopteran pests decreased, making them more effective in controlling these pests (War et al., 2016). Higher CO_2 levels make generalist predators more efficient and specialist predators less effective. In the parasitoid-host relationship, the time of emergence of the parasitoid is very important for effective infection. Parasitoids should emerge during the vulnerable phase of the host life cycle. Very early or very late emergence of parasitoids creates poor synchrony between the parasitoid and host, making the parasitoid incapable of attacking the host insect. Higher CO_2 levels delay the decomposition rates of plant debris, holding the pathogen for a longer time and helping in the overwintering of pathogens. Overwintering of pathogens results in higher inoculum levels in the early stages of the crop, thus promoting its survival and spread (Zhou et al., 2019). Higher CO_2 levels increase the spore production in fungal pathogens. In some crops, increased CO_2 concentration increases host plant resistance to pathogens. High carbohydrate concentration in plants promotes the survival of fungal pathogens like rust. Increased biomass, plant density, leaf wetness duration, light and water use efficiency, morphology, root exudates, and humid microclimate favor the infection and survival of pathogens. The consequences of increased CO_2 levels on plant pathogens cannot be generalized. The germination rate and germ tube growth of *Colletotrichum gloeosporioides* are delayed under higher CO_2 levels (Neshar et al., 2011). But once the fungus infects the plants, the development and sporulation are faster than at normal CO_2 levels. Controlled growth chambers, CO_2-filled tunnels, chambers with open tops, and air carbon enrichment systems are used to study the effect of CO_2 concentrations on plants, insects, and pathogens.

Fungal profile variations, microbial density, root exudates, and disease index vary under higher CO_2 levels. *Fusarium oxysporium* grown under different concentrations of CO_2 levels in phytotrons showed significant variation in the above-mentioned characters. Increased temperature and CO_2 levels will significantly affect the root architecture and the quantity and quality of root exudates in the rhizosphere in rocket plants affected by *Fusarium oxysporium*. The chemical composition of the root exudates determines the biological interactions of the plants with soil-borne pathogens, specifically the root-feeding fungal agents.

5.4 IMPACT OF ELEVATED O_3 LEVELS IN THE ATMOSPHERE ON INSECT PEST AND DISEASE INCIDENCES

The intense use of fossil fuels increases the O_3 concentration in the troposphere. Every year, the O_3 concentration is increasing by 0.5–2.5%, and it is expected that O_3 levels in the troposphere will reach 60 nmol/mol by 2050. O_3 is a highly phytotoxic compound. O_3 enters the plants through gaseous exchange and promotes the synthesis of reactive oxygen species (ROS). ROS affects the cellular components and mechanisms, biochemical processes, and physiological activities of the plants. Exposure of plants to O_3, and the resulting ROS accumulation disturbs major physiological processes and affects the expression of many genes, particularly genes involved in the synthesis of salicylic acid and jasmonic acid (Kumar et al., 2023). When the plants are suddenly exposed to higher O_3 levels, they show symptoms similar to a plant's response to pest and disease incidence. Mostly, hypersensitive cell death occurs during sudden exposure to higher O_3 levels. The same genes that are activated during pest and disease attacks are activated during O_3 exposure as well. The interaction between the plant growth regulators, jasmonic acid, salicylic acid, ethylene, and abscisic acid is also similar during biotic stresses and O_3 exposure (El Sabagh et al., 2021). During pathogenic infections, the linolenic acid content of plants increases, which in turn increases the synthesis of jasmonic acid. During O_3 exposure, the lipid profile of plants is disturbed, and linolenic acid level increases, and synthesis of jasmonic acid is initiated. O_3-induced ethylene production triggers the biosynthesis of ABA, which is actively involved in stomatal movement. O_3-triggered changes in plants seriously interfere with plant defense mechanisms against pathogens and make the host more resistant to the pathogens. During an infection, the pathogens break the defense mechanisms of the host and cause infection. During the process of evolution, the plants develop new mechanisms of defense and the pathogens develop new ways to break the host defense mechanisms. The changes in the genetic makeup of plants and pathogens develop through an antagonistic evolutionary process. O_3-induced changes in the host plant mechanisms make the pathogens less efficient in initiating the infection process.

When soybean plants grown under higher O_3 levels were artificially infected with soybean mosaic virus (SMV), the multiplication rate and spread of the disease are much reduced (Belgin et al., 2008). However, this induced resistance is lost during the later stages of crop growth, and an infection pattern was observed. When the plants are subjected to O_3 treatment, ROS like H_2O_2 and O_2^-. The ROS triggers the genes responsible for the defense against pathogens and initiates the pathways of JA, SA, ET, and ABA (El Sabagh et al., 2021). When the O_3-treated plants were infected with SMV later, the plants already have the biochemicals needed for fighting against the virus and hence the establishment and spread of the virus is greatly affected.

The time of onset of ROS synthesis is the important factor in deciding the resistance of a host against a pathogen.

As O_3 treatment propounds the synthesis of ROS chemicals, the host plants show increased resistance to the virus. Similar results were obtained when O_3-treated tobacco plants were treated with tobacco mosaic virus, *Arabidopsis* infected with *Pseudomonas syringae*. O_3 exposure to plants affects the phenylpropanoid biosynthetic pathway which induces the formation of flavonoids and other antioxidants. These antioxidants deactivate the ROS chemicals and suppress the resistant mechanisms. As O_3 treatment affects the phenylpropanoid pathway, the resistance of the host towards the pathogen increases. Increased O_3 levels trigger the genes corresponding to the synthesis of flavonoids. These flavonoids in turn increase the concentration of phytoanticipins and phytoalexins. These chemicals exhibit strong antimicrobial properties and hinder the infection of a broad spectrum of pathogens including SMV. Although O_3 induces resistance against SMV, the overall production of the crop is reduced. O_3 at very high concentration induces hypersensitive response and results in cell death. Both pathogen entry and O_3 exposure induce ROS production, but ROS induced by pathogens acts in a controlled manner whereas ROS due to O_3 reacts without any pattern.

Comparing the effects of CO_2 and O_3 on host plant resistance, CO_2 has a more prominent role than O_3. When the transcriptomes of soybean plants grown under elevated CO_2 and O_3 levels and normal environment, later infected with Japanese beetles were studied, the number of transcripts was found to be higher in plants grown in elevated CO_2 and O_3 environment (Casteel et al., 2008).

5.5 CASE STUDIES

Various case studies were made on the effects of climate change on insect pest and disease incidence and a few are discussed here.

5.5.1 THE DESERT LOCUST

Concern over the post-corona situation has increased due to the Desert Locust (*Schistocerca gregaria*) swarm's eastward migration in the Indian states of Rajasthan, Uttar Pradesh, and Madhya Pradesh. According to FAO Locust Watch, storm Amphan, which originated in the Bay of Bengal, is believed to be the cause of the strong westerly winds that are driving the movement.

The heavy rains (20–59%) in all central and northern Indian states may also foster the right environment for the locust swarm to flourish near the second week of April, reports of the first swarm encounter were made near Jaisalmer, Rajasthan. Six Indian states—Rajasthan, Gujarat, Uttar Pradesh, Madhya Pradesh, Maharashtra, and Tamilnadu have been impacted by the swarms since May 2020 (Press Bureau of India, 2020; Kumar et al., 2020).

An established function for wind, temperature, and vegetation in the desert locust's movement has been suggested by earlier investigations on its behavior. After arriving in a region with recent widespread heavy rains and temperatures between 20 and 35°C, the migration came to a complete stop. The female digs into the ground and detects sufficient soil moisture to lay hundreds of eggs in egg-pods. After depositing, the swarm picks up its migration once more and continues laying eggs hundreds of kilometers from the initial site in the second and third waves. The eggs may go through latent stages and remain for three months depending on the temperature and moisture of the soil. The incubation duration was shortened by the higher temperature and moisture content.

The reverse migration of these locusts during the monsoonal wind reversal in the afflicted states may result in a second and third wave of invasion. The accelerating long-distance migration caused by the changing environment can be traced back as the motivation behind the invasion.

5.5.2 THE FALL ARMYWORM

Spodoptera frugiperda, often known as the fall armyworm (FAW), is a destructive insect pest that is a member of the Noctuidae family in the Lepidoptera order. It is a polyphagous pest that harms economically significant farmed cereal crops like maize, rice, sorghum, cotton, and different vegetable crops, which has an effect on food security in the long run. The FAW feeds on the plant's leaves, stems, and reproductive organs. It is indigenous to the Americas' tropical and subtropical climates. One of the frequent pests of maize in South and North America is FAW, which was initially discovered in America. It was initially discovered in Africa in 2016 and has since spread to over 30 nations in tropical and southern Africa, including Madagascar and the Seychelles.

The first signs appear when the larval stage makes various-sized papery windows in the leaves, which causes severe defoliation of plants, the presence of fecal contents, and later-stage effects on plant growth and development.

This insect moves in a manner akin to an army on the march, wreaking havoc and destroying crops in its route. The FAW is extremely destructive, and CABI estimates that if control measures are not taken, the pest might result in a loss of 6.1 billion US dollars in just African nations. The fall armyworm's larval stage is the most harmful to crops and the most destructive in the natural world. Depending on the stage of the plant's growth, the FAW larvae can be seen on the young leaves, leaf whorls, tassels, and cob of infected maize plants. Numerous factors must be taken into account in order to calculate the loss attributable to FAW. The number of pests, the timing of infection, the natural enemies and diseases of the pest that are present at that time, and the health (nutritional and moisture) state of the plant all have a role in crop infestation caused by pests.

Baudron (Baudron et al., 2019) reported that when there is 26.4% to 55.9% of pest incidence in maize then there is a yield reduction of 11.57%. Chimweta (Chimweta et al., 2019) revealed that the 25–50% damage of leaf, silk, and tassel results in 58% of yield reduction while 55–100% of severity at the period of mid to late whorl stage caused up to 73% of yield loss. Kiprop reported a 12.5 to 30% loss in the country's economy in Malawi. During the reporting period, there were yield losses of 3.2 million tons in Tanzania, 13.91 million tons in Uganda, and 30.54 million tons in Ethiopia. FAW in Kenya affected 250,000 ha of agricultural land or 11% of the nation's total area planted to maize. Similar to this, FAW calculated that maize production losses in Ghana and Zambia were 45% and 40%, respectively. In twelve African nations, including Ghana and Zambia, losses from FAW were projected to be between 8.5 and 21 million tons or between $250 and 630 million USD, in the absence of control measures. According to estimates, FAW in India spread to 10 states and adversely affected 170,000 acres of maize crops. This insect has been reported to affect 80,000 hectares of land in Yunnan province, China, where it has been proven to harm crops like maize, sorghum, sugarcane, and ginger. In China, a total of 11,192.17 acres have been impacted, and maize covers 98.6% of the affected area. The same was true for more than 10,000 acres of maize in four regions of Cambodia. In Bangladesh, FAW was found in 8 regions, 22 districts, and 71 administrative regions, with an infestation rate ranging from 0.5 to 32%.

Similar reports of this worm infestation include about 10,000 ha of land in Indonesia, 16,200 ha of land in Myanmar, and 46,000 ha of land in Vietnam. According to FAW, the yield loss in Thailand will be between 25–40%, costing between 130 and 260 million dollars. According to Beshir et al., (2019), a deadly bug with a ravenous appetite for crops like maize and others might have a significant negative impact on the farmers and economy of Nepal. The potential crop loss in maize is expected to reach 100% if this pest is not handled correctly since the climatic conditions in Nepal are favorable for the formation of this pest's population. After the eggs hatch, the FAW infection in maize begins to show symptoms. The characteristic FAW symptoms include papery windows with ragged edges and an oblong to round look on the leaves, which cause the leaves to become loose and separate from the plants. Due to the voracious feeding habits of the larval instars, considerable defoliation and an abundance of fecal waste left on the plant can be seen at the severe stage. Crops eventually stop growing and developing, which prevents the formation of cobs and tassels. First and second-instar infestations have windowpane-shaped translucent patches, while third to sixth-instar infestations show larger, elongated holes. The FAW's feces eventually appear as sawdust-like debris on the leaves or in the maize funnel.

The likelihood of this insect spreading around the world and severely reducing agricultural productivity is high. An integrated management approach is needed to control the fall armyworm, and it's critical to identify the controlling mechanism and conduct field inspections at the first signs of a pest infestation. The incidence of pests can be decreased by implementing an awareness program through advisory services regarding pest identification, damage symptoms, and management methods using the proper procedures. Smallholder farmers all around the world may rely heavily on locally accessible resources and techniques; hence, it is vital to make them aware of the fall armyworm control procedures. The precautions can support global stability by reducing the occurrence and damage caused by insect invasions. It might be suggested to take joint activity that is essential for managing the fall armyworm.

5.5.3 THE EMERALD ASH-BORER

Agrilus planipennis belongs to the order Coleopteran and the family Buprestidae. Natural resource managers are increasingly concerned with the detection, prevention, and long-term management of invasive species (Aukema et al., 2010). This is especially true for newly introduced phloem-feeding insects, which can quickly kill trees. The emerald ash borer (EAB) was first discovered in North America in 2002 near Detroit, Michigan, and Windsor, Ontario, Canada (Cappaert et al., 2005). This metallic wood-boring beetle primarily spreads through human activity in northeastern North America and feeds on the phloem tissue of ash trees (Fraxinus Linnaeus species, Oleaceae). One of the biggest challenges in successfully managing EAB is early detection of infestations. There are three main reasons why determining low EAB densities can be difficult: (1) Ash species are widespread in many eastern North American ecosystems, but they have recently experienced a decline at the landscape level (Sinclair and Griffiths, 1994), making it hard to pinpoint the causes of symptoms. Infested trees at low population levels do not show external signs of EAB colonization until they have been infected for several years. The delimitation techniques used in this outlier infestation were considered successful. Winter is the optimal time to survey severely affected trees. These trees can be located, cut down, and debarked. Girdling one or more nearby trees to serve as a trap and attract adults emerging from those trees in the following flying season would be an additional precaution. The trap trees can then be cut down and destroyed the next winter. This work provides an approach for delimitation with real-time evaluation of EAB presence, along with a case study on the use of SLAM techniques in an outlier infestation.

KEYWORDS

- **climate change**
- **host-plant resistance**
- **molecular mechanism**
- **pest attack**
- **plant defense**

REFERENCES

Abarca, M., & Spahn, R. (2021). Direct and indirect effects of altered temperature regimes and phenological mismatches on insect populations. *Current Opinion in Insect Science, 47*, 67–74.

Aukema, J. E., McCullough, D. G., Holle, B. V., Liebhold, A. M., Britton, K., & Frankel, S. J. (2010). Historical accumulation of nonindigenous forest pests in the continental United States. *Bioscience, 60*(11), 886–897.

Bajracharya, A. S. R., Mainali, R. P., Bhat, B., Bista, S., Shashank, P. R., & Meshram, N. M. (2016). The first record of South American tomato leaf miner, Tuta absoluta (Meyrick 1917) (Lepidoptera: Gelechiidae) in Nepal. *Journal of Entomology and Zoology Studies, 4*(4), 1359–1363.

Baudron, F., Zaman-Allah, M. A., Chaipa, I., Chari, N., & Chinwada, P. (2019). Understanding the factors influencing fall armyworm (*Spodoptera frugiperda* J. E. Smith) damage in African smallholder maize fields and quantifying its impact on yield: A case study in Eastern Zimbabwe. *Crop Protection, 120*, 141–150.

Belgin, D. D., Aldea, M., Neil, B. F. O., Benitez, M., Li, M., Clough, S. J., & DeLucia, E. H. (2008). Elevated ozone alters soybean-virus interaction. *Molecular Plant Microbe Interactions, 21*(10), 1297–1308.

Beshir, A. R., Shrestha, H. K., & Pradhan, B. (2019). Destructive pest alert: Fall armyworm (FAW) is paving its way to Nepal. Retrieved from: maize.org.

Cappaert, D., McCullough, D., & Poland, T. (2005). The upside of the emerald ash borer catastrophe: A feast for woodpeckers. In Mastro and Reardon (Eds.), *Proceedings of the Emerald Ash Borer Research and Technology Development Meeting* (pp. 69–70).

Casteel, C. L., Neil, B. F. O., Zavala, J. A., Belgin, D. D., Berenbaum, M. R., & DeLucia, E. H. (2008). Transcriptional profiling reveals elevated CO_2 and elevated O_3 alter resistance of soybean to Japanese beetle. *Plant, Cell and Environment, 31*(4), 419–434.

Chandra, A. K., Joshi, A., Tripathi, A., Kumar, A., Pandey, S., Singh, A., Lal, D., Bharati, A., Adhikari, S., & Dinkar, V. (2022). Climate-resilience maize: Heat stress, signaling, and molecular interventions. *Journal of Plant Growth Regulation, 42*(4), 1–18.

Chimweta, M., Nyakudya, I. W., Jimu, L., & Mashingaidze, A. B. (2019). Fall armyworm [Spodoptera frugiperda (J.E. Smith)] damage in maize: Management options for flood recession cropping smallholder farmers. *International Journal of Pest Management, 66*(2), 142–154.

Crowley, T. J. (2000). Causes of climate change over the past 1000 years. *Science, 289*(5477), 270–277.

Csorba, A. B., Fora, C. G., Bálint, J., Felföldi, T., Szabó, A., Máthé, I., Loxdale, H. D., Kentelky, E., Nyaradi, I. I., & Balog, A. (2022). Endosymbiotic bacterial diversity of corn leaf aphid, *Rhopalosiphum maidis* Fitch (Hemiptera: Aphididae) associated with maize management systems. *Microorganisms, 10*(5), 939.

Du Plessis, H., Schlemmer, M. L., & Van den Berg, J. (2020). The effect of temperature on the development of *Spodoptera frugiperda* (Lepidoptera: Noctuidae). *Insects, 11*(4), 228.

Eastburn, D. M., McElrone, A. J., & Belgin, D. D. (2011). Influence of atmospheric and climate change on plant-pathogen interactions. *Plant Pathology, 60*(1), 54–69.

El Sabagh, A., Islam, M. S., Hossain, A., Iqbal, M. A., Mubeen, M., Waleed, M., Reginato, M., Battaglia, M., Ahmed, S., Rehman, A., & Arif, M. (2022). Phytohormones as growth regulators during abiotic stress tolerance in plants. *Frontiers in Agronomy, 4*, 4.

Jhou, Y.-S., Poovendhan, S., Huang, L.-H., & Tsai, C.-W. (2021). Host acceptance and plant resistance: A comparative behavioral study of *Myzus persicae* and *Acyrthosiphon pisum*. *Insects, 12*(11), 975. https://doi.org/10.3390/insects12110975.

Juroszek, P., Racca, P., Link, S., Farhumand, J., & Kleinhenz, B. (2020). Overview on the review articles published during the past 30 years relating to the potential climate change effects on plant pathogens and crop disease risks. *Plant Pathology, 69*(2), 179–193.

Kumar, M., Joshi, J., Prithivi, R. V., Chandresh, S. B., & Birari, V. V. (2020). Desert locust (*Schistocerca gregaria* F.) outbreak in Gujarat. *Agricultural and Food E-Newsletter* (Article I.D. 30260).

Kumar, P., Singh, B., Rajak, S., Pandey, S., & Pati, P. K. (2023). Dynamics of reactive oxygen species and lignin biosynthesis during leaf spot disease of *Withania somnifera* (L.) Dunal. *Plant Biology, 25*(5), 757–770. https://doi.org/10.1111/plb.13548.

Mallikarjuna, S., Rachappa, V., Sreenivas, A. G., Teggelli, R. G., Hanchinal, S. G., & Rathod, P. S. (2020). Climate change variables effect on variation in damage and development of *Helicoverpa armigera* (Hubner) (Lepidoptera: Noctuidae) and its host plant chickpea. *Journal of Experimental Zoology India, 23*(1), 381–386.

Nesher, I., Minz, A., Kokkelink, L., Tudzynski, P., & Sharon, A. (2011). Regulation of pathogenic spore germination by CgRac1 in the fungal plant pathogen *Colletotrichum gloeosporioides*. *Eukaryotic Cell, 10*(8), 1122–1130. https://doi.org/10.1128/EC.00321-10.

Pan, L. L., Miao, H., Wang, Q., Walling, L. L., & Liu, S. S. (2021). Virus-induced phytohormone dynamics and their effects on plant–insect interactions. *New Phytologist, 230*(4), 1305–1320.

Pandey, S., Muthamilarasan, M., Sharma, N., Chaudhry, V., Dulani, P., Shweta, S., Jha, S., Mathur, S., & Prasad, M. (2019). Characterization of DEAD-box family of RNA helicases in tomato provides insights into their roles in biotic and abiotic stresses. *Environmental and Experimental Botany, 158*, 107–116.

Pandey, S., Sahu, P. P., Kulshreshtha, R., & Prasad, M. (2018). Role of host transcription factors in modulating defense response during plant-virus interaction. In B. L. Patil (Ed.), *Genes, Genetics and Transgenics for Virus Resistance in Plants* (pp. 25–54). Caister Academic Press.

Pandey, S., Sharma, N., & Prasad, M. (2020). Role of RNA-interacting proteins in modulating plant-microbe interactions. *Advances in Genetics, 105*, 67–94.

Press Bureau of India. (2020). https://pib.gov.in/PressReleasePage.aspx?PRID=1627276.

Quesada, T., Lucas, S., Smith, K., & Smith, J. (2019). Response to temperature and virulence assessment of *Fusarium circinatum* isolates in the context of climate change. *Forests, 10*(1), 40. https://doi.org/10.3390/f10010040.

Saroj, S. (2019). Effect of climate change on agricultural insect pest. *Acta Scientific Agriculture, 3*(12), 74–80.

Sinclair, W. A., & Griffiths, H. M. (1994). Ash yellows and its relationship to dieback and decline of ash. *Annual Review of Phytopathology, 32*(1), 49–60.

Skendzic, S., Zovko, M., Zivkovic, I. P., Lesic, V., & Lemic, D. (2021). Impact of climate change on agricultural insect pests. *Insects, 12*(5), 440.

Sood, P., Pandey, S., & Prasad, M. (2021). Biotechnological advances and advanced mutation breeding techniques for crop plants. In *Crop Improvement* (pp. 139–165). CRC Press.

Tonnang, H. E., Sokame, B. M., Abdel-Rahman, E. M., & Dubois, T. (2022). Measuring and modelling crop yield losses due to invasive insect pests under climate change. *Current Opinion in Insect Science, 50*, 100873.

War, A. R., Taggar, G. V., War, M. Y., & Hussain, B. (2016). Impact of climate change on insect pests, plant chemical ecology, tritrophic interactions and food production. *International Journal of Clinical and Biological Sciences, 1*(2), 16–29.

Zhou, Y., Van Leeuwen, S. K., Pieterse, C. M., Bakker, P. A., & Van Wees, S. C. (2019). Effect of atmospheric CO_2 on plant defense against leaf and root pathogens of Arabidopsis. *European Journal of Plant Pathology, 154*, 31–42.

Zhou, Y., Vroegop-Vos, I., Schuurink, R. C., Pieterse, C. M. J., & Van Wees, S. C. M. (2017). Atmospheric CO_2 alters resistance of Arabidopsis to *Pseudomonas syringae* by affecting abscisic acid accumulation and stomatal responsiveness to coronatine. *Frontiers in Plant Science, 8*, 700. https://doi.org/10.3389/fpls.2017.00700.

CHAPTER 6

Reactive Oxygen Species (ROS) and Biotic Stress: Evolving Roles Under Climate Change

VINAY KUMAR DHIMAN,[1] DEVENDRA SINGH,[2] VIVEK KUMAR DHIMAN,[3] SUNIL KUMAR VERMA,[2] and HIMANSHU PANDEY[3]

[1]*Department of Basic Sciences, Dr. Yashwant Singh Parmar University of Horticulture and Forestry, Solan, Himachal Pradesh, India*

[2]*Faculty of Biotechnology, Institute of Biological Sciences and Technology, Shri Ramswaroop Memorial University, Barabanki, Uttar Pradesh, India*

[3]*Department of Biotechnology, Dr. Yashwant Singh Parmar University of Horticulture and Forestry, Solan, Himachal Pradesh, India*

ABSTRACT

Climate change poses an impact on global weather patterns which has caused uncertainties in crop yields. The harsh environmental conditions affect the physiology and metabolism of plants and restrict their growth and productivity. Also, decreased soil fertility, pest attacks, and microbial infection have also contributed to stress. These stresses cause cellular dysfunction through the production of ROS species. Plants have evolved defense mechanisms to deal with such stresses by activating a series of signaling pathways that induce defense genes causing the production of antioxidants, enzymes, accumulation of stress hormones, etc., leading to defense action. Understanding the effect of climate change on plants is of utmost importance to producing a climate-resilient crop and ensuring food security by maintaining crop productivity and nutritional value.

Climate Change and Biotic Factors: A Molecular Approach.
Ashutosh Singh, Saurabh Pandey, and Amarjeet Kumar, (Eds.)

6.1 INTRODUCTION

The variation in climate has become the greatest challenge of our times, and its repercussions are being felt across the world. As global temperatures continue to rise, researchers are uncovering new information about how climate change is impacting a host of interconnected ecosystems and systems. One area that is particularly under scrutiny is the effect of climate change on ROS formation in different plants as well as the stress-associated responses (Mittler et al., 2022). Climate stress environment refers to both abiotic as well as biotic stresses that severely affect the growth and development stages, productivity, and survival of organisms. The term includes global stressors (such as changing earth's climate) as well as local stressors (like environmental pollution) that can act singly or in combination to produce negative impacts (Singh et al., 2022a; Chauhan et al., 2023). Organisms have evolved a variety of mechanisms for survival during extreme environmental stress, but these may be inadequate in the face of rapid change or multiple stresses. For example, plants can acclimatize to gradual changes in temperature or light intensity, but not to sudden changes such as those caused by wildfires. Similarly, animals can migrate to new areas to escape harsh conditions, but only if suitable habitat is available and migration routes are not blocked by human activities (Harvy et al., 2022).

There have been concerns related to the growing negative effect of climate change on plant growth and development. For example, higher temperatures can lead to reduced photosynthesis and water use efficiency, while increasing carbon dioxide levels may enhance photosynthesis but also lead to greater water use (Corlett and Westcott, 2013). Extreme weather conditions lead to unexpected droughts and floods, causing damage to commercial food crops and disrupting their production systems. Biotic stresses such as pest attacks and diseases are also likely to become more prevalent under climate change. Pests may benefit from warmer temperatures, while diseases may spread more easily in hotter, wetter conditions (Skendžić et al., 2021). Changes in rainfall patterns can also adversely affect agricultural production by affecting the timing and amount of water available for crops (Calzadilla et al., 2013). In this article, we will explore the evolving roles of ROS and biotic stress under climate change, offering insights into why these processes matter and what actions can be taken to mitigate them.

6.2 WHAT IS ROS?

The ROS, or reactive oxygen species, are by-product molecules produced during several metabolic processes such as during the light-dependent reactions of photosynthesis. Electrons are transferred through a series of electron carriers, resulting in the formation of ROS, for instance, superoxide (O_2^-), hydrogen peroxide (H_2O_2) (Table 6.1). The electron transport chain (ETC) in the mitochondria also generates ROS as a byproduct of normal cellular respiration. Environmental stressors, such as high and low temperatures, drought, and exposure to UV radiation, can also lead to their production, and several other metabolic reactions taking place in the cell also lead to their production (Karuppanapandian et al., 2011). These ROS can damage cells and cause oxidative stress, which is linked to the origin of many chronic diseases. However, recent research has shown that ROS may also contribute to plant immunity and stress tolerance. In plants, ROS are produced in response to a range of biotic and abiotic stresses, including pathogens and herbivores, within extreme weather conditions. These molecules help the plant to mount an effective defense against the stressor. For example, ROS can help to kill viruses and bacteria, or to the production of defensive compounds like phytoalexins (Singh et al., 2022b). Figure 6.1 represents the ROS found in various plants.

TABLE 6.1 Members of ROS Species (Das & Roychoudhury, 2014)

ROS	Description	Examples
Superoxide	A high ROS formed by the one-electron reduction of molecular oxygen	$O_2^{\bullet-}$
Hydrogen Peroxide	A milder ROS that is formed by the dismutation of superoxide	H_2O_2
Hydroxyl Radical	ROS (a highly reactive) is formed by the Fenton reaction	$OH^{\bullet}$
Singlet Oxygen	ROS (a highly reactive) is formed by the photosensitization of molecular oxygen	$^1O_2$
Peroxyl Radical	ROS (a highly reactive) is formed by the addition of a hydrogen atom to hydrogen peroxide	$RO_2^{\bullet}$

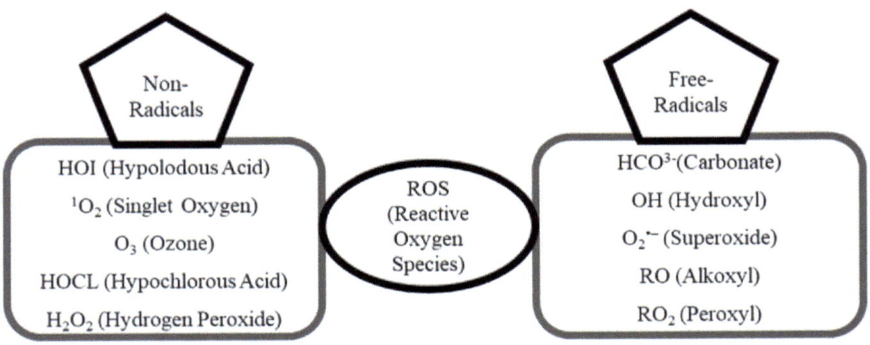

FIGURE 6.1 A sketch representing the ROS found in various plants.

Under climate change, the frequency, as well as severity of stresses (biotic and abiotic), is expected to increase. This means that plants will need to produce more ROS in order to cope with the added stressors. However, too much ROS can be detrimental to plant health. Therefore, it is important for plants to have the ability to regulate ROS production in response to changing environmental conditions (Araujo et al., 2015; Xie et al., 2019). In general, ROS are being produced routinely during normal plant cell metabolism, but excess ROS can lead to oxidative stress, which can damage cellular components and biomolecules such as different lipids, proteins, and cellular DNA. To defend themselves against oxidative stress, plants have evolved a number of antioxidant defense mechanisms (Sharma et al., 2023).

In order to adapt to stresses, plants have several antioxidants which act as defense mechanisms to protect themselves from oxidative stress caused by the production of ROS. These mechanisms consist of:

- **Enzymatic antioxidants:** Plants produce several antioxidant defense enzymes viz. SOD (superoxide dismutase), CAT (catalase), and POD (peroxidase) eliminate or neutralize ROS by converting them into less harmful substances. This can be explained as follows:

  Superoxide dismutase catalyzes the conversion of dioxide anion i.e., O_2^- (superoxide) to H_2O_2 (hydrogen peroxide) and O_2 (oxygen).

  SOD catalyzes the reaction: $2\ O_2^- + 2\ H^+ \rightarrow H_2O_2 + O_2$

  CAT and POD both enzymes neutralize H_2O_2 by converting it into water.

  CAT catalyzes the reaction: $2\ H_2O_2 \rightarrow 2\ H_2O + O_2$

  Peroxidase (POD): Peroxidase catalyzes the breakdown of Perhydroxic acid i.e., H_2O_2 (hydrogen peroxide) into H_2O (water) and O_2 (oxygen) or converts it to other less reactive substances, in the presence of a substrate such as phenols, alcohols, or amines.

  POD catalyzes the reaction: H_2O_2 + Substrate → Product + H_2O

  SOD, CAT, and POD enzymes are important for removing hydrogen peroxide, which is a relatively stable ROS, from the cells, preventing it from causing damage to the cell. It is stated that, although these enzymes are important for neutralizing ROS, they also generate some ROS as a byproduct of their enzymatic reaction. However, the amount generated is usually low and can be neutralized by other antioxidant mechanisms (Rajput et al., 2021). Alternatively, genes that have evolved to encode and biosynthesize specific antioxidant enzymes, i.e., glutathione S-transferase (GST), glutathione reductase (GR), and ascorbate peroxidase (APX), for enhancing the antioxidant defense in plants. These enzymes are also involved in the detoxification of hydrogen peroxide and other ROS by scavenging them.

- **Non-enzymatic antioxidants:** These antioxidants like ascorbate (vitamin C) are also present in plants, glutathione, and tocopherol (vitamin E) that can scavenge ROS and prevent them from causing damage (Ahmad et al., 2009). For example:

  Ascorbic acid (vitamin C) acts as a reducing agent, donating electrons to neutralize ROS such as hydrogen peroxide (H_2O_2).

  Ascorbic acid reacts with H_2O_2 to form water and dehydroascorbic acid:

  H_2O_2 + Ascorbic acid → Water + Dehydroascorbic acid

- **Phenolic compounds:** Phenolic compounds such as flavonoids, tannins, and anthocyanins are also important antioxidants in plant cells. They act as free radical scavengers and also prevent the oxidation of important biomolecules (Olszowy, 2019). For example:

  Flavonoids can neutralize O_2^- by reacting with it, forming a flavonoid radical:

  Flavonoid + O_2^- → Flavonoid radical + O_2

- **Metal-chelating compounds:** Some plants contain metal-chelating compounds such as metallothionein's that can bind and detoxify potentially harmful metal ions that can generate ROS (Hassinen et al., 2011).

- **Protein and DNA repair systems:** DNA and protein repair systems exist in the cell that repairs any damage due to ROS (Zhang et al., 2008).

  These antioxidant defense mechanisms work together to shield the plant from oxidative stress, allowing it to survive and thrive in its environment. These are just a few examples of the many reactions involved in the antioxidant defense mechanisms in plants. Overall, these mechanisms work to neutralize and detoxify ROS, preventing them from causing damage to the plant's cells (Devendra & Vishnu, 2021; Kumar et al., 2022).

6.3 WHAT IS BIOTIC STRESS?

Biotic stress is environmental stress that arises from the interactions between an organism and its biotic surroundings (Table 6.2). The term can refer to both negative and positive interactions, but is typically used in the context of negative interactions, such as those that lead to disease or death (Suzuki et al., 2014). Climate change is causing an increase in biotic stresses on plants and animals around the world. One of the major ways climate change is doing this is by altering the distribution of species. As temperatures rise and habitats shift, many species are forced into new areas where they may not be well-adapted. This can lead to competition with other species for resources, and increased exposure to pests and diseases. Another way climate change is causing more biotic stress is through changes in the timing of seasonal events (Surówka et al., 2020). For example, earlier springs can disrupt the synchrony between plants blooming and their pollinators being active. This can reduce pollination rates, leading to lower seed production and ultimately reduced populations (Forrest, 2015).

TABLE 6.2 Types of Biotic Stresses in Plants

Biotic Stress	Role	Examples	References
Pathogens	Microorganisms that cause disease in plants	Bacteria, Fungi, Viruses, Phytoplasmas	Miller et al. (2017)
Herbivores	Animals that feed on plant tissues	Insects, Mites, Snails, Rodents	Oerke, (2006)
Parasitic Plants	Plants that obtain nutrients from other plants	Striga, Cuscuta	Hegenauer et al. (2017)
Competition	Interference of one organism's survival or reproduction by another organism	Allelopathy, Root competition	Gawronska & Golisz, (2006)
Nematodes	Worms (very small in size) that nourish on plant roots	Root-knot (plant-parasitic) nematode, Cyst nematode	Kumar & Yadav, (2020)

The increased prevalence of biotic stresses under climate change is a serious concern because it can lead to population declines and even extinctions. To conserve biodiversity from climate change impact, it will be important to identify which species are most vulnerable to biotic stresses and take steps to protect them (Devendra Singh & Agarwal, 2021; Kumar, Rai et al., 2022). Plants have evolved to protect themselves from biotic stress such as interactions with other organisms, i.e., pathogens, herbivores, and competition with other plants (Shahzad et al., 2021). Some of the ways plants cope with biotic stress include:

- **Physical Barriers:** Many plants have developed protective outer layers modified with thickened cuticles, trichomes, and lignin to protect themselves from biotic stressors. These barriers can make it difficult for pathogens to penetrate the plant's surface and for herbivores to eat the plant's leaves (Belete, 2018).
- **Chemical Defense:** Plants produce numerous chemical compounds that can deter or kill pathogens and herbivores. These chemicals can act as toxins, repellents, or attractants for natural enemies of the pathogens or herbivores (Levin, 1976).
- **Induced Resistance:** Plants can also mount a defense response after sensing the presence of a pathogen or herbivore. This type of resistance induces activation of the defense-related genes as well as the production of defense-related proteins (Frost, 2008).
- **Symbiotic Association:** Some plants form symbiotic associations with microorganisms, for instance the association of mycorrhizal fungi, rhizobia, and endophytic bacteria that can protect the plant from pathogens and enhance nutrient uptake (Badri, 2009).
- **Systemic Acquired Resistance (SAR):** a systemic response that occurs when a plant is exposed to a pathogen or herbivore in one part of the plant. The plant will activate its defense mechanisms systemically throughout the whole plant, making it more resistant to future attacks (Fu and Dong, 2013).
- **Defense Signaling Pathways:** Plants have evolved complex defense signaling pathways, like the ethylene, jasmonate, and salicylic acid pathways, which are activated in response to biotic stressors and help coordinate the plant's defense response (Verma et al., 2016).
- **Allelopathy:** Plants can also suppress the growth of nearby plants by releasing chemicals that inhibit germination or growth, known as allelopathy (Šoln et al., 2022).
- Overall, plants have developed stimuli in response to cellular changes caused by biotic stress, which trigger defense mechanisms balancing their homeostasis, allowing them to survive and thrive in such environments.

6.4 HOW DO ROS AND BIOTIC STRESS INTERACT UNDER CLIMATE CHANGE?

Plants get exposed to varied environmental stresses, both natural and physical, which in turn produce cellular reactive oxygen species (ROS). Plants have evolved complex defense mechanisms that balance such reactive molecules. ROS and biotic stress are two important drivers of change in the natural world, and their interaction is likely to be a key determinant of how ecosystems respond to climate change (Surówka et al., 2020). There are indications that climate change has already affected the distribution and abundance of ROS-producing organisms, as well as the rate at which they produce ROS.

These changes are likely to have cascading effects on biotic stressors (Table 6.3) since many of them are sensitive to changes in ROS levels (Chauhan et al., 2022). For example, increased ROS production may lead to greater plant mortality than herbivory, as well as increased competition for resources among plants. In turn, these effects could alter community composition and function, with potentially far-reaching consequences for ecosystem services and human welfare (Garg & Manchanda, 2009).

TABLE 6.3 Effects of Biotic Stress from Environmental Changes

Mechanism	Effects	Biotic stress response	References
Changes in temperature	Increased temperatures can lead to variations in the range and dispersal of pests and pathogens, allowing them to expand into new areas and increasing their population density.	The mountain pine beetle's range has been able to grow as a result of warmer temperatures resulting in the devastation of pine forests in western North America.	Creeden et al. (2014)
Changes in precipitation	Drought or flooding can affect the plant's growth stages leaving them vulnerable to susceptible to biotic stressors.	In some regions, drought-stressed plants are more likely to be attacked by pests and pathogens. On the other hand, flooding can also increase the susceptibility of plants to biotic stressors, like fungal and bacterial infections, as well as reduce the oxygen supply to the roots, leading to hypoxia	Mariani & Ferrante, (2017) Szczepaniec & Finke, (2019)
Changes in atmospheric CO_2	Elevated atmospheric CO2 can change the chemistry of plants, making them more palatable to herbivores and less resistant to pathogens.	Research has shown that elevated level of CO_2 has resulted in an increased level of protein content within some plants, making them more attractive to herbivores.	Robinson et al. (2012)

TABLE 6.3 *(Continued)*

Mechanism	Effects	Biotic stress response	References
Changes in phenology	Climate variation affects the periodic events or cycle of plant growth and development, leading to mismatches between plants and their associated herbivores and pollinators.	Earlier springs due to warming temperatures can cause plants to flower before their pollinators are active, leading to reduced seed production.	Kudo & Ida, (2013)
Changes in biodiversity	Climate has also an impact on changing the diversity and distribution of plant species, altering the interactions between plants and their associated biotic stressors.	As temperatures rise, some floral species adapt to higher elevated zones or latitudes which in turn shift the interactions between plants and pests, pathogens within their vicinity.	Sharma, (2014)

The changing climate is altering the way in which plants produce and respond to ROS. ROS production is integral during the metabolic process of plant growth and development stages, but under certain conditions, for instance, drought or heat stress, ROS can potentially harm plant cells. The ability of plants to tolerate these stresses relies solely on the homeostasis between ROS and antioxidants within cells (Cassia et al., 2018). Changes in the climate alter the production of ROS in plants in several ways. For example, rising temperatures may increase the rate of photosynthesis, leading to increased production of ROS. Alternatively, changes in precipitation patterns could lead to drought conditions that would decrease the availability of water for plants, resulting in increased ROS (Reddy et al., 2004).

In addition to changes in production, climate change is also expected to alter the way in which plants respond to ROS. For example, higher temperatures could result in increased oxidative stress, leading to damage to plant cells. Also, high degrees of rainfall could lead to waterlogging or flooding, which could also lead to increased oxidative stress (Hasanuzzaman et al., 2012). The evolving roles of ROS and biotic stress under climate change are sophisticated and not fully understood. However, the changing climate will have a significant impact on the way in which plants produce and respond to ROS (Das et al., 2022).

6.5 BIOTECHNOLOGICAL APPROACHES AGAINST BIOTIC STRESS AND ROS

In recent years, the use of genomics and proteomics approaches to combat biotic stress has resulted in an increase in ROS within plants and has become increasingly prevalent. Genomic approaches involve the study of gene sequences, while proteomic approaches involve the study of proteins. These techniques have enabled researchers to identify genes or proteins that play a part in biotic stress responses and ROS production in plants. Through this knowledge, scientists are able to develop strategies for beneficial and healthier plant growth by manipulating these genes or proteins. By studying the encoding patterns of these genes and their expressed proteins, we can better understand how plants respond to biotic stress and ROS (Das et al., 2022).

In the context of biotic stress, genomics can be used to identify genes that are intrinsic to a plant's defense mechanisms against various pathogens and pests. For example, researchers can use genomic techniques to identify resistance genes that confer immunity to specific pathogens. This knowledge can be vitally used to develop more resistant crop varieties through breeding or genetic engineering (Van Esse et al., 2020). Proteomics, on the other hand, is the analysis of an organism's entire set of proteins, including their interactions, function, and structure. In the context of biotic stress, proteomics can be used to identify proteins responsible for the plant's response to pathogens and pests. For example, researchers can use proteomic techniques to identify the proteins that are responsible for the production of phytohormones, which play an essential part in the plant's defense response (Wu et al., 2013; Khatabi et al., 2019). Most importantly, these biotechnological approaches can also be used to study the mechanisms that plants use to detoxify ROS. For example, by utilizing these techniques, one can identify genes and proteins that are responsible for the formation of crucial antioxidant enzymes, such as catalase and superoxide dismutase, which neutralize ROS. Additionally, it can identify genes and proteins active in the repair of ROS-induced damage (Gupta et al., 2019).

The transgenic approach can be used to enhance antioxidant defense in plants by introducing genes responsible for encoding the enzymes actively participating in the synthesis or scavenging of antioxidants (Wani et al., 2016). The introduction of genes that encode the antioxidant enzymes like SOD or CAT is able to stimulate the detoxification process in plants against ROS, which is responsible for causing damage to plant cells. Additionally, the introduction of genes that encode the biosynthesis of antioxidants like Vitamin E, Vitamin C, and β-carotene can also enhance the plant's antioxidant defense (Dhiman et al., 2021; Verma et al., 2022).

There are several examples of transgenic plants that have been shown to increase tolerance against biotic stress caused by pathogens and pests. One example is transgenic rice plants that have been developed to resist the bacterial pathogen *X. oryzae* pv. Oryzae, which causes bacterial blight disease in rice (Zeng et al., 2015). These transgenic plants express a gene encoding an enzyme called ribonucleases (RNases), which can degrade the RNA of the pathogen and inhibit its growth. Another example is transgenic tobacco plants that have been developed to be resistant to the tobacco mosaic virus (TMV), which causes severe damage to tobacco crops (Zhu et al., 2009). These transgenic plants express a gene encoding a viral coat protein that mimics the viral coat protein of TMV and interferes with viral replication. Additionally, transgenic tomato and potato plants have been developed to be resistant to PLRV (potato leafroll virus) and PVY (potato virus Y) respectively, by expressing a viral coat protein gene. Transgenic cotton plants have been developed to be resistant to the cotton bollworm (*Helicoverpa armigera*) by expressing a gene encoding the Bt toxin, which is toxic to the bollworm (Wu et al., 2008). This transgenic cotton variety was developed to be resistant to a major cotton pest known as the bollworm, which is responsible for causing substantial yield losses in cotton crops. The Bt cotton variety expresses the same gene which is present in the bacterium *Bacillus thuringiensis* (Bt), which encodes for the Bt toxin. This toxin is toxic to the bollworm, and thus protects the cotton plant from damage (Dhiman et al., 2022; Kalia et al., 2020). In several countries, Bt cotton is widely cultivated due to its promising outcome in effectively reducing the use of chemical pesticides with increasing yields. Similarly, Bt corn has been developed to be resistant to the European corn borer, a major pest that causes substantial yield loss in corn crops (Siegfried & Hellmich, 2012). Like Bt cotton, Bt corn expresses a gene from *Bacillus thuringiensis* that expresses the Bt toxin, which is toxic to the European corn borer, protecting the corn plants from damage (Dohare et al., 2021; Negi et al., 2021; Pandey et al., 2022). Transgenic papaya has been developed to be resistant to the papaya ringspot virus (PRSV), which severely harms papaya crops (Azad et al., 2014). The transgenic papaya expresses a gene from the virus itself, which can interfere with viral replication and protect the papaya plants from damage. Transgenic soybeans have been developed to be tolerant to the herbicide glyphosate, which can be used to control weeds in soybean crops (Bonny, 2008). Transgenic rice plants have been developed to be more tolerant to oxidative stress caused by highly intense light. These transgenic plants express a gene encoding a protein called HsfA3, which can help regulate the plant's response to high light intensity and reduce the damage caused by ROS (Singh & Grover, 2008). Another example is transgenic tomato plants that have been developed to be more tolerant to oxidative stress due to high temperature (Krishna et al., 2019). These transgenic plants express a gene encoding a protein called HsfA3, which can enhance the plant's ability to resist higher temperatures and reduce the damage caused by ROS. Transgenic tobacco plants expressing the enzyme catalase (CAT) have been developed for improved tolerance to oxidative stress facilitated by high light intensity and salt stress (Lee et al., 2007; Li et al., 2023). These transgenic plants showed an increased level of CAT enzyme, which can neutralize hydrogen peroxide and reduce the damage caused by ROS. Transgenic *Arabidopsis thaliana* plants expressing the enzyme peroxidase (POD) have been developed to cope with oxidative stress caused by high light intensity and salt stress (Suzuki et al., 2013). These transgenic plants showed an increased level of POD enzyme, which can neutralize hydrogen peroxide and reduce the damage caused by ROS. Transgenic plants expressing the enzyme glutathione reductase (GR) have been developed to enhance oxidative stress tolerance due to salt stress as well as high light intensity (Le Martret et al., 2011; Yousuf et al., 2012). The GR enzyme helps in the detoxification of hydrogen peroxide and other ROS by scavenging them. Transgenic plants expressing the gene encoding for the enzyme ascorbate peroxidase (APX) have been developed to improve tolerance to oxidative stress caused by high light intensity and salt stress (Singh et al., 2022; Singh et al., 2022; Verma et al., 2022). The APX enzyme helps in the detoxification of hydrogen peroxide molecules and other ROS by scavenging them. Transgenic plants expressing the enzyme glutathione S-transferase (GST) have been developed to improve oxidative stress tolerance against extreme light intensity and salt stress. The GST enzyme helps in the detoxification of hydrogen peroxide and other ROS by scavenging them. Transgenic plants have been shown to be effective in enhancing tolerance to oxidative stress. The development and commercialization of these plants is a complex process that requires extensive testing and evaluation to ensure safety and environmental compatibility (Li et al., 2023). In summary, genomics and proteomics are powerful biotechnological tools that can be used to study the mechanisms that plants use to respond to biotic stress and ROS. By identifying the genes and proteins involved in these processes, researchers can develop strategies to improve crops' resistance to pathogens and pests and to increase the tolerance of crops to ROS (Singh et al., 2022; Verma et al., 2022).

6.6 CONCLUSION

In summary, ROS and biotic stress have evolved to become increasingly important components of the climate change puzzle. With increasing temperatures due to global warming, it is essential that we understand how these reactions are influencing plant physiology and productivity in order to accurately predict future trends in crop production. Additionally, the development of adaptive strategies like those mentioned here may prove beneficial for mitigating the impacts of oxidative damage caused by ROS and other abiotic stresses under a changing climate.

KEYWORDS

- **biotic stress**
- **climate changes**
- **crop yields**
- **ROS**

REFERENCES

Ahmad, P., Jaleel, C. A., Azooz, M. M., & Nabi, G. (2009). Generation of ROS and non-enzymatic antioxidants during abiotic stress in plants. *Botany Research International, 2*(1), 11–20.

Araujo, S. S., Beebe, S., Crespi, M., Delbreil, B., Gonzalez, E. M., Gruber, V., & Patto, M. C. V. (2015). Abiotic stress responses in legumes: Strategies used to cope with environmental challenges. *Critical Reviews in Plant Sciences, 34*(1–3), 237–280.

Azad, M., Kalam, A., Amin, L., & Sidik, N. M. (2014). Gene technology for papaya ringspot virus disease management. *The Scientific World Journal, 2014*, 768038.

Badri, D. V., Weir, T. L., van der Lelie, D., & Vivanco, J. M. (2009). Rhizosphere chemical dialogues: Plant–microbe interactions. *Current Opinion in Biotechnology, 20*(6), 642–650.

Belete, T. (2018). Defense mechanisms of plants to insect pests: From morphological to biochemical approach. *Trends Tech. Sci. Res, 2*(2), 555584.

Bonny, S. (2008). Genetically modified glyphosate-tolerant soybean in the USA: Adoption factors, impacts and prospects. A review. *Agronomy for Sustainable Development, 28*, 21–32.

Calzadilla, A., Rehdanz, K., Betts, R., Falloon, P., Wiltshire, A., & Tol, R. S. (2013). Climate change impacts on global agriculture. *Climatic Change, 120*, 357–374.

Cassia, R., Nocioni, M., Correa-Aragunde, N., & Lamattina, L. (2018). Climate change and the impact of greenhouse gasses: CO_2 and NO, friends and foes of plant oxidative stress. *Frontiers in Plant Science, 9*, 273.

Chauhan, D., Singh, D., Pandey, H., Khan, Z., Srivastava, R., Dhiman, V. K., & Dhiman, V. K. (2023). Impact of transcription factors in plant abiotic stress: A recent advancement for crop improvement. In *Plant Transcription Factors* (pp. 271–286). Academic Press.

Chauhan, V., Kaushal, D., Dhiman, V. K., Kanwar, S. S., Singh, D., Dhiman, V. K., & Pandey, H. (2022). An insight in developing carrier-free immobilized enzymes. *Frontiers in Bioengineering and Biotechnology, 10*, 794411. https://doi.org/10.3389/fbioe.2022.794411.

Corlett, R. T., & Westcott, D. A. (2013). Will plant movements keep up with climate change? *Trends in Ecology & Evolution, 28*(8), 482–488.

Creeden, E. P., Hicke, J. A., & Buotte, P. C. (2014). Climate, weather, and recent mountain pine beetle outbreaks in the western United States. *Forest Ecology and Management, 312*, 239–251.

Das, K., & Roychoudhury, A. (2014). Reactive oxygen species (ROS) and response of antioxidants as ROS-scavengers during environmental stress in plants. *Frontiers in Environmental Science, 2*, 53.

Das, R., Deb, P., Pandey, H., Shyam, P., & Singh, D. (2022). Botanical synthesis of silver nanoparticles (AgNPs) and its antifungal effect against *Alternaria porri* causing purple blotch of onion: An in vitro and natural epiphytic study. *Journal of Agriculture and Food Research, 10*, 100390. https://doi.org/10.1016/j.jafr.2022.100390.

Devendra Singh, D., & Agarwal, V. (2021). Screening of antimicrobial, anti-quorum sensing activity and cytotoxicity of origanum oil against gram-positive and gram-negative bacteria. *Biomedicine, 41*(3), 599–603. https://doi.org/10.51248/.v41i3.1200.

Devendra, S., & Vishnu, A. (2021). A review on the COVID-19, its history, diagnostic approaches, role of herbs and current world scenario. *Biomedicine, 41*(2). https://doi.org/10.51248/.v41i2.1032.

Dhiman, V. K., Chauhan, V., Kanwar, S. S., Singh, D., & Pandey, H. (2021). Purification and characterization of actinidin from *Actinidia deliciosa* and its utilization in inactivation of α-amylase. *Bulletin of the National Research Centre, 45*(1), 213. https://doi.org/10.1186/s42269-021-00673-0.

Dhiman, V. K., Singh, D., Pandey, H., Chauhan, D., Dhiman, V. K., & Pandey, D. (2022). Mycoremediation of agricultural waste for the cultivation of edible mushrooms. In T. Aftab (Ed.), *Sustainable Management of Environmental Contaminants* (pp. 471–482). Springer International Publishing. https://doi.org/10.1007/978-3-031-08446-1_18.

Dohare, S., Singh, D., Sharma, D., & Agarwal, V. (2021). The effect of *Staphylococcus epidermidis* cells on *Pseudomonas aeruginosa*-associated virulence factors. *Journal of Applied Biology & Biotechnology, 9*(6), 122–127. https://doi.org/10.7324/JABB.2021.96015.

Forrest, J. R. (2015). Plant–Pollinator interactions and phenological change: What can we learn about climate impacts from experiments and observations? *Oikos, 124*(1), 4–13.

Frost, C. J., Mescher, M. C., Carlson, J. E., & De Moraes, C. M. (2008). Plant defense priming against herbivores: Getting ready for a different battle. *Plant Physiology, 146*(3), 818–824.

Fu, Z. Q., & Dong, X. (2013). Systemic acquired resistance: Turning local infection into global defense. *Annual Review of Plant Biology, 64*, 839–863.

Garg, N., & Manchanda, G. (2009). ROS generation in plants: Boon or bane? *Plant Biosystems, 143*(1), 81–96.

Gawronska, H., & Golisz, A. (2006). Allelopathy and biotic stresses. In *Allelopathy: A Physiological Process with Ecological Implications* (pp. 211–227). Netherlands: Springer.

Gupta, S., Dong, Y., Dijkwel, P. P., Mueller-Roeber, B., & Gechev, T. S. (2019). Genome-wide analysis of ROS antioxidant genes in resurrection species suggest an involvement of distinct ROS detoxification systems during desiccation. *International Journal of Molecular Sciences, 20*(12), 3101.

Harvey, J. A., Tougeron, K., Gols, R., Heinen, R., Abarca, M., Abram, P. K., & Chown, S. L. (2022). Scientists' warning on climate change and insects. *Ecological Monographs, 93*(1), e1553.

Hasanuzzaman, M., Hossain, M. A., da Silva, J. A. T., & Fujita, M. (2012). Plant response and tolerance to abiotic oxidative stress: Antioxidant defense is a key factor. In *Crop Stress and Its Management: Perspectives and Strategies* (pp. 261–315).

Hassinen, V. H., Tervahauta, A. I., Schat, H., & Kärenlampi, S. O. (2011). Plant metallothionein's–metal chelators with ROS scavenging activity? *Plant Biology, 13*(2), 225–232.

Hegenauer, V., Körner, M., & Albert, M. (2017). Plants under stress by parasitic plants. *Current Opinion in Plant Biology, 38*, 34–41.

Kalia, M., Singh, D., Sharma, D., Narvi, S., & Agarwal, V. (2020). *Senna alexandriana* Mill as a potential inhibitor for quorum sensing-controlled virulence factors and biofilm formation in *Pseudomonas aeruginosa* PAO1. *Pharmacognosy Magazine, 16*(72), 802. https://doi.org/10.4103/pm.pm_315_20.

Karuppanapandian, T., Moon, J. C., Kim, C., Manoharan, K., & Kim, W. (2011). Reactive oxygen species in plants: Their generation, signal transduction, and scavenging mechanisms. *Australian Journal of Crop Science, 5*(6), 709–725.

Khatabi, B., Gharechahi, J., Ghaffari, M. R., Liu, D., Haynes, P. A., McKay, M. J., & Salekdeh, G. H. (2019). Plant–microbe symbiosis: What has proteomics taught us? *Proteomics, 19*(16), 1800105.

Krishna, R., Karkute, S. G., Ansari, W. A., Jaiswal, D. K., Verma, J. P., & Singh, M. (2019). Transgenic tomatoes for abiotic stress tolerance: Status and way ahead. *3 Biotech, 9*, 1–14.

Kudo, G., & Ida, T. Y. (2013). Early onset of spring increases the phenological mismatch between plants and pollinators. *Ecology, 94*(10), 2311–2320.

Kumar, P., Rai, S., Verma, S. K., Prakash, P. S., & Chitara, D. (2022). Classification, mode of action and uses of various immunomodulators. In R. K. Kesharwani, R. K. Keservani, & A. K. Sharma (Eds.), *Immunomodulators and Human Health* (pp. 3–38). Springer Nature Singapore. https://doi.org/10.1007/978-981-16-6379-6_1.

Kumar, P., Verma, S. K., Rai, S., Prakash, P. S., & Chitara, D. (2022). Role of homocysteine metabolism in cardiovascular diseases. In G. P. Dubey, K. Misra, R. K. Kesharwani, & R. P. Ojha (Eds.), *Homocysteine Metabolism in Health and Disease* (pp. 257–276). Springer Nature Singapore. https://doi.org/10.1007/978-981-16-6867-8_14.

Kumar, Y., & Yadav, B. C. (2020). Plant-parasitic nematodes: Nature's most successful plant parasite. *International Journal of Research and Review, 7*(3), 379–386.

Le Martret, B., Poage, M., Shiel, K., Nugent, G. D., & Dix, P. J. (2011). Tobacco chloroplast transformants expressing genes encoding dehydroascorbate reductase, glutathione reductase, and glutathione-S-transferase exhibit altered antioxidant metabolism and improved abiotic stress tolerance. *Plant Biotechnology Journal, 9*(6), 661–673.

Lee, Y. P., Kim, S. H., Bang, J. W., Lee, H. S., Kwak, S. S., & Kwon, S. Y. (2007). Enhanced tolerance to oxidative stress in transgenic tobacco plants expressing three antioxidant enzymes in chloroplasts. *Plant Cell Reports, 26*, 591–598.

Levin, D. A. (1976). The chemical defenses of plants to pathogens and herbivores. *Annual Review of Ecology and Systematics, 7*(1), 121–159.

Li, J., Yu, B., Ma, C., Li, H., Jiang, D., Nan, J., & Li, H. (2023). Functional characterization of sugar beet M14 antioxidant enzymes in plant salt stress tolerance. *Antioxidants, 12*(1), 57.

Li, J., Yu, B., Ma, C., Li, H., Jiang, D., Nan, J., & Li, H. (2023). Functional characterization of sugar beet M14 antioxidant enzymes in plant salt stress tolerance. *Antioxidants, 12*(1), 57.

Mariani, L., & Ferrante, A. (2017). Agronomic management for enhancing plant tolerance to abiotic stresses—Drought, salinity, hypoxia, and lodging. *Horticulturae, 3*(4), 52.

Miller, R. N. G., Costa Alves, G. S., & Van Sluys, M. A. (2017). Plant immunity: Unravelling the complexity of plant responses to biotic stresses. *Annals of Botany, 119*(5), 681–687.

Mittler, R., Zandalinas, S. I., Fichman, Y., & Van Breusegem, F. (2022). Reactive oxygen species signaling in plant stress responses. *Nature Reviews Molecular Cell Biology, 23*(10), 663–679.

Negi, S., Singh, D., Pandey, H., & Singhal, R. K. (2021). Biofortification for high Fe and Zn in various Poaceae crops by using different molecular breeding and biotechnological approaches. *Plant Physiology Reports, 26*(4), 636–646. https://doi.org/10.1007/s40502-021-00625-w.

Oerke, E. C. (2006). Crop losses to pests. *The Journal of Agricultural Science, 144*(1), 31–43.

Olszowy, M. (2019). What is responsible for antioxidant properties of polyphenolic compounds from plants? *Plant Physiology and Biochemistry, 144*, 135–143.

Pandey, H., Singh, D., Dhiman, V. K., Dhiman, V. K., & Pandey, D. (2022). Extremophiles and their application in bioremediation. In A. B. Gunjal, R. Thombre, & J. A. Parray (Eds.), *Advances in Environmental Engineering and Green Technologies* (pp. 188–206). IGI Global. https://doi.org/10.4018/978-1-7998-9144-4.ch009.

Rajput, V. D., Singh, R. K., Verma, K. K., Sharma, L., Quiroz-Figueroa, F. R., Meena, M., & Mandzhieva, S. (2021). Recent developments in enzymatic antioxidant defense mechanism in plants with special reference to abiotic stress. *Biology, 10*(4), 267.

Reddy, A. R., Chaitanya, K. V., & Vivekanandan, M. (2004). Drought-induced responses of photosynthesis and antioxidant metabolism in higher plants. *Journal of Plant Physiology, 161*(11), 1189–1202.

Robinson, E. A., Ryan, G. D., & Newman, J. A. (2012). A meta-analytical review of the effects of elevated CO_2 on plant–Arthropod interactions highlight the importance of interacting environmental and biological variables. *New Phytologist, 194*(2), 321–336.

Shahzad, A., Ullah, S., Dar, A. A., Sardar, M. F., Mehmood, T., Tufail, M. A., & Haris, M. (2021). Nexus on climate change: Agriculture and possible solution to cope future climate change stresses. *Environmental Science and Pollution Research, 28*, 14211–14232.

Sharma, H. C. (2014). Climate change effects on insects: Implications for crop protection and food security. *Journal of Crop Improvement, 28*(2), 229–259.

Sharma, K., Kumar, M., & Chandran, D. (2023). Ameliorative effects of phenolics in oxidative stress management in plants. In *Plant Phenolics in Abiotic Stress Management* (pp. 369–390). Singapore: Springer Nature Singapore.

Siegfried, B. D., & Hellmich, R. L. (2012). Understanding successful resistance management: The European corn borer and Bt corn in the United States. *GM Crops & Food, 3*(3), 184–193.

Singh, A., & Grover, A. (2008). Genetic engineering for heat tolerance in plants. *Physiology and Molecular Biology of Plants, 14*, 155–166.

Singh, D., Dhiman, V. K., Pandey, H., Dhiman, V. K., & Pandey, D. (2022a). Crosstalk between salicylic acid and auxins, cytokinins and gibberellins under biotic stress. In *Auxins, Cytokinins and Gibberellins Signaling in Plants* (pp. 249–262). Cham: Springer International Publishing.

Singh, D., Pandey, H., Dhiman, V. K., Bang, N. T. H., Dhiman, V. K., Singh, V., & Pandey, D. (2022b). Secondary metabolite engineering for plant immunity against various pathogens. In *Metabolic Engineering in Plants* (pp. 123–143). Singapore: Springer Nature Singapore.

Singh, D., Pandey, H., Dhiman, V. K., Thakur, K., Sharma, I., Dhiman, V. K., Singh, V., & Pandey, D. (2022). Assessment of environmental impacts of metal/metalloid pollution on plants. In *Metals Metalloids Soil Plant Water Systems* (pp. 217–232). Elsevier. https://doi.org/10.1016/B978-0-323-91675-2.00022-6.

Singh, D., Pandey, H., Shrivastava, N. K., Das, R., & Singh, V. (2022). Green synthesized gold and silver nanoparticles for antimicrobial applications. In C. Baskar, S. Ramakrishna, & A. Daniela La Rosa (Eds.), *Encyclopedia of Green Materials* (pp. 1–12). Springer Nature Singapore. https://doi.org/10.1007/978-981-16-4921-9_254-1.

Singh, V., Pandey, H., Misra, V., Tiwari, V., Srivastava, P., & Singh, D. (2022a). Hypolipidemic effect of [6]-gingerol-loaded Eudragit polymeric nanoparticles in high-fat diet-induced rats and gamma scintigraphy evaluation of gastric-retention time. *Journal of Applied Pharmaceutical Science, 12*(6), 156–163. https://doi.org/10.7324/JAPS.2022.120615.

Skendžić, S., Zovko, M., Živković, I. P., Lešić, V., & Lemić, D. (2021). The impact of climate change on agricultural insect pests. *Insects, 12*(5), 440.

Šoln, K., Klemenčič, M., & Koce, J. D. (2022). Plant cell responses to allelopathy: From oxidative stress to programmed cell death. *Protoplasma, 259*(5),1111–1124.

Surówka, E., Rapacz, M., & Janowiak, F. (2020). Climate change influences the interactive effects of simultaneous impact of abiotic and biotic stresses on plants. In *Plant Ecophysiology and Adaptation under Climate Change: Mechanisms and Perspectives I: General Consequences and Plant Responses* (pp. 1–50).

Suzuki, N., Miller, G., Sejima, H., Harper, J., & Mittler, R. (2013). Enhanced seed production under prolonged heat stress conditions in *Arabidopsis thaliana* plants deficient in cytosolic ascorbate peroxidase 2. *Journal of Experimental Botany, 64*(1), 253–263.

Suzuki, N., Rivero, R. M., Shulaev, V., Blumwald, E., & Mittler, R. (2014). Abiotic and biotic stress combinations. *New Phytologist, 203*(1), 32–43.

Szczepaniec, A., & Finke, D. (2019). Plant-vector-pathogen interactions in the context of drought stress. *Frontiers in Ecology and Evolution, 7*, 262.

Van Esse, H. P., Reuber, T. L., & van der Does, D. (2020). Genetic modification to improve disease resistance in crops. *New Phytologist, 225*(1), 70–86.

Verma, S. K., & Hussain, R. (2020). Comparative analysis of growth kinetics, concentration of sugar and alcohol tolerance of various yeast strains isolated from five different sources. *International Journal of Research and Review, 02*(08), 2582–5208.

Verma, S. K., & Kumar, P. (2022a). Biofuel policies in India: An assessment of policy barriers. In A. K. Rathoure & S. M. Khade (Eds.), *Advances in Environmental Engineering and Green Technologies* (pp. 44–64). IGI Global. https://doi.org/10.4018/978-1-6684-5269-1.ch004.

Verma, S. K., Singh, V., & Singh, D. (2022). Analysis of isolated yeast strains from different sources for bio-ethanol production. *International Journal of Research and Review, 2*(3), 100–104. https://doi.org/10.5281/zenodo.7067346.

Verma, V., Ravindran, P., & Kumar, P. P. (2016). Plant hormone-mediated regulation of stress responses. *BMC Plant Biology, 16*, 1–10.

Wani, S. H., Sah, S. K., Hossain, M. A., Kumar, V., & Balachandran, S. M. (2016). Transgenic approaches for abiotic stress tolerance in crop plants. In *Advances in Plant Breeding Strategies: Agronomic, Abiotic and Biotic Stress Traits* (pp. 345–396).

Wu, J., Luo, X., Wang, Z., Tian, Y., Liang, A., & Sun, Y. (2008). Transgenic cotton expressing synthesized scorpion insect toxin AaHIT gene confers enhanced resistance to cotton bollworm (*Heliothis armigera*) larvae. *Biotechnology Letters, 30*, 547–554.

Wu, L., Wang, S., Chen, X., Wang, X., Wu, L., Zu, X., & Chen, Y. (2013). Proteomic and phytohormone analysis of the response of maize (*Zea mays* L.) seedlings to sugarcane mosaic virus. *PLoS One, 8*(7), e70295.

Xie, X., He, Z., Chen, N., Tang, Z., Wang, Q., & Cai, Y. (2019). The roles of environmental factors in regulation of oxidative stress in plant. *BioMed Research International, 2019*, 9732325.

Yousuf, P. Y., Hakeem, K. U. R., Chandna, R., & Ahmad, P. (2012). Role of glutathione reductase in plant abiotic stress. In *Abiotic Stress Responses in Plants: Metabolism, Productivity and Sustainability* (pp. 149–158).

Zeng, X., Tian, D., Gu, K., Zhou, Z., Yang, X., Luo, Y., & Yin, Z. (2015). Genetic engineering of the *Xa10* promoter for broad-spectrum and durable resistance to *Xanthomonas oryzae* pv. *oryzae*. *Plant Biotechnology Journal, 13*(7), 993–1001.

Zhang, Q., Pan, J., Zhao, C., Wang, Y., Jia, Z., & Zheng, R. (2008). Non-enzymatic fast repair of DNA oxidative damage might also exist in cells. *Cell Biology International, 32*(6), 654–662.

Zhu, C. X., Song, Y. Z., Yin, G. H., & Wen, F. J. (2009). Induction of RNA-mediated multiple virus resistance to potato virus Y, tobacco mosaic virus and cucumber mosaic virus. *Journal of Phytopathology, 157*(2), 101–107.

CHAPTER 7

Molecular Mechanism of Melatonin Interaction on Biotic Stress Tolerance in Horticulture Crops

M. CHITHRA, S. SENTHILKUMAR, R. S. AFIYA, and S. MANIVANNAN

Department of Horticulture, Central University of Tamil Nadu, Thiruvarur, Tamil Nadu, India

ABSTRACT

Biotic stress causes tremendous damage to agriculture and horticulture crops worldwide. In general, plants have an inbuilt resistance mechanism to several biotic stress factors via phytohormones. These hormones aid in the enhancement of the pathogen-associated molecular patterns (PAMPs), synthesis of secondary metabolites, polyphenol oxidase (PPO), and plant resistance (R) proteins within the plant system that trigger immunity against insects and pathogen infestation. Among several phytohormones, ethylene, jasmonic acid, salicylic acid (SA), and melatonin are major key regulators in boosting the immune/defense mechanism in the plant system. On the other hand, the practice of using several synthetic pesticides and fungicides by the plant growers regulates the infestations of different insects and disease-causing pathogens and also causes negative impacts on soil fertility, crop production, and productivity. It also causes a carcinogenic effect in humans when consuming the products. Hence, to address these issues, we can consider an alternative approach by using eco-friendly plant hormones/phytohormones. Melatonin, an eco-friendly and economical phytohormone, has tremendous impacts on stimulating crop growth and development, including a defense mechanism against the ill effects of various biotic stresses. However, the molecular mechanisms of melatonin that govern the hormonal networks are

Climate Change and Biotic Factors: A Molecular Approach.
Ashutosh Singh, Saurabh Pandey, and Amarjeet Kumar, (Eds.)

unknown. We discuss how hormone signaling could be manipulated and how the host-pathogen interaction improves the resistance of crops to different biotic factors in this chapter.

7.1 INTRODUCTION

Under favorable environmental conditions, plants' growth and development are encouraged. The plants are said to be under stress if there is any variation from their usual behavior. Stress can be classified into two categories: biotic and abiotic stress. The metabolic, physiological, and molecular activities of plant systems are affected by both stresses, and these effects may cause a direct and indirect impact on production and productivity. "Biotic stress factor" refers to stress brought on by any pathogen, insect, or other living thing or biological system. Abiotic stress can be brought on by environmental factors including excessive or low light, temperature, unpredictable rainfall, uneven rainfall distribution, etc. Here, we are limited to focusing on biotic stress. To deal with both pressures, a variety of synthetic compounds are used. Despite this, it has a number of negative consequences on plants and the environment. Consider this point of view: "Phyto-hormones" are substances produced by plants that control or alter physiological activity, even in the presence of stress. (Saqib et al., 2022; Eun et al., 2019).

The commencement of growth of the plants can be regulated through Phyto-hormones. Some of the plant hormones have great potential against different biotic and abiotic ill effects. The plant hormones melatonin, ethylene, jasmonate, and salicylic acid (SA) are crucial in controlling plant immune reactions. Some other plant/phytohormones are also extensively studied to regulate/control plant growth. The phytohormones viz., auxin, ABA, cytokinin, gibberellins, and brassinosteroids, have also recently come to light as important regulators of plant immunity.

Melatonin is a good antioxidant molecule in plants and is both environmentally beneficial and biodegradable. The fundamental purpose of an oxygen scavenger is to safeguard membrane structure and maintain the proper balance of ROS metabolism. Fruits treated with exogenous melatonin had significantly higher disease resistance because they produced enough protective enzyme activities, such as superoxide dismutase (SOD), peroxidase (POD), catalase (CAT), phenylalanine ammonia-lyase (PAL), polyphenol oxidase (PPO), chitinase (CHI), and -1,3-glucanase. In this study, we discuss brand-new findings about the regulatory functions of melatonin against disease-causing pathogens and discuss how intricate relationships between other hormones for controlling different biotic stress in different horticulture crops.

7.2 MELATONIN FUNCTION AND ITS ROLE IN PLANTS

The main effect of melatonin is decreasing oxidative stress and it can trigger the immunity/resistance gene against different biological pathogens. Melatonin can also help to change the circadian rhythms in plants. Tryptophan is the common ancestor of both melatonin and auxin, making them both kinds of indole amines. As a result of its function, in *Chenopodium rubrum*, previous research suggested that melatonin controls the circadian rhythms of the plant (Kolar et al., 2009). Additionally, the treatment of melatonin in *Chenopodium rubrum* had an impact on flower growth during the early photoperiod (Kolar et al., 2003). In soybean, the size of the leaf, height of the plant, size of the pod, and seed production all significantly improved after being treated with melatonin (Wei et al., 2015). After melatonin treatment, the slowed down chlorophyll deterioration in the leaves of barley (Arnao et al., 2009). Similarly, in *Lupinus albus*, enhancement of the formation of etiolated cotyledons was higher when exposed to melatonin (Wen et al., 2016).

7.3 MELATONIN BIO-SYNTHESIS

Light, changes in temperature, high radiation, and intermittent and inadequate rainfall are just a few of the factors that might affect the plants to produce melatonin (Zeng et al., 2022). Tryptophan is the main precursor of melatonin. Due to tryptamine 5-hydroxylase (T5H), the tryptophan is changed into tryptamine followed by another enzyme, tryptamine 5-hydroxylase (T5H), to synthesize serotonin (Posmyk et al., 2009). From serotonin, the conversion of N-acetyl serotonin is mediated by serotonin N-acetyltransferase (SNAT) and arylalkyl amine N-acetyltransferase (AANAT). The synthesis of melatonin (HIOMT) from N-acetyl serotonin is done by N-acetylserotonin methyltransferase (ASMT)/hydroxy indole-O-methyltransferase. In the second pathway, 5-methoxy-tryptamine is converted from serotonin by the HIOMT enzyme, and melatonin is produced by the enzyme SNAT from 5-methoxy-tryptamine (Choi et al., 2017). In the reverse conversion of melatonin, N-acetyl-serotonin is changed into serotonin (Lee et al., 2018).

7.4 SIGNALING OF MELATONIN AND ITS RECEPTORS IN PLANTS

Several studies have revealed that melatonin is present in a wide range of plant species and is essential for regulating various physiological processes. These processes include growth stimulation, root induction, seed germination stimulation, increased photosynthetic efficiency, reduction of leaf water/CO_2 levels, and slowing down ripening/senescence processes. The specific role and signaling mechanism of Phyto-melatonin are still largely unknown due to our limited understanding of its receptor. *A. thaliana* contains the first Phyto-melatonin receptor ($CAND_2$/PMTR1), which was recently discovered by Wei et al. (2018). They found that Ca^{2+} flow dynamics also mediate H_2O_2 generation through the $CAND_2$/PMTR1-dependent Phyto-melatonin signaling. They proposed that Phyto-melatonin acts as a novel Phyto-hormone through the H_2O_2 and Ca^{2+} signaling transduction cascade mediated by $CAND_2$/PMTR1.

MAPK cascade may need to be activated by melatonin receptors, according to some earlier results that did not support melatonin-induced H_2O_2 generation and stomatal closure (Jagodzik et al., 2018). The $CAND_2$ acts as a melatonin receptor according to Lee et al. (2020). The $CAND_2$ is a protein that is apart from the plasma membrane and is located in the cytoplasm. Both mutant lines of SALK 071,302 ($CAND_2$–1) and SALK 068848 in *A. thaliana* proceed with the investigation of the role of $CAND_2$–2. They found that the melatonin-mediated defense gene induction (such as GST1) and activation of mitogen-activated protein kinase (MAPK) were altered with mutant lines of $CAND_2$ compared to the wild-type Col-0. The plant melatonin receptor $CAND_2$ has increased awareness of the overall role of melatonin as a powerful antioxidant (Reina et al., 2018).

7.5 THE MECHANISMS THROUGH WHICH PLANTS PROTECT THEMSELVES FROM BIOTIC STRESSES

Naturally, plants have highly developed immune systems to handle the different biotic stresses in various fruit crops. Several plants have physical defenses like waxes, thick cuticles, and distinctive trichomes that help keep insects and pathogens away. Plants also create chemical compounds to protect themselves against diseases (Zaynab et al., 2018). Additionally, two mechanisms that trigger defense reactions enable plants to sense pathogens. The most crucial factor is the pattern recognition receptor, which can identify

PAMPs such as quorum sensing, flagellin, bacterial lipopolysaccharides, fungal chitin, and peptidoglycans. This essential defense process is called PAMP-triggered immunity (PTI) (Monaghan et al., 2012).

The plant resistance (R) proteins are synthesized in the second pathway of the immune system; they aid in identifying the precise pests' or infections' effectors (Avr proteins) and trigger the defensive/immune system in plants via effector-triggered immunity (ETI). This will lead to the programming of cell death and hypersensitive reactions (HR) in infected cells according to Sopel et al. (2012).

7.6 REGULATORY ASPECTS OF MELATONIN FOR STRESSES DUE TO BIOTIC FACTORS

Naturally, the plants express different aspects of defense mechanisms to cope with and acclimatize to stressful situations. Biotic stressors annually result in enormous economic losses of crucial crops. Worldwide, these pests directly reduce agricultural productivity by 20% to 40% (Oreck et al., 1999; Teng, 1987). Different management strategies involve controlling insects, pests, and pathogens using various cultural methods, chemical methods, biological control, and integrated pest management. All of these strategies have the potential to alter microenvironments that are suitable for pests and have an impact on their prevalence (Wyckhuys et al., 2019).

Chemical control is giving quick results to control the different biotic organisms; hence, the producers mostly prefer this method. Since it has immediate results against target pests and pathogens, it causes unfavorable environmental and health effects (Lamichhane et al., 2018). The chemical control method is very costly and also more harmful to human beings and the surrounding environment. The main disadvantage of this method is the long residual effect on the crop plant itself, which invites some other undesirable insect pests in crop production and productivity. Hence, there is an urgent need to focus on environmentally friendly, biodegradable, less toxic, more cost-effective, and no longer residual effects of plant-based chemicals. Melatonin already has many desirable effects as mentioned above in terms of increasing all desired characteristics. Numerous phytohormones that are produced by the plant itself are discussed above. Melatonin is a key immune-regulating plant molecule that protects against viruses, bacteria, fungal infections, and nematodes, among others.

7.6.1 MELATONIN AS A PLANT PATHOGEN ALLEVIATING AGENT

Melatonin is an excellent antioxidant that is effectively protected from damage via scavenging ROS/RNS (Reactive oxygen species and Reactive nitrogen species). Melatonin remains a stress-relieving agent according to several research statements (Arnao et al., 2019). The expression of several genes is changed when plants are exposed to melatonin in addition to being an antioxidant. When dealing with challenging circumstances and assuming reinforcement against plant stress, these regulatory changes are advantageous. As a result, melatonin can help to improve physiological functions like seed and plant germination, increased photosynthetic efficiency, somatic uptake, enhanced plant growth, stimulate plant rooting, osmo-regulation, anti-senescence, improve all metabolic activity, as well as the regulation of plant hormones (Arnao et al., 2019; Arnao et al., 2018). Plants may use the eco-friendly chemical melatonin to protect themselves against biotic stress in an economical manner. Because of its immune-modulating, antioxidant, anti-inflammatory, and neuro-protective properties in plants and animals, melatonin has proven to be a successful therapeutic approach for the prevention of microbial illnesses (Vielma et al., 2014; Regodon et al., 2005).

7.6.2 ANTIVIRAL CONSEQUENCES OF MELATONIN

The external application of melatonin helps to reduce the number of viral RNA and its concentration in *Solanum lycopersicum* and *Nicotiana glutinosa* seedlings (100 mM, twice). According to Zhao et al. (2019), the high concentration of SA in the NO-dependent pathway is responsible for the beneficial effect of melatonin. Additionally, melatonin effectively eliminated the apple stem grooving virus (ASGV) in the shoots of "Gala" apple (Chen et al., 2019). This finding shows that melatonin has antiviral properties (Table 7.1).

7.6.3 MELATONIN BIOACTIVITY AGAINST BACTERIA

Melatonin has proven to have effective antibacterial activity against phytopathogens. *Mycobacterium TB* (H37Rv strain) was also strongly inhibited by melatonin application (Oreck et al., 2006). Melatonin can also activate MAPK cascades to promote SA production in *Arabidopsis thaliana* that has been infected with *Pseudomonas syringae* pv. tomato DC3000 (Liu et al., 2016). In watermelon and *Arabidopsis*, melatonin regulates the ETI- and PTI-associated genes (Table 7.2).

TABLE 7.1 Melatonin's Impact on the Plant Viral Pathogen System

Plant Name	Pathogen Name	Melatonin Dosage (μM)	Delivery Method	Involved Mechanism	Resulting Effect	References
Nicotiana glutinosa and *Solanum lycopersicum*	Tobacco mosaic virus (TMV)	100 (twice)	Irrigation	The increased expression of PR1 and PR5 genes through high Salicylic acid (SA) concentrations	Decrease the virus concentration in infected plants	Zhao, et al., 2019
Malus domestica	Apple stem grooving virus (ASGV)	15	Through shoot proliferation	No longer recognized	Elimination of virus from infected shoot tips.	Chen, et al., 2019

TABLE 7.2 Melatonin's Advantages for Plants Harboring Bacterial Infections

Plant Name	Name of the Pathogen	Dosage of Melatonin (μM)	Application Method	Mechanism	Effect	References
Arabidopsis thaliana, Nicotiana benthamiana	*Pseudomonas syringae*	1 or 10	Foliar application	The regulation and the defense genes PR1, PR5, and PDF1.2 should be expressed and upregulated in increasing order.	The pathogen proliferation should be prevented.	Lee et al., 2014
A. thaliana	*P. syringae*	10	Application of treatments through leaf	Pathogenesis-related (PR) genes via synchronized SA and ET signaling	Upregulate pathogen resistance	Lee et al., 2015
A. thaliana	*P. syringae*	10	Direct application on to the medium	Activities of vacuolar invertase and CWI (VI)	Callose formation and cell-wall strengthening during infection	Zhao et al.,2015

7.6.4 ANTIFUNGAL PROPERTIES OF MELATONIN

In the plant-fungus interaction, melatonin improved the resistance against *Botrytis cinerea* in tomato by regulating H_2O_2 synthesis and the jasmonic acid signaling pathway (Liu et al., 2019). Melatonin also prevented all microorganisms from growing in the apple juice during storage (4–12 hours), including mold, yeast, and bacteria (Zhang et al., 2018). By scavenging defense-related ROS in the affected fruits, it was found that melatonin increased resistance to *Penicillium digitatum*-caused green mold disease in citrus fruit, in contrast to the previous studies stated (Lin et al., 2019). In all cucurbits, gene expression should be changed, such as ETI- and PAMP-mediated responses increasing resistance to soil-borne oomycetes, powdery mildew, and foliar diseases (Mandal et al., 2018) (Table 7.3).

7.7 MELATONIN AND OTHER PHYTOHORMONE INTERACTIONS

Researchers have documented that Phyto melatonin interacts with other plant hormones like auxin, gibberellin, cytokinin, ABA, Ethylene, etc., as well as traditional hormones like auxin and gibberellin because of their shared molecular structure.

7.7.1 MELATONIN-AUXIN INTERACTION

Gravitropism, rooting capacity, and growth capacity are the three factors that were looked at for melatonin and the auxin IAA. The similarity of melatonin and IAA compounds in lupin plants was studied (Hernandez-Ruiz et al., 2004). Research has revealed that melatonin induces vegetative growth through a distinct method (Arnao et al., 2017). When melatonin levels are high, both IAA and melatonin impede development. However, recent research has shown that melatonin controls how signaling components, including auxin regulators, receptors, and tiny auxin-upregulated RNA genes, function to induce growth pathways (Tan et al., 2019).

7.7.2 MELATONIN AND GIBBERELLIC ACIDS (GAS) INTERACTION

The soluble GA receptor (GID genes), which forms a complex with the GA and DELLA proteins and prevents DELLAs from obstructing GA signaling, was increased. Hence, strong GA signaling is produced, which promotes

TABLE 7.3 The Function of Exogenous Melatonin in the Interaction Between Plants and Fungi

Plant Name	Pathogen Name	Melatonin Dosage (_M)	Delivery Method	Involved Mechanism	Resulting Effect	References
Malus Prunifolia	*Diplocarpon Mali*	50–500	Fertigation	Sustain intracellular H2O2 levels N The actions of enzymes involved in plant defense	reducing the effects of disease damage resistance to fungus decrease of lesions	Yin et al., 2013
Musa acuminata	*Fusarium oxysporum*	100	Foliar and root spray	By controlling the MaHSP90s gene's expression, resistance is produced.	The disease resistance was enhanced abundantly after the application of melatonin.	Wei et al., 2017
Fragaria Ananassa	*Botrytis cinerea* and *Rhizopus Stolonifer*	100	Fruits were dipped in treatment.	N Antioxidant enzyme activity, H2O2 levels	In stored strawberry, the decay and deterioration should be reduced.	Aghdam et al., 2017
Citrullus Lanatus	*Podosphaera xanthii* and *Phythophthora capsici*	100	Foliar application	Increased action of PTI- and ETI-associated genes	Enhanced Disease resistance	Mandal et al., 2018
A. thaliana	—	1 x 10^{-4} or 100	The seedlings were dipped in solution.	In abscisic acid (ABA), ET, SA, and JA pathways some of the important genes should be upregulated functionally and structurally.	Improve resistance against biotic stress	Weeda et al., 2014

seedling growth (Tan et al., 2019). In apple plants, melatonin therapy increased GA levels and cyclin synthesis (Mao et al., 2020). In *Arabidopsis*, increasing expression of blooming locus C and decreasing flowering locus T expression, melatonin administration prevented the floral transition in terms of reproductive development (Shi et al., 2015). Melatonin has been shown to increase GA levels in response to various stressors, but it is doubtful if it directly affects GA levels (Arnao et al., 2013).

7.7.3 INTERACTIONS BETWEEN ABSCISIC ACID AND MELATONIN (ABA)

In the melatonin-ABA interaction, melatonin can cause a decrease in ABA content. The ABA production enzymes were downregulated as a result of altered Abscisic acid signaling element control. The 9-cis-epoxy carotenoid dioxygenase (NCED), a key enzyme in ABA biosynthesis, is frequently down-regulated by melatonin therapy, while the ABA catabolism genes (CYP707 mono oxygenase) are up-regulated, causing a considerable drop in ABA levels. Examples of this include hickory, apples, Chinese cabbage, and cucumbers (Arnao et al., 2021). In mango, the downregulation of pectin-modifying enzymes including NCED, ACS, and ACO results in a decrease in ripening and softening, as well as a low synthesis of ABA and ethylene after treatment with melatonin.

7.7.4 INTERACTION BETWEEN MELATONIN AND ETHYLENE

Melatonin affects the ripening process by controlling RIN, CNR, NOR, and AP2a. Melatonin causes tomatoes to express genes responsible for carotenoid production. The ACO and ACS gene activity was upregulated by melatonin (EIL1, EIL3, and ERF2). The scientific evidence of proteomic analysis states that melatonin has a significant effect on a number of proteins related to the ripening pathways in tomato. Postharvest markers on strawberries improved after melatonin application. The application of melatonin raised fruit quality indices, antioxidant content, and bioactive chemical levels while extending fruit shelf life (Zahedi et al., 2020).

7.7.5 SALICYLIC ACID (SA) WITH MELATONIN INTERACTION

SA controls how plants respond to biological stress (Nabi, R.B.S et al., 2021). The application of melatonin typically causes a rise in SA, JA, and ethylene production. According to past research, melatonin is essential for plants' defense against biotic stress (Arnao et al., 2018). The higher count of *Pseudomonas syringae* DC3000 had higher levels of SA and melatonin in infected *Arabidopsis* plants. SNAT knockout mutants had decreased melatonin and SA levels and were more vulnerable to pathogen assaults (Lee et al., 2015). When melatonin was exogenously applied, plants produced increased SA and NO as a defense against the tobacco mosaic virus. Melatonin is one of the best hormones for reducing viral RNA (Zhao et al., 2019).

7.7.6 INTERACTION BETWEEN MELATONIN AND JASMONIC ACID (JA)

Jasmonic acid and Methyl jasmonate (MeJA) are two important plant hormones. They can be referred to as jasmonates (JA). They have influence over many variables that affect how plants grow, develop, and respond to stress. Jasmonates cause an upregulation of genes, metabolites, and proteins, and are involved in the plant defense system (Yang et al., 2019). According to recent studies, the connection between melatonin and JA is extremely complex. Melatonin therapies, for instance, have an effect on JA levels in studies of abiotic stress, but this effect is not immediately apparent. In B. napus, melatonin decreases JA production under salt stress. Moreover, the signaling pathway of JA should be blocked by the synthesis of JAZ proteins, which leads to blocking the signaling pathway of JA. Plant growth and tolerance to salt are improved as a result of the reduced response to JA (Tan et al., 2019).

7.8 MOLECULAR MECHANISM OF MELATONIN WITH OTHER ENZYMES

Melatonin is a superb antioxidant chemical found in plants. By triggering antioxidant mechanisms including non-enzymatic antioxidants and enzymatic antioxidants, the exogenous application of this melatonin can successfully decrease the rate of ROS levels in fruits and delay the ripening process (Onik et al., 2021; Wang et al., 2020; Liu et al., 2020). It eliminates ROS and increases the activity of enzymes like SOD, CAT, and POD that scavenge ROS (Cao et al., 2017; Liu et al., 2019; Zhang, et al., 2018). SOD is a key antioxidant enzyme that is found in many different types of organisms and catalyzes the dismutation of oxygen into molecular oxygen and hydrogen.

In contrast to SOD, CAT and APX can convert H_2O_2 to H_2O and O_2 in cells, while CAT and POD can dismutate O_2 to O_2 and H_2O_2 (Mittler et al., 2004; Zhu et al., 2015). All eukaryotes and prokaryotes include GR, also known as flavoprotein oxidoreductase. It is a crucial antioxidant enzyme for the removal of active oxygen from plant tissues, which effectively lowers the amount of H_2O_2 in plants and increases their resilience to environmental challenges.

7.9 ROLE AND MOLECULAR APPROACHES OF MELATONIN-MEDIATED DEFENSE IN FRUIT CROPS

The antioxidant capacity of fresh-cut pears was significantly enhanced by melatonin treatment by raising the total phenolic content, the capacity to scavenge DPPH and 2,2'-Azinobis (3-ethylbenzothiazoline-6-sulfonic acid Ammonium Salt) (ABTS), and the activities of PAL and CHS, as well as by lowering the levels of MDA and H_2O_2 and ascorbic acid. This reduced browning and the growth of microorganisms (Zheng et al., 2019). Zezhen et al., claimed in 2022 to have described further molecular melatonin techniques. The price of post-harvest grapes is significantly impacted by *Botrytis cinerea*. The grapes were given injections of 0.02, 0.2, and 2 mmol/L melatonin, after injections of *B. cinerea* solution 48 hours later. Melatonin exhibited no observable effects on the mycelium growth and spore germination of *B. cinerea*. On the other hand, post-harvest melatonin administration decreased malondialdehyde formation, prevented a rise in cell membrane permeability, and boosted the generation and accumulation of total phenols and flavonoids. Fruits treated with the chemical also showed considerably higher activity levels of protective enzymes such as superoxide dismutase (SOD), peroxidase (POD), catalase (CAT), phenylalanine ammonia-lyase (PAL), polyphenol oxidase (PPO), chitinase (CHI), and 1,3-glucanase.

A successful melatonin therapy (600 mol/L for 1–2 hours) was found to delay the fruit ripening and reduce the anthracnose incidence in guava fruit, according to Si Lin Fan et al., 2022. Melatonin substantially increased the antioxidant capacity and decreased oxidative damage such as superoxide anions, malondialdehyde, and hydrogen peroxide in the fruit. Moreover, it boosted the overall antioxidant capacity and improved both non-enzymatic antioxidants and enzymatic antioxidants. Superoxide dismutase, ascorbate peroxidase, catalase, and glutathione reductase all experienced large increases in activity when melatonin was present. Melatonin was responsible for maintaining the amounts of ascorbic acid and total flavonoids. PPO helps turn the polyphenols in plants into quinones, which helps to stop pathogen growth and lessen harmful pathogen infection. Grapes with melatonin treatments produced more total phenols and flavonoids, which in turn significantly raised the PAL, POD, and PPO activities. In peach fruit, melatonin administration successfully increased CAT, SOD, APX, and POD activities while decreasing MDA, O_2, and hydrogen peroxide content. It also preserved membrane integrity and delayed postharvest senescence (Gao et al., 2016).

The *Candidatus Liberibacter asiaticus* is the disease's putative cause in citrus, and *Diaphorina citri* is its purported carrier. Huanglongbing (HLB) is a devastating illness that affects the fruit worldwide. The *D. citri* infestation and Ca. *L. asiaticus* infection significantly improved the external melatonin content in Valencia sweet orange and upregulated the gene expression of CsTDC, CsT5H, CsSNAT, CsASMT, and CsCOMT, which are involved in its production. Salicylic acid has a correlation with melatonin induction, even though trans-jasmonic acid does not. According to Yasser Nehela and Nabil Killinya, more melatonin increased the endogenous levels of the stress-related phytohormones trans-jasmonate, auxins, abscisic acid, and salicylates, as well as the transcript levels of their biosynthetic genes. It's interesting to see that the Ca. *L. asiaticus* titer in the affected leaves appeared to have decreased after melatonin delivery.

Exogenous melatonin increased the activity of antioxidants enzymes like PAL, C4H, and 4CL and also increased the rate of accumulation of non-enzyme antioxidants such as flavonoids and phenolics in litchi, which made them resistant to *Peronophythora litchii* (Zhang et al., 2021). Some important enzyme activities, such as SOD, CAT, POD, PAL, and PPO, were increased by melatonin. These enzymes are basically defense-related. The *Macerozyme guilliermondii* Y-1 effectively increased as well in apple fruit, increasing the rate of non-enzyme antioxidants (TP and lignin) and the total antioxidant capacity (T-AOC), which led to a reduction in the incidence of disease (Sun et al., 2021).

7.10 ROLE AND MOLECULAR INTERACTIONS OF MELATONIN IN VEGETABLE CROPS

Production of cucumbers is seriously threatened by downy mildew, which is caused by *Pseudoperonospora cubensis*. The effects of pre-treatment with melatonin on cucumber resistance to downy mildew. The antioxidant capacity of those plants after inoculation was improved by the application of 100 mol L1 melatonin, which increased antioxidant enzyme activity and antioxidant gene expression. This pre-treatment successfully maintained the membrane's integrity by reducing the amount of malondialdehyde and relative electrolytic leakage. The gene expression and enzyme activity of this pathogen were also improved by melatonin. The nitrogen metabolism was also increased after inoculation. These responses show how melatonin supports fruit yields and strengthens cucumber resistance to downy mildew (Zeyu Liu et al., 2019). Following inoculation, the ability to metabolize nitrogen also increased. These results demonstrate how melatonin promotes fruit production and increases cucumber resistance to downy mildew (Zeyu Liu et al., 2019). Melatonin considerably raised the expression of CAT by factors of 1.12 and 1.81, while it greatly elevated the expression of cAPX by factors of 0.68 and 1.38. PPO activity rose throughout time, with levels being regularly greater than those of PAL, and occasionally up to five times higher. These results illustrated the pivotal function PPO serves in plants' defenses against DM. Melatonin effectively increased the phenylpropanoid pathway enzyme activities, such as 4CL, PAL, and POD, which help to improve the rate of accumulation of total phenolics, lignin, and flavonoids in cherry tomato fruit. S. Li and associates 2019. Melatonin aids a vital role in signaling elements and it has a variety of physiological activities in tomato plants' responses to biotic stressors. The possible impact of melatonin on fruit defense against *Botrytis cinerea* and also associated with molecular and physiological mechanisms. As a result, discovered that melatonin administration boosted the resistance against disease, with 50 M being the ideal concentration. The defense-related enzyme activities are improved by melatonin therapy and increased antioxidant enzyme activity while decreasing hydrogen peroxide (H_2O_2) concentration. According to Chunxue Liu et alresearch.'s (2019), melatonin therapy elevated the levels of methyl jasmonate (MeJA), promoted the expression of SlLoxD, SlAOC, and SlPI II, and lowered the expression of SlMYC2 and SlJAZ1. By controlling H_2O_2 levels and the JA signaling pathway, melatonin contributed to increasing resistance against *Botrytis cinerea* in tomato.

7.11 CONCLUSION

Melatonin has recently received a lot of interest due to its unique benefits in controlling plant development and increasing plant tolerance to diverse biotic stresses. Melatonin has recently drawn a lot of attention due to its useful role in the regulation of development and growth of plants. The potential role of melatonin is greatest in terms of being a regulatory molecule in circadian rhythms and photoperiodicity.

The plant's circadian oscillator can alter a number of biological processes. These processes cover many different periodic (photoperiodic/seasonal) cycle-related processes, including protein stability, metabolic control, and many more. Even in the presence of biotic and abiotic stress circumstances, the strong circadian rhythm influences photosynthesis and agricultural yield and promotes plant growth. The primary effects of melatonin on growth and development are discussed in the current review, along with how melatonin and PGPR treatments jointly alter plants' ability to withstand various environmental challenges. We also discuss how melatonin and other phytohormones interact under stress and reveal that various plants have melatonin receptors and signaling. More investigation is needed to better understand how melatonin influences plant infection by viruses, nematodes, and insects.

KEYWORDS

- **biotic stress**
- **hormone-pathogen interaction**
- **horticulture crop**
- **melatonin**
- **molecular mechanism**
- **PAMP**
- **phytohormone**
- **PPO**
- **resistance mechanism**
- **signaling pathways**

REFERENCES

Aghdam, M. S., & Fard, J. R. (2017). Melatonin treatment attenuates postharvest decay and maintains nutritional quality of strawberry fruits (*Fragaria* × *anannasa* cv. Selva) by enhancing GABA shunt activity. *Food Chemistry, 221*, 1650–1657.

Arnao, M. B., & Hernández-Ruiz, J. (2009). Protective effect of melatonin against chlorophyll degradation during the senescence of barley leaves. *Journal of Pineal Research, 46*(1), 58–63.

Arnao, M. B., & Hernández-Ruiz, J. (2017). Growth activity, rooting capacity, and tropism: Three auxinic precepts fulfilled by melatonin. *Acta Physiologiae Plantarum, 39*, 1–9.

Arnao, M. B., & Hernández-Ruiz, J. (2019a). Melatonin and reactive oxygen and nitrogen species: A model for the plant redox network. *Melatonin Research, 2*(3), 152–168.

Arnao, M. B., & Hernández-Ruiz, J. (2019b). Melatonin: A new plant hormone and/or a plant master regulator? *Trends in Plant Science, 24*(1), 38–48.

Arnao, M. B., & Hernández-Ruiz, J. (2021). Melatonin as a regulatory hub of plant hormone levels and action in stress situations. *Plant Biology, 23*, 7–19.

Cao, J. J., Yu, Z. C., Zhang, Y., Li, B. H., Liang, W. X., & Wang, C. X. (2017). Control efficiency of exogenous melatonin against post-harvest apple grey mold and its influence on the activity of defensive enzymes. *Plant Physiology Journal, 53*(1753), 1760.

Choi, G. H., Lee, H. Y., & Back, K. (2017). Chloroplast overexpression of rice caffeic acid O-methyltransferase increases melatonin production in chloroplasts via the 5-methoxytryptamine pathway in transgenic rice plants. *Journal of Pineal Research, 63*(1), e12412.

Fan, S., Xiong, T., Lei, Q., Tan, Q., Cai, J., Song, Z., & Zhu, X. (2022). Melatonin treatment improves postharvest preservation and resistance of guava fruit (*Psidium guajava* L.). *Foods, 11*(3), 262.

Fortunati, E., Mazzaglia, A., & Balestra, G. M. (2019). Sustainable control strategies for plant protection and food packaging sectors by natural substances and novel nanotechnological approaches. *Journal of the Science of Food and Agriculture, 99*(3), 986–1000.

Hernandez-Ruiz, J., Cano, A., & Arnao, M. B. (2004). Melatonin: A growth-stimulating compound present in lupin tissues. *Planta, 220*, 140–144.

Kolář, J., Johnson, C. H., & Macháčková, I. (2003). Exogenously applied melatonin (N-acetyl-5-methoxytryptamine) affects flowering of the short-day plant *Chenopodium rubrum*. *Physiologia Plantarum, 118*(4), 605–612.

Lamichhane, J. R., Osdaghi, E., Behlau, F., Köhl, J., Jones, J. B., & Aubertot, J. N. (2018). Thirteen decades of antimicrobial copper compounds applied in agriculture. A review. *Agronomy for Sustainable Development, 38*, 1–18.

Lee, H. Y., Byeon, Y., & Back, K. (2014). Melatonin as a signal molecule triggering defense response against pathogen attack in Arabidopsis and tobacco. *Journal of Pineal Research, 57*(3), 262–268.

Lee, H. Y., Byeon, Y., Tan, D. X., Reiter, R. J., & Back, K. (2015). Arabidopsis serotonin N-acetyltransferase knockout mutant plants exhibit decreased melatonin and salicylic acid levels resulting in susceptibility to an avirulent pathogen. *Journal of Pineal Research, 58*(3), 291–299.

Li, C., Wang, P., Wei, Z., Liang, D., Liu, C., Yin, L., & Ma, F. (2012). The mitigation effects of exogenous melatonin on salinity-induced stress in *Malus hupehensis*. *Journal of Pineal Research, 53*(3), 298–306.

Li, Z., Zhang, S., Xue, J., Mu, B., Song, H., & Liu, Y. (2022). Exogenous melatonin treatment induces disease resistance against *Botrytis cinerea* on post-harvest grapes by activating defense responses. *Foods, 11*(15), 2231.

Lin, Y., Fan, L., Xia, X., Wang, Z., Yin, Y., Cheng, Y., & Li, Z. (2019). Melatonin decreases resistance to postharvest green mold on citrus fruit by scavenging defense-related reactive oxygen species. *Postharvest Biology and Technology, 153*, 21–30.

Liu, C., Chen, L., Zhao, R., Li, R., Zhang, S., Yu, W., & Shen, L. (2019). Melatonin induces disease resistance to *Botrytis cinerea* in tomato fruit by activating jasmonic acid signaling pathway. *Journal of Agricultural and Food Chemistry, 67*(22), 6116–6124.

Mandal, M. K., Suren, H., Ward, B., Boroujerdi, A., & Kousik, C. (2018). Differential roles of melatonin in plant-host resistance and pathogen suppression in cucurbits. *Journal of Pineal Research, 65*(3), e12505.

Mao, J., Niu, C., Li, K., Chen, S., Tahir, M. M., Han, M., & Zhang, D. (2020). Melatonin promotes adventitious root formation in apple by promoting the function of MdWOX11. *BMC Plant Biology, 20*(1), 1–11.

Mittler, R., Vanderauwera, S., Gollery, M., & Van Breusegem, F. (2004). Reactive oxygen gene network of plants. *Trends in Plant Science, 9*(10), 490–498.

Monaghan, J., & Zipfel, C. (2012). Plant pattern recognition receptor complexes at the plasma membrane. *Current Opinion in Plant Biology, 15*(4), 349–357.

Nabi, R. B. S., Rolly, N. K., Tayade, R., Khan, M., Shahid, M., & Yun, B. W. (2021). Enhanced resistance of atbzip62 against *Pseudomonas syringae* pv. *tomato* suggests negative regulation of plant basal defense and systemic acquired resistance by AtbZIP62 transcription factor. *International Journal of Molecular Sciences, 22*(21), 11541.

Nehela, Y., & Killiny, N. (2020). Melatonin is involved in citrus response to the pathogen huanglongbing via modulation of phytohormonal biosynthesis. *Plant Physiology, 184*(4), 2216–2239.

Oerke, E. C. (1994). Estimated crop losses due to pathogens, animal pests and weeds. In *Crop Production and Crop Protection* (pp. 535–597). Elsevier Science Publishing, New York, NY.

Onik, J. C., Wai, S. C., Li, A., Lin, Q., Sun, Q., Wang, Z., & Duan, Y. (2021). Melatonin treatment reduces ethylene production and maintains fruit quality in apple during postharvest storage. *Food Chemistry, 337*, 127753.

Posmyk, M. M., & Janas, K. M. (2009). Melatonin in plants. *Acta Physiologiae Plantarum, 31*, 1–11.

Posmyk, M. M., Bałabusta, M., Wieczorek, M., Sliwinska, E., & Janas, K. M. (2009). Melatonin applied to cucumber (*Cucumis sativus* L.) seeds improves germination during chilling stress. *Journal of Pineal Research, 46*(2), 214–223.

Saqib, S., Nazeer, A., Ali, M., Zaman, W., Younas, M., Shahzad, A., & Nisar, M. (2022). Catalytic potential of endophytes facilitates synthesis of biometallic zinc oxide nanoparticles for agricultural application. *BioMetals, 35*(5), 967–985.

Shi, H., Tan, D. X., Reiter, R. J., Ye, T., Yang, F., & Chan, Z. (2015). Melatonin induces class A1 heat-shock factors (HSFA 1s) and their possible involvement of thermotolerance in Arabidopsis. *Journal of Pineal Research, 58*(3), 335–342.

Spoel, S. H., & Dong, X. (2012). How do plants achieve immunity? Defense without specialized immune cells. *Nature Reviews Immunology, 12*(2), 89–100.

Sun, C., Huang, Y., Lian, S., Saleem, M., Li, B., & Wang, C. (2021). Improving the biocontrol efficacy of *Meyerozyma guilliermondii* Y-1 with melatonin against postharvest gray mold in apple fruit. *Postharvest Biology and Technology, 171*, 111351.

Sun, Y., Liu, Z., Lan, G., Jiao, C., & Sun, Y. (2019). Effect of exogenous melatonin on resistance of cucumber to downy mildew. *Scientia Horticulturae, 255*, 231–241.

Tan, X., Long, W., Zeng, L., Ding, X., Cheng, Y., Zhang, X., & Zou, X. (2019). Melatonin-induced transcriptome variation of rapeseed seedlings under salt stress. *International Journal of Molecular Sciences, 20*(21), 5355.

Teng, P. S. (1987). *Crop Loss Assessment and Pest Management* (No. BOOK). American Phytopathological Society.

Vielma, J. R., Bonilla, E., Chacín-Bonilla, L., Mora, M., Medina-Leendertz, S., & Bravo, Y. (2014). Effects of melatonin on oxidative stress, and resistance to bacterial, parasitic, and viral infections: A review. *Acta Tropica, 137*, 31–38.

Wang, S. Y., Shi, X. C., Wang, R., Wang, H. L., Liu, F., & Laborda, P. (2020). Melatonin in fruit production and postharvest preservation: A review. *Food Chemistry, 320*, 126642.

Weeda, S., Zhang, N., Zhao, X., Ndip, G., Guo, Y., Buck, G. A., & Ren, S. (2014). Arabidopsis transcriptome analysis reveals key roles of melatonin in plant defense systems. *PloS One, 9*(3), e93462.

Wei, W., Li, Q. T., Chu, Y. N., Reiter, R. J., Yu, X. M., Zhu, D. H., & Chen, S. Y. (2015). Melatonin enhances plant growth and abiotic stress tolerance in soybean plants. *Journal of Experimental Botany, 66*(3), 695–707.

Wei, Y., Hu, W., Wang, Q., Zeng, H., Li, X., Yan, Y., & Shi, H. (2017). Identification, transcriptional and functional analysis of heat-shock protein 90s in banana (*Musa acuminata* L.) highlight their novel role in melatonin-mediated plant response to *Fusarium* wilt. *Journal of Pineal Research, 62*(1), e12367.

Wen, D., Gong, B., Sun, S., Liu, S., Wang, X., Wei, M., & Shi, Q. (2016). Promoting roles of melatonin in adventitious root development of *Solanum lycopersicum* L. by regulating auxin and nitric oxide signaling. *Frontiers in Plant Science, 7*, 718.

Wyckhuys, K. A., Hughes, A. C., Buamas, C., Johnson, A. C., Vasseur, L., Reymondin, L., & Sheil, D. (2019). Biological control of an agricultural pest protects tropical forests. *Communications Biology, 2*(1), 10.

Yang, J., Duan, G., Li, C., Liu, L., Han, G., Zhang, Y., & Wang, C. (2019). The crosstalks between jasmonic acid and other plant hormone signaling highlight the involvement of jasmonic acid as a core component in plant response to biotic and abiotic stresses. *Frontiers in Plant Science, 10*, 1349.

Yin, L., Wang, P., Li, M., Ke, X., Li, C., Liang, D., Wu, S., Ma, X., Li, C., Zou, Y., & Ma, F. (2013). Exogenous melatonin improves Malus resistance to Marssonina apple blotch. Journal of pineal research, 54(4), 426–434. https://doi.org/10.1111/jpi.12038.

Zahedi, S. M., Hosseini, M. S., Abadía, J., & Marjani, M. (2020). Melatonin foliar sprays elicit salinity stress tolerance and enhance fruit yield and quality in strawberry (*Fragaria* × *ananassa* Duch.). *Plant Physiology and Biochemistry, 149*, 313–323.

Zaynab, M., Fatima, M., Abbas, S., Sharif, Y., Umair, M., Zafar, M. H., & Bahadar, K. (2018). Role of secondary metabolites in plant defense against pathogens. *Microbial Pathogenesis, 124*, 198–202.

Zeng, W., Mostafa, S., Lu, Z., & Jin, B. (2022). Melatonin-mediated abiotic stress tolerance in plants. *Frontiers in Plant Science, 13*, 847175.

Zhang, H., Liu, X., Chen, T., Ji, Y., Shi, K., Wang, L., & Kong, J. (2018). Melatonin in apples and juice: Inhibition of browning and microorganism growth in apple juice. *Molecules, 23*(3), 521.

Zhao, L., Chen, L., Gu, P., Zhan, X., Zhang, Y., Hou, C., & Wang, Q. C. (2019). Exogenous application of melatonin improves plant resistance to virus infection. *Plant Pathology, 68*(7), 1287–1295.

Zheng, H., Liu, W., Liu, S., Liu, C., & Zheng, L. (2019). Effects of melatonin treatment on the enzymatic browning and nutritional quality of fresh-cut pear fruit. *Food Chemistry, 299*, 125116.

Zhu, X. M., Shi, X. P., Yong, D. J., Zhang, Y., Li, B. H., Liang, W. X., & Wang, C. X. (2015). Induction of resistance against *Glomerella cingulata* in apple by endophytic actinomycetes strain A-1. *Plant Physiology Journal, 51*(6), 949–954.

PART III

Advanced Tools and Methods for Climate Resilience Crops Under Biotic Stress

CHAPTER 8

The Exploitation of Next-Generation Breeding Strategies to Build Biotic Stress Resilience in Crops

PRASANTA KUMAR MAJHI,[1] TANMAYA KUMAR BHOI,[2] IPSITA SAMAL,[3] DEEPAK KUMAR MAHANTA,[4] PARTHA PRATIM BEHERA,[5] MANARANJAN SENAPATI,[6] J. KOMAL,[7] K. JAGADEESH,[8] YAGNAYANI SENAPATI,[9] and SUBRAT KUMAR SENAPATI[10]

[1]Department of Plant Breeding and Genetics, Regional Research and Technology Transfer Station (RRTTS), Odisha University of Agriculture and Technology (OUAT), Keonjhar, Odisha, India

[2]Forest Protection Division, ICFRE–Arid Forest Research Institute (AFRI), Jodhpur, Rajasthan, India

[3]Department of Entomology, Faculty of Agriculture, Sri Sri University, Cuttack, Odisha, India

[4]Department of Entomology, Dr. Rajendra Prasad Central Agricultural University, Pusa, Bihar, India

[5]Department of Plant Breeding and Genetics, Assam Agricultural University, Jorhat, Assam, India

[6]Division of Genetics, ICAR–Indian Agricultural Research Institute (IARI), Pusa Campus, New Delhi, India

[7]Department of Entomology, Navsari Agricultural University, Navsari, Gujarat, India

[8]Department of Genetics and Plant Breeding, Professor Jayashankar Telangana State Agricultural University, Hyderabad, Telangana, India

Climate Change and Biotic Factors: A Molecular Approach.
Ashutosh Singh, Saurabh Pandey, and Amarjeet Kumar, (Eds.)

[9]*Department of Plant Breeding and Genetics, College of Agriculture, Odisha University of Agriculture and Technology (OUAT), Bhubaneswar, Odisha, India*

[10]*Department of Floriculture and Landscaping, Odisha University of Agriculture and Technology (OUAT), Bhubaneswar, Odisha, India*

ABSTRACT

The emerging field of molecular genetics and related technologies is an avenue for selecting new crop species. By assisting in the insertion of target genes into the genotype that is indeed adapted, molecular markers are used in rapid parental genome recovery using marker-assisted backcross breeding (MABB). In QTL mapping, a suitable marker connected to the desired trait is sought after. The goal of this pyramiding of genes technique is to create a genotype that combines the advantageous features inherited from many parents. Due to recent developments in transgenic technologies, we can now modify crops for biotic resistance in a targeted manner that has minimal impact on soil microbes and the environment. A new generation of genome-sequencing tools and the identification of large-scale gene polymorphisms have made the emerging science, which we refer to as "IPM-omics," commercially viable for pest-insect systems. By combining NGS data with high-throughput polymorphism detection methods, we can learn more about pest populations and make more informed pest management decisions. The field of plant stress research extensively uses Machine Learning and its offshoot Deep Learning. The four types of plant stress that can be identified, categorized, quantified, and predicted have all been studied using these techniques (ICQP). The most common use of transcriptome and genome studies is to predict phenotypic features before they manifest. Thus, these next-generation breeding strategies can build climate resilience in crops against biotic stresses.

8.1 INTRODUCTION

Around 11,500 years ago, nomadic man gradually learned how to cultivate cereal and root crops and live a settled existence based on farming. By 2,000 years ago, agriculture supported the majority of the world's population.

The yield, harvest index, and quality features of cultivated plants have been modified to meet human requirements and preferences using a combination of artificial and natural selection techniques since their domestication. Crop productivity depends incredibly on edaphoclimatic conditions. The persistent climate change, driven by the human-caused increase in greenhouse gases such as CO_2, CH_4, etc., and characterized by a rise in global temperature and changes in precipitation patterns, is a major risk to farming. Agricultural intensification and irrigation have already begun to degrade land, and climate change may hasten this process (Kissoudis et al., 2016). The global production of maize and wheat is anticipated to decrease by 3.8% and 5.5%, respectively, because of the effects of global warming. United Nations climate scientists released their first report as part of the Intergovernmental Panel on Climate Change (IPCC), warning that a temperature increase of 1.5 degrees Celsius would likely occur by 2040. Feeding the world's expanding population by 2050, agricultural output worldwide must increase by 60% per year from 2005–07 to 2050. This includes increases of 77% in poor countries and 24% in wealthy nations (Malhi et al., 2021).

The disease triangle shows how the host plant, the pathogen, and the environment all have a role in the progression of a disease in a plant. Insect and pest population dynamics are primarily influenced by temperature, but due to the wide variety of insect species, pathogen races, and host plants, we can expect a wide range of reactions to climate change in various regions. Changes in climate may cause alterations in their range, generation time, overwintering success, invading insect species, plant diseases spread by insects, and how they affect their hosts and natural foes (Skendi et al., 2021). Activation of latent pathogen races due to global warming could trigger an unexpected pandemic by shifting the geographic distribution of existing diseases. Warmer temperatures combined with saturated soil produce an environment that is ideal for the growth of soilborne pathogens and foliar diseases. It may also lead to the development of new races and diseases in an ecosystem (Hunjan & Lore, 2020). Extreme temperatures and weather variability may reduce the flowering period leading to a decrease in the nectar and pollen availability to honey bees, thereby hampering pollination and fertilization, as well as a reduction in bee population due to unavailability of food.

Taking the right steps at the right time, like using adapted cultivars with different sowing times and different ways of employing fertilizer and water, etc., can help us deal with climate change and may even make it easier for crops to grow. To counteract the detrimental effects of climate change on crop productivity, it is essential to breed climate-resilient crops. Increasing crop productivity by creating climate-resilient superior genotypes is one way that next-generation breeding technologies can help the world's food supply meet future demands (Razzaq et al., 2021). Current improvements in genomic-assisted breeding (GAB) methods have made it possible to create fully labeled crop pan-genomes that show the whole landscape of genetic diversity (GD) and bring back the lost gene repertoire of a species. The crop pan-genomes offer fresh ways to utilize this distinctive genetic material or variant in order to enhance breeding efforts. Therefore, the present chapter highlighted different next-generation breeding approaches including Marker-Assisted Backcross Breeding, QTL mapping, Multiomics approaches, Transgenic approaches, genome editing, artificial intelligence, and bioinformatics tools to develop biotic stress resilience in crops.

8.2 NEXT-GENERATION BREEDING INNOVATION WITH MOLECULAR MARKERS

8.2.1 QTL MAPPING FOR TARGET TRAITS

Quantitative features, which include the vast majority of economic significance, are typically controlled by a large number of genes. According to this, there are various ways in which quantitative traits and qualitative traits differ. A large number of major and minor genes affect quantitative traits, while qualitative qualities are typically determined by only one gene or a few genes (oligogenic). Two, for quantitative traits, the effect of many genes is small and additive, while for qualitative traits, the effect of a gene is large. Three, environmental factors have a significant impact on quantitative traits. Four, these traits are typically recorded as observable variations with continuous or distributed normality. The genomic regions governing these quantitative traits are known as Quantitative trait locus (QTLs). To identify and utilize these loci, two strategies are followed: one is family mapping (linkage mapping) and the other is population-based mapping (association mapping) (Singh & Singh, 2015). Many disease and pest resistance traits have been mapped in various crops, as shown in Table 8.1 for major crops. Compounding the linkage after using several mapping functions, including the Haldane mapping operation, which assumes that interference doesn't happen, and Kosambi mapping operation, which assumes that interference does happen, yields the genetic distance between the marker and the trait. Below, we outline the standard prerequisites and methodology for QTL mapping.

TABLE 8.1 Different Biotic Stress Characterized Using Molecular Mapping Procedure

Sl. No.	Crop	Traits	Property of Inheritance	Number of Qtls/ Genes Reported	References
1.	Rice	BLB	Monogenic	46	Fiyaz et al. (2022)
2.		Blast	Mostly Polygenic	118	Kalia & Rathour, (2019)
3.		Sheath Blight	Polygenic	≈ 112	Senapati et al. (2022)
4.		Brown plant hopper	Mostly monogenic	37	Muduli et al. (2021)
5.		White backed plant hopper	Both Monogenic and Polygenic	23	Makkar et al. (2019)
6.	Wheat	Stem rust	Monogenic	≈ 60	Randhawa et al.(2019)
7.		Yellow rust	Monogenic	78	Randhawa et al. (2019)
8.		Leaf rust	Monogenic	77	Randhawa et al. (2019)
9.		powdery mildew	Monogenic	73	Krishnappa et al. (2022)
10.		*Fusarium* Head Blight	Polygenic	7	Randhawa et al. (2019)
11.		Wheat blast	Both Monogenic and Polygenic	8	Randhawa et al. (2019)
12.		Nematodes	Both Monogenic and Polygenic	≈ 60	Awol et al. (2021)
13.	Mustard	White rust	Mostly monogenic	≈ 6	Inturrisi et al. (2021)
14.	Chickpea	Wilt (*Fusarium*)	Polygenic	≈ 2	Roorkiwal et al. (2020)

8.2.1.1 SUITABLE PARENTAL LINES THAT SHOULD BE PERFECTLY CONTRAST FOR TRAITS OF INTEREST

In general, for any biotic stress, the source of resistance must be identified first. Then only resistance-governing genomic areas can be mapped. As parents, suitable resistant and susceptible lines are employed to create mapping populations. The two parents should be easily crossable and should contrast significantly with regard to the disease or pest in question.

8.2.1.2 SUITABLE MAPPING POPULATION (F_2, RIL DH, NIL, ETC.) FROM THE PARENTAL LINES

Several forms of mapping populations have previously been established during studies; we must employ them based on our objective. For preliminary mapping or mapping heterotic QTLs, the F_2 and $F_{2:3}$ populations are acceptable. For precisely mapping the QTLs in different environments over a number of years, RIL populations are favored. If your objective is to fine-map the resistance QTL, you can choose amongst NIL, $RILF_2$, and $NILF_2$ populations.

8.2.1.3 ROBUST PHENOTYPING IN MULTIPLE ENVIRONMENTS OR IN DIFFERENT SEASONS

Phenotypic observation of the mapping population is the key to finding the QTLs for resistance. Protocols for accurate phenotyping have previously been established for a number of disease and pest screenings in important cereal and pulse crops. The majority of diseases are scored on a 0 to 9 scale, where 0 indicates immunity (0% infection) and 9 indicates high susceptibility (>90% infection). Artificial inoculation of a pathogen at a particular growth stage to produce the symptoms and maintain the microclimatic conditions that are conducive to growing the pathogen more rapidly is the key to disease response screening. The growing environment can be controlled in the field or in hot spot areas, particularly for that pests will help in accurate phenotyping.

8.2.1.4 CHECKING POLYMORPHISM BETWEEN TWO PARENTS BY USING SUITABLE GENETIC MARKERS

To check for polymorphism between the mapping population's parents, a suitable genetic marker is used. Simple sequence repeats (SSRs) and single nucleotide polymorphisms (SNPs) are the most common markers used today because of their high throughput, reliability, and ease of use. In crops with limited molecular development, first-generation genetic markers (RFLP, RAPD) are predominantly used. Two parents are checked with a large number of markers, and only clear-cut polymorphic markers are selected. Polymorphism percentage can vary from genomic region to genomic region as well as from crop to crop; rice has an average polymorphic percentage of 10%, meaning that screening roughly 400 markers will yield approximately 40 polymorphic markers (in the case of SSRs).

8.2.1.5 GENOTYPING OF THE WHOLE MAPPING POPULATION

The mapping population must be completely genotyped with polymorphic markers between the parents. Prior to further analysis, markers should be scored, segregation distortion markers deleted, and the number of missing markers kept to a minimum.

8.2.1.6 SUITABLE STATISTICAL ANALYSIS TO FIND THE PROBABLY LINKED MARKER

Single QTL mapping and multiple QTL mapping are the two most prevalent strategies. In both instances, the primary objective is the correlation between the marker segregation pattern and disease/pest response data. If the trait segregates normally, it can be used directly; if necessary, it can be transformed using a suitable data transformation mechanism. Both Single Marker Analysis (SMA) and Simple Interval Mapping (SIM) have benefits because they are easy to use, but they also have some drawbacks. For example, in SMA, the genetic distance between a marker and a trait cannot be determined, and in SIM, a ghost QTL may be isolated. Some of the QTL mapping methods that can be used to avoid these problems are composite interval mapping, multiple interval mapping, and Bayesian techniques. IciMapping and QTL Mapper are only two of the many software programs available for mapping QTL.

8.2.2 GENOME-WIDE ASSOCIATION STUDIES (GWAS)

Association mapping, also known as linkage disequilibrium (LD) mapping or population-based mapping, and genome-wide association studies (GWAS) are the most common types of linkage mapping. The founding principle of GWAS is to estimate the LD between the trait and the markers. In general, we don't require a mapping population here; rather, we go for phenotyping of a large random germplasm set or multiparent population like (NAM or MAGIC population). Here, historical recombination events are considered to aid in locating marker-trait associations (MTAs) that are more directly associated with the target features. It can find QTLs with considerably smaller effects than linkage mapping, which is a benefit. The genotyping coverage should be as comprehensive as feasible, encompassing the entire genome of crops (mostly SNPs are used). Population structure and kinships are the two parameters that influence MTA robustness, which can be minimized by estimating the Q matrix and K matrix respectively. As the number of markers increases, the likelihood of a spurious association increases, which must be adjusted by applying Bonferroni correction or measuring the false discovery rate (Bonferroni correction is the robust one). Frequently, TASSEL and

GAPIT, an R-based tool, are used to calculate the MTAs. Principal component analysis (PCA)-based matrix is nowadays used as a covariate in GAPIT as a replacement of K matrix and Q matrix.

8.2.3 AB-QTL MAPPING

QTLs can be detected and transferred into a superior background using state-of-the-art approaches such as advanced backcross QTL (Tanksley & Nelson, 1996). Marker-assisted backcross breeding (MABB) is a method for rapidly increasing the frequency of desirable traits in a popular cultivar's offspring. But if the gene or QTL is not identified, there is a risk of linkage drag. The AB QTL strategy involves crossing the desired donor parent with the most popular elite cultivar as the recipient parent. The populations are advanced to further generations like BC_2 or BC_3 with stringent phenotyping. Now the BC_2F_2 or BC_3F_2 are used in the QTL mapping. The final output of this method is the QTL-NIL which is totally identical to the elite cultivars with an additional feature of disease resistance. Some major advantages of AB QTL mapping are: it reduces the linkage drag, only two to three generations are required to isolate the NILs, and the pleiotropic effect of QTL can be identified.

8.2.4 MARKER-ASSISTED BACKCROSS APPROACH AND GENE PYRAMIDING

Recently, molecular markers have become increasingly popular as a trustworthy and efficient supplement to conventional plant breeding for the development of superior crop varieties. Molecular markers include RAPD, ISSR, SSR, and STS based on polymerase chain reaction as well as RFLP based on Southern blot hybridization. Some PCR-based markers, like AFLP and CAPS, employ restriction digestion either before or after amplification to narrow down a gene's potential variants. Single nucleotide polymorphism (SNP) uses crop species DNA sequence data to create the latest marker technology. Using a predetermined set of heritable (foreground selection) markers, the desired gene or genes can be transferred from one species to another. MAS uses a large number of randomly placed or mapped markers with whole genome coverage to swiftly obtain the recurrent parent genome in backcross breeding (Prabhu et al., 2009). Improved crop resilience to biotic and abiotic stresses is a common goal of marker-assisted backcross breeding (MABB) nowadays. Recent climate change has increased the risk of many biotic and abiotic stressors. Dixit et al. (2020) used marker-assisted forward breeding (MAFB) to develop high-yielding climate-resilient rice with genes for bacterial leaf blight (BLB) (*Xa4, Xa5, Xa13, Xa21*), blast (*Pi9*), gall midge (GM) (*Gm4, Gm8*), brown planthopper (BPH) (Bph3, Bph17), and drought tolerance (*qDTY1.1* and *qDTY3.1*). Dixit et al. (2020) observed that ILs containing seven to ten genes outperformed the recurring parent Swarna+drought in yield (1.0 t per ha) and resistance to blast, BLB, BPH, and GM without sacrificing grain quality in non-stress situations.

Pandit et al. (2021) used marker-assisted backcross breeding to find that Maudamani offspring with the Sub1 and GW5 (wide-grain) traits were more submergence-resistant and yielded more grain. Background screening made it up to 96.875% easier to retrieve the genome of a parent from a progeny who carried the target QTL in the BC_3 pyramided line. The BC_3F_2 generation tracked the number of recipient parent homozygous progenies for the target QTLs *Sub1* and GW5 (wide-grain) alleles and the yield component QTL *OsSPL14*. The seven pyramided lines beat both parents in terms of grain yield and tolerance to 14 days of submersion. Bruchids (*Callosobruchus spp.*) are a threat to mung beans, which makes the beans taste bad and cause a lot of damage. Hence, comprehending how resistance works at the molecular level will make it easier to find resistance genes or QTLs using the existing genome sequence. These genes or QTLs can then be introduced into the cultivated variety using the MABB strategy (Majhi et al., 2022). Wu et al. (2022) used MABB approach to transfer the bruchid resistance gene VrPGIP2 from V2802 to Kamphaeng Saen 1, a popular cultivar. Genetic selection determined the background, and VrBR-SSR013 and DMB-SSR158 determined the foreground in this experiment. The VrPGIP2 resistance gene in three highly inbred BC3F6 lines (R67-11, R67-22, and R92-15) boosts resistance.

In order to ensure agricultural sustainability, gene pyramiding, marker-assisted selection (MAS), and other approaches have sped up the production of extremely accurate, long-lasting resistant/tolerant lines (Dormatey et al., 2020). Direct genotypic selection employing DNA markers that co-segregate with the genes of interest avoids phenotypic selection issues. The introduction of genetic markers has had a dramatic impact on plant breeding and genetics. As a way to develop new, pest-resistant plant varieties, this method may prove effective (Bhoi et al., 2019). For the most part, gene stacking has not been fully utilized in order to enhance quality and build biotic stress resistance in the most important agricultural crops. To boost crop tolerance to environmental stresses for long-term advantage, contemporary agriculture is employing successful gene pyramiding tactics (Dormatey et al., 2020). Pyramiding depends on the quantity of transferred genes, the variety within genes and flanking markers, the number of selected populations in each breeding generation, the features of genes and germplasms, and breeders' ability to identify target genes. For the next few decades, the evolution of stress-tolerant rice varieties will rely on gene pyramiding, which has been made more precise and dependable by modern breeding procedures like the marker-assisted backcrossing approach (Haque et al., 2021).

8.3 MULTI-OMICS APPROACHES IN BIOTIC STRESS MANAGEMENT

In recent decades, the application of various "omics" technologies has significantly contributed to the advancement of plant-system technologies. The development of Next-generation sequencing (NGS) has led to the emergence of a new generation of omics approaches such as genomics, transcriptomics, and proteomics, which have proved successful in plant research. Besides, several omics studies such as metabolomics, ionomics, and phenomics have also been widely used for plant protection purposes.

8.3.1 STUDYING THE GALL-INDUCING INSECTS

Oates et al. (2016) studied the development of insect galls explored using omics-level approaches. The study suggested that the secretion of effector proteins and the disturbance of phytohormonal balance could be the potential mechanisms involved in the development of galls. However, a complete understanding of these mechanisms and their functioning within the system is still lacking. Further research is necessary to address the unanswered questions before utilizing this information for developing strategies aimed at breeding or engineering plants with better resistance.

8.3.2 UNDERSTANDING THE COMPLEX VIRUS-VECTOR INTERACTION

Genome sequence data serves as the foundation for the analysis of multiple 'omics data types, including transcriptomics, proteomics, and metabolomics (Kukurba and Montgomery, 2015). The detection and measurement of gene expression at the RNA level during particular developmental stages or in response to changes in physiological conditions is known as transcriptomics and is a frequently utilized 'omics method (Wang et al., 2010). This method can be used to quantify a number of RNA species, including viral RNA, short RNAs, long non-coding RNAs, and mRNA (Brault et al., 2010). According to Thorpe et al. (2016), the study of these RNA species can reveal potential links between particular genes and/or their gene products and various phases of insect development. For instance, differences in gene expression during viral acquisition, analysis of the saliva of several aphid species that serve as vectors, insect responses to various host plants, or studies of the molecular mechanisms behind pesticide resistance (Christodoulides et al., 2017). The development of instruments with increased sensitivity is now possible thanks to developments in "omics technologies" (Yang et al., 2013). These instruments are also becoming more readily available and affordable. The identification of RNA, protein, or metabolite species at the individual or small population of cells level is now possible thanks to emerging technologies such as single-cell transcriptomics (Ziegenhain et al., 2018), laser capture microdissection (Zhu et al., 2016), and MALDI-imaging (Vrkoslav et al., 2010). The development of expression atlas resources for model insects has been aided by these developments.

8.3.3 UNDERSTANDING METAMORPHOSIS AND INSECT DEVELOPMENT

Small RNA molecules known as microRNAs (miRNAs) are essential for controlling gene expression at the post-transcriptional level and have been shown to increase the complexity of biological pathways (Bartel, 2009; Nouzova et al., 2018). Recent research has investigated the miRNA repertoire of an insect's *corpora allata* (CA), the gland responsible for juvenile hormone (JH) synthesis. Results from this study showed significant differences in miRNA profiles between the pupal and adult stages of *Aedes aegypti* CA, with varying JH synthesis rates (Nouzova et al., 2018). Additional research has indicated the role of the bantam miRNA in regulating JH titers (Qu et al., 2017) in *Drosophila melanogaster*, highlighting the potential importance of post-transcriptional control mechanisms, such as miRNA suppression of gene expression, in regulating JH synthesis rates in response to environmental cues (Perez-Hedo et al., 2013). Further research using omics techniques as well as functional studies using RNA interference (RNAi) or CRISPR-Cas9 technology may identify new molecules or key players involved in the regulation of CA activity (Duvall et al., 2017). Additionally, epigenetic factors could potentially affect JH synthesis in social insects and other animals (Noriega, 2004). In conclusion, while early studies in each omics field have yielded promising results, there is still much to learn about the complex regulation of JH synthesis (Ernst et al., 2015).

8.3.4 DECIPHERING INSECT-ASSOCIATED MICROBIAL COMMUNITY

The term "meta-omics" was coined by De Clerck et al. (2015) to describe a set of techniques used for the analysis of organism communities. Metagenomics, meta-transcriptomics, meta-proteomics, and metabolomics are some of these approaches. A sample's collection of genomes can be identified using metagenomics (Malacrin, 2018). Using a short nucleotide fragment known as a barcode (e.g., 16S, 18S, ITS, or COI) for identification, metabarcoding, a more focused version of metagenomics, focuses on the taxonomical reconstruction of biological communities in a particular sample (McCarthy et al., 2011). In addition to information on relative taxonomic abundance and the presence of specific genes, metagenomic and metabarcoding methods can both provide a qualitative assessment of the diversity of organisms in a sample. But all they tell us about is who's there (taxonomic reconstruction) and what's going on (gene identification). With the use of methods like PICRUSt for bacteria and FUNGuild (Jia et al., 2013) for fungus (Engel et al., 2012), the functional role of recreated communities may be predicted. Meta-omics tools have been used in several studies across various host taxa to describe the insect-associated microbial community, including *Pentalonia nigronervosa* (Coquerel) (Peroh, 2015), *Lutzomyia longipalpis* (Lutz & Neiva) (Handelsman et al., 1998), and Lutzomyia intermedia (Lutz & Neiva) (Abdelfattah et al., 2018).

8.3.5 TARGETING OMICS-BASED HOLISTIC PEST MANAGEMENT APPROACH THROUGH SYSTEM APPROACH AND IPM-OMICS

The resistance of insect pests to insecticides has become a major concern in crop management. Integrated pest management (IPM) aims to slow the development of resistance and reduce insect damage by using natural predators and chemical insecticides sparingly. However, the implementation of IPM has not improved significantly over the years. To address this issue, a "systems" perspective called IPM-omics has been developed for the management of cowpea and other crop pests in West African regions, which utilizes emerging genomic technologies. IPM-omics entails finding polymorphisms in insect populations, using them to analyze collections of individual insects collected from the field, overlaying this information with other data sets, and improving pest management decisions. It also involves using information and communication technologies to more effectively manage pests (Bello-Bravo et al., 2010). High-throughput polymorphism detection systems and next-generation sequencing (NGS) data can be utilized to gain new insights into pest populations and make more informed pest management decisions. The "systems" view of organism communities in IPM-omics has the potential to improve IPM approaches by enhancing the evaluation, adaptation, and optimization of pest management strategies. However, extension agents, governmental organizations, and non-governmental organizations (NGOs) must have access to and the capacity to implement the IPM research findings through information and communications technologies in order for any resulting IPM program to be implemented successfully (Agunbiade et al., 2012).

8.4 TRANSGENIC APPROACHES TO DEVELOP BIOTIC STRESS RESISTANCE CROP PLANT

Biotic constraints reduce agricultural yields significantly and degrade crop quality, and they may constitute a risk to food and nutritional security (McKenna et al., 2020; Douglas, 2018). They cause significant yield loss and quality deterioration of crops. The total yield loss due to these biotic stress factors is around (17–30)% (Savary and Willocquet, 2020). Limited genetic variability and inherent barriers to cross-breeding among existing species make traditional breeding methods for crop improvement ineffective. Mutational breeding and somaclonal variation technologies enhance genetic variability, but their effectiveness is erratic. However, plant genetic engineering and tissue culture techniques are considered the most advanced and productive breeding technologies in crop improvement programs for biotic stress resistance (Wijerathna-Yapa, 2017). The most recent advancements in transgenic technology, according to Kamthan et al. (2016), offer a direct and directional technique to improve crops for biotic resistance while having little impact on beneficial soil microorganisms and the environment. In light of this, genetic engineering provides a number of different options for choosing resistance genes from a variety of sources to introduce them into plants to impart resistance to various biotic stresses (Tohidfar & Khosravi, 2015). The strategy enhances plant breeding by expanding the range of genes and germplasm available for crop incorporation and reducing the time required to develop new varieties and hybrids (Abebe & Tafa, 2022). Through recombinant DNA technology, resistance genes from different species are combined to overcome genetic hurdles. Foreign genes can be inserted into food plants through electroporation, gene gun or particle bombardment, floral dip, and Agrobacterium-mediated transformation (Yu et al., 2022). Anjanappa and Gruissem (2020) state that transgenic crop cultivars are created by cloning desired genes, constructing expression vectors, genetically transforming recipient cultivars, and identifying transgenic lines. These crop cultivars enhance or introduce new desirable traits. Transgenic technology can alter or remove undesirable genes to modify their traits in individuals (Zhang et al., 2017).

8.4.1 TRANSGENICS FOR INSECT-PEST TOLERANCE

Transgenic technology offers an alternative and new way to control pests that is good for the environment, efficient, long-lasting, and good for yield. Introducing the *Bacillus thuringiensis* bacterial gene Bt is a well-known way to make plants resistant to insects. Other genes used to control insects include Cowpea trypsin inhibitor (CpTI), Alpha-amylase inhibitor (AI), snowdrop lectin (Galanthus nivalis agglutinin: GNA), protease inhibitor II gene (Pin II), etc.

These transgenic crops are mostly resistant to lepidopteran, coleopteran, and a hemipteran class of insects and are successfully utilized in cotton, maize, potato, rice, soybean, sugarcane, cowpea, eggplant, and tomato varieties and hybrids for commercial use. This leads to a reduction in pesticide usage and an increase in yield (Tilgam et al., 2021).

In maize, multiple *Bt* genes (*Cry2Ab2, Cry1F, Cry1A.105, Cry35Ab1, Cry34Ab1,* and *Cry3Bb1*) were incorporated and showed differential interactions and expression against Colorado potato beetle and western corn rootworm (Xiao & Wu, 2019). Calles-Torrez et al. (2018) found that western corn rootworms caused less damage to *Cry3Bb1* + *Cry34/35Ab1* maize crosses than to Bt maize with either *Cry3Bb1* or *Cry34/35Ab1* protein alone. The effectiveness of the *cry1Ab* gene in controlling the bean pod borer in cowpea led to less damage to pods and seeds in West Africa (Addae et al., 2020).

TABLE 8.2 Improvement of Crops Through Transgenic Approaches

Sl. No.	Crop	Stress	Gene in action	References
1.	Cotton	Pink and army bollworm	Eagle-2 variety, *Cry3Bb1* and *Cry3* protein expression.	Zafar et al. (2022)
2.	Tobacco	*Spodoptera litura*	Vigna mungo protease inhibitor (*VmPI*)	Mudiyappanayar and Koundal (2020)
3.	Potato	Colorado potato beetle	*cry3A*	Salehian et al. (2021)
4.	Sugarcane	*Chilo infuscatellus*	*Vip3A*	Riaz et al. (2020)
5.	Maize	*S. frugiperda*	*Cry1Ab*, *Vip3Aa20*	Eghrari et al. (2021)
6.	Rice	*C. suppressalis* and *C. medinalis*	*Cry1Ab* and *Vip3A* fusion protein	Xu et al. (2018)
7.	Cowpea	Aphid-borne mosaic potyvirus	Expressing the Viral coat protein gene	Mundembe et al. (2009)
8.	Tobacco	Viral disease (CMV)	*dsRNA* resistant to CMV-O and CMV-Y	Ntui et al. (2014)
9.	Tomato	*Phytophthora infestans*	*MYB49* enhances resistance	Cui et al. (2018)
10.	Mustard	Fungal pathogens	Overexpression of *NPR1*	Ali et al. (2017)
11.	Wheat	Bacterial disease	*Arabidopsis EF-Tu* receptor	Schoonbeek et al. (2015)

8.4.2 TRANSGENICS FOR DISEASE RESISTANCE

About 15% of food yields are lost because of fungi and bacteria, and another 3% is lost because of viruses (Abebe et al., 2022). The pathogen-derived resistance (PDR) method was used to make plants resistant to harmful viruses (Mundembe et al., 2009). Mundembe et al. (2009) said that the viral coat protein gene in the modified plant slowed down the spread of disease (Prins et al., 2008; Yu et al., 2022). The finding of RNAi (RNA interference) technology was a major step forward in developing transgenic plants that are resistant to viruses. Yu et al. (2022) describe how viruses can be controlled by using dsDNA to control their housekeeping genes, such as the coat protein (CP), the replication protein (Rep), and the moving protein (MP). A lot of work went into making transgenic plants resistant to different fungal diseases in crop plants. The natural genes in these transgenic plants that make them immune to pathogenic fungi were turned up to 11. Albert et al. (2015) found that when *Sclerotinia sclerotiorum* or *Phytophthora infestans* infected potatoes, adding the immune receptor gene *AtRLP23* made the pathogens take up less space and caused the tumors to be smaller. According to Song et al. (2018) and Du et al. (2015), the SmELR from wild potatoes and the immune receptor gene *SlVe1* from tomatoes both improved resistance to *Phytophthora infestans* in potatoes and Verticillium wilt in tobacco and cotton, respectively. These include the introduction of resistance genes from *Solanum commersonii* to improve bacterial wilt resistance in potatoes (Boschi et al., 2017), *Arabidopsis* EFR expression, and the incorporation of antibacterial proteins and bacteriophages in transgenic plants to check a large number of bacterial diseases in crop plants (Luo et al., 2021). Table 8.2 displays the progress made in creating transgenic crop plants. Although genetic engineering has many benefits, it also has several drawbacks that have been acknowledged. The lack of efficient transformation and regeneration mechanisms, particularly for monocots, which make up the majority of cereal crops grown worldwide, is one of the barriers to plant genetic engineering. Furthermore, because full plants cannot be grown from transformable cells in monocots and legumes, the success of genetic engineering in these plant species is limited. Combining traditional and molecular breeding programs with gene alteration approaches for creating crop plants with a tolerance to biotic stress is predicted to result in a high yield.

8.5 GENOME EDITING IN MANAGING BIOTIC STRESS

8.5.1 CRISPR/CAS IN DISEASE MANAGEMENT

CRISPR/Cas gene editing has shown great potential in improving the biotic stress tolerance of crops. Global agricultural output losses due to viral, fungal, and bacterial infections are estimated to be between 20–40% (Walker, 1984), and developing pathogen-resistant crops can help to address food crises (Borrelli et al., 2018). Studies have shown that CRISPR/Cas9 knockout can lead to plants resistant to bacterial, fungal, and viral diseases and insect pests (Chen et al., 2019). Powdery mildew-resistant wheat cultivars have been created by employing CRISPR/Cas9 to remove TaMLO alleles (Wang et al., 2014). The production of powdery mildew-resistant crops has also been facilitated by the use of CRISPR/Cas9 to knock out SlMLO and VvMOL3 in tomato and grape (Wan et al., 2020), as well as a mutation in SlPMR4 in tomato (Santillán Martínez et al., 2020). Rice blast resistance was increased by modifying OsERF922 and OsSEC3A genes with CRISPR/Cas9 and inducing Bsr-d1 and Pi21 mutations in rice (Ma et al., 2018). Knocking out miR482b and miR482c genes from tomato using CRISPR/Cas9 has led to the development of late blight-resistant tomato (Hong et al., 2021). In addition, CRISPR/Cas9-based knocking down of *OsSWEET13*, a gene that encodes a sucrose transporter, has led to bacterial blight-resistant rice plants (Silva et al., 2021). Targeted modification of the CsLOB1 promoter *EBEPthA4* gene with CRISPR/Cas9 has led to the development of *Xanthomonas citri* subspecies citri-resistant citrus plants (Peng et al., 2017). Geminivirus-resistant *Nicotiana benthamiana* and *Arabidopsis* BSCTV have been developed by targeting genome coding and non-coding sgRNA/Cas9 (Ji et al., 2015). In addition, delivering sgRNA molecules to N. benthamiana plants overexpressing Cas9 using a tobacco rattle virus (TRV) vector demonstrated mixed infection immunity using an sgRNA targeting conserved sequences from various viral strains (Ali et al., 2015). Additionally, MeMV, Cotton Leaf Curl Kokhran Virus (CLCuKoV), and TYLCV severe and mild strain coding and non-coding regions have been targeted using several CRISPR/Cas9 techniques (Ali et al., 2016). Since RNA viruses require plant host factors like eIF4E, eIF(iso)4E, and eIF4G to survive, potyvirus-resistant cucumber plants have been developed by editing two eIF4E sites using CRISPR/Cas9 (Chandrasekaran et al., 2016). Rice becomes more resistant to plant hoppers and stem borers when OsCYP71A1 is knocked out because this affects serotonin production and raises salicylic acid levels (Lu et al., 2018). Additionally, it was discovered that the gmcdpk38 mutant with Hap3 deletion using CRISPR/Cas9 was extremely resistant to common cutworm (Li et al., 2022).

8.5.2 CRISPR-CAS GENOME EDITING IN AGRICULTURE FOR THE MANAGEMENT OF INSECT PESTS

8.5.2.1 CRISPR-CAS GENOME EDITING IN INSECTS

Numerous research has demonstrated the efficacy of modifying insect genomes using the CRISPR-Cas system. For instance, the yellow gene in Drosophila fruit flies was modified, and the *BmBLOS2* gene in silkworms was targeted. The gossypol detoxifying gene cluster *CYP6AE* was deleted from H. armigera, increasing the insect's sensitivity to phytotoxin. In addition, the use of CRISPR-Cas to target genes such as *CpomOR1, Orco,* and *OR16* resulted in reduced fertility and fecundity in insects, affected mating behavior and host plant choice, and prevented males from detecting pheromone signals and controlling insect populations. The technology has also been used to remove growth genes in insects such as *Spodoptera litura, Spodoptera frugiperda,* and *Plutella xylostella,* resulting in abnormal gonads, disarmed prolegs, and body segment functions. These results show the potential of *CRISPR-Cas* gene editing technology for managing and controlling insects (Hafeez et al., 2019; Garczynski et al., 2017; Chang et al., 2017; Bi et al., 2016; Koutroumpa et al., 2016; Li et al., 2016; Wang et al., 2016; Wang et al., 2013; Gratz et al., 2013).

8.5.2.2 CRISPR-CAS-MEDIATED GENE DRIVE IN INSECT PEST MANAGEMENT

The CRISPR-Cas genome editing technology can be used to create a gene drive that can spread modified genes through an insect population, leading to alterations in their traits. However, gene disruption caused by *CRISPR-Cas* gene drives may result in sterility or mortality in targeted insect species, causing their population collapse and even extinction over several generations. This approach can be utilized for insect pest control by releasing modified insect strains that possess undesirable traits like lethality, infertility, biased sex ratios, or insecticidal sensitivity. By deleting genes that make insects resistant to Bt toxins, for instance, the sustainable management of Bt resistance in *H. armigera* can be achieved. Additionally, by disrupting the X chromosome, the *CRISPR-Cas* gene drive can change the male-female ratio over time by producing more male progeny and fewer females. As a result, *CRISPR-Cas* gene drives have the potential to revolutionize the control of insect pests, albeit careful thought should be given to their possible ecological effects (McFarlane et al., 2018; Jin et al., 2018).

8.5.2.3 CRISPR-CAS TECHNOLOGY IN GENOME EDITING OF CROP PLANTS

It is possible to improve plant quality using CRISPR-Cas technology, enhancing the crops' tolerance to biotic and abiotic challenges (Wang et al., 2020). One application of CRISPR-Cas in plants is modifying the production of enzymes that either repel or attract insect pests and predators (Books, 2019). For example, transgenic plants modified to produce the aphid-produced sesquiterpene hydrocarbon (E)-farnesene (Ef) have been found to inhibit the feeding ability of other hosts and attract the parasitic wasp *Diaeretiella rapae*, which effectively controls the aphid population (Beale et al., 2006). Another way the CRISPR-Cas system can improve plant resistance to insects is by enhancing the plant's natural defense mechanisms. For instance, editing two beta-1–3 glucanase genes in barley using CRISPR-Cas9 rendered the phloem sap inaccessible to the aphid *Rhopalosiphum padi* (Bassett et al., 2013).

8.5.2.4 UTILIZATION OF CROP WILD RELATIVES (CWRS) IN INSECT-PEST RESISTANCE THROUGH CRISPR-CAS APPROACH

The use of CRISPR-Cas systems for gene editing can aid in addressing regulatory issues related to the introduction of foreign genes in transgenic plants. Wild crop cousins frequently have low yields despite being resistant to biotic and abiotic stress. Through domestication and breeding, crop yields have increased, but they have become more susceptible to insect pests. CRISPR-Cas9 editing can be used to modify or delete genes that make crops vulnerable to insects, or introduce beneficial traits from wild relatives to create insect-resistant cultivars (Rathinam et al., 2019). The process involves two steps. First, insect-resistant crops can be bred from wild relatives. Using CRISPR-Cas editing, genes that provide resistance to insects can be identified and edited to create insect-resistant crop varieties. For example, the wild tomato *Solanum pimpinellifolium* has a built-in resistance to spider mites and other arthropod insects (Rakha et al., 2017), and the resistance genes can be modified after testing against related insects to create crop varieties that are resistant to insects (Li et al., 2016). By over-expressing or silencing genes using CRISPR-Cas editing, sequence variation can be introduced to produce insect-resistant phenotypes in commercially important crops. However, the effectiveness of this approach is yet to be fully proven.

8.6 ROLE OF RNAI IN BIOTIC STRESS MANAGEMENT

RNAi pest control is superior because it is efficient, simple to implement, adaptable, targeted, and reliable. For studying gene function and biological management of agricultural insect pests and pathogens, it is a powerful reverse genetic tool when used properly (Andrade and Hunter, 2016).

8.6.1 SYSTEMIC RNAI IN INSECTS

According to research by Gordon and Waterhouse (2007), RdRP does not amplify SiRNA molecules in humans or insects, but it does in plants, nematodes, and fungi. Unexpectedly, *Tribolium castaneum* studies (Tomoyasu et al., 2008) showed that the insect responded to systemic RNAi even in the absence of RdRP. Moreover, despite the absence of canonical RdRPs in insect genomes, a subunit of RNA polymerase II that exhibits RdRP-like activity has been found in *Drosophila melanogaster*, playing a role in RNAi and transposon suppression, as reported by Richards et al. (2008). These findings suggest that unknown mechanisms that resemble RdRP may be responsible for RNAi amplification in insects. Additionally, RNAi's systemic nature in insects is evidenced by its effects being observed across developmental stages and generations in some insect species.

8.6.2 INSECT CELLULAR UPTAKE OF DSRNA

In addition to bacteria, plants, fungi, insects, and even human parasitic insects, RNAi signals are transmitted both within and between the various biological kingdoms (Knip et al., 2014). In insects, siRNAs can act as cellular messengers derived either from the environment or other tissues. RNAi occurring in a single cell is called "cell autonomous RNAi," whereas RNAi occurring in different cells or tissues is called "systemic RNAi" (Baum and Roberts, 2014). Environmental and systemic RNAi have been found to affect insect behavior (Prentice et al., 2015). Insect systemic silencing is attributed to the uptake of dsRNA via trans-membrane channels and endocytosis (Joga et al., 2016). Proteins SID-1 and SID-2 are required for the uptake of substances across membranes (Cappelle et al., 2016). According to in silico analysis, *Tribolium castaneum* SID-1 homologs resemble *C. elegans* tag-130 more than SID-1. However, systemic RNAi in C. elegans does not require the *tag-130* gene. This suggests that insects can pick up the silent signal without the necessity for SID-1. In the absence of Sid orthologs, for instance, *T. castaneum* and mosquitoes show a robust systemic RNAi response, indicating that they use a distinct dsRNA absorption pathway (Tomoyasu et al., 2008). Furthermore, SID-1-like channel proteins and receptor-mediated endocytosis are used by the Colorado potato beetle, *Leptinotarsa decemlineata*, to take up dsRNA (Cappelle et al., 2016).

8.7 ARTIFICIAL INTELLIGENCE FOR BIOTIC STRESS RESILIENCE

8.7.1 MACHINE LEARNING AND DEEP LEARNING

Machine learning (ML) and its variant Deep Learning (DL) are used in a wide range of artificial intelligence applications in the study of plant stress. These methods have been used to analyze the four categories of plant stress for identification, classification, quantification, and prediction (Singh et al., 2016). The most popular application of ML and DL algorithms for analyzing genomic and transcriptome data is the prediction of phenotypic traits prior to their expression (Singh et al., 2021). This contrasts with the majority of published image-processing phenotyping research, which uses these methods to detect and classify stress symptoms. The use of traditional statistics has been broadened by ML and DL techniques for modeling non-linear systems like biological processes (Osama et al., 2015). Therefore, both methods are useful for processing the study of plant responses to the occurrence of biotic and abiotic stressors. Plant science, plant stress, plant disease, plant resistance, and machine learning or deep learning are all terms that could be used in a search. By identifying the algorithms utilized, the ICQP paradigm category on which each study is based, the stressors evaluated, and the datasets analyzed, it provides a classification of current studies that use ML approaches to explain plant stress.

The quality of supervised algorithms is influenced by the following factors: data sources, feature extraction from the data, output variable selection, and learning techniques (Silva et al., 2019). The Support Vector Machine (SVM), Random Forest (RF), and Convolutional Neural Networks (CNN) are the most frequently discussed methods for faithfully simulating plant stress responses using artificial neural networks. The principal application of these techniques was the detection and categorization of stress-related symptoms in massive image databases. Machine learning algorithms have also been used to predict a variety of other stress-dependent features, including disease resistance gene expression (Hiddar et al., 2021), transcription factor expression (Song et al., 2020), yield (Shook et al., 2021), growth (Yasrab et al., 2021), and specialized metabolites biosynthesis (García-Pérez et al., 2021). In situations where traditional machine learning (ML) approaches have failed, new, high-performance artificial neural networks (ANN) were required since AI systems are becoming more complex as a result of the growing amount of data available to understand biological systems. The phrase "deep learning" refers to neural networks that are characterized by an increase in the number of layers and non-linear network changes, as well as an improvement in the effectiveness of the training process (Dargan et al., 2020).

Prior to the present DL peak, there was a long period of apathy, mostly because of hardware constraints and a lack of financing. However, as soon as more potent technology, particularly Graphics Processing Units, became available, these strategies were reevaluated (GPUs) (Nicodeme, 2020). These systems were quickly employed to process DL algorithms since they were first created to calculate three-dimensional visuals for video games and were shown to be effective parallel computing devices. DL is one of the most effective AI techniques for modeling big non-linear systems and analyzing a significant amount of experimental data because of the multicores of recent GPUs. With the aid of powerful software, DL can be applied to a variety of systems biology issues. Protein-protein interactions are widespread in biological systems and fundamental to deciphering the mechanism underlying physiological responses. Pavlopoulou et al. (2019) describe how computational models have been used to simulate such interactions. Millions of molecules in a library can be effectively screened using in silico predictions of protein-protein and protein-ligand interactions rather than time-consuming and expensive studies (Stokes et al., 2020). For the purpose of studying receptor binding during the perception of stressors in plant systems, discovering novel phytohormones, and developing elicitors that will provide the best possible responses for the control of hormesis, in silico prediction may provide a useful framework (Wang et al., 2021).

Given the exorbitant cost of high-performance GPUs, the widespread belief that DL requires an expensive processing infrastructure is not completely unfounded. However, multiple options exist for doing so on the cloud, making DL model training possible. In addition, several of these platforms offer a no-cost trial period, making cloud computing accessible to a wider audience. This strategy avoids the high cost, specialized storage space, and power requirements of a locally hosted computing cluster. Due to the lengthy training time required for a DL model to achieve optimal performance, these factors pose serious limitations on the feasibility of such computing. It is essential to consider both hardware characteristics and existing software to keep operating performance at an acceptable level while training ML networks. Since many existing frameworks are freely available, ML and DL algorithms can be used on a wide range of systems. Because of this, computer modeling has become widely used in agriculture (Khan and Al-Badi, 2020). BAIR and the community developed Caffe, a popular deep-learning framework. Keras, PyTorch, scikit-learn, and Theano are Python-based programs, as are high-level C^{++} and C applications like OpenCV (Bresilla et al., 2019). Google Brain's TensorFlow and Microsoft Cognitive Toolkit (CNTK) are two of the most prominent open-source DL systems. Finally, Matlab and Nvidia CUDA should be investigated for agricultural AI applications (Thomas et al., 2021).

Hormesis management improves stress tolerance, activates specialized metabolism, and increases crop output and quality (Vargas-Hernandez et al., 2017). However, it is a difficult and costly procedure to characterize hormetic curves for various species and to assess various stressor dosages to induce the desired physiological response. Furthermore, controlled elicitation experiments show that in order to evaluate the effects of low-dose stress and develop eustress management strategies, high-tech greenhouses or considerable crop growth are necessary (Valencia-Hernandez et al., 2022; Sáenz-de la et al., 2021). To make treatment recommendations that are likely to be successful, hundreds of different plant markers, including genotyping information, yield performance, and environmental information, must be taken into account. Due to its capacity to unearth intricate non-linear relationships between environmental factors and gene expression and to gain an understanding of gene networks and signaling pathways, DL would be a helpful tool for overcoming hormesis management limits (Khaki and Wang, 2019). Convolutional neural networks (CNN) are among the most frequently used biometric techniques in agriculture for assessing plant identification and morphology, development, disease, and environmental stress (Wu et al., 2019; Ma et al., 2018; Tausen et al., 2020; Durmu et al., 2017; Ferentinos, 2018; Wang et al., 2017). The structure of a CNN is similar to that of a data analysis matrix. The filters are set up to segment the input and collect specific data at certain points to help train the deep neural network (LeCun et al., 2015). Modern plant analysis methods may quickly find important stress-related factors with considerable sensitivity to identify hormetic oscillations.

8.7.2 HIGH-THROUGHPUT PHENOTYPING PLATFORMS (HTPP) FOR DISEASE & PEST MAPPING

Phenotyping is used differently in plant breeding and agriculture. While the primary use of phenotyping in crop improvement is to find plants with enhanced features, it is currently most frequently used in agricultural production for monitoring fertilizer needs and weed identification. Future pathogen and pest identification have the potential to completely transform precision agriculture (Mahlein et al., 2010). Phenotyping techniques might vary by crop, trait, stage of development, and available resources.

8.7.2.1 SATELLITE-BASED IMAGING

With multispectral spatial resolutions ranging from 1.24 m (WorldView-4) to 260 m (CBERS-2), satellite imagery is widely accessible. Because most plant breeding efforts are carried out on incredibly small pieces of land, better satellite imagery is required. Tattaris et al. (2016) used the Digital Globe WorldView-2 satellite to test satellite imaging for evaluating germplasm in a wheat breeding trial. They found that a plot size of 8.4 m x 2.4 m is adequate at such resolution, but a smaller plot size of 2 m x 0.8 m was insufficient for analysis. A cutting-edge satellite called WorldView-3 is now in use. Its multispectral resolution is 1.24 meters, and its panchromatic nadir spatial sensor has a GSD of 0.31 meters. On tiny plots of one square meter or less, thousands of genotypes are frequently tested in plant breeding programs' early selection cycles. Modern satellite sensors, such as WorldView-3, are not able to analyze such maps. Smaller trial plots can be evaluated using proximal phenotyping or remote sensing with unmanned aerial vehicles (UAVs), while larger trial plots benefit most from satellite imagery. In plant breeding, yield tests are often undertaken with larger plot sizes of several meters in width and length, especially in later generations of selections; satellite imagery may be a useful substitute for these kinds of studies.

8.7.2.2 UNMANNED AERIAL VEHICLES (UAVS)

UAVs can be broadly divided into four categories: fixed-wing systems, rotorcraft, parachutes, and blimps (Sankaran et al., 2015). Rotorcraft typically fly between 10 and 200 meters above the ground, giving them a far greater spatial resolution and shorter GSD than satellites, which hover at an altitude of over 700 kilometers. The sensor size, camera focal length, and distance to the ground object all have an impact on the orthomosaic's final spatial resolution. Unmanned aerial vehicles (UAVs) have undergone significant testing in preparation for application in plant breeding to assess various traits in different crops. Successful breeding efforts have also employed fixed-wing gliders. At a height of 150 meters, a multispectral camera placed on a fixed-wing glider was flown to assess the genetic diversity of breed wheat in response to decreased nitrogen availability. A team of researchers discovered this (Zaman-Allah et al., 2015). De Souza et al. (2017) measured the height of sugarcane plants from a fixed-wing glider equipped with an RGB-band camera at a height of 150 meters. A hexacopter equipped with a thermal camera was flown at a height of 25 meters to assess the black poplar's drought tolerance (Ludovisi et al., 2017). To digitally count maize plants, an octocopter equipped with an RGB camera was flown at a 50 m altitude (Gnädinger and Schmidhalter, 2017). A pheno-copter fitted with thermal, near-infrared, and RGB cameras was used to explore lodging in wheat crops, sugarcane canopy temperatures, and sorghum ground cover (Chapman et al., 2014). These findings demonstrate the enormous potential of UAVs for application in plant breeding when taken as a whole. The selection of superior inbred lines and clones can be accelerated as the cost of UAVs continues to decline and new imaging and analysis techniques are developed, which might result in higher genetic advancements in the cropping business. A number of problems, including lens distortion, GSD, the overlap of the images taken while in flight, and the method used to estimate wide angles during image collections, can cause orthomosaics made from UAV footage to have significant flaws. Error reduction and orthomosaic estimation are thus essential for performing quantitative analysis, or even just for identifying the objects in the image.

8.7.2.3 PROXIMAL PHENOTYPING

Plant phenotypes are studied by proximal phenotyping, which makes use of sensors and ground-based vehicles. Phenotyping platforms can range from handheld sensors to moving vehicles to fixed towers to cable suspensions (Deery et al., 2014). Common uses for portable sensors in agriculture include measuring plant height, nitrogen content, leaf area, and chlorophyll fluorescence. Maize's resistance to drought was measured using a handheld infrared thermometer (Keener and Kircher, 1983). For application in plant breeding, a variety of mobile platforms are being created and tested. These kinds of platforms are frequently used to measure plant biomass, leaf area index, count plants, measure plant height, and evaluate early vigor and maturity. Mobile platforms can be moved around manually or with the help of a motor. The proximal sensing cart (PSC), which was constructed using two bicycles as the wheels and a metal frame, had three infrared thermometers, two digital cameras, a positioning GPS receiver, and two data loggers (White and Conley, 2013). PSC was applied in the camelina, cotton, wheat, and barley seedling nurseries. The creation of motorized moveable platforms has greatly increased the mobility and comfort of field workers. The phenomobile has a real-time kinematic (RTK) GPS system that is accurate to within 2 cm, a hydraulic driving system, three LiDAR sensors, four RGB stereo cameras, an infrared thermometer, a light source on an adjustable boom, and an infrared thermometer. In a wheat breeding study, Deery et al. (2014) reported successful usage of phenomobile for measuring canopy height, counting wheat canopy spikes, monitoring canopy temperature, and assessing plant stress. Three active spectrum sensors, one passive hyperspectral sensor, and RTK-GPS location are all included with the PhenoTrac 4 tractor. PhenoTrac 4 was used in wheat and barley breeding nurseries to assess plant nitrogen uptake, grain production, and plant dry weights (Barmeier and Schmidhalter, 2016; 2017). It is frequently important to assess plants during breeding experiments for a variety of traits, including their morphology, phenology, tolerance to abiotic stress factors, host resilience to diseases and pests, and general health. Therefore, a phenotyping platform benefits from having the capacity to evaluate numerous features simultaneously. This technique reduces costs, saves time, and gets rid of any mistakes made by several people performing multiple phenotyping rounds on the same plots. With the use of affordable handheld phenotyping instruments, several hundred plots can be examined. However, employing portable sensors is less viable for a big breeding facility due to the time and labor required as well as the errors caused by temporal variation in data.

8.8 BIOINFORMATICS TOOLS TO PREDICT ACCURACY FOR BIOTIC STRESS

The climate uncertainties have enhanced the impact of stress factors, resulting in significant yield losses. Among these, the major ones are biotic stressors such as diseases and insect pests. This has increased the need for novel methods to overcome these challenges and achieve sustainable yields. At this crossroads, the utilization of vast resources of bioinformatic databases and tools is crucial. Several crop-specific genomic databases are available, such as Wheatgenome, Maizegdb, Peanutbase, etc., and some of them are even family-based, such as Solenomics, etc., as listed in Table 8.3. After a genome has been sequenced, the next step is annotation, which provides context regarding the genes' structure and function (Salzberg et al., 2019). Tools like B2GO and PSAT can be used for annotation. There are numerous databases available, but the most up-to-date one is RGdb 4.0, which contains data on 199 reference resistance genes and 586,652 potential resistance genes derived from 182 sequenced proteomes.

TABLE 8.3 Some of the Crop-Specific Databases

Sl. No	URL	Importance
1.	https://www.wheatgenome.org/	Genomic Data Wheat
2.	https://www.maizegdb.org/	Genomic Data of Maize
3.	https://www.cottongen.org/	Genomic Data of Cotton
4.	https://peanutbase.org/home	Genomic Data of Groundnut
5.	https://www.pulsedb.org/	Genomic Data of Pulses
6.	https://solgenomics.net/	Genomic Data of Solanaceae species
7.	http://www.plantgdb.org/	Genomic data of Multispecies

Information on Pathogen Receptor Genes (PRGs) is available in PRGdb, an open-access database (Joan Calle García et al., 2021). One of the primary goals of plant breeding is the characterization of different disease resistance proteins of plants, which is crucial for the defeat of aggressive pathogens (Hofberger et al., 2014). RFPDR is a bioinformatics-based tool that can be used as a supplement to other similar resources by providing data on proteins that confer resistance to disease (Simon et al., 2022).

In silico analysis is an approach that can exploit existing genomic data resources with the help of computer science for a better understanding of the plant process. The utility of DNA markers has gained pace.

The current most popular markers are SSR and SNP types; however, their development costs are prohibitive. Tools such as polymorph prediction, SSRIT, SNP detector, and many others can be used to mine available genomic datasets, such as ESTs, for these markers (Table 8.4).

TABLE 8.4 Bioinformatic Tools for In-Silico Analysis

Sl. No.	Tool	URL	Application
1.	Polymorph Predict	http://webtom.cabgrid.res.in/polypred/index.php	Finds SSR repeats
2.	Prot-Param tool	https://web.expasy.org/protparam/	Physical-chemical parameters of proteins
3.	KASPspoon	https://github.com/AlsammanAlsamman/Spoon	PCR analysis
4.	-	http://insilico.ehu.es/	Experiments with complete genomes
5.	Plant care	https://bioinformatics.psb.ugent.be/webtools/ plantcare/html/	Analysis of promoter sequence
6.	SSRIT	https://www.southgreen.fr/glossary/term/650	Finds SSR repeats
7.	SNP detector	http://lpg.nci.nih.gov/	Finds SNP
8.	Synbrowse	http://www.synbrowse.org/	Comparative genomic tool

Future insights into the role of acyl-CoA-binding proteins (ACBPs) in biotic stress resistance were predicted by in silico analysis of data from Soy Base (https://www.soybase.org/) and the Soybean eFP Browser (http://bar.utoronto.ca/efpsoybean/cgi-bin/efpWeb.cgi). The application of insecticides has been one of the most used measures for controlling insect pests. The monotonous use of insecticides resulted in the macroevolution of the insect population and is challenging the effectiveness of chemical control (Denholm, 2002; Hemingway, 2002). As a reliable measure for identifying resistance, mutation frequency measurement in insect populations at the field level became important (Feng et al., 2015). To enhance efficiency, an RNA-sequence analysis-based pipeline named "ACE" was developed to predict insect acetylcholinesterase mutations (Guo et al., 2017). Comparative genomics is one branch that provides a holistic idea about how crops evolved and genes are conserved over a period. Comparative genomic analysis using in-silico tools like SynBrowse, SYBIL, Vista, etc. The application of the above and many other bioinformatic tools can complement several other breeding approaches for biotic stress.

8.9 CONCLUDING REMARKS AND WAY FORWARD

The discovery and utilization of next-generation sequencing breeding strategies played a key role in unraveling the information tangled in the form of the complex nucleotide sequence. Advancement of science resulted in easy accessibility of those technologies at affordable costs. This augmented their usage, during which a huge amount of information, i.e., genomic data, is generated. Yet, storage databases were developed because there was a demand for such a service. Once the data has been collected and organized, specialists will require specialized software to analyze the collected information and decipher the hidden meanings. This has led to the development of numerous bioinformatics programs. Bioinformatics is one such field, as it uses the data produced by different molecular experiments to inform plant breeding and provide answers to numerous mysteries. This progress has allowed for the collection of massive amounts of information on nucleotide and amino acid sequences, protein structure and function, and gene expression levels, among other fields of biological science. The correlations between genetics and environment are studied by repeatedly conducting yield trials in various locations. Due to the dispersed nature of many research sites, it may be difficult to conduct local or drone-based observations without resorting to satellite imagery. There is likely to be a rise in the use of satellite imagery in breeding programs for the purpose of assessing multi-location yield trials as a result of the steadily decreasing cost of such services.

KEYWORDS

- **biotic stress**
- **climate resilience**
- **machine learning**
- **marker-assisted backcross breeding**
- **next generation breeding**
- **QTL mapping**

REFERENCES

Abdelfattah, A., Malacrinò, A., Wisniewski, M., Cacciola, S. O., & Schena, L. (2018). Metabarcoding: A powerful tool to investigate microbial communities and shape future plant protection strategies. *Biological Control, 120*, 1–10.

Abebe, A., & Tafa, Z. (2022). A review on: Impact of genetic engineering in biotic stresses resistance crop breeding. *International Journal of Research and Applied Sciences in Biotechnology, 9*(2), 243–250.

Addae, P. C., Ishiyaku, M. F., Tignegre, J. B., Ba, M. N., Bationo, J. B., Atokple, I. D., & Pittendrigh, B. R. (2020). Efficacy of a cry1Ab gene for control of *Maruca vitrata* (Lepidoptera: Crambidae) in cowpea. *Journal of Economic Entomology, 113*(2), 974–979.

Agunbiade, T., Steele, L., Coates, B., Gassmann, A., Margam, V., Ba, M., & Pittendrigh, B. (2012). IPM-omics: From genomics to extension for integrated pest management of cowpea. In O. Boukar, O. Coulibaly, C. Fatokun, K. Lopez, & M. Tamò (Eds.), *Improving Livelihoods in the Cowpea Value Chain Through Advancements in Science: Proceedings of the 5th World Cowpea Research Conference* (pp. 231–248).

Albert, I., Böhm, H., Albert, M., Feiler, C. E., Imkampe, J., Wallmeroth, N., & Nürnberger, T. (2015). An RLP23–SOBIR1–BAK1 complex mediates NLP-triggered immunity. *Nature Plants, 1*(10), 1–9.

Ali, S., Mir, Z. A., Tyagi, A., Mehari, H., Meena, R. P., Bhat, J. A., & Grover, A. (2017). Overexpression of NPR1 in *Brassica juncea* confers broad spectrum resistance to fungal pathogens. *Frontiers in Plant Science, 8*, 1693.

Ali, Z., Ali, S., Tashkandi, M., Zaidi, S. S., & Mahfouz, M. M. (2016). CRISPR/Cas9-mediated immunity to geminiviruses: Differential interference and evasion. *Scientific Reports, 6*, 26912. doi:10.1038/srep26912.

Andrade, C. E., & Hunter, W. B. (2016). RNA interference natural gene-based technology for highly specific pest control (HiSPeC). In I. Y. Abdurakhmonov (Ed.), *RNA Interference* (pp. 391–409). InTech.

Anjanappa, R. B., & Gruissem, W. (2021). Current progress and challenges in crop genetic transformation. *Journal of Plant Physiology, 261*, 153411.

Awol, S., İmren, M., ALİ, M. A., Toumi, F., Paulitz, T., & DABABAT, A. A. (2021). Genetic resistance of wheat towards plant-parasitic nematodes: Current status and future prospects. *Biotech Studies, 30*(1), 43–62.

Azlan, N. S., Guo, Z. H., Yung, W. S., Wang, Z., Lam, H. M., Lung, S. C., & Chye, M. L. (2021). In silico analysis of acyl-CoA-binding protein expression in soybean. *Frontiers in Plant Science, 12*, 646938.

Barmeier, G., & Schmidhalter, U. (2016). High-throughput phenotyping of wheat and barley plants grown in single or few rows in small plots using active and passive spectral proximal sensing. *Sensors, 16*, 1860.

Barmeier, G., & Schmidhalter, U. (2017). High-throughput field phenotyping of leaves, leaf sheaths, culms and ears of spring barley cultivars at anthesis and dough ripeness. *Frontiers in Plant Science, 8*, 1920.

Bartel, D. P. (2009). MicroRNAs: Target recognition and regulatory functions. *Cell, 23*, 215–233.

Bassett, A. R., Tibbit, C., Ponting, C. P., & Liu, J. L. (2013). Highly efficient targeted mutagenesis of *Drosophila* with the CRISPR/Cas9 system. *Cell Reports, 4*, 220–228.

Baum, J. A., & Roberts, J. K. (2014). Progress towards RNAi-mediated insect pest management. *Advances in Insect Physiology, 47*, 249–295.

Beale, M. H., Birkett, M. A., & Bruce, T. J. (2006). Aphid alarm pheromone produced by transgenic plants affects aphid and parasitoid behavior. *Proceedings of the National Academy of Sciences of the United States of America, 103*(27), 10509–10513.

Bello-Bravo, J., Diaz, R., Venugopal, S., Viswanathan, M., & Pittendrigh, B. R. (2010). Expanding the impact of practical scientific concepts for low-literate learners through an inclusive and participatory virtual knowledge ecosystem. *Journal of World Universities Forum, 3*, 147–164.

Bhoi, T. K., Majhi, P. K., Prabhulinga, T., & Samal, I. (2019). Gene Pyramiding: A Tool for Insect Resistance Management. In A. K. Singh (Ed.), *Advances in Biotechnology and Bioscience* (pp. 1–14). AkiNik Publications.

Bi, H. L., Xu, J., Tan, A. J., & Huang, Y. P. (2016). CRISPR/Cas9-mediated targeted gene mutagenesis in *Spodoptera litura. Insect Science, 23*(3), 469–477.

Borrelli, V. M. G., Brambilla, V., Rogowsky, P., Marocco, A., & Lanubile, A. (2018). The enhancement of plant disease resistance using CRISPR/Cas9 technology. *Frontiers in Plant Science, 9*, 1245.

Boschi, F., Schvartzman, C., Murchio, S., Ferreira, V., Siri, M. I., Galván, G. A., & Dalla-Rizza, M. (2017). Enhanced bacterial wilt resistance in potato through expression of *Arabidopsis* EFR and introgression of quantitative resistance from *Solanum commersonii. Frontiers in Plant Science, 8*, 1642.

Brault, V., Tanguy, S., Reinbold, C., Le Trionnaire, G., Arneodo, J., Jaubert-Possamai, S., Guernec, G., & Tagu, D. (2010). Transcriptomic analysis of intestinal genes following acquisition of Pea enation mosaic virus by the pea aphid *Acyrthosiphon pisum. Journal of General Virology, 91*, 802–808.

Bresilla, K., Perulli, G. D., Boini, A., Morandi, B., & Corelli Grappadelli, L. (2019). Single-shot convolution neural networks for real-time fruit detection within the tree. *Frontiers in Plant Science, 10*, 611.

Calles-Torrez, V., Knodel, J. J., Boetel, M. A., Doetkott, C. D., Podliska, K. K., Ransom, J. K., & Fuller, B. W. (2018). Transgenic Bt. corn, soil insecticide, and insecticidal seed treatment effects on corn rootworm beetle emergence, larval feeding injury, and corn yield in North Dakota. *Journal of Economic Entomology, 111*(1), 348–360.

Cappelle, K., de Oliveira, C. F., Van Eynde, B., Christiaens, O., & Smagghe, G. (2016). The involvement of clathrin-mediated endocytosis and two Sid-1-like transmembrane proteins in double-stranded RNA uptake in the Colorado potato beetle midgut. *Insect Molecular Biology, 25*, 315–323.

Chandrasekaran, J., Brumin, M., Wolf, D., Leibman, D., Klap, C., Pearlsman, M., & Morin, S. (2016). Development of broad virus resistance in non-transgenic cucumber using CRISPR/Cas9 technology. *Molecular Plant Pathology, 17*(7), 1140–1153. doi:10.1111/mpp.12375.

Chang, H., Liu, Y., & Ai, D. (2017). A pheromone antagonist regulates optimal mating time in the moth *Helicoverpa armigera*. *Current Biology, 27*, 1610–1615.

Chapman, S., Merz, T., Chan, A., Jackway, P., Hrabar, S., Dreccer, M., & Jimenez-Berni, J. (2014). Pheno-copter: A low-altitude, autonomous remote-sensing robotic helicopter for high-throughput field-based phenotyping. *Agronomy, 4*, 279–301.

Chen, J. S., Dagdas, Y. S., Kleinstiver, B. P., Welch, M. M., Sousa, A. A., Harrington, L. B., et al. (2017). Enhanced proofreading governs CRISPR-Cas9 targeting accuracy. *Nature, 550*(7676), 407–410.

Christodoulides, N., Van Dam, A. R., Peterson, D. A., Frandsen, R. J. N., Mortensen, U. H., Petersen, B., & Hardy, N. B. (2017). Gene expression plasticity across hosts of an invasive scale insect species. *PLOS ONE, 12*(5), e0176956.

Cui, J., Jiang, N., Zhou, X., Hou, X., Yang, G., Meng, J., & Luan, Y. (2018). Tomato MYB49 enhances resistance to *Phytophthora infestans* and tolerance to water deficit and salt stress. *Planta, 248*, 1487–1503.

Dargan, S., Kumar, M., Ayyagari, M. R., & Kumar, G. (2020). A survey of deep learning and its applications: A new paradigm to machine learning. *Archives of Computational Methods in Engineering, 27*(4), 1071–1092.

De Clerck, C., Fujiwara, A., Joncour, P., Léonard, S., Félix, M.-L., Francis, F., & Massart, S. (2015). A metagenomic approach from aphid's hemolymph sheds light on the potential roles of co-existing endosymbionts. *Microbiome, 3*, 63.

De Souza, C. H. W., Lamparelli, R. A. C., Rocha, J. V., & Magalhães, P. S. G. (2017). Height estimation of sugarcane using an unmanned aerial system (UAS) based on structure from motion (SfM) point clouds. *International Journal of Remote Sensing, 38*(8–10), 2218–2230.

Deery, D., Jimenez-Berni, J., Jones, H., Sirault, X., & Furbank, R. (2014). Proximal remote sensing buggies and potential applications for field-based phenotyping. *Agronomy, 4*, 349–379.

Denholm, I., Devine, G. J., & Williamson, M. S. (2002). Evolutionary genetics. Insecticide resistance on the move. *Science, 297*(5590), 2222–2223.

Dixit, S., Singh, U. M., Singh, A. K., Alam, S., Venkateshwarlu, C., Nachimuthu, V. V., & Kumar, A. (2020). Marker assisted forward breeding to combine multiple biotic-abiotic stress resistance/tolerance in rice. *Rice, 13*, 1–15.

Dormatey, R., Sun, C., Ali, K., Coulter, J. A., Bi, Z., & Bai, J. (2020). Gene pyramiding for sustainable crop improvement against biotic and abiotic stresses. *Agronomy, 10*(9), 1255.

Douglas, A. E. (2018). Strategies for enhanced crop resistance to insect pests. *Annual Review of Plant Biology, 69*, 637–660.

Du, J., Verzaux, E., Chaparro-Garcia, A., Bijsterbosch, G., Keizer, L. C., Zhou, J., & Vleeshouwers, V. G. (2015). Elicitin recognition confers enhanced resistance to *Phytophthora infestans* in potato. *Nature Plants, 1*(4), 1–5.

Durmuş, H., Güneş, S. E. O., & Kırcı, M. (2017). Disease detection on the leaves of tomato plants by using deep learning. In *Proceedings of the 2017 6th International Conference on Agro-Geoinformatics* (pp. 1–5). IEEE.

Duvall, L. B., Basrur, N. S., Molina, H., McMeniman, C. J., & Vosshall, L. B. (2017). A peptide signaling system that rapidly enforces paternity in the *Aedes aegypti* mosquito. *Current Biology, 27*, 3734–3742.

Eghrari, K., Oliveira, S. C., Nascimento, A. M., Queiroz, B., Fatoretto, J., de Souza, B. H. S., & Môro, G. V. (2022). The implications of homozygous vip3Aa20-and cry1Ab-maize on *Spodoptera frugiperda* control. *Journal of Pesticide Science, 95*(1), 115–127.

Engel, P., Martinson, V. G., & Moran, N. A. (2012). Functional diversity within the simple gut microbiota of the honey bee. *Proceedings of the National Academy of Sciences of the United States of America, 109*, 11002–11007.

Ernst, U. R., Van Hiel, M. B., Depuydt, G., Boerjan, B., De Loof, A., & Schoofs, L. (2015). Epigenetics and locust life phase transitions. *Journal of Experimental Biology, 218*, 88–99.

Feng, X., Yang, C., Yang, Y., Li, J., Lin, K., Li, M., et al. (2015). Distribution and frequency of G119S mutation in ace-1 gene within *Anopheles sinensis* populations from Guangxi, China. *Malaria Journal, 14*, 470.

Ferentinos, K. P. (2018). Deep learning models for plant disease detection and diagnosis. *Computers and Electronics in Agriculture, 145*, 311–318.

Fiyaz, R. A., Shivani, D., Chaithanya, K., Mounika, K., Chiranjeevi, M., Laha, G. S., & Sundaram, R. M. (2022). Genetic improvement of rice for bacterial blight resistance: Present status and future prospects. *Rice Science, 29*(2), 118–132.

García-Pérez, P., Zhang, L., Miras-Moreno, B., Lozano-Milo, E., Landin, M., Lucini, L., & Gallego, P. P. (2021). The combination of untargeted metabolomics and machine learning predicts the biosynthesis of phenolic compounds in *Bryophyllum* medicinal plants. *Plants, 10*, 2430.

Garczynski, S. F., Martin, J. A., Griset, M., Willett, L. S., Cooper, W. R., Swisher, K. D., & Unruh, T. R. (2017). CRISPR/Cas9 editing of the codling moth (*Lepidoptera: Tortricidae*) CpomOR1 gene affects egg production and viability. *Journal of Economic Entomology, 110*, 1847–1855.

Gnädinger, F., & Schmidhalter, U. (2017). Digital counts of maize plants by unmanned aerial vehicles (UAVs). *Remote Sensing, 9*, 544.

Gotelli, N. J., Ellison, A. M., & Ballif, B. A. (2012). Environmental proteomics, biodiversity statistics and food-web structure. *Trends in Ecology and Evolution, 27*, 436–442.

Gratz, S. J., Cummings, A. M., Nguyen, J. N., Hamm, D. C., Donohue, L. K., Harrison, M. M., & O'Connor-Giles, K. M. (2013). Genome engineering of *Drosophila* with the CRISPR RNA-guided Cas9 nuclease. *Genetics, 194*, 1029–1035.

Guo, D., Luo, J., Zhou, Y., et al. (2017). ACE: An efficient and sensitive tool to detect insecticide resistance-associated mutations in insect acetylcholinesterase from RNA-Seq data. *BMC Bioinformatics, 18*, 330.

Hafeez, M., Liu, S., Jan, S., Shi, L., Fernández-Grandon, G. M., Gulzar, A., & Wang, M. (2019). Knock-down of gossypol inducing cytochrome P450 genes reduced deltamethrin sensitivity in *Spodoptera exigua* (*Hübner*). *International Journal of Molecular Sciences, 20*, 2248.

Handelsman, J., Rondon, M. R., Brady, S. F., Clardy, J., & Goodman, R. M. (1998). Molecular biological access to the chemistry of unknown soil microbes: A new frontier for natural products. *Chemistry and Biology, 5*, 245–249.

Haque, M. A., Rafii, M. Y., Yusoff, M. M., Ali, N. S., Yusuff, O., Datta, D. R., & Ikbal, M. F. (2021). Recent advances in rice varietal development for durable resistance to biotic and abiotic stresses through marker-assisted gene pyramiding. *Sustainability, 13*, 10806.

Hemingway, J., Field, L., & Vontas, J. (2002). An overview of insecticide resistance. *Science, 298*(5591), 96–97.

Hiddar, H., Rehman, S., Lakew, B., Verma, R. P. S., Al-Jaboobi, M., Moulakat, A., & Amri, A. (2021). Assessment and modeling using machine learning of resistance to scald (*Rhynchosporium commune*) in two specific barley genetic resources subsets. *Scientific Reports, 11*, 15967.

Hofberger, J. A., Zhou, B., Tang, H., Jones, J. D., & Schranz, M. (2014). A novel approach for multi-domain and multi-gene family identification provides insights into evolutionary dynamics of disease resistance genes in core eudicot plants. *BMC Genomics, 15*, 966.

Hong, Y., Meng, J., He, X., Zhang, Y., Liu, Y., Zhang, C., et al. (2021). Editing miR482b and miR482c simultaneously by CRISPR/Cas9 enhanced tomato resistance to *Phytophthora infestans*. *Phytopathology, 111*(6), 1008–1016.

Huang, Y., Chen, Y., & Zeng, B. (2016). CRISPR/Cas9 mediated knockout of the *abdominal-A* homeotic gene in the global pest, diamondback moth (*Plutella xylostella*). *Insect Biochemistry and Molecular Biology, 75*, 98–106.

Inturrisi, F. C., Barbetti, M. J., Tirnaz, S., Patel, D. A., Edwards, D., & Batley, J. (2021). Molecular characterization of disease resistance in *Brassica juncea*: The current status and the way forward. *Plant Pathology, 70*(1), 13–34.

Jagdish, M., & Koundal, K. R. (2020). Constitutive expression of protease inhibitor gene isolated from black gram (*Vigna mungo* L.) confers resistance to *Spodoptera litura* in transgenic tobacco plants. *Indian Journal of Biotechnology, 19*(2), 94–101.

Ji, X., Si, X., Zhang, Y., Zhang, H., Zhang, F., & Gao, C. (2018). Conferring DNA virus resistance with high specificity in plants using virus-inducible genome-editing system. *Genome Biology, 19*(1), 197.

Jia, S., Zhang, X., Zhang, G., Yin, A., Zhang, S., Li, F., & Tala, et al. (2013). Seasonally variable intestinal metagenomes of the red palm weevil (*Rhynchophorus ferrugineus*). *Environmental Microbiology, 15*, [Epub ahead of print].

Jin, L., Wang, J., & Guan, F. (2018). Dominant point mutation in a tetraspanin gene associated with field-evolved resistance of cotton bollworm to transgenic Bt cotton. *Proceedings of the National Academy of Sciences of the United States of America, 115*(46), 11760–11765.

Joan Calle, G., Anna, G., Andreu, P. G., Alfonso, S. V., Ciro Gianmaria, A., Daniela, D., & Maria Raffaella Ercolano. (2021). PRGdb 4.0: An updated database dedicated to genes involved in plant disease resistance process. *Nucleic Acids Research, 50*(D1), D1483–D1490.

Joga, M. R., Zotti, M. J., Smagghe, G., & Christiaens, O. (2016). RNAi efficiency, systemic properties, and novel delivery methods for pest insect control: What we know so far. *Frontiers in Physiology, 7*, 553.

Kalia, S., & Rathour, R. (2019). Current status on mapping of genes for resistance to leaf- and neck-blast disease in rice. *3 Biotech, 9*, 1–14.

Kamthan, A., Chaudhuri, A., Kamthan, M., & Datta, A. (2016). Genetically modified (GM) crops: Milestones and new advances in crop improvement. *Theoretical and Applied Genetics, 129*(9), 1639–1655.

Keener, M. E., & Kircher, P. L. (1983). The use of canopy temperature as an indicator of drought stress in humid regions. *Agricultural Meteorology, 28*, 339–349.

Khaki, S., & Wang, L. (2019). Crop yield prediction using deep neural networks. *Frontiers in Plant Science, 10*, 621.

Khan, A. I., & Al-Badi, A. (2020). Open-source machine learning frameworks for industrial Internet of Things. *Procedia Computer Science, 170*, 571–577.

Knip, M., Constantin, M. E., & Thordal-Christensen, H. (2014). Transkingdom cross-talk: Small RNAs on the move. *PLoS Genetics, 10*(12), e1004602.

Koutroumpa, F. A., Monsempes, C., Francois, M. C., Cian, A. D., Royer, C., Concordet, J. P., & Jacquin-Joly, E. (2016). Heritable genome editing with CRISPR/Cas9 induces anosmia in a crop pest moth. *Scientific Reports, 6*, 29620.

Krishnappa, G., Tyagi, B. S., Gupta, V., Gupta, A., Venkatesh, K., Kamble, U. R., & Singh, G. (2022). Wheat breeding. In G. Tyagi, B. S. and V. Gupta, A. Gupta, K. Venkatesh, U. R. Kamble, and G. Singh (Eds.), *Fundamentals of Field Crop Breeding* (pp. 39–111). Springer.

Kukurba, K. R., & Montgomery, S. B. (2015). RNA sequencing and analysis. *Cold Spring Harb. Protoc., 2015*, 951–969.

Langille, M. G. I., Zaneveld, J., Caporaso, J. G., McDonald, D., Knights, D., Reyes, J. A., & Knight, R. (2013). Predictive functional profiling of microbial communities using 16S rRNA marker gene sequences. *Nat. Biotechnol., 31*(9), 814–821.

LeCun, Y., Bengio, Y., & Hinton, G. (2015). Deep Learning. *Nature, 521*, 436–444.

Li, J., Zhu, L., & Hull, J. J. (2016). Transcriptome analysis reveals a comprehensive insect resistance response mechanism in cotton to infestation by the phloem feeding insect *Bemisia tabaci*. *Plant Biotechnol. J., 14*(10), 1956–1975.

Li, X., Hu, D., Cai, L., Wang, H., Liu, X., Du, H., et al. (2022). Calcium-dependent protein kinase38 regulates flowering time and common cutworm resistance in soybean. *Plant Physiol., 190*(1), 480–499.

Li, Y., Zhang, J., Chen, D., Yang, P., Jiang, F., Wang, X., & Kang, L. (2016). CRISPR/Cas9 in locusts: Successful establishment of an olfactory deficiency line by targeting the mutagenesis of an odorant receptor co-receptor (Orco). *Insect Biochem. Mol. Biol., 79*, 27–35.

Lu, H. P., Luo, T., Fu, H. W., Wang, L., Tan, Y. Y., Huang, J. Z., et al. (2018). Resistance of rice to insect pests mediated by suppression of serotonin biosynthesis. *Nat. Plants, 4*(6), 338–344.

Ludovisi, R., Tauro, F., Salvati, R., Khoury, S., Mugnozza Scarascia, G., & Harfouche, A. (2017). UAV-based thermal imaging for high-throughput field phenotyping of black poplar response to drought. *Front. Plant Sci., 8*, 1681.

Luo, M., Xie, L., Chakraborty, S., Wang, A., Matny, O., Jugovich, M., & Ayliffe, M. (2021). A five-transgene cassette confers broad-spectrum resistance to a fungal rust pathogen in wheat. *Nat. Biotechnol., 39*(5), 561–566.

Ma, J., Chen, J., Wang, M., Ren, Y., Wang, S., Lei, C., et al. (2018). Disruption of OsSEC3A increases the content of salicylic acid and induces plant defense responses in rice. *J. Exp. Bot., 69*(5), 1051–1064.

Ma, W., Qiu, Z., Song, J., Li, J., Cheng, Q., & Zhai, J. (2018). A deep convolutional neural network approach for predicting phenotypes from genotypes. *Planta, 248*, 1307–1318.

Mahlein, A. K., Steiner, U., Dehne, H. W., & Oerke, E. C. (2010). Spectral signatures of sugar beet leaves for the detection and differentiation of diseases. *Precis. Agric., 11*, 413–431.

Majhi, P. K., Bhoi, T. K., Mogali, S. C., Shiv, A., Sahoo, K. C., & Saini, V. (2022). Advances in Molecular Breeding for Bruchid (*Callosobruchus* spp.) Resistance in Mungbean [*Vigna radiata* (L.) Wilczek]: A Review. *Legume Res- An Int. J., 45*(8), 933–941.

Makkar, G. S., Bhatia, D., Suri, K. S., & Kaur, S. (2019). Insect resistance in Rice (*Oryza sativa* L.): Overview on current breeding interventions. *International Journal of Tropical Insect Science, 39*, 259–272.

Malacrinò, A. (2018). Meta-omics tools in the world of insect-microorganism interactions. *Biol., 7*(4), 50.

Mayr, A., Klambauer, G., Unterthiner, T., & Hochreiter, S. (2016). DeepTox: Toxicity Prediction Using Deep Learning. *Front. Environ. Sci., 3*, 80.

McCarthy, C. B., Diambra, L. A., & Rivera Pomar, R. V. (2011). Metagenomic analysis of taxa associated with *Lutzomyia longipalpis*, vector of visceral leishmaniasis, using an unbiased high-throughput approach. *PLoS Negl. Trop. Dis., 5*(11), e1304.

McFarlane, G. R., Whitelaw, C. B. A., & Lillico, S. G. (2018). CRISPR-based gene drives for pest control. *Trends Biotechnol., 36*, 130–133.

McKenna, T. P., Koziol, L., Bever, J. D., Crews, T. E., & Sikes, B. A. (2020). Abiotic and biotic context dependency of perennial crop yield. *PloS ONE, 15*(6), e0234546.

Muduli, L., Pradhan, S. K., Mishra, A., Bastia, D. N., Samal, K. C., Agrawal, P. K., & Dash, M. (2021). Understanding brown planthopper resistance in Rice: Genetics, biochemical and molecular breeding approaches. *Rice Sci., 28*(6), 532–546.

Mundembe, R., Matibiri, A., & Sithole-Niang, I. (2009). Transgenic plants expressing the coat protein gene of cowpea aphid-borne mosaic potyvirus predominantly convey the delayed symptom development phenotype. *Afr. J. Biotechnol., 8*(12), 2682–2690.

Nguyen, N. H., Song, Z., Bates, S. T., Branco, S., Tedersoo, L., Menke, J., & Kennedy, P. G. (2016). FUNGuild: An open annotation tool for parsing fungal community datasets by ecological guild. *Fungal Ecol., 20*, 241–248.

Nicodeme, C. (2020). Build Confidence and Acceptance of AI-Based Decision Support Systems-Explainable and Liable AI. In *Proceedings of the 2020 13th International Conference on Human System Interaction (HSI),* Tokyo, Japan, 6–8 June, 2020; IEEE: Piscataway, NJ, USA, pp. 20–23.

Noriega, F. G. (2004). Nutritional regulation of JH synthesis: A mechanism to control reproductive maturation in mosquitoes? *Insect Biochem. Molec. Biol., 34*, 687–693.

Nouzova, M., Etebari, K., Noriega, F. G., & Asgari, S. (2018). A comparative analysis of corpora allata-corpora cardiaca microRNA repertories revealed significant changes during mosquito metamorphosis. *Insect Biochem. Mol. Biol., 96*, 10–18.

Nouzova, M., Rivera-Pérez, C., & Noriega, F. G. (2018). Omics approaches to study juvenile hormone synthesis. *Current Opinion Insect Sci., 29*, 49–55.

Ntui, V. O., Kynet, K., Khan, R. S., Ohara, M., Goto, Y., Watanabe, M., & Mii, M. (2014). Transgenic tobacco lines expressing defective CMV replicase-derived dsRNA are resistant to CMV-O and CMV-Y. *Mol. Biotechnol., 56*, 50–63.

Oates, C. N., Denby, K. J., Myburg, A. A., Slippers, B., & Naidoo, S. (2016). Insect gallers and their plant hosts: From omics data to systems biology. *Inter. J. Mol. Sci., 17*(11), 1891.

Osama, K., Mishra, B. N., & Somvanshi, P. (2015). Machine learning techniques in plant biology. In B. N. Mishra (Ed.), *PlantOmics: The Omics of Plant Science* (pp. 731–754). Springer.

Pandit, E., Pawar, S., Barik, S. R., Mohanty, S. P., Meher, J., & Pradhan, S. K. (2021). Marker-assisted backcross breeding for improvement of submergence tolerance and grain yield in the popular rice variety 'Maudamani.' *Agron., 11*, 1263.

Pavlopoulou, A., Karaca, E., Balestrazzi, A., & Georgakilas, A. G. (2019). In silico phylogenetic and structural analyses of plant endogenous danger signaling molecules upon stress. *Oxidative Med. Cell. Longev., 2019*(2), 1–14.

Peng, A., Chen, S., Lei, T., Xu, L., He, Y., Wu, L., & Yao, Q. (2017). Engineering canker resistant plants through CRISPR/Cas9-targeted editing of the susceptibility gene CsLOB1 promoter in citrus. *Plant Biotechnol. J., 15*(12), 1509–1519.

Perez-Hedo, M., Rivera-Perez, C., & Noriega, F. G. (2013). The Insulin/TOR signal transduction pathway is involved in the nutritional regulation of juvenile hormone synthesis in *Aedes aegypti*. *Insect Biochem. Mol. Biol., 43*, 495–500.

Peršoh, D. (2015). Plant-associated fungal communities in the light of meta'omics. *Fungal Divers., 75*, 1–25.

Prabhu, K. V., Singh, A. K., Basavaraj, S. H., Cherukuri, D. P., Charpe, A., Krishnan, S. G., & Singh, V. P. (2009). Marker assisted selection for biotic stress resistance in wheat and rice. *Indian J. Genet. Plant Breed., 69*(04), 305–314.

Prentice, K., Pertry, I., Christiaens, O., Bauters, L., Bailey, A., Niblett, C., & Smagghe, G. (2015). Transcriptome analysis and systemic RNAi response in the African sweet potato weevil. *PLoS ONE, 10*(3), e0115336.

Prins, M., Laimer, M., Noris, E., Schubert, J., Wassenegger, M., & Tepfer, M. (2008). Strategies for antiviral resistance in transgenic plants. *Mol. Plant Pathol., 9*(1), 73–83.

Qu, Z., Bendena, W. G., Nong, W., Siggens, K. W., Noriega, F. G., Kai, Z. P., & Bendena, W. G. (2017). MicroRNAs regulate the sesquiterpenoid hormonal pathway in *Drosophila* and other arthropods. *Proc. Biol Sci., 284*(1869), 20171827.

Rakha, M., Bouba, N., Ramasamy, S., Regnard, J. L., & Hanson, P. (2017). Evaluation of wild tomato accessions (*Solanum spp.*) for resistance to two-spotted spider mite (*Tetranychus urticae* Koch) based on trichome type and acylsugar content. *Genet. Resour. Crop Evol., 64*, 1011–1022.

Randhawa, M. S., Bhavani, S., Singh, P. K., Huerta-Espino, J., & Singh, R. P. (2019). Disease resistance in wheat: Present status and future prospects. In S. K. Sarial & R. K. Behl (Eds.), *Disease Resistance in Crop Plants: Molecular, Genetic and Genomic Perspectives* (pp. 61–81).

Rathinam, M., Mishra, P., Mahato, A. K., Singh, N. K., Rao, U., & Sreevathsa, R. (2019). Comparative transcriptome analyses provide novel insights into the differential response of Pigeonpea (*Cajanus cajan* L.) and its wild relative (*Cajanus platycarpus* (Benth.) Maesen) to herbivory by *Helicoverpa armigera*. *Plant Mol. Biol., 101*, 163–182.

Riaz, S., Nasir, I. A., Bhatti, M. U., Adeyinka, O. S., Toufiq, N., Yousaf, I., & Tabassum, B. (2020). Resistance to *Chilo infuscatellus* (Lepidoptera: Pyraloidea) in transgenic lines of sugarcane expressing *Bacillus thuringiensis* derived Vip3A protein. *Molecular Biology Reports, 47*(4), 2649–2658.

Richards, S., Gibbs, R. A., Weinstock, G. M., Brown, S. J., Denell, R., et al. (2008). The genome of the model beetle and pest *Tribolium castaneum*. *Nature, 452*, 949–955.

Rischbeck, P., Elsayed, S., Mistele, B., Barmeier, G., Heil, K., & Schmidhalter, U. (2016). Data fusion of spectral, thermal and canopy height parameters for improved yield prediction of drought stressed spring barley. *European Journal of Agronomy, 78*, 44–59.

Roorkiwal, M., Bharadwaj, C., Barmukh, R., Dixit, G. P., Thudi, M., Gaur, P. M., & Kumar, S. (2020). Integrating genomics for chickpea improvement: Achievements and opportunities. *Theoretical and Applied Genetics, 133*, 1703–1720.

Sáenz-de la, O. D., Morales, L. O., Strid, Å., Torres-Pacheco, I., & Guevara-González, R. G. (2021). Ultraviolet-B exposure and exogenous hydrogen peroxide application lead to cross-tolerance toward drought in *Nicotiana tabacum* L. *Physiologia Plantarum, 173*, 666–679.

Salehian, H., Rahnama, H., Dezhsetan, S., & Babaei, S. (2021). Constitutive expression of *cry3A* gene in transgenic potato plants for resistance to Colorado Potato Beetle (CPB). *Potato Research, 64*(4), 667–679.

Salzberg, S. L. (2019). Next-generation genome annotation: We still struggle to get it right. *Genome Biology, 20*(1), 1–3.

Sanfaçon, H. (2015). Plant translation factors and virus resistance. *Viruses, 7*(7), 3392–3419.

Sankaran, S., Khot, L. R., Espinoza, C. Z., Jarolmasjed, S., Sathuvalli, V. R., Vandemark, G. J., & Pumphrey, M. O. (2015). Low-altitude, high-resolution aerial imaging systems for row and field crop phenotyping: A review. *European Journal of Agronomy, 70*, 112–123.

Santillán Martínez, M. I., Bracuto, V., Koseoglou, E., Appiano, M., Jacobsen, E., Visser, R. G. F., & Govers, F. (2020). CRISPR/Cas9-targeted mutagenesis of the tomato susceptibility gene *PMR4* for resistance against powdery mildew. *BMC Plant Biology, 20*(1), 284.

Savary, S., & Willocquet, L. (2020). Modeling the impact of crop diseases on global food security. *Annual Review of Phytopathology, 58*, 313–341.

Schoonbeek, H. J., Wang, H. H., Stefanato, F. L., Craze, M., Bowden, S., Wallington, E., & Ridout, C. J. (2015). Arabidopsis EF-Tu receptor enhances bacterial disease resistance in transgenic wheat. *New Phytologist, 206*(2), 606–613.

Senapati, M., Tiwari, A., Sharma, N., Chandra, P., Bashyal, B. M., Ellur, R. K., & Singh, A. K. (2022). *Rhizoctonia solani* Kühn pathophysiology: Status and prospects of sheath blight disease management in rice. *Frontiers in Plant Science, 13*, 881116.

Shi, W., Syrenne, R., Sun, J. Z., & Yuan, J. S. (2010). Molecular approaches to study the insect gut symbiotic microbiota at the 'omics' age. *Insect Science, 17*, 199–219.

Shook, J., Gangopadhyay, T., Wu, L., Ganapathy Subramanian, B., Sarkar, S., & Singh, A. K. (2021). Crop yield prediction integrating genotype and weather variables using deep learning. *PLoS ONE, 16*, e0252402.

Silva, C. J., van den Abeele, C., Ortega-Salazar, I., Papin, V., Adaskaveg, J. A., Wang, D., & Thomma, B. P. (2021). Host susceptibility factors render ripe tomato fruit vulnerable to fungal disease despite active immune responses. *Journal of Experimental Botany, 72*(7), 2696–2709.

Silva, J. C. F., Teixeira, R. M., Silva, F. F., Brommonschenkel, S. H., & Fontes, E. P. (2019). Machine learning approaches and their current application in plant molecular biology: A systematic review. *Plant Science, 284*, 37–47.

Simón, D., Borsani, O., & Filippi, C. V. (2022). RFPDR: A random forest approach for plant disease resistance protein prediction. *PeerJ, 10*, e11683.

Singh, A., Ganapathy Subramanian, B., Singh, A. K., & Sarkar, S. (2016). Machine learning for high-throughput stress phenotyping in plants. *Trends in Plant Science, 21*, 110–124.

Singh, A., Jones, S., Ganapathy Subramanian, B., Sarkar, S., Mueller, D., Sandhu, K., & Nagasubramanian, K. (2021). Challenges and opportunities in machine-augmented plant stress phenotyping. *Trends in Plant Science, 26*, 53–69.

Singh, B. D., & Singh, A. K. (2015). Marker-assisted plant breeding: Principles and practices. In S. D. Tanksley & J. C. Nelson (Eds.), *Advanced Backcross QTL Analysis: A Method for the Simultaneous Discovery and Transfer of Valuable QTLs from Unadapted Germplasm into Elite Breeding Lines* (pp. 191–203). *Theoretical and Applied Genetics, 92*(2).

Song, Q., Lee, J., Akter, S., Rogers, M., Grene, R., & Li, S. (2020). Prediction of condition-specific regulatory genes using machine learning. *Nucleic Acids Research, 48*, e62.

Song, Y., Liu, L., Wang, Y., Valkenburg, D. J., Zhang, X., Zhu, L., & Thomma, B. P. (2018). Transfer of tomato immune receptor Ve1 confers Ave1-dependent *Verticillium* resistance in tobacco and cotton. *Plant Biotechnology Journal, 16*(2), 638–648.

Stokes, J. M., Yang, K., Swanson, K., Jin, W., Cubillos-Ruiz, A., Donghia, N. M., & Bloom, Z. (2020). A Deep Learning Approach to Antibiotic Discovery. *Cell, 180*, 688–702.

Tattaris, M., Reynolds, M. P., & Chapman, S. C. (2016). A direct comparison of remote sensing approaches for high-throughput phenotyping in plant breeding. *Frontiers in Plant Science, 7*, 1131.

Tausen, M., Clausen, M., Moeskjær, S., Shihavuddin, A., Dahl, A. B., Janss, L., & Andersen, S. U. (2020). Greenotyper: Image-based plant phenotyping using distributed computing and deep learning. *Frontiers in Plant Science, 11*, 1181.

Thomas, G., Balocco, S., Mann, D., Simundsson, A., & Khorasani, N. (2021). Intelligent agricultural machinery using deep learning. *IEEE Instrumentation & Measurement Magazine, 24*, 93–100.

Thorpe, P., Cock, P. J., & Bos, J. (2016). Comparative transcriptomics and proteomics of three different aphid species identifies core and diverse effector sets. *BMC Genomics, 17*, 172.

Tilgam, J., Kumar, K., Jayaswal, D., Choudhury, S., Kumar, A., Jayaswall, K., & Saxena, A. K. (2021). Success of microbial genes based transgenic crops: Bt and beyond Bt. *Molecular Biology Reports, 48*(12), 8111–8122.

Tohidfar, M., & Khosravi, S. (2015). Transgenic crops with improved resistance to biotic stresses: A review. *Biotechnology, Agronomy and Society and Environment, 19*(1), 62–70.

Tomoyasu, Y., Miller, S. C., Shuichiro, T., Schomeier, M., Grossmann, D., & Bucher, G. (2008). Exploring systemic RNA interference in insects: A genome-wide survey for RNAi genes in Tribolium. *Genome Biology, 9*, 1–22.

Turnbaugh, P. J., & Gordon, J. I. (2008). An invitation to the marriage of metagenomics and metabolomics. *Cell, 134*, 708–713.

Valencia Hernandez, J. A., Solano Alvarez, N., Rico Rodriguez, M. A., Rodriguez Ontiveros, A., Torres Pacheco, I., Rico Garcia, E., & Guevara Gonzalez, R. G. (2022). Eustressic dose of cadmium in soil induces defense mechanisms and protection against *Clavibacter michiganensis* in tomato. *Journal of Plant Growth Regulation, 1111*(1111), 1111.

Vargas Hernandez, M., Macias-Bobadilla, I., Guevara-Gonzalez, R. G., Romero-Gomez, S. D. J., Rico-Garcia, E., Ocampo-Velazquez, R. V., Alvarez-Arquieta, L. d. L., & Torres-Pacheco, I. (2017). Plant hormesis management with biostimulants of biotic origin in agriculture. *Frontiers in Plant Science, 8*, 1762.

Vrkoslav, V., Muck, A., Cvacka, J., & Svatos, A. (2010). MALDI imaging of neutral cuticular lipids in insects and plants. *Journal of the American Society for Mass Spectrometry, 21*, 220–231.

Walker, P. T. (1984). Quantification and economic assessment of crop losses due to pests, diseases and weeds. In *Advancing Agricultural Production in Africa: Cabs First Scientific Conference, Arusha, Tanzania,* 12–18 February.

Wan, D. Y., Guo, Y., Cheng, Y., Hu, Y., Xiao, S., Wang, Y., et al. (2020). CRISPR/Cas9-mediated mutagenesis of VvMLO3 results in enhanced resistance to powdery mildew in grapevine. *Horticulture Research, 7*, 116.

Wang, G., Sun, Y., & Wang, J. (2017). Automatic image-based plant disease severity estimation using deep learning. *Computational Intelligence and Neuroscience, 2017*, 1–8.

Wang, J., Ma, H., Zuo, Y., Yang, Y., & Wu, Y. (2020). CRISPR-mediated gene knockout reveals nicotinic acetylcholine receptor (nAChR) subunit α6 as a target of spinosyns in *Helicoverpa armigera. Pest Management Science, 76*, 2925–2931.

Wang, J., Zhang, H., Wang, H., Zhao, S., Zuo, Y., Yang, Y., & Wu, Y. (2016). Functional validation of cadherin as a receptor of Bt toxin Cry1Ac in *Helicoverpa armigera* utilizing the CRISPR/Cas9 system. *Insect Biochemistry and Molecular Biology, 76*, 11–17.

Wang, X. W., Luan, J. B., Li, J. M., Bao, Y. Y., Zhang, C. X., & Liu, S. S. (2010). De novo characterization of a whitefly transcriptome and analysis of its gene expression during development. *BMC Genomics, 11*, 400.

Wang, Y., Li, Z., Xu, J., Zeng, B., Ling, L., You, L., et al. (2013). The CRISPR/Cas system mediates efficient genome engineering in *Bombyx mori. Cell Research, 23*, 1414–1416.

Wang, Y., Zhou, M., Zou, Q., & Xu, L. (2021). Machine learning for phytopathology: From the molecular scale towards the network scale. *Briefings in Bioinformatics, 22*, bbab037.

White, J. W., & Conley, M. M. (2013). A flexible, low-cost cart for proximal sensing. *Crop Science, 53*, 1646–1649.

Wijerathna-Yapa, A. (2017). Transgenic plants: Resistance to abiotic and biotic stresses. *Journal of Agricultural and Environmental International Development, 111*(1), 245–275.

Wu, C., Zeng, R., Pan, J., Wang, C. C., & Liu, Y. J. (2019). Plant phenotyping by deep-learning-based planner for multi-robots. *IEEE Robotics and Automation Letters, 4*, 3113–3120.

Wu, K., Shirk, P. D., Taylor, C. E., Furlong, R. B., Shirk, B. D., Pinheiro, D. H., & Siegfried, B. D. (2018). CRISPR/Cas9 mediated knockout of the abdominal-A homeotic gene in fall armyworm moth (*Spodoptera frugiperda*). *PLoS ONE, 13*(12), e0208647.

Wu, R., Zhang, Q., Lin, Y., Chen, J., Somta, P., Yan, Q., Xue, C., Liu, J., Chen, X., & Yuan, X. (2022). Marker-assisted backcross breeding for improving Bruchid (*Callosobruchus* spp.) resistance in mung bean (*Vigna radiata* L.). *Agronomy, 12*, 1271.

Xiao, Y., & Wu, K. (2019). Recent progress on the interaction between insects and *Bacillus thuringiensis* crops. *Philosophical Transactions of the Royal Society B: Biological Sciences, 374*(1767), 20180316.

Xu, C., Cheng, J., Lin, H., Lin, C., Gao, J., & Shen, Z. (2018). Characterization of transgenic rice expressing fusion protein Cry1Ab/Vip3A for insect resistance. *Scientific Reports, 8*(1), 1–8.

Yang, N., Xie, W., Jones, C. M., Bass, C., Jiao, X., Yang, X., Liu, B., Li, R., & Zhang, Y. (2013). Transcriptome profiling of the whitefly *Bemisia tabaci* reveals stage-specific gene expression signatures for thiamethoxam resistance. *Insect Molecular Biology, 22*, 485–496.

Yasrab, R., Zhang, J., Smyth, P., & Pound, M. P. (2021). Predicting plant growth from time-series data using deep learning. *Remote Sensing, 13*, 331.

Yu, H., Wang, Y., Fu, F., & Li, W. (2022). Transgenic improvement for biotic resistance of crops. *International Journal of Molecular Sciences, 23*(22), 14370.

Zafar, M. M., Mustafa, G., Shoukat, F., Idrees, A., Ali, A., Sharif, F., & Li, F. (2022). Heterologous expression of cry3Bb1 and cry3 genes for enhanced resistance against insect pests in cotton. *Scientific Reports, 12*(1), 10878.

Zaidi, S. S., Tashkandi, M., Mansoor, S., & Mahfouz, M. M. (2016). Engineering plant immunity: Using CRISPR/Cas9 to generate virus resistance. *Frontiers in Plant Science, 7*, 1673.

Zaman-Allah, M., Vergara, O., Araus, J. L., Tarekegne, A., Magorokosho, C., Zarco-Tejada, P. J., & Craufurd, P. (2015). Unmanned aerial platform-based multi-spectral imaging for field phenotyping of maize. *Plant Methods, 11*, 35.

Zhang, J., Khan, S. A., Heckel, D. G., & Bock, R. (2017). Next-generation insect-resistant plants: RNAi-mediated crop protection. *Trends in Biotechnology, 35*(9), 871–882.

Zhou, J., Peng, Z., Long, J., Sosso, D., Liu, B., Eom, J. S., et al. (2015). Gene targeting by the TAL effector PthXo2 reveals cryptic resistance gene for bacterial blight of rice. *The Plant Journal, 82*(4), 632–643.

Zhu, Y., Li, H., Bhatti, S., Zhou, S., Yang, Y., Fish, T., & Thannhauser, T. W. (2016). Development of a laser capture microscope-based single-cell-type proteomics tool for studying proteomes of individual cell layers of plant roots. *Horticulture Research, 3*, 16026.

Ziegenhain, C., Vieth, B., Parekh, S., Hellmann, I., & Enard, W. (2018). Quantitative single-cell transcriptomics. *Briefings in Functional Genomics, 17*, 220–232.

CHAPTER 9

Applications and Recent Advances in CRISPR-Cas Editing for Biotic Stress Management in Cereals

PRIYANKA AGARWAL,[1] AJAY KUMAR CHANDRA,[1] ALKA BHARATI,[2] RADHESHYAM KUMAWAT,[3] PRAMOD KUMAR PANDEY,[4] and AMARJEET KUMAR[5]

[1]*ICAR–Indian Agricultural Research Institute, New Delhi, India*

[2]*Scientist, ICAR–Central Agroforestry Research Institute, Jhansi, Uttar Pradesh, India*

[3]*Research Scholar, College of Post-Graduate Studies in Agricultural Sciences, Central Agricultural University (Imphal), Umiam, Meghalaya, India*

[4]*Assistant Maize Breeder, AICRP on Maize, College of Agriculture, Central Agricultural University, Kyrdemkulai, Meghalaya, India*

[5]*Assistant Professor, Genetics and Plant Breeding, College of Agriculture, Central Agricultural University (Imphal), Kyrdemkulai, Meghalaya, India*

ABSTRACT

CRISPR-Cas (Clustered Regularly Interspaced Short Palindromic Repeat/CRISPR-Associated Protein), the 21st-century technique, is a highly efficient gene editing technique that shows high competence for dealing with stress management. It's the new way to characterize genes. Both abiotic (cold stress, drought, salinity, etc.) and biotic stresses (bacteria, fungi, viruses) significantly affect agronomic traits such as plant growth, grain yield, and nutritional components. The most eco-friendly and long-term solution to this

Climate Change and Biotic Factors: A Molecular Approach.
Ashutosh Singh, Saurabh Pandey, and Amarjeet Kumar, (Eds.)

problem is to breed tolerant cultivars against these stresses. Emerging genetic engineering techniques offer new prospects for agricultural enhancement by applying precision genome editing for certain crop traits. The impact of this CRISPR/Cas technique has been visualized on cereal crops. Cereal crops are the staple food crops and constitute the major source of nutritional value for human life. This strategy has produced some of the biggest success stories about staple food crops, including rice, wheat, and maize. Precision plant breeding and creating non-transgenic cereals that can withstand the negative effects of climate change are expected to benefit greatly from the CRISPR-Cas system. The CRISPR-Cas methodology, combined with existing and future breeding technologies, can combat global food security.

9.1 INTRODUCTION

Cereals are recognized as staple food crops in both developed and emerging economies. They are the main source of fiber, protein, carbohydrates, and energy. They significantly contribute to the GDP of agriculture in many developed and developing nations and hence play a crucial role in the economy. There is an urgent need to increase staple crop yield to achieve global food security and feed the world's growing population in the future. Both biotic (fungi, bacteria, and viruses) and abiotic (drought, salt, heat, heavy metals, nutritional deficiency, and cold) stresses have a significant impact on the growth and yield of cereals (Chandra et al., 2022). Studies on functional genomics and molecular breeding have helped us understand and improve cereals' growth and yield under biotic and abiotic challenges. Depending on the cereal and the region, these pressures can impair the output of cereals by up to >75%. In the case of rice, most cultivars are susceptible to blast, sheath blight, sheath rot, brown leaf spot, bacterial leaf blight, tungro virus, bakanae disease, leaf scald, false smut, and grain discoloration (Shekhar et al., 2020). Approximately 50–100% of rice crop yield is estimated to be lost due to this disease worldwide (Fiyaz et al., 2022). Congruently, the major biotic stresses of wheat are stripe rust, stem rust, leaf rust, rust, karnal bunt, powdery mildew, leaf streak, black chaff, and head blight (Agarwal et al., 2021; Raghunandan et al., 2022; Mallick et al., 2022). Among them, yellow rust has caused significant yield losses in wheat worldwide (Rozewicz et al., 2021; Nyamgoud et al., 2022; Tambong, 2022). The continuous yield losses of major cereal crops may lead to food and nutritional insecurity concerns.

To address yield losses caused by different biotic stressors and factors, scientists have to urgently take a U-turn from conventional to modern plant breeding strategies. So far, several modern molecular tools and techniques are available, including marker-assisted breeding, GWAS, genomic selection (GS), Tilling, multi-omics, transgenic, and many others for the improvement of stress-resilient crops (Sood et al., 2019; Chandra et al., 2020). Aided by these high-throughput biotechnological techniques, genome editing can provide a more reliable and stable option for the development of biotic stress-resilient cereal crops. Since the genome editing tools like ZFNs, MNs, TALENs, and CRISPR/Cas system offer precise gene/genome editing events (Chandra et al., 2022b; Dongariyal et al., 2023). So, it can be a novel breakthrough in the field of stress biology research, particularly biotic stress breeding of major cereal crops.

To address yield losses caused by different biotic stressors and factors, scientists have to urgently take a U-turn from conventional to modern plant breeding strategies. So far, several modern molecular tools and techniques are available, including marker-assisted breeding, GWAS, genomic selection (GS), Tilling, multi-omics, transgenic, and many others for the improvement of stress-resilient crops (Sood et al., 2019; Chandra et al., 2020). Aided by these high-throughput biotechnological techniques, genome editing can provide a more reliable and stable option for the development of biotic stress-resilient cereal crops. Since the genome editing tools like ZFNs, MNs, TALENs, and CRISPR/Cas system offer precise gene/genome editing events (Chandra et al., 2022b; Dongariyal et al., 2023). So, it can be a novel breakthrough in the field of stress biology research, particularly biotic stress breeding of major cereal crops.

9.2 GENOME EDITING: A POTENTIAL TOOL FOR STRESS-RESILIENT CROP IMPROVEMENT

In the age of molecular biology, precise and effective genome alteration systems are made possible by genome editing at particular loci using cutting-edge molecular techniques and tools (Dongariyal et al., 2023). Meganucleases (MNs), transcription activator-like effector nucleases (TALENs), zinc finger nucleases (ZFNs), Cpf1, and clustered regularly interspaced short palindromic repeats (CRISPR)/CRISPR associated 9 (Cas9) are different classes of engineered nucleases (EN) that have been developed over the past 20 years (Figure 9.1). These EN are typically used as effective genome editing tools. The earliest endonucleases that can precisely detect and cleave up to 12 to 40 bp of recognition sequence were meganucleases (MNs) (Jurica et al., 1998). Eventually, other methods like TALENs and ZFNs emerged as the primary methods for altering the genome of target species. Contrastingly, ZFNs have targeted DNA cleavage proteins that are specifically engineered to cleave DNA sequences at certain places. Although, the major restrictions of these techniques are that the editing endonucleases require two different peptide linker proteins for their endonuclease activity (binding and site-specific cleavage) which were very difficult designs (Martin-Pizarro and Pose, 2018).

These problems forced researchers to develop more effective, practical, affordable, and advanced editing tools for focused progress. In view of this, the CRISPR/Cas9 system has provided Cas9 with more sophisticated and effective genome-editing platforms, expanding the possibilities for precise modification of genes or genomes. However, the main benefits of the CRISPR/Cas9 editing technology are that it is more effective, cost-effective, and simple. The invention of CRISPR/Cas9 has led to the development of a tool that is more versatile and precise, and most importantly, it can identify and cleave numerous loci of the target genes (Shivani et al., 2017; Dongariyal et al., 2023).

9.3 TOOLS AND TECHNOLOGY OF CRISPR SYSTEM

CRISPR is a family of repetitive DNA sequences which are spaced apart by fragments of variable sequences found in the genomes of archaea (84%) and bacteria (45%) (Bao et al., 2019). It was initially discovered downstream of the alkaline phosphatase isozyme gene in *Escherichia coli* in the 1990s (Ishino et al., 1987). For sequence-specific interference with invading nucleic acids, the CRISPR system makes use of short guide RNAs (gRNAs). CRISPR is an array of short, repeated sequences separated by highly unique spacers. The source of spacers and repeats are virus nucleic acid and plasmids. CRISPR/Cas, a group of CRISPR-associated proteins, has the power to identify, cut, and, finally, change phage DNA in a certain manner. The protein Cas has an enzyme activity and a specific function in DNA sequences, and CRISPR displays nuclease activity. Three steps are frequently used to summarize the CRISPR/Cas process. Firstly, distinct sequences are inserted into the CRISPR locus, known as spacer acquisition or adaptation. Thereafter, transcription of the CRISPR locus and processing of gRNA (expression or gRNA biogenesis), followed by target interference, detected and degraded by gRNA and Cas proteins (Liu et al., 2017). There are three classes (Classes I, II, and III) of the CRISPR/Cas system (Makarova et al., 2020). Each of these depends on the proteins and signature sequences to offer a unique mechanism for the immune process. Type I and III use various Cas proteins to assemble a complex with distinguishing proteins for binding

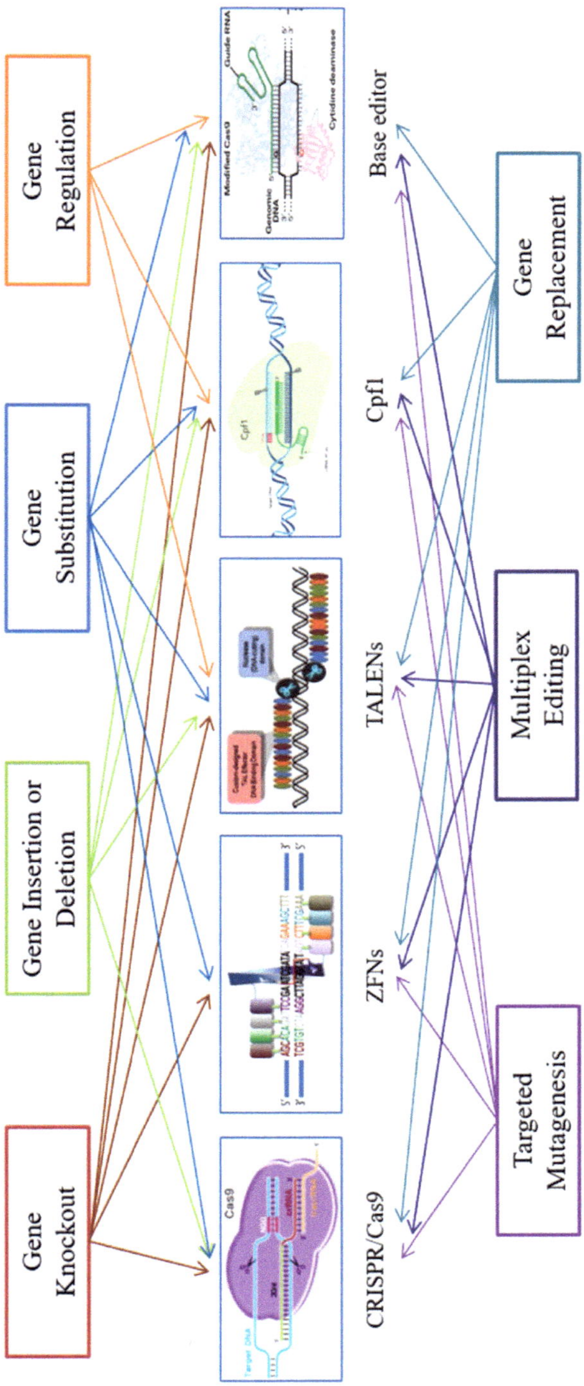

FIGURE 9.1 Various genome editing technologies and their possible role in stress-resilient breeding.

and targeting; however, class II systems commonly use a simplified protocol. The type II system, also known as CRISPR-associated protein 9 (Cas9), consists of a single multifunctional protein that recognizes RNA-guided target DNA and catalyzes their cleavage through the action of the RuvC and HNH domains (Zetsche et al., 2015). Cas1, 2, and 9 are the three signature proteins encompassing three subtypes and are usually used to classify the class II CRISPR/Cas systems (II-A, II-B, and II-C). The recently developed CRISPR/Cas12a and Cas12b fall into classes VA and VB, respectively. In addition, other effector nucleases, including Cas3, Cas8, Cas10, and Cas13, are rarely used for genome editing (Makarova et al., 2018; Xu et al., 2019).

9.3.1 CRISPR-CAS9 EXPRESSION SYSTEM

The single-stranded guide RNA (sgRNA) protein complex, which is active in CRISPR/Cas9-mediated genome editing, is the basis for this approach. A DNA endonuclease called Cas9 is derived from bacteria like *B. laterosporus, S. thermophiles, S. pyogenes,* and *S. aureus* (Bashar and Hanif, 2021). The most common editing process among these is *S. pyogenes*-mediated CRISPR/Cas9, in which the Cas9 endonuclease attaches to the CRISPR RNA (crRNA) and transactivates the CRISPR RNA (tracrRNA) (Steinert et al., 2015). The HNH domain and the RuvC-like domain are generally the two functional domains found in Cas9 nucleases. Target DNA sequences are precisely cleaved into their complementary and opposite strands by the HNH and RuvC-like domains, respectively. TracrRNA and crRNA, two naturally occurring small RNAs, can combine to create sgRNA and act as a guide sequence for screening target genes in a region with an adjacent protospacer adjacent motif (PAM) sequence. Cas9 activity is necessary for the CRISPR/Cas9 system to function properly, directed by sgRNA that needs a PAM to determine the target site (Mojica et al., 2009). However, five important steps are typically used to carry out genome editing using the CRISPR/Cas9 system. This includes (i) recognition of PAM sequence near or within the target gene, (ii) development of sgRNA homology to the target 'scaffold' sequence, (iii) insertion of sgRNA into a binary vector, (iv) transformation into a cell line or suitable host, followed by (v) expression profiling and selection of edited lines (Jaganathan et al., 2018). To effectively modify the complex genomes of plants and animals, CRISPR/Cas9 more frequently follows these sequential steps (Liu et al., 2017). The construction of vectors for delivery and expression is extremely important for effective editing, but it is extremely difficult in the majority of polyploid crops. Despite all these challenges, progress is on its way, and different binary vectors have been developed to accomplish effective delivery, target editing, and expression in important cereal crops like rice, wheat, maize, sorghum, barley, etc. Moreover, these findings can also demonstrate the capability of CRISPR/Cas9 to amend specific loci or genes in target plants.

9.4 EFFECT OF BIOTIC STRESS AND CONTROL PRACTICES IN MAJOR CEREAL CROP

Plants' biotic stress is caused by viruses, bacteria, fungi, nematodes, insects, arachnids, and weeds. Biotic stress agents directly deprive their hosts of essential nutrients, as opposed to abiotic stress, which is brought by environmental factors like heat, temperature, chilling, and drought. This reduces plant vigor and, in extreme cases, causes plant mortality. Different types of pathogens attack cereal crops and make them vulnerable to diseases. To get rid of this, pesticides, herbicides, and fungicides are used to control cereal diseases. These chemicals not only pollute the environment but also affect the metabolic pathway of cereals. Hence, the efficient and effective use of biotechnological tools like CRISPR/Cas technology can induce plant defense mechanisms against pathogens like bacteria, viruses, fungi, and insects (Figure 9.2). Given this, researchers have already targeted some biotic stress-responsive genes by CRISPR/Cas events, and in most cases, very encouraging results are reported (Table 9.1).

TABLE 9.1 Potential use of CRISPR/Cas9 Mediated Genome Editing System for Development of Biotic Stress-Resilient Cereal Crops

S. No.	Cereal crops	Disease or stress	Target gene	References
1.	Wheat	Powdery mildew	*TaMLO-A1*, *TaMLO-B1* and *TaMLO-D1*	Wang et al., 2014
2.	Rice	Bacterial blight	*OsSWEET13*	Zhou et al., 2015
3.	Rice	Blast	*OsERF922,* and *OsSEC3A*	Wang et al., 2016; Ma et al., 2018
4.	Wheat	Powdery mildew	*EDR1*	Zhang et al., 2017
5.	Rice	Rice tungro spherical virus	*eIF4G*	Macovei et al., 2018
6.	Rice	Plant hopper	*OsCYP71A1*	Lu et al., 2018
7.	Rice	Stem borer	*OsCYP71A1*	Lu et al., 2018
8.	Rice	Herbicide	*OsALS*	Kuang et al., 2020
9.	Maize	Herbicide	*ZmALS1* and *ZmALS2*	Jiang et al., 2020

9.5 POTENTIAL APPLICATION OF CRISPR/CAS9 FOR IMPROVEMENT OF BIOTIC STRESS-RESILIENT CROPS

9.5.1 RESISTANCE TO BACTERIAL DISEASES

Bacterial diseases are one of the most prevalent infections in agriculture, which are typically brought on by bacteria, thus reducing the growth and yield potential of cereal crops. The CRISPR technique is widely utilized in major cereals like rice to enhance resistance against bacterial blight, commonly reported in West Africa and Southeast Asia. When rice plants are infected with bacterial blight disease, it is reported that sucrose family (SWEET) transporters are involved in rice growth and yield. Hence, rice growth against bacterial blight disease can be enhanced by CRISPR/Cas9 edited *OsSWEET13* gene knockout (Zhou et al., 2015). In a different study, the CRISPR/Cas9 system was also used to alter SWEET genes (SWEET11, 13, and 14), which assisted in improving traits like height and panicle length (Oliva et al., 2019).

9.5.2 RESISTANCE TO FUNGAL DISEASES

Fungal disease is a kind of devastating disease in crops. Chemical fungicides are mainly used in agriculture to minimize yield and biomass loss due to fungal attacks. The environment and sustainable agriculture can be improved by developing fungicide-free cereals that are resistant to fungal infections. Among fungal diseases, plants with greater resistance to powdery mildew were produced by CRISPR/Cas9 through knocking out all three *TaMLO* (TaMLO-A1, B1, and D1) alleles in wheat (Wang et al., 2014). Furthermore, wheat with resistance to powdery mildew was also created by using the CRISPR/Cas9 system to edit the enhanced disease resistance1 (*EDR1*) gene (Zhang et al., 2017). Similarly, *the ethylene-responsive factor 22 (OsERF922)* was knocked out using CRISPR/Cas9 technology, leading to the development of rice resistant to blast (Wang et al., 2016). Additionally, improved rice cultivar resistance to blast disease was achieved through the CRISPR/Cas9 system by disrupting *OsSEC3A* (Ma et al., 2018).

9.5.3 RESISTANCE TO INSECTS AND PESTS

Pests' attacks are responsible for up to 20–40% of the loss in agricultural production worldwide (Douglas, 2018). In a recent FAO report, about 40% of the world's crops are lost to pests each year. According to Deutsch et al. (2018), as the weather has warmed up, losses due to pests and insects in agriculture have increased over the past few years. Yet, pesticide use on a large scale might harm the ecosystem; hence, safe and efficient technology is urgently needed to reduce the number of insects and pests that affect crops. When *OsCYP71A1* was disrupted through CRISPR/Cas9 to limit serotonin production, the amount of salicylic acid increased, which, however, increased the rendering resistance of rice against plant hoppers and stem borers (Lu et al., 2018). The *gmcdpk38* mutant with *Hap3* elimination via CRISPR/Cas9 demonstrated great resistance to common cutworms (Li et al., 2022).

9.5.4 RESISTANCE TO VIRUSES

Plant viruses pose a serious threat to several staple crops that are also economically significant. Single-stranded DNA (ssDNA), double-stranded DNA (dsDNA) viruses, double-stranded RNA (dsRNA), reverse-transcribing viruses, positive-sense single-stranded RNA (ssRNA+) viruses, and negative-sense single-stranded RNA (ssRNA) viruses are the six main viral groups affecting standing plants/crops as well as plant produce (Roossinck et al., 2015). In tropical Asia, tungro disease, which is caused by the combination of the tungro spherical virus and the tango bacilliform virus, severely limits the grain and biomass yield of rice. The translation initiation factor 4 gamma (*eIF4G*) gene has been shown to contribute to the management of tungro spherical viruses in rice (Lee et al., 2010). Correspondingly, rice lines resistant to the tungro spherical virus were developed by editing the *eIF4G* gene using CRISPR/Cas9 technology (Macovei et al., 2018).

9.5.5 RESISTANCE TO WEEDS

Weeds and insects also constitute a major biotic stress affecting cereals' growth and productivity. Cereals suffer the most damage from weeds. They make it harder for cereals to develop and flourish properly since there is more competition for resources like nutrition, space, sunlight, water, and fertilizers. Although using the same herbicide frequently can make weeds resistant. With genome editing techniques like base editing and prime editing, it might be possible to develop herbicide resistance in plants more effectively. Through base editing, we can insert point mutations into the genome of plants, assisting in the development of herbicide-resistant plants. It has been efficiently used in rice to amend the acetolactate synthase (*OsALS*) gene for herbicide-tolerant rice (Kuang et al., 2020). In a different study, the same *OsALS* gene was targeted by CBE (cytosine base editor), which enhanced rice growth even in nicosulfuron, imazapic, pyroxsulam, and fucarbazone treatment conditions (Zhang et al., 2020). Herbicide resistance has been developed particularly in cereals, like rice and wheat, via prime editing. Recently, Butt et al. (2020) introduced prime editing for cereals for the first time in rice (Figure 9.2).

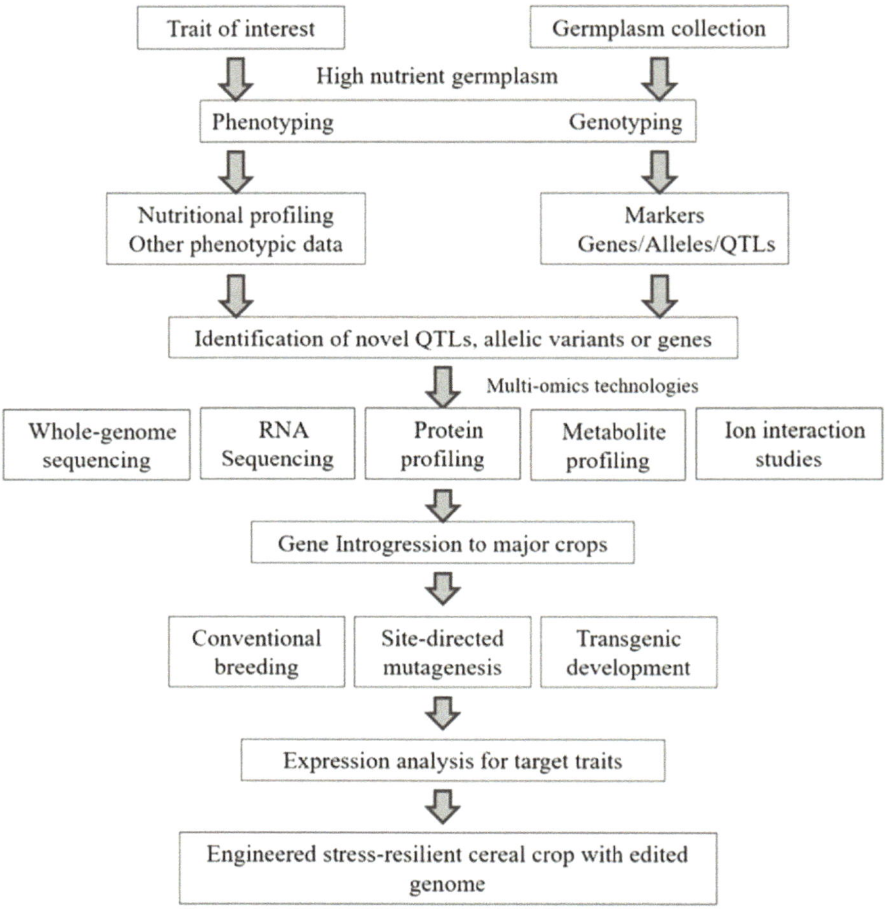

FIGURE 9.2 Multi-Omics-based approaches for improvement of stress-resilient traits in crops.

They have targeted the *OsALS* and *OsIPA1* genes in rice using prime editing. Among them, the *OsALS* gene (editing efficiency 26%) improved rice growth when exposed to the herbicide Bispyribac sodium. Jiang et al. (2020) used base editing to create two ZmALS genes (*ZmALS1* and *ZmALS2*) to create herbicide-tolerant maize with a greater editing efficiency (53.2%).

9.6 SOFTWARE AND PLANT DATABASES: CRISPR/CAS9-BASED STRESS-RESILIENT CROP IMPROVEMENT

This is the era of next-generation sequencing, genomics, and bioinformatic techniques, which have replaced conventional breeding approaches with high-throughput screening for crop improvement.

The integration of omics datasets of plant species such as wheat, rice, maize, millet, and sorghum is commonly available in the public domain. These available resources and datasets could be effectively utilized to design future stress-resilient crops (Kumar et al., 2018; Chandra et al., 2021). Recently, bioinformatics platforms have offered remarkable innovations to promote model research and applied plant species. These computational approaches, along with gene editing tools, could play a potential role in eliminating the global hunger problem, benefiting the agriculture sector. The widely available sequence information in various crops is a great resource and serves as the basis for the development of gene editing techniques (Table 9.2).

TABLE 9.2 Details of Genomics-Based Databases Available in Public Domains That Can be Used for Stress-Resilient Crop Improvement

S. No.	Genomics based database	Web links for search
1.	MaizeGDB	http://www.maizegdb.org/
2.	EnsemblPlants	http://plants.ensembl.org/
3.	Maize sequence database	http://www.maizesequence.org/
4.	autoSNPdb	http://autosnpdb.appliedbioinformatics.com.au/
5.	Brassica rapa genome database	http://brassicadb.org/
6.	European Molecular Biology Laboratory (EMBL) nucleotide sequence database	http://www.ebi.ac.uk/embl/
7.	PlantsDB	muenchen.de/plant/genomes.jsp
8.	The Plant Ontology	http://www.plantontology.org/
9.	Rice Genome Annotation Project	http://rice.plantbiology.msu.edu/
10.	SSR taxonomy tree	http://appliedbioinformatics.com.au/projects/ssrtaxono my/php/
11.	International Nucleotide Sequence Database Collaboration (INSDC)	http://www.insdc.org/
12.	Graingenes	http://wheat.pw.usda.gov/
13.	International Crop Information System (ICIS)	http://www.icis.cgiar.org
14.	Gramene	http://www.gramene.org/
15.	Brachypodiumdatabase	http://www.brachypodium.org/

TABLE 9.2 *(Continued)*

S. No.	Genomics based database	Web links for search
16.	SOL Genomics Network	(SGN) http://solgenomics.net/
17.	Phytozome	http://www.phytozome.net/
18.	Brassica rapagenome database	http://brassicadb.org/
19.	European bioinformatics institute, EnsEMBL plants	http://plants.ensembl.org/
20.	SoyBase	http://soybase.org/
21.	The Crop Expressed Sequence Tag database, CR-EST	http://pgrc.ipk-gatersleben.de/cr-est/
22.	The Triticeae Repeat Sequence Database (TREP)	http://wheat.pw.usda.gov/ITMI/Repeats/
23.	Wheat genome information	http://www.wheatgenome.info
24.	TAGdb	http://flora.acpfg.com.au/tagdb/
25.	Oryzabase	http://www.shigen.nig.ac.jp/rice/oryzabase/

9.7 CONCLUSION AND FUTURE DIRECTION

By 2050, the world's population is projected to exceed 10 billion, leading to increased food demand. To meet the needs of the growing population, sustainable and integrated agricultural production is essential. Both biotic and abiotic stress factors negatively impact the growth and production of all cereals. Developing new cultivars that can withstand biotic and abiotic challenges will greatly benefit sustainable agricultural production. Over the past few decades, various multi-omics approaches have been steadily developed and systematically applied in plant research as valuable tools for crop improvement. The recently developed CRISPR/Cas technology offers a simple, adaptable, and accurate protocol for developing novel stress-resistant crops. CRISPR/Cas9 technology shows great promise in introducing synthetic DNA sequences, such as promoters and transcriptional regulatory elements, to enhance plant performance and improve yield, quality, and resistance to biotic and abiotic challenges. The CRISPR/Cas9 genome editing method can generate new variants and has broad applications for enhancing crops quantitatively and qualitatively. With precise genome editing and transgene-free applications, CRISPR is expected to overcome significant barriers to crop improvement. In the scientific realm, CRISPR technology will play a crucial role in ensuring global food and nutritional security in the face of challenges like climate change and population growth.

KEYWORDS

- **biotic stress**
- **cereal crop**
- **CRISPR- Cas9**
- **crop improvement**
- **gene editing**
- **genetic engineering**
- **modern biotechnological method**
- **plant genomes**

REFERENCES

Agarwal, P., Jha, S. K., Sharma, N. K., Raghunanadan, K., Mallick, N., Niranjana, M., Saharan, M. S., & Singh, J. B. (2021). Identification of the improved genotypes with 2NS/2AS translocation through molecular markers for imparting resistance to multiple biotic stresses in wheat. *Indian Journal of Genetics and Plant Breeding, 81*, 28–34.

Bao, A., Burritt, D. J., Chen, H., Zhou, X., et al. (2019). The CRISPR/Cas9 system and its applications in crop genome editing. *Critical Reviews in Biotechnology, 39*, 321–336.

Bashar, K. K., & Hanif, M. A. (2021). Crop gene editing against biotic stresses via CRISPR/Cas9 tools: A review. *Archives of Phytopathology and Plant Protection, 54*, 1159–1181.

Butt, H., Rao, G. S., Sedeek, K., Aman, R., Kamel, R., & Mahfouz, M. (2020). Engineering herbicide resistance via prime editing in rice. *Plant Biotechnology Journal, 18*, 2370.

Chandra, A. K., Jha, S. K., Agarwal, P., Mallick, N., Murukan, N., & Vinod. (2022b). Leaf rolling in bread wheat (*Triticum aestivum* L.) is controlled by the upregulation of a pair of closely linked/duplicate zinc finger homeodomain class transcription factors during moisture stress conditions. *Frontiers in Plant Science, 13*, Article 1038881.

Chandra, A. K., Joshi, A., Tripathi, A., Kumar, A., Pandey, S., Singh, A., Lal, D., Bharati, A., Adhikari, S., & Dinkar, V. (2022a). Climate resilience maize: Heat stress, signaling, and molecular interventions. *Journal of Plant Growth Regulation, 42*, 6349–6366.

Chandra, A. K., Kumar, A., Bharati, A., Joshi, R., Agrawal, A., & Kumar, S. (2020). Microbial-assisted and genomic-assisted breeding: A two-way approach for the improvement of nutritional quality traits in agricultural crops. *3Biotech, 10*(2), 1–15.

Chandra, A. K., Pandey, D., Tiwari, A., Gururani, K., Agarwal, A., Dhasmana, A., & Kumar, A. (2021). Metal-based nanoparticles trigger the differential expression of key regulatory genes which regulate iron and zinc homeostasis mechanisms in finger millet. *Journal of Cereal Science, 100*, Article 103235.

Deutsch, C. A., Tewksbury, J. J., Tigchelaar, M., Battisti, D. S., Merrill, S. C., Huey, R. B., et al. (2018). Increase in crop losses to insect pests in a warming climate. *Science, 361*, 916–919.

Dongariyal, A., Chandra, A. K., Dongriyal, A., Kumar, A., & Sharma, P. (2023). Tending genome editing via CRISPR/Cas9-induced mutagenesis: Opportunity and challenges for yield, quality and nutritional improvement of fruit crops. *Scientia Horticulturae, 311*, Article 111790.

Douglas, A. E. (2018). Strategies for enhanced crop resistance to insect pests. *Annual Review of Plant Biology, 69*, 637–660.

Fiyaz, R. A., Shivani, D., Chaithanya, K., Mounika, K., Chiranjeevi, M., Laha, G. S., Viraktamath, B. C., Rao, L. S., & Sundaram, R. M. (2022). Genetic improvement of rice for bacterial blight resistance: Present status and future prospects. *Rice Science, 29*, 118–132.

Ishino, Y., Shinagawa, H., Makino, K., Amemura, M., & Nakata, A. (1987). Nucleotide sequence of the iap gene, responsible for alkaline phosphatase isozyme conversion in *Escherichia coli*, and identification of the gene product. *Journal of Bacteriology, 169*, 5429–5433.

Jaganathan, D., Ramasamy, K., Sellamuthu, G., Jayabalan, S., & Venkataraman, G. (2018). CRISPR for crop improvement: An updated review. *Frontiers in Plant Science, 9*, Article 985.

Jiang, Y. Y., Chai, Y. P., Lu, M. H., Han, X. L., Lin, Q., Zhang, Y., Zhang, Q., Zhou, Y., Wang, X. C., Gao, C., & Chen, Q. J. (2020). Prime editing efficiently generates W542L and S621I double mutations in two ALS genes in maize. *Genome Biology, 21*, Article 1–10.

Jurica, M. S., Monnat Jr, R. J., & Stoddard, B. L. (1998). DNA recognition and cleavage by the LAGLIDADG homing endonuclease I-CreI. *Molecular Cell, 2*, 469–476.

Kuang, Y., Li, S., Ren, B., Yan, F., Spetz, C., Li, X., Zhou, X., & Zhou, H. (2020). Base-editing-mediated artificial evolution of OsALS1 in planta to develop novel herbicide-tolerant rice germplasms. *Molecular Plant, 13*, 565–572.

Kumar, A., Gupta, S., Abrol, G., & Chandra, A. (2018). Deriving innovations through intervention of newer technologies in integrated manner for value addition and development of value-added products in agri-food-nutrition and health sectors. *Acta Scientiae Agriculturae, 2*, 78–79.

Lee, J. H., Muhsin, M., Atienza, G. A., Kwak, D. Y., Kim, S. M., De Leon, T. B., Angeles, E. R., Coloquio, E., Kondoh, H., Satoh, K., & Cabunagan, R. C. (2010). Single nucleotide polymorphisms in a gene for translation initiation factor (eIF4G) of rice (*Oryza sativa*) associated with resistance to Rice tungro spherical virus. *Molecular Plant-Microbe Interactions, 23*, 29–38.

Li, X., Hu, D., Cai, L., Wang, H., Liu, X., Du, H., et al. (2022). Calcium dependent protein kinase38 regulates flowering time and common cutworm resistance in soybean. *Plant Physiology, 190*, 480–499.

Liu, C., Zhang, L., Liu, H., & Cheng, K. (2017). Delivery strategies of the CRISPR-Cas9 gene-editing system for therapeutic applications. *Journal of Controlled Release, 266*, 17–26.

Lu, H. P., Luo, T., Fu, H. W., Wang, L., Tan, Y. Y., Huang, J. Z., et al. (2018). Resistance of rice to insect pests mediated by suppression of serotonin biosynthesis. *Nature Plants, 4*, 338–344.

Ma, J., Chen, J., Wang, M., Ren, Y., Wang, S., Lei, C., & Cheng, Z. (2018). Disruption of OsSEC3A increases the content of salicylic acid and induces plant defense responses in rice. *Journal of Experimental Botany, 69*, 1051–1064.

Macovei, A., Sevilla, N. R., Cantos, C., Jonson, G. B., Slamet-Loedin, I., Čermák, T., Voytas, D. F., Choi, I. R., & Chadha-Mohanty, P. (2018). Novel alleles of rice eIF4G generated by

CRISPR/Cas9-targeted mutagenesis confer resistance to Rice tungro spherical virus. *Plant Biotechnology Journal, 16*, 1918–1927.

Makarova, K. S., Wolf, Y. I., & Koonin, E. V. (2018). Classification and nomenclature of CRISPR/Cas systems: Where from here? *CRISPR Journal, 1*, 325–336.

Makarova, K. S., Wolf, Y. I., Iranzo, J., Shmakov, S. A., Alkhnbashi, O. S., Brouns, S. J., Charpentier, E., Cheng, D., Haft, D. H., Horvath, P., & Moineau, S. (2020). Evolutionary classification of CRISPR–Cas systems: A burst of class 2 and derived variants. *Nature Reviews Microbiology, 18*, 67–83.

Mallick, N., Jha, S. K., Agarwal, P., Kumar, S., et al. (2022). Marker-assisted transfer of leaf and stripe rust resistance from *T. turgidum* var. *durum* cv. Trinakria to wheat variety HD2932. *Frontiers in Genetics, 13*, Article 941287.

Martín-Pizarro, C., & Pose, D. (2018). Genome editing as a tool for fruit ripening manipulation. *Frontiers in Plant Science, 9*, 1415. https://doi.org/10.3389/fpls.2018.01415.

Mojica, F. J. M., Diez-Villasenor, C., Garcia-Martinez, J., & Almendros, C. (2009). Short motif sequences determine the targets of the prokaryotic CRISPR defense system. *Microbiology, 155*(3), 733–740. https://doi.org/10.1099/mic.0.023960-0.

Nyamgoud, S., Tyagi, S., Chandra, A. K., et al. (2022). Development and characterisation of leaf rust resistant *Triticum timopheevii* derived introgression lines in hexaploid wheat. *Indian Journal of Genetics and Plant Breeding, 82*(4), 1–10. https://doi.org/10.31018/ijgpb.2022.82001.

Oliva, R., Ji, C., Atienza-Grande, G., Huguet-Tapia, J. C., Perez-Quintero, A., Li, T., Eom, J. S., Li, C., Nguyen, H., Liu, B., & Auguy, F. (2019). Broad-spectrum resistance to bacterial blight in rice using genome editing. *Nature Biotechnology, 37*(12), 1344–1350. https://doi.org/10.1038/s41587-019-0266-0.

Raghunandan, K., Tanwar, J., Patil, S. N., Chandra, A. K., et al. (2022). Identification of novel broad-spectrum leaf rust resistance sources from Khapli wheat landraces. *Plants, 11*(11), 1965. https://doi.org/10.3390/plants11111965.

Różewicz, M., Wyzińska, M., & Grabiński, J. (2021). The most important fungal diseases of cereals—Problems and possible solutions. *Agronomy, 11*(4), 714. https://doi.org/10.3390/agronomy11040714.

Shekhar, S., Sinha, D., & Kumari, A. (2020). An overview of bacterial leaf blight disease of rice and different strategies for its management. *International Journal of Current Microbiology and Applied Sciences, 9*(9), 2250–2265. https://doi.org/10.20546/ijcmas.2020.909.281.

Sood, S., Joshi, D. C., Chandra, A. K., & Kumar, A. (2019). Phenomics and genomics of finger millet: Current status and future prospects. *Planta, 250*(3), 731–751. https://doi.org/10.1007/s00425-019-03237-2.

Steinert, J., Schiml, S., Fauser, F., & Puchta, H. (2015). Highly efficient heritable plant genome engineering using Cas9 orthologues from *Streptococcus thermophilus* and *Staphylococcus aureus*. *The Plant Journal, 84*(6), 1295–1305. https://doi.org/10.1111/tpj.13078.

Tambong, J. T. (2022). Bacterial pathogens of wheat: Symptoms, distribution, identification, and taxonomy. In R. A. Mahmood (Ed.), *Wheat* (pp. 1–22). IntechOpen. https://doi.org/10.5772/intechopen.102855.

Wang, F., Wang, C., Liu, P., Lei, C., Hao, W., Gao, Y., Liu, Y. G., & Zhao, K. (2016). Enhanced rice blast resistance by CRISPR/Cas9-targeted mutagenesis of the ERF transcription factor gene *OsERF922*. *PLOS ONE, 11*(4), e0154027. https://doi.org/10.1371/journal.pone.0154027.

Wang, Y., Cheng, X., Shan, Q., Zhang, Y., Liu, J., Gao, C., & Qiu, J. L. (2014). Simultaneous editing of three homoeoalleles in hexaploid bread wheat confers heritable resistance to powdery mildew. *Nature Biotechnology, 32*(9), 947–951. https://doi.org/10.1038/nbt.2969.

Xu, J., Hua, K., & Lang, Z. (2019). Genome editing for horticultural crop improvement. *Horticulture Research, 6*, 113. https://doi.org/10.1038/s41438-019-0171-1.

Zetsche, B., Gootenberg, J. S., Abudayyeh, O. O., Slaymaker, I. M., Makarova, K. S., Essletzbichler, P., Volz, S. E., Joung, J., van der Oost, J., Regev, A., Koonin, E. V., & Zhang, F. (2015). Cpf1 is a single RNA-guided endonuclease of a class 2 CRISPR-Cas system. *Cell, 163*(3), 759–771. https://doi.org/10.1016/j.cell.2015.09.038.

Zhang, R., Chen, S., Meng, X., Chai, Z., Wang, D., Yuan, Y., Chen, K., Jiang, L., Li, J., & Gao, C. (2021). Generating broad-spectrum tolerance to ALS-inhibiting herbicides in rice by base editing. *Science China Life Sciences, 64*(10), 1624–1633. https://doi.org/10.1007/s11427-020-1782-8.

Zhang, Y., Bai, Y., Wu, G., Zou, S., Chen, Y., Gao, C., & Tang, D. (2017). Simultaneous modification of three homeologs of *Ta EDR 1* by genome editing enhances powdery mildew resistance in wheat. *The Plant Journal, 91*(4), 714–724. https://doi.org/10.1111/tpj.13597.

Zhou, J., Peng, Z., Long, J., Sosso, D., Liu, B. O., Eom, J. S., Huang, S., Liu, S., Vera Cruz, C., Frommer, W. B., & White, F. F. (2015). Gene targeting by the TAL effector PthXo2 reveals cryptic resistance gene for bacterial blight of rice. *The Plant Journal, 82*(4), 632–643. https://doi.org/10.1111/tpj.12834.

CHAPTER 10

Recent Advances in CRISPR-Cas for Biotic Stress Management in Fruits

R. S. AFIYA, S. SENTHILKUMAR, M. CHITHRA, and S. MANIVANNAN

Department of Horticulture, School of Life Sciences, Central University of Tamil Nadu, Thiruvarur, Tamil Nadu, India

ABSTRACT

Global food demand and crop production are under severe threat as a result of rising population, depleting land resources, climate change, and global climate variation, as well as the appearance of new infectious diseases and pests. By capturing carbon dioxide from the atmosphere and storing carbon in their trunks, trees reduce global warming brought on by climate change. Due to the biological growth and carbon sequestration occurring in fruit orchards, they even have a significant impact on climate change. Fruit crops are vulnerable to phytopathogens, specifically viruses, germs, bacteria, microbes, fungi, nematodes, insects, pests, and weeds, which can cause biotic or living stress and more than 30–40% loss globally. These agents directly rob the nutrients from their host, which reduces plant vigor and, in severe cases, causes the host plant to die. Biotic stress is a significant factor in pre- and postharvest losses in the horticulture industry. To feed the growing population, developing crop varieties that are resistant to disease is crucial. Numerous technologies have been used to create crops that are resistant to disease, and genetic variation is a crucial part of resistance or immunity breeding. Creation of diversity in genes with the improvement of resistance against pathogenic genes is a necessary component to fully utilize the idea of resistance breeding. These facts have led to the utilization of various genetic code editing technologies to create genetic variations. The

Climate Change and Biotic Factors: A Molecular Approach.
Ashutosh Singh, Saurabh Pandey, and Amarjeet Kumar, (Eds.)

CRISPR/Cas9 (Clustered Regularly Interspaced Short Palindromic Repeats) immune mechanism, an RNA-driven technology for effective modification of structural genes, is one of the most recent scientific advances. Host plants have been found to contain numerous genetic codes, and transcription factors are negatively regulated, recessive, and increase the risk of disease, which may play a part in a pathogen-friendly interaction. The genome modification and genetic code editing potential of the CRISPR Cas9 setup in fruit species is still being explored. This chapter evaluates recent advancements in horticulture research utilizing CRISPR/Cas9 technology for editing genes, summarizes them, and considers how they might be used to advance resistance breeding and production, with a particular emphasis on fruit crops. This could support future developments in the use of CRISPR, a defense technology for pathogens, to increase fruit production and sustainability.

10.1 INTRODUCTION

Phytopathogens like viruses, bacteria, and fungi pose a serious threat to crop productivity and are responsible for 30 to 40% of crop losses worldwide. Developing crop varieties resistant to disease is crucial to feed the growing population. Fruits are among plants that are essential in promoting human health because they are packed with dietary fiber, vitamins, and minerals. Fruit growth, quality, and yield are threatened or hampered by biotic factors (pests, diseases, and environmental stressors like flooding and drought). When harmful microorganisms like bacteria, viruses, viroids, phytoplasma, and nematodes interact with plants, it creates a situation where plant growth is unsustainable. Internal or external interactions between microorganisms and plant tissues can cause symptoms of biotic stress like discoloration, growth retardation, decaying, or the development of localized lesions (Balodi et al., 2017). Insect pests have been causing significant problems with yield losses, posing a risk to agricultural productivity, similar to other biotic stress factors. Additionally, as a result of climate change, phytophagous pests may become more damaging to fruit crops (Sun et al., 2017). According to Wang et al., biotic stress causes significant postharvest losses as microorganisms infect leaves, ultimately reducing photosynthetic activity levels in all major crops.

Chemical pesticides have historically been used to control plant diseases, but this method frequently results in residual contamination, is beneficial to harmful insects, and can lead to insecticide immunity in vectors (Stelinski et al., 2012). Additionally, crop losses brought on by plant diseases and insect pests not only reduce crop productivity but also significantly jeopardize global food and nutritional security (Miller et al., 2009). Plant breeding has been used for creating disease- and insect-hardy crop varieties for thousands of years, but it is labor-intensive, time-consuming, stochastic, and the screening process is a very difficult procedure (Scheben et al., 2017). Back-crosses and selection are two common traditional breeding techniques for enhancing crop traits; however, these methods take a long time and require a lot of resources. Clustered regularly interspaced short palindromic repeats (CRISPR)-CRISPR-associated protein-9 (Cas9), a new breeding method, has the potential to rapidly and accurately improve a variety of characteristics, including nutritional qualities, resistance to disease, tolerance to environmental stress factors, yield, and quality in fruit crops. Due to its easy operation and high mutagenesis effectiveness, this process has been used to obtain novel genetic stock through gene-directed mutagenesis.

Genome editing (GE) can be crucial because it offers a more promising and environmentally friendly way to handle the situation. The target DNA undergoes base substitutions and/or insertions/deletions (indels) as a result of the GE technique. It uses a variety of methods, such as the current improvement of the CRISPR-associated nuclease 9 (Cas9) methodologies and the usage of transcriptional activator-like zinc finger nucleases (ZFNs), effector nucleases (TALENs), and transcriptional activator-like effector nucleases (ZFNs). The CRISPR/Cas9 methodology is highly straightforward, simpler with usage than TALENs and ZFNs because to ascertain the destination of the Cas9 nuclease, only one guide ribonucleic acid (gRNA) is required (Puchta et al., 1993; Upadhyay, 2021). Considering the numerous uses in fruit crops utilizing CRISPR/Cas9 genetic modifications, such as reducing living and non-living stress susceptibility (Nishitani et al., 2016; Malnoy et al., 2016; Jia et al., 2016), lengthening fruit shelf life (Hu et al., 2021), and altering plant architecture and flowering time (Ren et al., 2020; Shao et al., 2020; Charrier et al., 2019; Varkonyi-Gasic et al., 2019; Zhou et al., 2020; Varkonyi-Gasic et al., 2021). This review focuses on key approaches and discoveries for managing insects and diseases in plants while taking into account how well genetic manipulation with the CRISPR Cas9 system reduces biological stress factors in fruit crops.

10.2 EMERGENCE OF CRISPR AS THE BEST GENOME EDITING TOOL

Escherichia coli (E. coli) were found to contain CRISPR sequence in 1987 by Ishano et al., but their function was not described. In 2000, Mojica et al., discovered that CRISPR sequences are widespread in other microbes. The first practical implication or support for CRISPR's role in adaptive immunity was discovered in 2007. CRISPR for genome editing was proposed by Charpentier et al., in 2010. Zhang et al., performed the first eukaryotic cell Cas9 genome engineering demonstration in 2013. Nishimasu et al., discovered the crystal structure of the Cas9-gRNA complex in 2014. Jennifer Doudna and Emmanuelle Charpentier were honored in 2020 with the Nobel Prize in Chemistry for the formulation of the CRISPR/Cas9 genetic code modifying method.

10.3 BIOTIC STRESS AND CRISPR CAS9

Biotic stress in agriculture is a significant factor in pre- and postharvest losses. Plants lack an adaptive immune response, or the capacity to adjust to new diseases and remember prior infections, in contrast to vertebrates. Despite lacking innate immune responses, plants have developed a wide range of complex defenses against biotic stresses. The genetic code of the plant contains the genetic underpinnings of these defense mechanisms. To feed the growing population, developing crop varieties that are resistant to disease is crucial. Opposing abiotic or non-living stress, which is brought on by ecological factors like heat and water scarcity, biotic stress agents directly rob their host of nutrients, which reduces plant vigor and, in severe conditions, results in plant mortality. A number of technologies have been used to create crops that are resistant to disease. The novel CRISPR/CRISPR-related nuclease 9 (Cas9) methodologies are one of the most crucial tools for altering plant genomes to withstand biotic stress. In the form of destructive diseases, viruses cause greater losses in crop yields. Additionally, the relatively complex CRISPR/Cas9 process is indeed a rapid and effective means that can provide plants with virus resistance. The host genome is precisely modified by CRISPR/Cas9 against a particular virus. However, the rate of mutation and the tolerance response differs depending on the type of plant species used, the methods used, and the sites that are being targeted.

A prokaryotic immune system called CRISPR is found in the genetic code sequences of various archaea and bacteria (Grissa et al., 2007). With three types in each class, it can be roughly divided into two forms. Class I requires numerous transcription factors, and class II necessitates only one effector protein. Class I includes three distinct forms, namely I, III, and IV, which require various Cas proteins to actively participate in multiple tasks like processing pre-crRNA, spacer sequences-gRNAs, and cleavage. Class

II is composed of three sub-forms, namely II, V, and VI, which rely solely on a solitary Cas protein (Chen et al., 2012). Together with Cas proteins, trans-activating RNA (tracrRNA) and CRISPR-RNA (crRNA) bind to and cleave the retrovirus genetic code at the location of complementary crRNA sequences. Members of Class II and Type II groups include CRISPR/Cas9 (Jinek et al., 2012). In terms of versatility, simplicity, and relative affordability, the CRISPR/Cas9 system outperforms previous editing tools (TALEN, ZNFs, and meganucleases); genetic mutations in agriculture are now widely applicable (Samanta et al., 2016). CRISPR/Cas9-mediated modifications offer a unique opportunity for complex and polycistronic editing of various target genes with relatively high efficiency (Ma et al., 2015).

10.3.1 IMPROVEMENT OF BIOTIC STRESS RESISTANCE

To increase plant resistance to biotic stress, two methods have been used: modifying the antidefense pathway genes in fruit crops, or designing sgRNAs and attempting to target the genome. The target genetic code associated with the protein's binding over the crop-specific protein as "eukaryotic translation initiation factor 4E" is the crucial first step in plant Y virus infection (eIF4E). A crucial eIF4E mutation can impact how the virus interacts with plants and lead to plant resistance to this virus (Bastet et al., 2019). On cucumbers, CRISPR-Cas was used to observe the N0 and C0 ends of non-transgenic homozygous plants produced by eIF4E in the T3 generation; these plants possess protection against vein yellow virus in cucumber and mosaic virus in pumpkin, as well as resistance to PRSV-W in papaya (Chandrasekaran et al., 2016). Gemini virus's pathogenicity can be significantly diminished by CRISPR-Cas9 mutations that are produced in both its coding and non-coding locations. SgRNA-Cas9 constructs in Nicotiana benthamiana target the gemini virus known as beet severe curly top virus and prevent it from accumulating in the leaves (Ji et al., 2015). Noncoding-region mutations in the gemini virus are thought to weaken or even prevent the virus from replicating. Noncoding-region mutations produce fewer viral variants than coding-region mutations (Ali et al., 2016).

The majority of diseases brought on by fungi have the potential to severely reduce fruit crop production and quality. Examples include both downy and powdery mildews, which severely reduce the value of tomatoes (Borrelli et al., 2018). The Fe(II)-dependent superfamily of 2-oxoglutarate oxygenases, which includes *Arabidopsis thaliana* DMR6, is involved in maintaining salicylic acid homeostasis. Plants that over-express DMR6 are less susceptible to downy mildew (Zeilmaker et al., 2015). By using CRISPR-Cas9 to delete tomato homologous genes, the DMR6 mutation was found to be resistant to *Phytophthora, Pseudomonas syringae*, and Xanthomonas spp (Thomazella et al., 2016). A membrane-linked protein is encoded by the powdery mildew ailment sensitivity gene *Mlo1* (Mildew resistant locus 1). *Pseudomonas syringae*, a bacterial pathogen, sources leaf spot illnesses with fruits, which greatly reduce fruit and vegetable yield and sensory qualities. Changing the C-terminal jasmonate domain (JAZ2Djas) of JAZ2 repressors in *Arabidopsis thaliana* confers resistance to *Pseudomonas syringae* by CRISPR-Cas9 (Andrés et al., 2018).

10.3.2 PHYSIOLOGY OF CRISPR CAS9 FOR BIOTIC STRESS

A cell's reaction to biotic stress pathways is activated by infection through the signaling pathways ethylene/jasmonate and salicylic acid (Caarls et al., 2015). In responding to pathogen attack and to prevent the dissemination of infection, a small amount of dominance or expression, the changes trigger off, which are preceded by biochemical and anatomical modifications in fruit crops. At the infection site, phenolic and peroxide compounds are released, and callose, papillae, and lignin deposit accounts form as well (Caarls et al., 2015; Bai et al., 2016). A wide variety of transgenic wine grapes have been created over the past 20 years, that overexpress sequence homology and heterological mechanisms of the cell responses to biological factors inducing stress, such as pathogen-linked proteins (Xenopus laevis Mag2), antimicrobial peptides (RCC2 (RIP), VvNPR1.1, Chi 11 (VpPR4–1, Vvtl-1), transcription factors, and antimicrobial peptides (VvbZIP VpWRKY3; VvWRKY33; VvbZIP60; VvDOF3). The sensitivity was decreased by the genes for cell defense (VqJAZ4; pPGIP; nag70; VpEIRP1; RPW8.2; VvSNAT2; VpPUB23; VaHAESA; VaPUB), subordinate stress-related metabolites (Vst1; STS), false mildew, and gray mold. However, the usage of these variations in the industry comes with extra costs in terms of period and substance to safeguard vegetation integrating with ecosystems and human consumption (Capriotti et al., 2020; Yang et al., 2020; Hartung and Schiemann, 2014). Because it makes the introduction of novel varieties with resistance to pathogens and makes them market-safe, quick, and affordable, the usage of CRISPR/Cas9-mediated genetic modification has indisputable comparative advantages.

10.3.3 BIOTIC STRESS-RESISTANT (TARGET) GENE IDENTIFICATION

Finding appropriate target genes that can be changed without having a reduced impact on crop performance is a significant barrier to engineering resistance. Several techniques can be used to find potential target genes. For vulnerability factors stimulated through microbial or defense-oriented genes, minor RNAs resulting after the microorganism and commensurate multitude security genes or inferred after the host target pathogenicity reveals on the cross-kingdom RNAi. The candidate-focused types can be used through genetic engineering, which can also avoid the pathogen from changing the host..

10.4 MECHANISM OF PLANT GENOME EDITING USING CRISPR CAS9

The progress of crop resistance against one or more viruses and an understanding of the interactions between plant viruses were made possible by CRISPR/Cas-mediated genetic modification. Cas9 and a single guide RNA are part of a two-component system. RuvC and HNH nucleases are two areas of the monomeric RNA-guided DNA endonuclease Cas9, which cleaves balancing and non-complementary DNA features to produce blunted ends in the aimed DNA and frameshift mutations that disrupt base pairs (Makarova et al., 2015; Doudna and Charpentier, 2014). Exact beleaguered DNA is cut by Cas9 as directed by the sgRNA. The novel components Cas9, PAM (Protospacer Associated Motif), and single guide RNA (sgRNA) were found crucial for the identification and cleavage of specific DNA sites. The 20 bp double-stranded sgRNA is created when crRNA and tracrRNA combine. According to Jinek et al. (2012), the transcrRNA couples with the pre-crRNA sequence to produce double-stranded RNA, which is then cut by RNase III to produce complete crRNA (Jiang and Doudna, 2017). Cas9 endonuclease then binds with the mature crRNA (crRNA: tracrRNA). Various PAM sites are recognized by the Cas9 protein. gRNA binds to the targeted DNA after Cas9 identifies a PAM sequence, and a double-strand break (DSB) occurs three base pairs upstream of the PAM (Jinek et al., 2012). Due to specific variations performed by Cas9, the targeted site is activated, leading to the formation of two nuclease domains that can cleave both strands of the target DNA near the 3 nucleotides in front of the PAM, causing DSBs (Garneau et al., 2010).

- **CRISPR Locus:** Spacers range from 26 to 72 bp, while repeat sequences fall between 21 and 48 bp. Within a CRISPR locus, the well-conserved repeat length, repeat order, and spacer length can differ between CRISPR loci in the same or different genomes. According to the target sequence's direction, a conserved sequence linked to CRISPR loci is referred to as the leader upstream of the CRISPR.

- **Proto-spacer:** To identify and subsequently prevent phage multiplication, the spacer's nucleotide sequences must resemble a region of the phage genome known as a proto-spacer.
- **Cas9 Structure:** The six-domain Cas9 protein. The largest domain, REC1, is in charge of binding guide RNA. It is still unclear what the REC II domain does. The arginine-rich bridge helix is used to start the six domains of the Cas9 protein. Cleavage activity, too. The interacting domain that starts binding to the target DNA is called the PAM domain. Single-strand DNA is cut by the nucleases RuvC and HNH.
- Cas9 types include dCas9, Cas9 D10A, and wild-type Cas9. Specific double-stranded DNA sites can be cleaved by wild-type Cas9. DNA repair mechanisms cause insertion and deletion to happen. It facilitates precise mutation and replacement.

10.4.1 GENERAL WORKING OF CRISPR

Two essential molecules make up the CRISPR-Cas9 system, which alters (mutates) the DNA. The two DNA strands are first cut at a particular location in the genetic code by an enzyme called Cas9, which acts as a pair of restriction endonucleases allowing for the addition or deletion of DNA strands. The second component has a small RNA fragment called a guide RNA (gRNA), which is contained within a larger RNA scaffold and is made up of a pre-designed RNA sequence only about 20 base pairs long. The pre-designed sequences "guide" Cas9 to the proper region of the genome while the scaffold component binds to DNA. This ensures that the cut made by the Cas9 enzyme occurs at the proper place in the genome. The cell now attempts to repair the DNA after realizing that it needs to. Scientists can alter one or more genes in a cell's genome through the application of DNA repair machinery.

10.4.2 STEPS OF PLANT GENOME EDITING USING CRISPR-CAS9

10.4.2.1 CLEAVAGE MECHANISM

Three phases make up CRISPR-Cas9-mediated adaptive immunity. i) Incorporation through adaptation. (ii) CRISPR RNA transcription and new spacer expression at the CRISPR locus (crRNA). Target DNA sequences are located and destroyed during interference.

10.4.2.2 REPAIR MECHANISM

Double-strand breaks (DSBs) are engineered using either a homology-directed repair (HDR) system or a limited fidelity non-homologous end-joining (NHEJ) technique (Liu et al., 2019). A gene's repair site may experience frameshift mutation as a result of indels (insertions or deletions) brought on by the NHEJ. This results in premature codons in the mRNA, which eventually leads to the formation of a protein that is incomplete and non-functional. While HDR repair uses homologous single-stranded oligodeoxynucleotides as the template to precisely and specifically knock out point mutations at a specified area (Chiang et al., 2016). Although CRISPR/Cas DSBs can be engineered by both NHEJ and HDR, NHEJ is the more frequent method (Feng et al., 2021). The target site cleavage can be performed accurately and without mistakes by the Cpf1 and gRNA complex (Ledford, 2015).

By the error-prone non-homologous end joining (NHEJ) pathway, the CRISPR/Cas-induced double-strand breaks (DSBs) would cause a straightforward change in a particular sequence, that also frequently results in tiny insertions or deletions (InDels) that lead to frame-shift mutations and consequently loss of a specific function. Another option is to use homology-directed repair (HDR), which facilitates the supplement of huge external DNA in a preferred genomic locus or the alteration of a region of interest by introducing desired mutations. Single nucleotide mutations can be produced further by using cytosine-base editors (CBEs) in association with cytidine deaminase or adenine base editors (ABEs) along the ADAs. Prime editing, which uses a nCas9-fused reverse transcriptase and relies on gRNA for prime editing rather than the more common sgRNAs, can result in more pronounced mutations.

10.4.3 DESIGNING AND SELECTION OF SGRNA

Developing a powerful gRNA against that target and the choice of a suitable target site are critical to the success of a CRISPR experiment. Each gRNA is made up of two components: a trans-activating CRISPR RNA scaffold sequence and a CRISPR RNA with a spacer (tracer RNA). The protospacer in the target's PAM region must match the nucleotide sequence of the sgRNA's design. Some bacteria have an integrated CRISPR Cas9-like gene-modifying system that they use to respond to intruding pathogens like traditional systems. In order to recognize and defend against the virus the next time it attacks, the bacteria use CRISPR to trim out portions of the virus DNA and retain a small portion of it. This system was modified by scientists so that it could be applied to other plant and animal cells, including those of mice and people.

The following steps show how the CRISPR/Cas method for editing the plant genome operates:

1. Cas9 and gRNA cassette-carrying CRISPR/Cas9 construct such as 35sP-Cas9-35s and Ter: Ubiquitin 6/U3P. Selectable markers include hpt II, npt II)-U6/U3 T; 35sP: sgRNA with t-RNA scaffold;
2. Using a variety of conversion techniques (gene gun, Agrobacterium, bombardment, etc.), the CRISPR hypothesis is delivered into explants (cotyledonary nodes, calli, immature embryos, etc.) and modified plants are produced.
3. A target gene (susceptible/recessive gene) is found in a cell's genome.
4. The double strand breaks Cas9 with the target gene mechanism.
5. Gene dysfunction results in the emergence of bacterial, fungal, or viral aversion in plants.

10.4.4 SYSTEMS OF CRISPR COMPONENTS DELIVERY INTO PLANTS

The biggest challenge is safely and effectively delivering the genome editing system of CRISPR Cas9 to specific target cells. The delivery must be customized for the desired tissues in order to prevent undesirable off-target gene-modifying events. Physical or non-viral entry and viral techniques are the two categories of delivery methods. A few non-viral delivery methods include electroporation, microinjection, induced transduction, cell-penetrating peptide, hydrodynamic injection, mechanical cell deformation, lipofection, and DNA nanostructure. Lentivirus and adeno-associated viruses are examples of viral delivery techniques. PEG-mediated delivery, Agrobacterium-regulated delivery, nanoparticle-mediated delivery, bombardment-mediated delivery, floral dip, or pollen-tube pathway method are among the widely used delivery techniques. The protoplast is incubated with a plasmid comprising Cas9 and gRNA while also being exposed to PEG in PEG-mediated delivery. Cas9/gRNA ribonucleoproteins are frequently delivered using PEG-mediated delivery, which is its main advantage. The vector-less or DNA-free edited modified crops won't likely encounter any ethical or regulatory obstacles. For bombardment-mediated delivery of vector or Cas9/gRNA ribonucleoproteins in transformation or genetic code transfer, a device known as a "biolistic gun or gene gun" is necessary. Silver, Gold, or tungsten particles are frequently cast off as the carriers of Cas9/gRNA ribonucleoproteins or vectors. This delivery method's main advantage is that it does not require the CRISPR/Cas9 binary vector. Numerous explant types can change when substantial amounts of DNA and RNA are delivered. The foremost significance is the Cas9/gRNA ribonucleoprotein complex, which could be carried potentially. Agrobacterium-mediated delivery is the most widely used technique for CRISPR/Cas9 construct delivery for a variety of plant species. By transforming the binary vector, which includes Cas9 and the gRNA expression cassette, the Agrobacterium strain is produced. In the past, plasmids were either assorted with pollen or functional to the receptive stigma using the floral dip or pollen tube pathway methods, or they were directly applied to the stigma's surface.

10.5 RECENT ADVANCES FOR CRISPR CAS9

Recently, modifications to the CRISPR system have been created and tested to overcome the drawbacks of the standard CRISPR-Cas9 editing tool. Different and expanded PAM sequences found in Cas9 enzymes culled out from several supplementary bacteriological types can improve on-target specificity (Vats et al., 2019). The endonuclease NmeCas9 from Neisseria meningitidis, SaCas9 from Staphylococcus aureus, and Cas12a enzymes (Cpf1) from Prevotella and Francisella with distinct PAM specificities have been found to have superior properties to those of SpCas9.

10.5.1 CHIMERIC FUSION TECHNOLOGY: PROGRESSIONS IN CAS9-BASED EDITING

Specific protein fusion techniques have been discovered to enhance homology-directed repair (HDR), along with typical modifications made to efficiently use the CRISPR/Cas9 system in multiple ways (Marzec et al., 2020). Prime Editor is a unique protein that is anchored at target-specific sites and acts as a carrier for selected modifications or alterations. This is a fusion of the Cas9 nickase (which cuts single-stranded DNA instead of double-stranded DNA) and a reverse transcriptase protein.

10.5.2 CRISPR-BASED GENE EDITING WINS OUT OVER RNA INTERFERENCE

Many research teams around the world are using plants to demonstrate how CRISPR-based genetic code and RNA interference (RNAi) modification techniques work. The main difference between the two is that RNAi limits gene dominance (knockdown) only at the RNA level, while CRISPR totally calms genes at the DNA level (knockout). Ineffective RNAi giveaway of target genes principals to leaky expression of target mRNA due to poor reproducibility, leading to off-target effects in which other genes are unintentionally affected. However, CRISPR-based tools improve precision and reduce off-target effects of altered genotypes. The main difference between CRISPR and RNAi is that in most cases CRISPR causes a comprehensive loss of functional effect, whereas RNAi usually results in a hypomorphic effect leading to fractional loss of gene function. Therefore, compared to RNAi, CRISPR produces more robust and reliable phenotypes.

10.6 RECENT ACHIEVEMENTS IN BIOTIC STRESS RESISTANCE THROUGH CRISPR CAS9

10.6.1 FUNGAL DISEASES

Fungal pathogens are more potent disease-causing agents in many economically significant food crops. The highly polymorphic properties of the effector proteins naturally produced by pathogenic organisms prevent the currently employed approaches from achieving long-lasting resistance. The use of new disease control strategies such as CRISPR/Cas9 tools is essential for modern generation crop production. This CRISPR/Cas9 editing of the 'S' gene conferred broad-spectrum fungal disease resistance. The 'S' gene promotes infection and helps pathogens interact favorably. Mutations in the 'S' allele may favor 'S' gene-based resistance over recessively inherited 'R' gene-based resistance (Schie et al., 2014).

10.6.1.1 GRAPEVINE

Actinomycetes, downy mildew, and powdery mildew are the top three grape diseases that diminish yield in grapes and require annual preventive fungicide treatment of vineyards, while only damaging the surrounding ecosystem. It also distresses the value of the berries that are harvested. Powdery mildew, downy mildew, and gray mold are all instigated by *Plasmopara viticola* (Berk. and M.A. Curtis), Berl. De Toni, Erysiphe necator (pig.), Burrill and *Botritis cenerea* Fr. (Fizikova et al., 2021). The CRISPR/Cas9 methodology was developed by Wang et al. (2018) to display that the transcript influence gene *VvWRKY52* is required for vulnerability to gray mold in the cultivar 'Thompson seedless.' Using gRNAs targeting four sites, they generated 15 lines with biallelic mutations and 7 lines with single allelic mutations. Both large (5 bp) and small erasures occurred during editing events (Wang et al., 2018). Botrytis-resistant grape transformants were generated using *WRKY52* gene editing technology. A transcription factor of cellular response to biological stress, encoded by WRKY52, functions as a suppressor of the JA signaling pathway. Pathogenesis-related 4 (PR4) protein expression increased in 'Riesling' plants after inoculation, suggesting that PR4 may be involved in resistance to *P. viticola* (Kortekamp, 2006). PR proteins are known to be essential for plants to defend themselves against attack by pathogens. In many plants, chitinases and the chitin-binding protein PR4 are essential for defense against pathogens.

In contrast to the standard procedure of transforming Agrobacteria with plasmids for CRISPR/Cas9 proofreading and selection, Osakabe et al., recently designed guidelines for the delivery and adaptation of gene constructs using RNPs. However, this takes only 2–3 weeks (Osakabe et al., 2018). The integral membrane protein Mlo is thought to be a detrimental regulator of cellular defense mechanisms as a result of enhanced spreading and reorganization of the actin cytoskeleton in the account of biotic stress. In the context of MLO mutations, H_2O_2 accumulation, host cell death, papilla advancement, and cell wall apposition (CWA) are associated with MLO-mediated resistance to powdery mildew (Consonni et al., 2006; Wan et al., 2020). Based on the triple knockdown of mlo7, mlo6, and mlo11, the grape variety 'Brachetto' was shown to have a 77% increase in susceptibility to powdery mildew (Pessina et al., 2016). Cells produce H_2O_2 as a defense mechanism against pathogenic organisms that transport genes through stomata (Wu et al., 1997; Yu et al., 2020). DAB staining shows an accumulation of H_2O_2 in leaf tissue. As shown by Kortekamp (2003), he develops ROS when both infected and resistant grape cultivars are infected with downy mildew. *P. viticola* developed faster in her VvPR4b knockout line than in WT according to DAB staining results. Furthermore, this study showed that her VvPR4b in grapes is important for downy mildew resistance. Using CRISPR/Cas9, we successfully added the mutation resistance gene *MLO-7* to Chardonnay grape cultivars, resulting in enhanced confrontation with powdery mildew (Malnoy et al., 2016).

10.6.1.2 APPLE

Chen investigated the role of MdMYBPA1 and MdSYP121 in anthocyanin production and activation of disease resistance against *Botryosphaeria dothidea* using apple callus (Wang et al., 2018; He et al., 2018). Using CRISPR/Cas9 technology, the mutagenesis-resistant genes encoding DIPM-1, DIPM-2, and DIPM-4 were effectively integrated into the Golden Delicious apple cultivar to improve resistance to fire blight (Malnoy et al., 2016). Furthermore, Zhou et al., showed an associated surge in salicylic acid (SA) mediation in apple callus cells and quieting of *PR* gene expression and DeNot. The protein CNGC2 is thus shown by the authors to be a repressor of *B. dothidea* resistance. Therefore, they are not the best candidates for creating apple varieties that are pathogen-resistant. However, in addition to resistant strains, CNGC2 mutations also have some pleiotropic effects, such as reduced fertility, incapacitated pollen tube emergence, and shortened anthers (Zhou et al., 2020).

10.6.2 POST-HARVEST INFECTION

For the first time, CRISPR/Cas9 genome modification technology was used in the plant pathogens *P. digitatum* and *P. expansum*. The *wetA* gene in both species was successfully disrupted using CRISPR/Cas9 methods to characterize the phenotype of these mutants. However, there is still room for improvement.

10.6.3 BACTERIAL DISEASES

A practical method for producing disease-resistant cultivars is gene editing. The creation of disease-resistant varieties has proven to be a successful and environmentally friendly method for controlling bacterial plant diseases. Endogenous plant genes known as susceptible (S) genes promote pathogen growth, infection, and symptom development during colonization (Tripathi et al., 2020). Recessive resistance to plant diseases may be brought on by functionless genes. Plants with aversion or resistance that is S-gene targeted may have enduring protection. Deactivation of a targeted factor in the host is required for pathogen existence resulting in S gene-based resistance. It has been reported that S genes in crops can be modified to increase resistance to a specific pathogen and, in some cases, even broad-spectrum resistance (Kim et al., 2019).

Discovering and understanding the cellular pathways that lead to disease development is the basis for identifying potential proteins or genes that, when deleted, triggered, or overexpressed, have the anticipated impact. The targeted genes most likely to be affected by the presence of the pathogen can be engineered in a number of different ways. Comparative transcriptome analysis (RNA-seq) examines gene expression differences between vulnerable and resistant populations following pathogen exposure. Infection, there is one way. To induce resistance, CRISPR is currently used to activate the endogenous banana genes Vicilin, Leucine-Rich Repeat (LRR), Wall-Associated Kinase (Wak2 and Wak5), Pathogenesis-Related (PR), and Disease Resistance R (e.g., Mouse *RPM1* gene). Transcriptome analysis of BXW-susceptible cultivar 'Pisang Awak' and BXW-resistant diploid banana stem cell 'Musa balbisiana' (Tripathi et al., 2019).

10.6.3.1 CITRUS

Using Cas9 technology, it was possible to create resistance to *Xanthomonas citri* subsp. citri, which is an occasional organism that causes bacterial citrus canker. The promoter gene *CsLOB1* was chosen as a target because it contributes to cancer development in citrus. A comparison of developed transgenic lines with wild types showed improved resistance to citrus canker (Peng et al., 2017). Xcc bacterial strain also secretes the transcriptional activator-like (TAL) vector PthA4, which fuses with EBE in the promoter of the *CsLOB1* gene to stimulate the dominance of this citrus canker susceptibility gene (Hu et al., 2014).

10.6.3.2 GRAPES

Red Blotch Disease Viral Contamination (GRBV) and *Xylella fastidiosa* Wells (XF) both cause anthocyanin collection in grapes, which is directed by the movement of the MYBA7 transcription factor. The GRBV in grapes shows itself at first as red circular spots on the leaves. When the disease impacts grapes, it delays fruit ripening, leads to acid build-up, and lowers the levels of sucrose and anthocyanin levels in berries. Insects that feed on plant sap and develop in the xylem of the plants disperse the GRBV (Sunitha and Rock, 2020; Johnson, 2020). An increase in anthocyanins, a decrease in inorganic phosphate levels, and an accumulation of abscisic acid are all associated with the XF pathogenesis in grapes (Lim et al., 2007). In response to phosphate, sugar, and ultraviolet light, anthocyanins are known to accumulate. Transcriptional small interfering RNA locus 4 and miR828 can regulate this process (TAS4). Grapes produce small interfering RNAs (SiRNAs) called TAS4a/b/c that target MYBA5 and are differentially expressed. Rock and Sunitha selected gene deletion lines for TAS4 and MYBA7 (Sunitha and Rock, 2020).

The absence of TAS4c and MYBA5/6 repetitive loci or inducing environmental stress conditions may account for the apparent lack of anthocyanin accumulation in regenerated plants. Although determining the role of anthocyanins in PD and GRBV resistance/tolerance processes in modified plants was difficult due to the lack of visible pigment changes. Further investigation of transformation shoots by sampling PD- and GRBV-resistant plants seems very promising. XF growth has been shown to be inhibited by phenolic substances in the past, though the quantity of polyphenols that various cultivars produce in infection-related responses can vary markedly (Maddox et al., 2010).

10.6.3.3 DUNCAN GRAPEFRUIT

In Duncan grapefruit, EBE editing of one allele of the *CsLOB1* gene with CRISPR/Cas9 slightly reduced cancer symptoms (Jia et al., 2016). Another study showed that cancer resistance in Duncan grapefruit was also due to CRISPR/Cas9 modification of the CsLOB1 coding region (Jia et al., 2017).

10.6.3.4 ORANGE

Citrus canker resistance was found in Wanjincheng oranges by mutagenesis of EBEs from both alleles of the CsLOB1 promoter (Peng et al., 2017). The innate immune marker gene *CsWRKY22* triggered by the orange pathogen Wanjincheng was deleted using the CRISPR/Cas9 system, and the mutant plants showed significantly reduced susceptibility to citrus canker (Wang et al., 2019).

10.6.3.5 APPLE

By encapsulating Cas9 and guide RNA in RNP particles, the genes encoding the receptor protein kinases DIPM-1, DIPM-2, and DIPM-4 were rendered inactive (Malnoy et al., 2016). The effector protein DspA/E was produced by the bacterium *Erwinia amylovora* Burrill, which causes bacterial blight on apples and other fruit trees belonging to the Rosaceae family (Siamer et al., 2014). A promising technique for generating transformed plants that accelerates and facilitates the industrial introduction of an efficient variety is the combined FLP/FRT system. It was used to achieve MdDIPM-4 deletion; heat shock removed foreign DNA from the genome.

10.6.4 VIRAL DISEASES RESISTANCE THROUGH CRISPR CAS

Plant viruses are obligate parasites that must replicate and maintain their life cycle entirely within the machinery of their hosts. Significant host susceptibility (S) factors that promote viral infection and replication include coiling of genetic material and eukaryotic translational induction factors. Defending fruit crops from viral pathogens is one of the biggest agricultural challenges. The addition of Cas9 has enhanced plant breeding efforts and assisted in the development of varieties with hereditary resistance to plant viral infections. Complexes of Cas9 endonuclease and sgRNAs (single-guide RNAs) target specific viral and host genomes, preventing viral cleavage or altering the plant host genome, and are used to confer tolerance to viral DNA and RNA. Consequently, the virus is no longer able to replicate effectively.

10.6.4.1 MECHANISM OF VIRAL DISEASE RESISTANCE

The CRISPR/Cas9 technique uses a three-step process to identify and target the genetic material of a pathogen: I acquisition, (ii) expression, and (iii) interference (Marraffini and colleagues., 2010).

- **Acquisition:** Short fragments of invasive foreign virus or plasmid genetic code that are recognized and incorporated into the CRISPR locus. Short RNA (crRNA) is created by copying CRISPR loci and using it to target virus components with effector endonuclease genes using simple complementarity. The protospacer adjacent motif (PAM) is typically composed of a brief (2–5 bp) stretch of nucleotides that are retained and that produce the DNA fragment (spacer) for identification. Pathogen attacks are resistant to CRISPR-mediated immunity due to changes in viral genomes and PAM (Shan-e-Ali Zaidi et al., 2016).

- **Expression:** Together, the tracrRNA molecule and the Cas9 protein effectively transcribe the huge pre-CRISPR RNA (pre-crRNA) extracted after the CRISPR locus, changing it to various crRNAs (Sternberg et al., 2015). Through base complementarity, the tracrRNA joins the crRNA repeat area and speeds up pre-crRNA processing in the crRNA (Arora and Narula, 2017). Activated crRNAs engage a CRISPR-associated antiviral defense complex to recognize and base-pair a precise marked area in external DNA (CASCADE) (Deltcheva et al., 2011).
- DNA interference in the CRISPR/Cas9 technique needs just one Cas9 protein (Zetsche et al., 2015). The crRNA protects against pathogen attacks by directing the Cas9 protein to the foreign DNA's target site, where it will be destroyed during the interference step (Marraffini et al., 2010).

Steps

1. Use a plant chromosome with a *Cas9* gene and multiple gRNAs.
2. Cas9 protein synthesis and gRNA transcription to pre-crRNA.
3. The Cas9 and cr-RNA complex is formed.
4. Cas9 complex attaching to viral genome homologous sequences with PAM.
5. Viral genetic code entry, the rolling circle mode changes one-stranded DNA (ssDNA) into two-stranded DNA (dsDNA), and cr-RNA.
6. Chopped viral genome and dsDNA disruption.

10.6.4.1.1 Banana

The common banana streak virus (BSV) infection of B-genome plants, which kills the plant, is a major impediment to hybridization using *M. balbisiana* and its offshoots. Tripathy et al., prevented the transcription and/or translation of functional viral proteins by introducing mutations in integrative viral elements using CRISPR/Cas9 editing (Tripathi et al., 2019, 2020). gRNAs were created in banana to combat banana streak virus and specifically target the BSOLV and eBSOLV sequences. The CRISPR/Cas9 methodology was employed effectively to modify the incorporated pattern of the eBSOLV in the Gonja Manjaya, a plantain cultivar. The outcomes showed that precise deletions in bananas can be produced very effectively by multiplexing the CRISPR/Cas9 technology.

Although initial findings indicated stable changes and the silencing of eBSOLV reactivation, due to available sources spread through mealy bugs in field conditions, assessment of edited plants encourages various generations and with new BSV-affected plants is required (Tripathi et al., 2019).

10.6.4.1.2 Grapevine

A significant and pervasive degeneration of the grapevine is caused by GFLV. It results in significant grape production reduction (up to 75–80%), poor fruit appeal, quality, sensory properties, and a steadily worsening condition of infected vines that may ultimately result in plant death (Martelli & Savino, 1990; Martelli, 1993; Bovey et al., 1980). Vigne et al., In 2003, the potential of transgenic grapes dominant in the *GFLV CP* gene to promote expression of potential GFLV recombinants in the vineyard was investigated. They also examined the viability of grape leaf virus (GFLV) recombinants produced by transgenic vines expressing the dominant GFLV coat protein (*CP*) gene.

10.6.4.1.3 Apricot

The disease of sharka which is triggered by Plum Pox Virus (PPV), is one of the primary factors limiting Prunus production globally. ParPMC2 is a pseudogene that mediates ParPMC1's functional paralog ParPMC1's down-regulation by silencing, according to Zuriaga et al. (2018). Overall, the results strongly imply that the host susceptibility genes ParPMC1 and/or ParPMC2 are required for PPV infection and that their silencing could impart a trait of PPV resistance.

10.6.5 PEST INFESTATION

Insects that eat plants produce a fifth of the world's crops each year. Biotic stress threatens the quality and quantity of crops, affecting food security in all countries. Hunger and poverty are just two of the devastating effects of the yield loss caused by insect infestation. Insect pests have raised serious issues regarding yield losses that put agricultural productivity at risk, just like other biotic stress factors. In order to successfully use the technology, emerging novel advancements of CRISPR-Cas9 play a crucial role in editing, modification, as well as transformation techniques in plant species. A cutting-edge strategy to control insect pests would be to use genome editing to knock down important regulators.

10.6.5.1 MECHANISM OF PEST RESISTANCE BY CRISPR CAS9

The application of CRISPR/Cas9 for genome modification was explored after the discovery of the master mechanism cleavage recombination, which is homology dependent, and it quickly became clear that it is a reliable genetic code editing tool. The wide range of site-specific genome editing techniques using gene editing tools presents countless opportunities to influence critical agricultural traits such as biotic stress resistance (Mao et al., 2017; Rath et al., 2015; Jinek et al., 2012; Char et al., 2017). When it comes to breeding or development for resistance to pests and the addition of new genes by genetic code modification, biotechnology plays a crucial part in controlling insect pests to safeguard crops. Biotechnology also increases yields in other areas (Sun et al., 2017). Gene disruption caused by *CRISPR-Cas* system gene drives could cause mortality or sterility in the specified insect species, leading to population declines and even extinctions owing to enhanced and plain recessive or suppressed lethal changes (McFarlane et al., 2018). Genome modification can be used to check insect communities by manipulating the genes of both target plants and insects. Crop insect pests can be controlled by rendering them sterile, halting pesticide resistance, or, in the absence of sufficient R-genes, by generating de novo resistance. CRISPR-Cas can be applied in agriculture to check insect pests and thereby protect crops.

A two-step method that involves changing the specified genetic code in insects and releasing them back into the environment can be used to successfully edit an insect's genome (Koutroumpa et al., 2016). In the Li et al. (2016) study, the pattern of the guide-target RNA was altered to suppress the expression of the odor receptor and co-receptor gene (*Orco*). Due to the defective electrophysiological responses caused by the *orco* gene mutation, mutant locusts lack the ability to be attracted to aggregation pheromones in crowded conditions. Editing crucial plant-resistant or immunity genes, such as those that govern the target gene interactions with insect effectors and genes that confer resistance, is another way to improve the host's resistance to pests (R-genes). R genes assess a plant's vulnerability to insects and diseases, whereas S genes make plants susceptible to stress (Rathinam et al., 2019). Editing the R and S genes is a dependable method for conferring insect resistance in plant species. For their development, immunity, and behaviors, insects are known to rely on essential chemicals found in plants; these have been seen in rice insects (Lu et al., 2018). Genetic modification in plants has been utilized to provide resistance to both insect and non-insect pests by deleting the S-genes in the plants. Tryptamine, an amino acid, was transformed to serotonin in plants via the CRISPR-Cas technique, which also resulted in the deletion of the tryptamine 5-hydroxylase gene, encrypting or encoding the *CYP71A1* gene, which prevented plant hopper growth. Using CRISPR-Cas, at present we can rapidly change, eliminate, delete, duplicate, add, and substitute hereditary material and DNA at almost every target site in any crop or insect species in order to avail fruit crops resistant to insect pests. This methodology must be developed in order to create crop plants resistant to insect pests.

10.6.5.2 APPLICATIONS OF INSECT MANAGEMENT IN FRUITS USING CRISPR CAS

All over the world, the codling moth (*Cydia pomonella L.*) is one of the most destructive insect pests of pome fruit. Targeting specific genes to alter egg viability and production, the codling moth genome was modified by utilizing *CRISPR-Cas* gene-editing technology (CpomOR1). The *CpomOR1* gene product is an odorant receptor and belongs to the subfamily of pheromone receptors (Table 10.1).

TABLE 10.1 Target Genes For Insect Pest Control by CRISPR/Cas9

Phenotype related gene		**Male related genes**	**Female related genes**
Tyrosine hydroxylase	Genes related to Abnormal wing disc,	Double sex (DSX)	Egg-specific protein (Esp)
BLOS2	Tyrosine hydroxylase	White (w)	Sxl
Ecdysone-induced TF promoting	Yellow Troponin C (TpnC)	Transformer	Double Sex (DSX)
		MSL Clamp proteins which are chromatin linked	
Metabolism related gene	Segmentation related genes	Pheromone and odor-related genes	Insecticide resistance-related gene
Succinyl-CoA synthetase/ligase	Paired	Binding protein for Pheromone	Nicotinic acetylcholine receptor
P450	Wnt-1	Receptor and Co-receptors for Olfactory and Odorant	ABCA2
NPC1b	Abdominal-A		PSST
	Ago 1		Chitin synthase 1 (CHS1)

Changes, such as insertions and deletions, were found to have been successfully introduced (Garczynski et al., 2017). By altering the *Helicoverpa armigera* genetic code, Cadherin receptors are knocked down to work on *Cry1Ac* gene factor which promotes toxin tolerance (Wang et al., 2016). Insect genomes can be modified using Cry protein binding receptors to lessen plant vulnerability. Resistance to insects was targeted in crops through the CRISPR-Cas-mediated removal or loss of the targeted gene cluster CYP6AE has highlighted the importance of the toxin removal of a number of harmful phytochemicals and revealed the polyphagous insect cotton bollworm's (*H. armigera*) vulnerability to plant toxins (Wang et al., 2019).

In *H. armigera*, odorant receptor 16 (OR16) was knocked out using CRISPR-Cas9 technology, which pheromone signals and prevents males from mating with undeveloped females. This resulted in the overproduction of infertile eggs and assisted in insect control (Chang et al., 2017). Another technique for controlling insects involves using CRISPR-Cas9 to restrict genes contributing to growth, like the *abd-A* gene (Abdominal A), for a broad range of insects including tobacco cutworm (*Spodoptera litura*) by Bi et al., in 2016, fall armyworm (*Spodoptera frugiperda*) by Wu et al., in 2019, and Diamondback moth (*Plutella xylostella*) by Huang et al., in 2016. It produced gonads that weren't normal. Using CRISPR-Cas9, numerous additional genes in a variety of insect pests were modified.

10.6.5.3 CRISPR CAS9'S ROLE IN PLANT PATHOLOGY

Enhancing or editing the genome, which regulates a disease's susceptibility component for infections caused by bacteria and fungi, can be used to create disease-resistant cultivars. We can create incompatible interactions between host pathogens by editing the genome that controls host-pathogen interactions. It aids in enhancing the effectiveness of biological control agents. By modifying the genetic code that is able to account for virus reproduction as well as virulence, we can produce virus-free resistant cultivars.

10.7 FUTURE THRUST

CRISPR/Cas9 had a significant impact on the crops of the future, enabling them to fend off different pathogens due to its straightforwardness, dependability, and broad application possibilities. Lack of efficient regeneration techniques and hindered tissue culture protocols are additional bottlenecks. Thanks to its minimalism, durability, adaptability, and flexibility, the

CRISPR/Cas9 method has evolved over the past ten years and emerged as a helpful tool for creating future crops. The traits of horticulturally important or dominant crop species have been enhanced, and new medical applications have been developed through precise gene insertions, additions, duplications, replacements, deletions, and substitutions in crop genetic makeup. The CRISPR/Cas tool marked the beginning of an era in plant speed breeding and enhanced plant protection against many pathogens by suppressing pathogen-host interactions. Multiplex editing improves the ability to simultaneously target multiple genomic locations for mutational induction. Ultimately, it assists scientists in developing diverse next-generation varieties with disease resistance and improved trait values. The CRISPR tool has the potential to transform the future of horticulture by exploring the prospects of specific crop species and enhancing their nutritional efficiency, productivity consistency, stress tolerance from abiotic factors, and resistance to weak pests and pathogens. This technology may serve as a breeding tool for the upcoming generation to improve crop yields and potentially lead to the next agricultural revolution (Ahmad et al., 2020).

10.8 CONCLUSION

Given the continually expanding arsenal of CRISPR tools, it is obvious that genetic modification will help to accelerate the breeding and fostering of aversion or resistance despite any issues with GMO legislation. With the increased and expanded specificity of CRISPR systems, practically any preferred alteration will be possible. Pathogen resistance is a critical characteristic that can be enhanced with gene editing or modification in a species or variety of crops, and CRISPR/Cas is presently being employed to fight greening in citrus, a serious infection that continues to threaten sweet orange. CRISPR/Cas-instigated mutations or changes can help develop crop ideotypes or cultivars having resistance to diseases caused by pathogens when resistance reserves in natural or automatic fluctuations or wild relatives are in short supply. Theoretically, plant species that are more resistant would also require fewer pesticide applications, which is crucial because pesticides can also become scarcer or lose their effectiveness over time (Borel, 2017). However, we must re-evaluate how rapidly we need to produce food, and we value ecosystems in a credible, stable environment or sound ecosystem if we are to truly minimize the consumption of agricultural chemicals.

KEYWORDS

- **biotic stress**
- **CRISPR-Cas**
- **fruit crops**
- **genome editing**
- **plant defense**

REFERENCES

Ali, Z., Ali, S., Tashkandi, M., Zaidi, S. S., & Mahfouz, M. M. (2016). CRISPR/Cas9-mediated immunity to geminiviruses: Differential interference and evasion. *Scientific Reports, 6*, 26912. https://doi.org/10.1038/srep26912.

Andres, O., Garcia, I. S., León, N., & Szurek, B. (2018). Design of a bacterial speck resistant tomato by CRISPR/Cas9-mediated editing of SlJAZ2. *Plant Biotechnology Journal, 17*(3), 665–673. https://doi.org/10.1111/pbi.12926.

Arora, L., & Narula, A. (2017). Gene editing and crop improvement using CRISPR-Cas9 system. *Frontiers in Plant Science, 8*, 1932. https://doi.org/10.3389/fpls.2017.01932.

Bai, C., Capell, T., Berman, J., Medina, V., Sandmann, G., Christou, P., & Zhu, C. (2016). Bottlenecks in carotenoid biosynthesis and accumulation in rice endosperm are influenced by the precursor-product balance. *Plant Biotechnology Journal, 14*(1), 195–205. https://doi.org/10.1111/pbi.12370.

Bastet, A., Zafirov, D., Giovinazzo, N., Guyon-Debast, A., Nogué, F., Robaglia, C., & Gallois, J. L. (2019). Mimicking natural polymorphism in eIF4E by CRISPR-Cas9 base editing is associated with resistance to potyviruses. *Plant Biotechnology Journal, 17*(9), 1736–1750. https://doi.org/10.1111/pbi.13103.

Bi, H. L., Xu, J., Tan, A. J., & Huang, Y. P. (2016). CRISPR/Cas9-mediated targeted gene mutagenesis in *Spodoptera litura*. *Insect Science, 23*(3), 469–477. https://doi.org/10.1111/1744-7917.12276.

Borrelli, V. M. G., Brambilla, V., Rogowsky, P., Marocco, A., & Lanubile, A. (2018). The enhancement of plant disease resistance using CRISPR/Cas9 technology. *Frontiers in Plant Science, 9*, 1245. https://doi.org/10.3389/fpls.2018.01245.

Bovey, R., Gärtel, W., Hewitt, W. B., Martelli, G. P., & Vuittenez, A. (1990). Soil-borne viruses transmitted by nematodes. In R. Bovey, W. Gärtel, W. B. Hewitt, G. P. Martelli, & A. Vuittenez (Eds.), *Virus and Virus-like Diseases of Grapevines* (pp. 46–50). Editions Payot.

Caarls, L., Pieterse, C. M., & Van Wees, S. C. (2015). How salicylic acid takes transcriptional control over jasmonic acid signaling. *Frontiers in Plant Science, 6*, 170:1–170:11. https://doi.org/10.3389/fpls.2015.00170.

Capriotti, L., Baraldi, E., Mezzetti, B., Limera, C., & Sabbadini, S. (2020). Biotechnological approaches: Gene overexpression, gene silencing, and genome editing to control fungal and oomycete diseases in grapevine. *International Journal of Molecular Sciences, 21*(16), 5701. https://doi.org/10.3390/ijms21165701.

Chandrasekaran, J., Brumin, M., Wolf, D., Leibman, D., Klap, C., Pearlsman, M., Sherman, A., Arazi, T., & Gal-On, A. (2016). Development of broad virus resistance in non-transgenic cucumber using CRISPR/Cas9 technology. *Molecular Plant Pathology, 17*(7), 1140–1153. https://doi.org/10.1111/mpp.12375.

Chang, H., Liu, Y., & Ai, D. (2017). A pheromone antagonist regulates optimal mating time in the moth *Helicoverpa armigera. Current Biology, 27*(11), 1610–1615. https://doi.org/10.1016/j.cub.2017.04.028.

Char, S. N., Neelakandan, A. K., Nahampun, H., Frame, B., Main, M., Spalding, M. H., & Yang, B. (2017). An Agrobacterium-delivered CRISPR/Cas9 system for high-frequency targeted mutagenesis in maize. *Plant Biotechnology Journal, 15*(2), 257–268. https://doi.org/10.1111/pbi.12611.

Charrier, A., Vergne, E., Dousset, N., Richer, A., Petiteau, A., & Chevreau, E. (2019). Efficient targeted mutagenesis in apple and first-time edition of pear using the CRISPR-Cas9 system. *Frontiers in Plant Science, 10*, 40:1–40:12. https://doi.org/10.3389/fpls.2019.00040.

Chen, L. Q., Qu, X. Q., Hou, B. H., Sosso, D., Osorio, S., Fernie, A. R., & Frommer, W. B. (2012). Sucrose efflux mediated by SWEET proteins as a key step for phloem transport. *Science, 335*(6065), 207–211. https://doi.org/10.1126/science.1213351.

Chiang, T. W. W., Le Sage, C., Larrieu, D., Demir, M., & Jackson, S. P. (2016). CRISPR-Cas9 D10A nickase-based genotypic and phenotypic screening to enhance genome editing. *Scientific Reports, 6*, 1–7. https://doi.org/10.1038/srep36421.

Consonni, C., Humphry, M. E., Hartmann, H. A., Livaja, M., Durner, J., Westphal, L., Vogel, J., Lipka, V., Kemmerling, B., & Schulze-Lefert, P. (2006). Conserved requirement for a plant host cell protein in powdery mildew pathogenesis. *Nature Genetics, 38*(6), 716–720. https://doi.org/10.1038/ng1806.

Deltcheva, E., Chylinski, K., Sharma, C. M., Gonzales, K., Chao, Y., Pirzada, Z. A., Eckert, M. R., Vogel, J., & Charpentier, E. (2011). CRISPR RNA maturation by trans-encoded small RNA and host factor RNase III. *Nature, 471*(7340), 602–607. https://doi.org/10.1038/nature09886.

Doudna, J. A., & Charpentier, E. (2014). The new frontier of genome engineering with CRISPR-Cas9. *Science, 346*(6213), 1258096. https://doi.org/10.1126/science.1258096.

Feng, Y. L., Liu, S. C., & Chen, R. (2021). Target binding and residence: A new determinant of DNA double-strand break repair pathway choice in CRISPR/Cas9 genome editing. *Journal of Zhejiang University Science B, Biomedicine & Biotechnology, 22*(1), 73–86. https://doi.org/10.1631/jzus.B2000456.

Garczynski, S. F., Martin, J. A., Griset, M., Willett, L. S., Cooper, W. R., Swisher, K. D., & Unruh, T. R. (2017). CRISPR/Cas9 editing of the codling moth (*Lepidoptera: Tortricidae*) CpomOR1 gene affects egg production and viability. *Journal of Economic Entomology, 110*(4), 1847–1855. https://doi.org/10.1093/jee/tox174.

Garneau, J. E., Dupuis, M. E., Villion, M., Romero, D. A., Barrangou, R., Boyaval, P., Fremaux, C., Horvath, P., Magadan, A. H., & Moineau, S. (2010). The CRISPR/Cas bacterial immune system cleaves bacteriophage and plasmid DNA. *Nature, 468*(7320), 67–71. https://doi.org/10.1038/nature09523.

Grissa, I., Vergnaud, G., & Pourcel, C. (2007). CRISPR finder: A web tool to identify clustered regularly interspaced short palindromic repeats. *Nucleic Acids Research, 35*(suppl_2), W52–W57. https://doi.org/10.1093/nar/gkm360.

Hartung, F., & Schiemann, J. (2014). Precise plant breeding using new genome editing techniques: Opportunities, safety and regulation in the EU. *The Plant Journal, 78*(5), 742–752. https://doi.org/10.1111/tpj.12578.

He, X., Huo, Y., Liu, X., Zhou, Q., Feng, S., Shen, X., Li, B., Wu, S., & Chen, X. (2018). Activation of disease resistance against *Botryosphaeria dothidea* by downregulating the expression of MdSYP121 in apple. *Horticulture Research, 5*, 24. https://doi.org/10.1038/s41438-018-0027-6.

Hu, C., Sheng, O., Deng, G., He, W., Dong, T., Yang, Q., Dou, T., Li, C., Gao, H., & Liu, S. (2021). CRISPR/Cas9-mediated genome editing of MaACO1 (*aminocyclopropane-1-carboxylate oxidase1*) promotes the shelf life of banana fruit. *Plant Biotechnology Journal, 19*(3), 654–656. https://doi.org/10.1111/pbi.13479.

Hu, Y., Zhang, J., Jia, H., Sosso, D., Li, T., Frommer, W. B., Yang, B., White, F. F., Wang, N., & Jones, J. B. (2014). Lateral organ boundaries 1 is a disease susceptibility gene for citrus bacterial canker disease. *Proceedings of the National Academy of Sciences USA, 111*(15), E521–E529. https://doi.org/10.1073/pnas.1402126111.

Huang, Y., Chen, Y., & Zeng, B. (2016). CRISPR/Cas9 mediated knockout of the abdominal-A homeotic gene in the global pest, diamondback moth (*Plutella xylostella*). *Insect Biochemistry and Molecular Biology, 75*, 98–106. https://doi.org/10.1016/j.ibmb.2016.07.004.

Ji, X., Zhang, H., Zhang, Y., Wang, Y., & Gao, C. (2015). Establishing a CRISPR–Cas-like immune system conferring DNA virus resistance in plants. *Nature Plants, 1*, 15144. https://doi.org/10.1038/nplants.2015.144.

Jia, H., Orbovic, V., Jones, J. B., & Wang, N. (2016). Modification of the PthA4 vector binding elements in Type I CsLOB1 promoter using Cas9/sgRNA to produce transgenic Duncan grapefruit alleviating XccDpthA4.3 infection. *Plant Biotechnology Journal, 14*(5), 1291–1301. https://doi.org/10.1111/pbi.12485.

Jia, H., Zhang, Y., Orbovic, V., Xu, J., White, F. F., Jones, J. B., & Wang, N. (2017). Genome editing of the disease susceptibility gene CsLOB1 in citrus confers resistance to citrus canker. *Plant Biotechnology Journal, 15*(7), 817–823. https://doi.org/10.1111/pbi.12678.

Jiang, F., & Doudna, J. A. (2017). CRISPR–Cas9 structures and mechanisms. *Annual Review of Biophysics, 46*, 505–529. https://doi.org/10.1146/annurev-biophys-062215-010822.

Jinek, M., Chylinski, K., Fonfara, I., Hauer, M., Doudna, J. A., & Charpentier, E. (2012). A programmable dual-RNA–guided DNA endonuclease in adaptive bacterial immunity. *Science, 337*(6096), 816–821. https://doi.org/10.1126/science.1225829.

Jinek, M., Chylinski, K., Fonfara, I., Hauer, M., Doudna, J. A., & Charpentier, E. (2010). A programmable dual-RNA–guided DNA endonuclease in adaptive bacterial immunity. *Science, 337*(6096), 816–821. https://doi.org/10.1126/science.1225829.

Johnson, K. B. (2010). Pathogen refuge: A key to understanding biological control. *Annual Review of Phytopathology, 48*, 141–160. https://doi.org/10.1146/annurev-phyto-073009-114418.

Kim, Y. A., Moon, H., & Par, C. J. (2019). CRISPR/Cas9-targeted mutagenesis of Os8N3 in rice to confer resistance to *Xanthomonas oryzae* pv. *oryzae*. *Rice, 12*(1), 67. https://doi.org/10.1186/s12284-019-0348-3.

Kortekamp, A. (2006). Expression analysis of defense-related genes in grapevine leaves after inoculation with a host and a non-host pathogen. *Plant Physiology and Biochemistry, 44*(1), 58–67. https://doi.org/10.1016/j.plaphy.2006.01.007.

Koutroumpa, F. A., Monsempes, C., Francois, M. C., Cian, A. D., Royer, C., Concordet, J. P., & Jacquin-Joly, E. (2016). Heritable genome editing with CRISPR/Cas9 induces anosmia in a crop pest moth. *Scientific Reports, 6*, 29620. https://doi.org/10.1038/srep29620.

Ledford, H. (2015). Alternative CRISPR system could improve genome editing. *Nature, 526*(7571), 17. https://doi.org/10.1038/526017a.

Li, Y., Zhang, J., Chen, D., Yang, P., Jiang, F., Wang, X., & Kang, L. (2016). CRISPR/Cas9 in locusts: Successful establishment of an olfactory deficiency line by targeting the mutagenesis of an odorant receptor co-receptor (Orco). *Insect Biochemistry and Molecular Biology, 79*, 27–35. https://doi.org/10.1016/j.ibmb.2016.10.002.

Lin, H., Doddapaneni, H., Takahashi, Y., & Walker, M. A. (2007). Comparative analysis of ESTs involved in grape responses to Xylella fastidiosa infection. *BMC Plant Biology, 7*, 8. https://doi.org/10.1186/1471-2229-7-8.

Liu, M., Rehman, S., Tang, X., Gu, K., Fan, Q., Chen, D., & Ma, W. (2019). Methodologies for improving HDR efficiency. *Frontiers in Genetics, 9*, 691. https://doi.org/10.3389/fgene.2018.00691.

Lu, H. P., Luo, T., Fu, H. W., Wang, L., Tan, Y. Y., Huang, J. Z., & Lou, Y. G. (2018). Resistance of rice to insect pests mediated by suppression of serotonin biosynthesis. *Nature Plants, 4*, 338–344. https://doi.org/10.1038/s41477-018-0131-7.

Ma, X., Zhang, Q., Zhu, Q., Liu, W., Chen, Y., Qiu, R., & Lin, Y. (2015). A robust CRISPR/Cas9 system for convenient, high-efficiency multiplex genome editing in monocot and dicot plants. *Molecular Plant, 8*(8), 1274–1284. https://doi.org/10.1016/j.molp.2015.04.007.

Maddox, C. E., Laur, L. M., & Tian, L. (2010). Antibacterial activity of phenolic compounds against the phytopathogen Xylella fastidiosa. *Current Microbiology, 60*(1), 53–58. https://doi.org/10.1007/s00284-009-9486-7.

Makarova, K. S., Wolf, Y. I., Alkhnbashi, O. S., Costa, F., Shah, S. A., Saunders, S. J., & Horvath, P. (2015). An updated evolutionary classification of CRISPR–Cas systems. *Nature Reviews Microbiology, 13*, 722–736. https://doi.org/10.1038/nrmicro3569.

Malnoy, M., Viola, R., Jung, M. H., Koo, O. J., Kim, S., Kim, J. S., & Nagamangala Kanchiswamy, C. (2016). DNA-free genetically edited grapevine and apple protoplast using CRISPR/Cas9 ribonucleoproteins. *Frontiers in Plant Science, 7*, 1904. https://doi.org/10.3389/fpls.2016.01904.

Mao, Y., Botella, J. R., & Zhu, J. K. (2017). Heritability of targeted gene modifications induced by plant-optimized CRISPR systems. *Cellular and Molecular Life Sciences, 74*(6), 1075–1093. https://doi.org/10.1007/s00018-016-2402-1.

Marraffini, L. A., & Sontheimer, E. J. (2010). CRISPR interference: RNA-directed adaptive immunity in bacteria and archaea. *Nature Reviews Genetics, 11*(3), 181–190. https://doi.org/10.1038/nrg2749.

Martelli, G. P. (1993). Handbook for detection and diagnosis: Grapevine degeneration–Fanleaf. In G. P. Martelli (Ed.), *Grafttransmissible Diseases of Grapevines* (pp. 9–18). FAO.

Martelli, G. P., & Savino, V. (1990). Fanleaf degeneration. In R. Pearson & A. C. Goheen (Eds.), *Compendium of Grape Diseases* (pp. 48–49). APS Press.

Marzec, M., Brąszewska-Zalewska, A., & Hensel, G. (2020). Prime Editing: A new way for genome editing. *Trends in Cell Biology, 30*(4), 257–259. https://doi.org/10.1016/j.tcb.2020.01.002.

McFarlane, G. R., Whitelaw, C. B. A., & Lillico, S. G. (2018). CRISPR-based gene drives for pest control. *Trends in Biotechnology, 36*(2), 130–133. https://doi.org/10.1016/j.tibtech.2017.09.010.

Nishitani, C., Hirai, N., Komori, S., Wada, M., Okada, K., Osakabe, K., & Osakabe, Y. (2016). Efficient genome editing in apple using a CRISPR/Cas9 system. *Scientific Reports, 6*, 31481. https://doi.org/10.1038/srep31481.

Osakabe, Y., Liang, Z., Ren, C., Nishitani, C., Osakabe, K., Wada, M., & Poli, M. (2018). CRISPR-Cas9-mediated genome editing in apple and grapevine. *Nature Protocols, 13*(12), 2844–2863. https://doi.org/10.1038/s41596-018-0069-1.

Paula de Toledo Thomazella, D., Brail, Q., Dahlbeck, D., & Staskawicz, B. (2016). CRISPR-Cas9 mediated mutagenesis of a DMR6 ortholog in tomato confers broad-spectrum disease resistance. *BioRxiv*. https://doi.org/10.1101/064824.

Peng, A., Chen, S., Lei, T., Xu, L., He, Y., Wu, L., & Zou, X. (2017). Engineering canker-resistant plants through CRISPR/Cas9-targeted editing of the susceptibility gene CsLOB1 promoter in citrus. *Plant Biotechnology Journal, 15*(12), 1509–1519. https://doi.org/10.1111/pbi.12738.

Pessina, S., Lenzi, L., Perazzolli, M., Campa, M., Dalla Costa, L., Urso, S., Valè, G., Salamini, F., Velasco, R., & Malnoy, M. (2016). Knockdown of MLO genes reduce susceptibility to powdery mildew in grapevine. *Horticulture Research, 3*, 16016:1–16016:9. https://doi.org/10.1038/hortres.2016.16.

Pompili, V., Dalla Costa, L., Piazza, S., Pindo, M., & Malnoy, M. (2020). Reduced fire blight susceptibility in apple cultivars using a high-efficiency CRISPR/Cas9-FLP/FRT-based gene editing system. *Plant Biotechnology Journal, 18*, 845–858. https://doi.org/10.1111/pbi.13230.

Puchta, H., Dujon, B., & Hohn, B. (1993). Homologous recombination in plant cells is enhanced by in vivo induction of double strand breaks into DNA by a site-specific endonuclease. *Nucleic Acids Research, 21*(22), 5034–5040. https://doi.org/10.1093/nar/21.22.5034.

Rath, D., Amlinger, L., Rath, A., & Lundgren, M. (2015). The CRISPR-Cas immune system: Biology, mechanisms and applications. *Biochimie, 117*, 119–128. https://doi.org/10.1016/j.biochi.2015.06.025.

Rathinam, M., Mishra, P., Mahato, A. K., Singh, N. K., Rao, U., & Sreevathsa, R. (2019). Comparative transcriptome analyses provide novel insights into the differential response of Pigeonpea (*Cajanus cajan* L.) and its wild relative (*Cajanus platycarpus* (Benth.) Maesen) to herbivory by *Helicoverpa armigera* (Hübner). *Plant Molecular Biology, 101*, 163–182. https://doi.org/10.1007/s11103-019-00904-w.

Ren, C., Guo, Y., Kong, J., Lecourieux, F., Dai, Z., Li, S., & Liang, Z. (2020). Knockout of VvCCD8 gene in grapevine affects shoot branching. *BMC Plant Biology, 20*, 47:1–47:8. https://doi.org/10.1186/s12870-020-2274-8.

Samanta, M. K., Dey, A., & Gayen, S. (2016). CRISPR/Cas9: An advanced tool for editing plant genomes. *Transgenic Research, 25*, 561–573. https://doi.org/10.1007/s11248-016-9943-6.

Scheben, A., Wolter, F., Batley, J., Puchta, H., & Edwards, D. (2017). Towards CRISPR/Cas crops–bringing together genomics and genome editing. *New Phytologist, 216*, 682–698. https://doi.org/10.1111/nph.14702.

Shan-e-Ali Zaidi, S., Mansoor, S., Ali, Z., Tashkandi, M., & Mahfouz, M. M. (2016). Engineering plants for Geminivirus resistance with CRISPR/Cas9 system. *Trends in Plant Science, 21*, 279–281. https://doi.org/10.1016/j.tplants.2015.11.009.

Shao, X., Wu, S., Dou, T., Zhu, H., Hu, C., Huo, H., & Bi, F. (2020). Using CRISPR/Cas9 genome editing system to create MaGA20ox2 gene-modified semi-dwarf banana. *Plant Biotechnology Journal, 18*, 17–19. https://doi.org/10.1111/pbi.13245.

Siamer, S., Guillas, I., Shimobayashi, M., Kunz, C., Hall, M. N., & Barny, M. A. (2014). Expression of the bacterial type III effector DspA/E in Saccharomyces cerevisiae down-regulates the sphingolipid biosynthetic pathway leading to growth arrest. *Journal of Biological Chemistry, 289*(27), 18466–18477. https://doi.org/10.1074/jbc.M114.552459.

Stelinski, L. L., Rogers, M. E., & Tiwari, S. (2012). Biochemical basis of organophosphate and carbamate resistance in Asian Citrus Psyllid. *Journal of Economic Entomology, 105*(2), 540–548. https://doi.org/10.1603/EC11382.

Sternberg, S. H., LaFrance, B., Kaplan, M., & Doudna, J. A. (2015). Conformational control of DNA target cleavage by CRISPR–Cas9. *Nature, 527*, 110–113. https://doi.org/10.1038/nature15544.

Sternberg, S. H., Redding, S., Jinek, M., Greene, E. C., & Doudna, J. A. (2014). DNA interrogation by the CRISPR RNA-guided endonuclease Cas9. *Nature, 507*, 62–67. https://doi.org/10.1038/nature13011.

Sun, D., Guo, Z., Liu, Y., & Zhang, Y. (2017). Progress and prospects of CRISPR/Cas systems in insects and other arthropods. *Frontiers in Physiology, 8*, 608. https://doi.org/10.3389/fphys.2017.00608.

Sunitha, S., & Rock, C. D. (2020). CRISPR/Cas9-mediated targeted mutagenesis of TAS4 and MYBA7 loci in grapevine rootstock 101–14. *Transgenic Research, 29*, 355–367. https://doi.org/10.1007/s11248-020-00199-4.

Tripathi, J. N., Ntui, V. O., Ron, M., Muiruri, S. K., Britt, A., & Tripathi, L. (2019). CRISPR/Cas9 editing of endogenous banana streak virus in the B genome of *Musa* spp. overcomes a major challenge in banana breeding. *Communications Biology, 2*, 46:1–46:11. https://doi.org/10.1038/s42003-019-0292-5.

Tripathi, L., Ntui, V. O., & Tripathi, J. N. (2020). CRISPR/Cas9 genome editing of banana for disease resistance. *Current Opinion in Plant Biology, 56*, 118–126. https://doi.org/10.1016/j.pbi.2020.04.005.

Upadhyay, S. K. (2021). *Genome Engineering for Crop Improvement*. John Wiley & Sons.

Varkonyi-Gasic, E., Wang, T., Cooney, J., Jeon, S., Voogd, C., Douglas, M. J., Pilkington, S. M., Akagi, T., & Allan, A. C. (2021). Shy Girl, a kiwifruit suppressor of feminization, restricts gynoecium development via regulation of cytokinin metabolism and signaling. *New Phytologist, 230*, 1461–1475. https://doi.org/10.1111/nph.17219.

Varkonyi-Gasic, E., Wang, T., Voogd, C., Jeon, S., Drummond, R. S. M., Gleave, A. P., & Allan, A. C. (2019). Mutagenesis of kiwifruit Centroradialis-like genes transforms a climbing woody perennial with long juvenility and axillary flowering into a compact plant with rapid terminal flowering. *Plant Biotechnology Journal, 17*, 869–880. https://doi.org/10.1111/pbi.13021.

Vats, S., Kumawat, S., Kumar, V., Patil, G. B., Joshi, T., Sonah, H., Sharma, T. R., & Deshmukh, R. (2019). Genome editing in plants: Exploration of technological advancements and challenges. *Cells, 8*(11), 1386. https://doi.org/10.3390/cells8111386.

Wan, D. Y., Guo, Y., Cheng, Y., Hu, Y., Xiao, S., & Wang, Y. (2020). CRISPR/Cas9-mediated mutagenesis of VvMLO3 results in enhanced resistance to powdery mildew in grapevine (*Vitis vinifera*). *Horticulture Research, 7*, 116:1–116:14. https://doi.org/10.1038/s41438-020-00350-7.

Wang, H., Shi, Y., Wang, L., Liu, S., Wu, S., Yang, Y., Feyereisen, R., & Wu, Y. (2018). CYP6AE gene cluster knockout in *Helicoverpa armigera* reveals role in detoxification of phytochemicals and insecticides. *Nature Communications, 9*, 4820. https://doi.org/10.1038/s41467-018-07273-4.

Wang, J., Zhang, H., Wang, H., Zhao, S., Zuo, Y., Yang, Y., & Wu, Y. (2016). Functional validation of cadherin as a receptor of Bt toxin Cry1Ac in *Helicoverpa armigera* utilizing the CRISPR/Cas9 system. *Insect Biochemistry and Molecular Biology, 76*, 11–17. https://doi.org/10.1016/j.ibmb.2016.07.005.

Wang, L., Chen, S., Peng, A., Xie, Z., He, Y., & Zou, X. (2019). CRISPR/Cas9-mediated editing of *CsWRKY22* reduces susceptibility to *Xanthomonas citri* subsp. *citri* in Wanjincheng orange (*Citrus sinensis* (L.) Osbeck). *Plant Biotechnology Reports, 13*(1), 1–10. https://doi.org/10.1007/s11816-019-00556-x.

Wang, N., Qu, C., Jiang, S., Chen, Z., Xu, H., Fang, H., Su, M., Zhang, J., Wang, Y., & Liu, W. (2018). The proanthocyanidin-specific transcription factor MdMYBPA1 initiates anthocyanin synthesis under low-temperature conditions in redfleshed apples. *Plant Journal, 96*(1), 39–55. https://doi.org/10.1111/tpj.14014.

Wang, X. (2018). CRISPR/Cas9-mediated efficient targeted mutagenesis in grape in the first generation. *Plant Biotechnology Journal, 16*, 844–855. https://doi.org/10.1111/pbi.12846.

Wu, G. (1997). Activation of host defense mechanisms by elevated production of H_2O_2 in transgenic plants. *Plant Physiology, 115*, 427–435. https://doi.org/10.1104/pp.115.2.427.

Wu, K., Shirk, P. D., Taylor, C. E., Furlong, R. B., Shirk, B. D., Pinheiro, D. H., & Siegfried, B. D. (2018). CRISPR/Cas9 mediated knockout of the *abdominal-A* homeotic gene in fall armyworm moth (*Spodoptera frugiperda*). *PLoS ONE, 13*(12), e0208647. https://doi.org/10.1371/journal.pone.0208647.

Yang, S., Shi, Y., Zou, L., Huang, J., Shen, L., Wang, Y., Guan, D., & He, S. (2020). Pepper CaMLO6 negatively regulates *Ralstonia solanacearum* resistance and positively regulates high temperature and high humidity responses. *Plant Cell Physiology, 61*(7), 1223–1238. https://doi.org/10.1093/pcp/pcaa060.

Yu, Y. H. (2020). Grape (*Vitis davidii*) VdGATA2 functions as a transcription activator and enhances powdery mildew resistance via the active oxygen species pathway. *Scientia Horticulturae, 267*, 109327. https://doi.org/10.1016/j.scienta.2020.109327.

Zeilmaker, T., Ludwig, N. R., Elberse, J., Seidl, M. F., Berke, L., Van Doorn, A., Schuurink, R. C., Snel, B., & Van den Ackerveken, G. (2015). Downy mildew resistant 6 and DMR6-LIKE OXYGENASE 1 are partially redundant but distinct suppressors of immunity in Arabidopsis. *Plant Journal, 81*(2), 210–222. https://doi.org/10.1111/tpj.12721.

Zetsche, B., Gootenberg, J. S., Abudayyeh, O. O., Slaymaker, I. M., Makarova, K. S., Essletzbichler, P., Volz, S. E., Joung, J., Van Der Oost, J., & Regev, A. (2015). Cpf1 is a single RNA-guided endonuclease of a class 2 CRISPR-Cas system. *Cell, 163*(3), 759–771. https://doi.org/10.1016/j.cell.2015.09.038.

Zhou, H., Bai, S., Wang, N., Sun, X., Zhang, Y., Zhu, J., & Dong, C. (2020). CRISPR/Cas9-mediated mutagenesis of MdCNGC2 in apple callus and VIGS-mediated silencing of MdCNGC2 in fruits improve resistance to *Botryosphaeria dothidea*. *Frontiers in Plant Science, 11*, 575477:1–11. https://doi.org/10.3389/fpls.2020.575477.

Zhou, J., Wang, G., & Liu, Z. (2018). Efficient genome-editing of wild strawberry genes, vector development and validation. *Plant Biotechnology Journal, 16*(10), 1868–1877. https://doi.org/10.1111/pbi.12925.

CHAPTER 11

The Role of Lipoxygenase in Crop Storage Tolerance and Biotic Stress Management

ROOHI SHARMA,[1] RAMAN TIKORIA,[1] MOHD. ALI,[1] DEEPAK KUMAR,[1] PARKIRTI,[1] KRITIKA KAUSHAL,[1] UMARA RAHMANI,[2] SUPRIYA PANDEY,[3] PUJA OHRI,[1] VISHAL DINKAR,[4] and ASHISH KUMAR SINGH[5]

[1]*Department of Zoology, Guru Nanak Dev University, Amritsar, Punjab, India*

[2]*Department of Plant Pathology, Sardar Vallabhbhai Patel University of Agriculture and Technology, Meerut, Uttar Pradesh, India*

[3]*Govind Ballabh Pant National Institute of Himalayan Environment, Kosi-Katarmal, Almora, Uttarakhand, India*

[4]*ICAR–Central Institute of Temperate Horticulture, Rangreth, Srinagar, Jammu and Kashmir, India*

[5]*Crop Protection Division, ICAR-VPKAS, Almora, Uttarakhand, India*

ABSTRACT

Natural enzymes called lipoxygenases (LOXs) are present in a vast array of animal and plant species. LOXs can be redox enzymes that contain manganese, non-heme iron, or non-sulfur iron dioxygenases. One of the key elements impacting the viability and durability of crop seeds is the deoxygenation of polyunsaturated fatty acids (PUFA) possessing a minimum of one cis, cis-1,4-pentadiene that is transformed into fatty acid hydroperoxides by LOXs, which assists in crop storage. Linolenic acid, a component in the production

Climate Change and Biotic Factors: A Molecular Approach.
Ashutosh Singh, Saurabh Pandey, and Amarjeet Kumar, (Eds.)

of jasmonic acid (JA), is oxygenated by LOX, allene oxide synthase (AOS), and allene oxide cyclase (AOC) to produce 12-oxo-phytodienoic acid. Jasmonic acid also takes part in numerous physio-biochemical processes such as fruit development and ripening, senescence, and seed germination. These *LOX* genes have climacteric contributions in the defense responses to abiotic and biotic stress. Abiotic stress includes wounding, UV rays, severe temperatures, oxidative stress, salt, drought, and flood, whereas biotic stress includes insects, pests, and pathogenic attacks. It has been hypothesized that LOXs have a protective role in host-parasite interactions with the action of enzymes that definitely includes AOS, AOC, and hydroperoxide lyase (HPL). LOXs assist in seed germination, delay fruit senescence, and maintain the robustness and viability of the seed. These ubiquitous enzymes are the obvious research subjects when biologists examine how plants respond to stressors. In this chapter, we addressed the numerous elements of the potential applications of this knowledge to enhance crop storage and emphasized the work that has been achieved in recognizing the position of LOXs in biotic stress management. This chapter would further help readers learn more about LOXs in a dynamic environment and offer fresh ideas for applying this information to increase biotic stress tolerance.

11.1 INTRODUCTION

With the increase in population, the food demand is increasing rapidly, with global demand rising to 70% more food by the year 2050 to feed the population of around 9.1 billion (Godfray et al., 2010). The crop loss due to biotic stresses (like pathogens, pests, weeds, herbivores, and microbes) and abiotic stresses (extremely low and high temperatures, humidity and rainfall, wounds, UV rays, drought and dehydration, flooding, salinity, oxidative stress) is huge. Accounting for the world's expanding population's nutritional needs is still a major challenge; however, as per estimates, greater than a third of food crops are damaged or invested as waste in post-harvest activities, including spoilage and spillage, quality issues, and nutritional value loss, commercial loss, and loss in seed viability (Kumar and Kalita, 2017). Sub-Saharan Africa alone faces losses of around USD 4 billion every year, according to a World Bank analysis (Zorya, 2011).

The mild to deadly pathogenic (fungi, bacteria, viruses, rats, cockroaches, etc.) crop diseases also lower crop performance and yield to a great extent. In the USA, roughly 90% of the main illnesses of the crops grown are caused by soil-borne plant pathogens (Pinkerton et al., 2000). Definitive diagnosis of a specific disease is challenging because of the resemblance in indicators including seedlings damping-off, root darkening, leaf blight, root rot, stunted growth, drooping, discoloration, bark breaking, and twigs or branches necrotic lesions which then makes the disease tougher to control (Nene et al., 2012; Steele and Gibbs, 2018). Environmental factors have numerous implications on the cellular processes and plant responses to external stimuli. Depending on their length and severity, some of the factors may inflict stress (Farooq et al., 2009). Growth and development are negatively impacted by frosts, extreme heat and cold, excessive humidity, acidity, saltiness, and acidity of the lithosphere, soils, pesticide pollution, and other factors. These factors also limit metabolism, alter its pattern, and need internal energy to combat the adverse consequences of stress (Shahzad et al., 2021). The mechanical damage to the crop may be the degradation of photosynthetic tissues, and in severe forms, defoliation may happen (Pandey et al., 2017). The heavy destruction of floral and foliar parts like buds, shoots, flowers, fruits, and seeds can also be seen. Pests are boring into the stems and tunneling through them, causing the disruption of sap flow, physical weakening of the stem, and ultimately, stem breakage (Srivastava and Subramaniam, 2016).

Despite the preharvest and onset-harvesting time troubles, the postharvest loss is estimated to be up to 80%, while it may reach around 20–40% in African regions where humidity, moisture, and temperature are the most crucial factors along with pathogens (Fox and Fimeche 2013; Abass, 2014). Throughout the supply chain for foods, storage is very crucial, and various studies have found that this is likewise the stage at which losses are the highest. Several regions only cultivate seasonal crops, and after harvest, grains are kept in storage for brief or extended time frames as food stocks, and also seeds are kept for the following season (Kiaya 2014; Majumder et al., 2016). Traditional methods (by drying grains in matka, Kothi, sanduk, patila, gagar, and some crops buried under the soil) have been overtaken by modern methods (granaries, metal cans, jute bags, polythenes, plastic containers) for crop storage but still, the economic and biological losses are immense (Grover and Singh 2013; Hell et al., 2000). Under improper storage, the crops may also undergo massive destruction by insect infestation (30–40%) and mycotoxins (25–40%) (Dubey et al., 2012; Haider and Rai, 2021).

A constructive way to enhance food supply, ease the burden on the environment, end hunger, and promote farmers is to reduce postharvest losses, especially in less developed countries (Boxall 2001; Stathers et al., 2020). While combating stress and crop loss by external methods is a tough phenomenon. The use of weedicides, insecticides, and other chemical pesticides is not a sustainable option; therefore, biological control measures are more efficient to scrap such consequences where they have an important role in the management of crops (Bardin et al., 2015; Sarker et al., 2021). Microbes, metabolites, biofertilizers, bioinsecticides, and biomolecules are among those effective biological tools (Barbu et al., 2021). Likewise, other essential biological metabolites, the plant, animal, and microbial Lipoxygenases (LOXs), are the key players in mitigating both biotic and abiotic stresses. Plant Lipoxygenases are helpful in ameliorating biotic stress and crop storage and management where plants rely heavily on LOXs (Kuhn, 2015). Lipoxygenase is a group of dioxygenase enzymes with two domains (C- and N-terminal) and sizes varying from 90–110 kDa with multiple functions (Singh et al., 2022). Plant lipoxygenases, a group of dioxygenases that do not contain heme iron or sulfur, play an integral role in physiological plant processes by oxidizing various lipids, particularly biotic and abiotic pressures (Vishwanath et al., 2020). Plant LOXs are placed into two important groups, viz., 9-LOXs and 13-LOXs (Upadhyay et al., 2019).

The oxidation process is catalyzed by the LOX proteins of the polyunsaturated fatty acid (PUFA) (linoleic acid and linolenic acid) into unsaturated fatty acid hydroperoxides, which then results in the production of oxylipin during pathogenic or pest attacks in particular. The compounds of the PUFA oxidation lipoxygenase pathway play a prominent role among some of the aspects that contribute to the development of adaptive mechanisms, essentially the signaling system properties such as detection, signal enhancement, and transformation, and ultimately defense gene expression (Vishwanath et al., 2020). It has been postulated that the LOX pathway has a significant role in plant defense against biting-chewing herbivores via important oxylipins such as 12-OPDA by the action of 13-LOXs, 10-OPDA by the action of 9-LOXs, the positional isomer of 12-OPDA, and 10-oxo-11-phytoenoic acid (10-OPEA). LOXs cause an antiherbivore oxidative change in response to herbivory (Wari et al., 2022; Crozier, 2000). This harms the herbivore by direct and indirect oxidative damage and transmits signals that cause the plant to defend itself. A vital plant stress hormone that aids in defense against diseases and herbivores is jasmonic acid (Shrestha et al., 2021). In *Cucumis sativus*, during day five following blossoming, 12 genes were discovered to have a variable expression in the moment of fruit development as well as have distinct expression designs in the exocarp, endocarp, and pulp. The outcomes of the expression study revealed that the *LOX* gene family in

Cucumber responded to many treatments, indicating that they have a broad function in fruiting, hormonal reactions, and abiotic stress (Yang et al., 2012).

11.2 LIPOXYGENASE ENZYME IN PLANTS

Lipoxygenase (LOX) is a ubiquitous oxidoreductase enzyme in both flora and fauna, and it participates in many physiological processes (Brash 1999; Babenko et al., 2015). Bohn and Haas (1928) discovered LOX as a carotene-oxidizing enzyme, while in 1932, it was first reported to be present in soybean plants (Andre and Hou, 1932). Craig (1936) recognized it as a 'fat oxidase' due to the higher oxygen content than carbon dioxide (CO_2) in the oxygen vicinity in white lupine (*Lupinus albus*). It was purified and crystallized from soybean in Theorell's laboratory in 1947. LOXs were also isolated from numerous plant species, including Thale cress (*A. thaliana*), rice, maize, grapes, popular tree (*Populus spp.*), tea (*Camellia sinensis*), peach, radish, cotton, and chili (Viswanath et al., 2020).

LOXs are multipurpose enzymes and catalyze three different reactions, namely dioxygenase, hydroperoxidase, and leukotriene synthase activities (Andreou and Feussner, 2009). LOXs are also known as "signal" enzymes due to the similarity in amino acid (aa) sequence, tertiary structure, and interactions of the N-domain with the C2-like domain of α-toxin from *Clostridium perfringens* (Horn et al., 2015). The carbon mostly positioned at the second position of domain-containing proteins is implicated in signaling pathways, and membrane traffic catalysis is carried out in the state corresponding to membrane structures (Smith and Murphy, 2016). Soybean lipoxygenases, through trypsin digestion, form a 60kDa fragment of LOX-1 (mini-LOX). It has been determined that this soybean section LOX-1, which consists of 15 amino acids and is located between the final section of the folded conformation in the N-terminus and the very first spiral of the catalytic C-terminus, can be broken down by trypsin through the rupture of the link between lysine-277 and serine-278 residues. This mini-LOX is less stable but more active, acidic, and hydrophobic in activity than LOX-1 itself (Maccarrone et al., 2001).

11.3 LOX DIVERSITY

LOXs are a class of non-heme, non-sulfur, or manganese (Mn)-containing dioxygenases that have a molecular weight of 90–110 kDa and are made up of two different domains (Singh et al., 2022). The two different helically arranged catalytic domains of LOX have a molecular weight of 25–30 kDa and 55–65 kDa at the N-terminus and C-terminus, respectively. Two barrels with four antiparallel plates comprise 160 amino acid residues that make up the LOX N-terminal. The C-terminal region contains an active iron (Fe^{3+}) site, 23 helices, two antiparallel plates, and 693 amino acid residues. The active state of LOX is a high-spin oxidized Fe^{3+}, whereas the inactive state is a high-spin Fe^{2+}. This activated iron is coordinated by five amino acid residues in the active site of liquid oxygen, and the H_2O molecule is the sixth ligand (Zhang et al., 2021).

Plant LOXs are largely divided into two major groups, i.e., 9-LOXs and 13-LOXs, which vary according to whether carbon is at position 9 or position 13 of the C-18 polyunsaturated fatty acids is oxidized in the substrate (Viswanath et al., 2020). Furthermore, LOXs have been split into Type-I and Type-II subfamilies depending on similarities in their sequences and structure. The proteins of Type-I LOXs show minimal resemblance in sequences (approximately 35%) and not a single chloroplast transit protein, whereas those of Type-II LOXs possess substantial sequence similarity (at least 75%) and a chloroplast transit peptide (Song et al., 2016). Currently, it is understood that Type-I contains members of the 9-LOXs subfamily, and Type-I and Type-II contain members of the 13-LOXs subfamily (Wang et al., 2019). Recently, it was shown that LOX-specific inhibitors might cause the conversion of 9-LOX to 13-LOX. The unique patterned primary conserved motifs of 9-LOX of TV/K (lysine motif) in *TdLOX2* may have replaced the conventional motif TV/R (arginine motif) of 9-LOX, possessing substantial amino acid protein sequence comprising *TdLOX2* in explaining this location-specificity (Menga and Trono, 2020). New *LOX* gene families are continuously characterized and expanding the plant LOXs (Umate 2011, Chen et al., 2015; Upadhyay et al., 2018; Shrestha et al., 2021). Numerous of these *LOX* gene families have various expression patterns, including the 14 LOX proteins in rice and the 6 *LOX* gene structure in *Arabidopsis* (Umate, 2011). The primary plant oxylipins generated from 13-LOX 18:3 are 12-oxo-phytodienoic acid (12-OPDA) and jasmonic acid (JA), whereas those derived from 9-LOX 18:3/18:2 are 10-oxo-11-phytoenoic acid (10-OPEA), 10-OPDA, and also the death acids (DAs) (Shi et al., 2020).

Different research studies have demonstrated the critical functions *LOX* genes play in plant growth, maturation, senescence, and response to biotic and abiotic stress (Andreou and Feussner, 2009, Figure 10.1). Two 9-LOXs (AtLOX1 and AtLOX5) and four 13-LOXs (AtLOX2, AtLOX3, AtLOX4, and AtLOX6) genes are present in *Arabidopsis* (RoyChowdhury et al., 2016). Pathogen attack, abscisic acid (ABA), and methyl jasmonate (MeJA) all upregulate AtLOX1, which is crucial for stomatal function during earlier stages of protection in vegetation (El-Mogy et al., 2019). It has been thoroughly documented how *LOX* genes of plants react to biotic as well as abiotic stressors. For instance, durum wheat TdLpx-A2 plays a significant function in abiotic stress response by blocking the increased output of reactive oxygen species (ROS) (Menga and Trono, 2020). Additionally, numerous findings have shown quantifiable characteristic markers associated with various LOX expressions where LOX activity was observed under water shortage. ChlMLOX may be involved in non-photochemical chlorophyll quenching, which protects the photosynthetic structure from ROS oxidation, according to the favorable relationship between ChlMLOX functionality and carotenoids (Permyakova et al., 2017). By facilitating a JA biosynthesis route and controlling stomatal activities to lessen water transpiration, CmLOX10 greatly boosts the drought resilience of oriental melons (Xing et al., 2020). According to the numerous findings of translational pattern analysis techniques, *LOX* genes may be connected to metabolic activities and the reaction to abiotic or biotic stressors (Arif et al., 2020). Furthermore, the *LOX* gene has been demonstrated to respond to heat variations. Only under cold stress conditions, GhLOX18 gets activated in cotton, whereas mostly GhLOX genes show changed expression in response to heat (Shaban et al., 2018). Tomato LOX1, LOX3, and LOX5 genes were shown to exclusively respond to heat variation pressure, whereas LOX11 only responded to chilling pressure (Upadhyay et al., 2019).

Linoleic and linolenic acids are naturally occurring LOX substrates in plant cells, as are fatty acid esters and unsaturated fatty acids with cis and trans-pentadiene structures (Singh et al., 2022). At various sites, different sources of LOX substrates lead to oxygenation and the consequent production of their byproducts. Linoleic acid may be oxidized to produce hydroperoxide at 6 or 13-lipo-hydroperoxide, whereas linolenic acid can be oxidized to produce 13-hydroperoxide (Shi et al., 2022). The LOX-1 subtype is reactive with a substance that has solubility in water, especially linoleyl sulfate. It has been discovered that LOX-2 and LOX-3, to a greater extent, are efficient against triglycerides and methyl esters of fatty acids in comparison to the free fatty acids, while LOX-2 works on substances for esterification to generate 9 or 13-hydroperoxide.

While a pH between 4.5 and 8 is thought to be suitable for LOX action, individual plant species may require a different range. JA, 12-oxophytodienoic acid (12-OPDA), and 13-hydroxyoctadecatrienoic acid (13-HODE) are only a few examples of the several 13-LOX-derived substances known with a good reputation for controlling the plant defensive behavior (Lim et al., 2015). A crucial defensive molecule, JA, controls the plant's ability to fight necrotrophic infections (Figure 11.1).

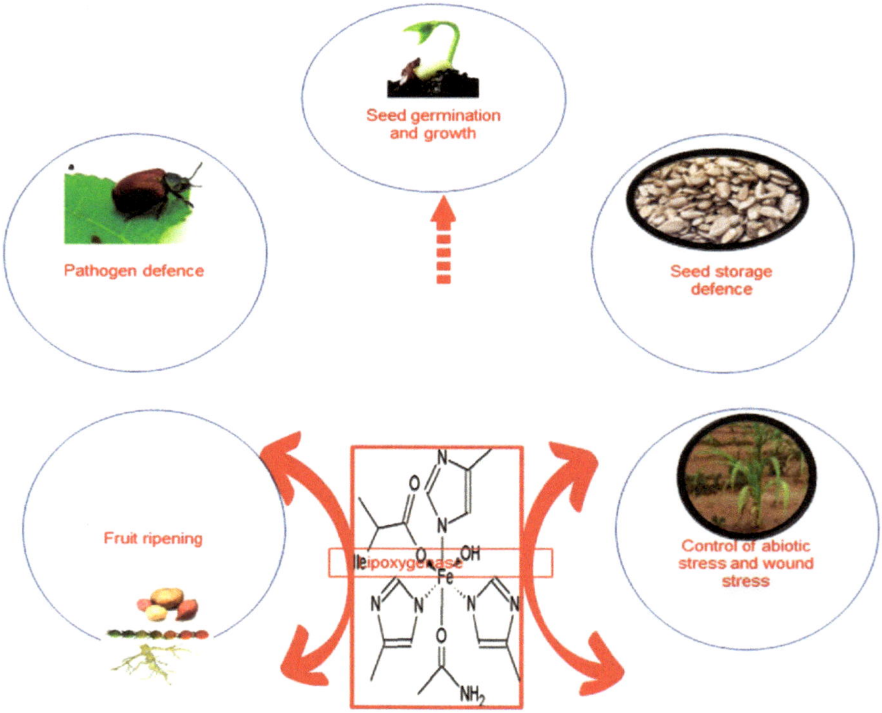

FIGURE 11.1 Application of plant LOXs in the development and growth of plants.

In addition, tobacco leaves infested with an avirulent strain of *Pseudomonas syringae* pv. *phaseolicola* exhibit increased 13-*LOX* gene expression and buildup of JA concentration (Weichert et al., 1999). The function of 9-LOXs in defense systems against biotic and abiotic stressors is unknown. While most of the LOXs that have been studied so far are soluble cytosolic enzymes, some are found in the mitochondria, chloroplasts, or vacuoles. In soybean, lipoxygenases have been discovered to contribute to the partitioning of nitrogen and assimilate, and they appear to be controlled according to the quantity of nitrogen in the plant in ways that are both tissue-specific and developmentally controlled (Hayward et al., 2016; Song et al.,

2016). Some LOX isoforms play a crucial role in manufacturing fatty acid hydroperoxides destined for JA, which activates the genes in plants involved in the wound-healing process (Singh et al., 2022).

11.4 LOXS AND THEIR METABOLITES IN THE DEVELOPMENT OF VARIOUS STRESS RESPONSES

The developmental evolution of plant responses to biotic and abiotic stressors involves numerous defense-related pathways (Kaur et al., 2022; Sharma et al., 2022). Pathogens and abiotic pressure patterns influence the signaling cascade's structure, generating and activating anti-stress chemicals (Banothu and Uma 2022; Sharma et al., 2023). Signaling molecules like JA, MeJA, ABA, salicylic acid (SA), cytokinins, ethylene, and other metabolites can function alone or interact with one another in a synergistic or antagonistic way (Bittner et al., 2022; Kumar and Ohri, 2022). Plants are naturally entangled in and they can use a multitude of signaling networks, enabling them to adapt to more effective defensive strategies when fighting biotic and abiotic challenges.

Several investigations have revealed that LOX activity is triggered by ozone effects, mechanical injuries, hyperthermia, elicitors, and stress outcomes (Cole et al., 2012; Lõhelaid et al., 2015; Kang et al., 2021). Furthermore, the activation of numerous key enzyme-coding genes enhances LOX metabolism. There is enough data to support that the LOX lipid membrane transformation route represents a significant and independent signaling mechanism (Babenko et al., 2017; Singh et al., 2022). The initial signal contact stimulation of plasma membrane receptor molecules causes a membrane-bound protein molecule to propagate down the signal cascade, just like with other signaling pathways. The LOX pathway's signals are amplified by using Ca^{2+} and calmodulin ions in autocatalytic cycles. Linoleic and linolenic acids, used as substrates by LOXs, are released from membrane phospholipids as a result of phospholipase-A being activated by Ca^{2+} and calmodulin (Kućko et al., 2022). MeJAs produce desaturase-related proteins like traumatin, JA, oxylipins, and aldehydes volatiles under autocatalytic cycle conditions, which catalyze the conversion of linoleic acid to linolenic acid (Thakur and Udayashankar, 2019). Next, LOX can activate protein kinases and stimulate the expression of a large number of genes whose by-products are thought to be engaged in defense processes (Kućko et al., 2022; Liburkin-Dan et al., 2022; Figure 11.2). The following chemicals appear to express the protein genes involved in the development of stress tolerance, even though the molecular mechanisms relating to oxylipin-mediated gene activation are not fully understood.

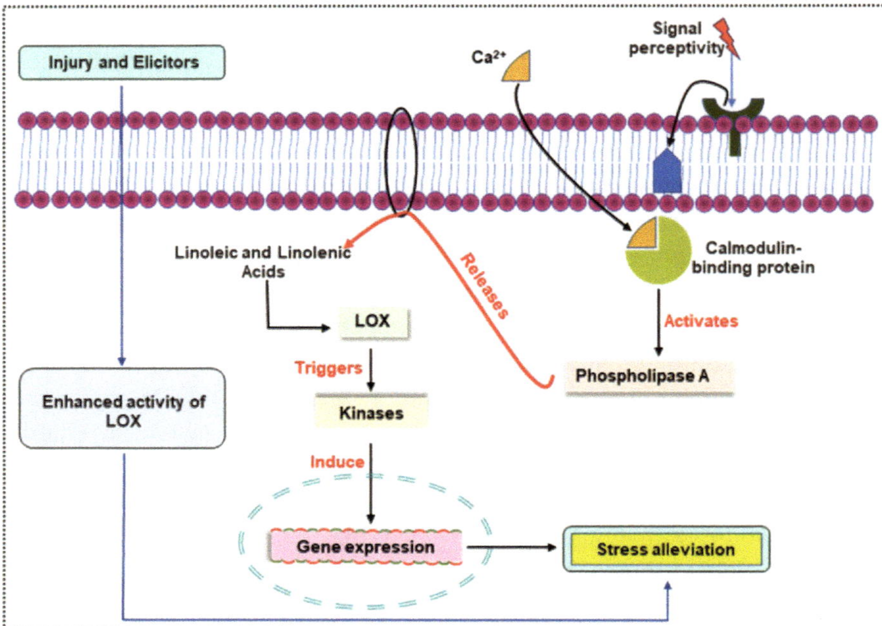

FIGURE 11.2 Lipoxygenase signaling pathway underplant stress.

Jasmonic acid (JA) is among the most well-studied oxylipins in the LOX signaling system. As a discerning inhibitor of polypeptide synthesis, JA and MeJA contribute to alkaloid secretion and accumulation and increase the formation of proteinase inhibitors (Szeliga et al., 2019). JA increases plant disease resistance by creating active oxygen molecules and causing the creation of several pathogenesis-related defense molecules (Liu et al., 2019; Zhu et al., 2023).

For example, water stress in canola was reduced by spraying 0.5 mM JA at various NaCl concentrations (40, 70, and 100 mM). Salt stress increased the ratio of root length to root weight while decreasing lateral and main root growth, as well as the roots' heaviness, density, diameter, and moisture content. Under different salt-stress levels, JA foliar spray boosted overall root growth but reduced dry shoot weights and primary root development, with no change in secondary root development or root water content (Farhangi-Abriz et al., 2019).

Phospholipase A is triggered by Ca^{2+} and calmodulin and releases linoleic and linolenic acids derived from membrane phospholipids that serve as substrates for LOXs. Furthermore, LOX can activate protein kinases and increase the expression of certain genes, producing byproducts that have also been linked to defense systems.

Similarly, using JA, crucial functions of *CmLOX10* in the control of drought tolerance in oriental melon seedlings were determined, where stomatal closure and *CmMYC2*-mediated feedback were regulated by JA signaling (Szeliga et al., 2019).

Furthermore, in polyunsaturated fatty acid oxidation, LOX plays a crucial role. Their by-products serve as precursors for subsequent events that generate bioactive oxylipins and are associated with a broad range of developmental processes, including the ability to resist stress (Lu et al., 2020; Barbaś et al., 2022). For instance, 18:3 is utilized by putative lipid droplets-localized 13-LOX to produce 13-hydroperoxyoctadecatrienoic acid, a precursor for 'Responsive To Dehydration20'-mediated 13-hydroxyoctadecatrienoic acid production (Shimada et al., 2015). Additionally, 13-hydroperoxyoctadeca-trienoic acid can be used as a substrate by enzymes such as allene oxide cyclase, hydroperoxide lyase, and allene oxide synthase to generate a variety of oxylipins with anti-stress effects (Ul Hassan et al., 2015; Babenko et al., 2017). In line with this, it has been found that overexpressing a gene for 13-LOX in *Hordeum vulgare* reduces the fertility of green peach aphids when they feed on the resulting lines, indicating the potential for modifying LOXs for crop improvement (Losvik et al., 2017).

Similarly, in response to abiotic circumstance challenges and plant growth modulators application, the expression study of 12 cucumber *LOX* genes demonstrated their differential transcript in reply to multiple treatments, demonstrating their varied activities in hormone and abiotic stress responses (Yang et al., 2012). In another experiment, methyl jasmonates increased the synthesis of harmful proteins in maize and triggered plant defense against *Ostriniafurnacalis* (Zhang et al., 2015). By upregulating the enzymes 3-ketoacyl-CoA thiolase B, LOX 2, LOX 5, and omega 6-fatty acid desaturase, wounding stress increases the synthesis of JA in *Daucus carota* (Jacobo-Velázquez et al., 2015). Additionally, a variety of secondary metabolites produced by jasmonates, the end products of intricate metabolic routes with branches, actively contribute to plant's abiotic and biotic stress resistance.

Moreover, the interaction of oxylipins with other biologically active components of signaling pathways, particularly phytohormones, is now well understood. For instance, abscisic acid positively influenced the LOX activity in injured plants (Saha et al., 2021; Chen et al., 2022). Applications of abscisic acid have been shown to increase stress tolerance, LOX activity, JA level, peroxide oxidation of membrane phospholipids, and JA in damaged rice leaves (Roychoudhury et al., 2009). The abscisic acid levels and LOX transcripts in water-stressed plants were positively correlated (Deluc et al., 2009). JA, abscisic acid, and LOX activities are also increased by mechanical damage (Huan et al., 2022; Wei et al., 2022). Melatonin-pretreated oat seedlings, when compared to control seedlings under salt stress, showed increased expression levels of the mitogen-activated protein kinase genes *Asmap1* and *Aspk11*, lipid peroxidase genes *LOX* and *peroxygenase,* a chlorophyll biosynthesis gene, and genes encoding *NAC, MYB, WRKY1*, and *WRKY3* transcription factors except DREB2 (Gao et al., 2019). Likewise, the activity of lipoxygenase alcohol acyl transferase and alcohol dehydrogenase in *Pyrus ussuriensis* fruits was increased after ethylene treatment, leading to the improvement of aroma in cold-storage conditions. Gracilariopsis lemaneiformis Gllox1 and Gllox2 expression are also stimulated by high temperature, jasmonic acid, and salicylic acid treatment (Hou et al., 2018). Therefore, it is evident that JA, SA, and ethylene communicate with LOX to positively control a number of plant physio-biochemical pathways that elicit a response to stress.

11.5 LOXS UNDER CROP STORAGE

LOX has been found in different cellular sections that generate signaling molecules involved in different functions, such as propagation, proliferation, and development of seeds. JA and ABA production, fruit ripening, senescence, cell death, wounding, and response against abiotic and biotic stress factors. The vital roles of LOXs in overall plant growth, storage, and stress management in different crops are summarized in Table 11.1.

11.5.1 LOXS IN SEED GERMINATION AND GROWTH

LOXsin seeds serve as vegetative retention proteins (VSPs) and have a crucial function in seedling development, including maturation, senescence, pest resistance, reaction to injury, and transport of lipids during germination (Ehrhardt-Brocardo et al., 2022). Many new LOXs are induced during the early stages of seedling growth in various plant species, including pea, watermelon, french bean, lupine, finger millet, mung bean, carrot, lentil, barley, cucumber, *Arabidopsis*, wheat, sunflower, and *Brassica napus* (Singh et al., 2022). However, the three LOX isoforms present in mature seeds vanish early in the germination of seeds in soybean, resulting in lipid degradation through the LOX-independent process during seed germination (Singh et al., 2022; Viswanath et al., 2020). Proteome profiling studies of rice seedlings and soybean seedlings revealed that lipids were degraded via a LOX-dependent pathway and that LOX may have also assisted in the quick removal of reactive oxygen species (ROS) during the reserve recruitment during soybean seedling development (Mangrauthia et al., 2016; Han et al., 2013). After germination, a significant LOX L-4 developed in soybean cotyledons, converting LA into 9(S)-hydroperoxy-10(E), 12(Z)-octadecadienoic acid, and 13S-HPOD. The *OsLOX2*gene in rice affects seed lifetime during storage and germination (Huang et al., 2014). Inappropriate *OsLOX2* gene suppression can postpone the aging process throughout storage without disrupting regular germination. Recent reports show that transgenic plants for antisense suppression of *LOX3* gene expression in rice endosperm lead to a marked massive drop in LOX transcript levels and that 9-HPOD from LA (Xu et al., 2015).

11.5.2 LOX ON PHYTOHORMONES

Plant hormones JA, SA, ABA, and nitric oxide all express the *LOX* gene. LOX is up-regulated when maize plants are exposed to SA, MeJA, and ABA (Tayyab et al., 2020). Similarly, in mature *Arabidopsis* plants, ABA and MeJA upregulated the expression of *LOX1* gene. However, it was found that *LOX2* expression rises in the leaves and flowers, whereas LOX6 expression declines in the seeds, roots, and stems (Viswanath et al., 2020; Montillet et al., 2013). Phytohormones such as auxin, kinetin, epibrassinolide, gibberellin, and SA inhibited LOX activities most under experiments. However, auxin and gibberellin simultaneously improved the quantity of 9-*LOX* gene transcripts, but other phytohormones decreased the LOX activity (Lemeza et al., 2010). Similarly, the first enzymes that are involved in Jasmonic acid biosynthesis are 13-lipoxygenases. There are four 13-lipoxygenase genes in *Arabidopsis*. The 13-lipoxygenases LOX2, LOX3, and LOX4 catalyze the transformation of 13(S)-hydroperoxide from linolenic acid. Among these, LOX3 and LOX4 interact with WRKYs and have W-boxes [the DNA sequence (C/T)TGAC(T/C)] in their promoters (Jiao et al., 2022).

11.5.3 LOX IN SENESCENCE AND FRUIT RIPENING

Various research studies showed that an increase in LOX activity is a typical characteristic of cellular activities in senescence in plants. Senescence is characterized by increased LOX activity, which generates superoxide radicals. It results in a rise in levels of lipid peroxidation. LOX is the primary controller of lipid peroxidation and may be a factor in the membrane damage that occurs during senescence just before senescence. LOX activity was dramatically increased in the cytoplasm and microsomes of senescing bean cotyledons in iris versicolor flowers (Ahmad and Tahir, 2018). When aerial spraying of 2,4-dichlorophenoxyacetic acid (2,4-D) in immature and mature leaves of peas and 13-LOX in *Alstroemeria peruviana* floral tissues, enhanced accumulation of lipid hydroperoxides leads to electrolyte leakage and membrane semi-permeability loss, which ultimately cause senescence (Elufisan et al., 2023; Viswanath et al., 2020). Transgenic *Arabidopsis* expressing persimmon DkLOX3 revealed LOX's precise expression pattern and function during senescence (Huetal., 2015). The over-expression of DkLOX3 in *Arabidopsis* leaves during senescence led to more yellowing with chlorophyll loss, elevated membrane permeability for electrolyte leakage, higher MDA, superoxide radicals (O_2 and H_2O_2), and dead cell content compared to wild-type plants. Comparative proteomic investigation of maize showed that LOX levels increased during the beginning of leaf senescence (Wu et al., 2018). In addition to senescence, LOX is important for regulating quality and flavor in a variety of climacteric and non-climacteric fruits throughout the ripening process (Schiller et al., 2015). Similarly, hexanol and hexyl esters, which depend on hexanal production via LOX activity on fatty lipids, are released in greater amounts autonomously during apple ripening (Contreras et al., 2016). *PpLOX1* and *PpLOX4* transcription increased in peach plants, playing an important role in the generation of ethylene during the ripening process and producing volatile flavor compounds (Zhang et al., 2010). The silencing of TomloxB in tomato increased post-harvest life during fruit ripening (León-García et al., 2017). However, by applying exogenous chlorogenic acid to apple pulp, delayed ripening and senescence of the fruit pulp were observed to have lower levels of LOX activity (Xi et al., 2016).

11.5.4 LOX IN WOUND STRESS

After wounding, different free radicals – ROS, lipid-derived carbonyl, and malondialdehyde (MDA) – are produced via the LOX pathway, which exacerbates membrane damage, resulting in localized cell death typically associated with hypersensitive reaction (HR). In pea seedlings, LOX is shown to have a significant role in the oxidative burst caused by wounding. It was discovered that extracellular LOX may be crucial for fatty acid oxidation and that it also produces reactive electrophiles that control signaling reactions (Viswanath et al., 2020). Singlet oxygen ($^1O_2$) is created during wounding through the LOX process, which takes part in the oxidation of either lipids or proteins. Catechol, a LOX inhibitor, demonstrated this method of ($^1O_2$) production at the site of a wound in *Arabidopsis* leaves, resulting in ($^1O_2$) synthesis and proliferation being the only option for chloroplasts (Viswanath et al., 2020; Li and Kim, 2021). The LOX found in chloroplasts in most terrestrial plants include C6-aldehydes like (Z)-3-hexenal, (E)-2-hexenal, (E)-2-hexanol, and n-hexanal, which are implicated in the plant defensive response to wound stress (Liang et al., 2022). C6-aldehydes are typically present in small levels, but under herbicide or wounding damage circumstances, LOX triggers the fast production of these chemicals in plants, controlling plant defense by inducing responses to herbivore and wound-induced defense mechanisms. CsLOX2, 3, and 4 mediate the generation of C6-aldehydes and alcohols in the tea plant, while CsLOX6 and CsLOX7 synthesize JA (Zhu et al., 2018). However, runner bean has four *LOX* genes viz., PcLOXA, B, C, and D. Of these, PcLOXA, B, and D were stimulated by injury and take part in the development of JA production (Mazur et al., 2018).

11.5.5 LOXS UNDER ABIOTIC STRESSES

Few investigations have unraveled the significance of plant LOX during abiotic trauma through reduced H_2O_2 accumulation, lipid peroxidation, and elevated levels of RD20, RD29A, RD29B, P5CS, COR15A, and DREB2A expression. The *CaLOX1* plays a critical role in osmotic, high salinity, and water deficit stress responses (Gao et al., 2019). With an increase in germination rate, root growth, up-regulation of stress-responsive gene expression, including RD22, RD29A, RD29B, and NCED3, and a decrease in H_2O_2 and superoxide radical build-up, DkLOX3 exhibits greater resistance to osmotic, high salinity, and water deficiency stress (Li et al., 2022). Under drought and nickel stress, the total LOX activity concentrations in forage millet were markedly elevated, inducing the antioxidative enzymes and reducing the influences of these stresses (Kotapati et al., 2017). Kumar et al. (2022) found that LOX affected lipid peroxidation during heavy metal stress in *Arabidopsis*. Similarly, the pea under cadmium and selenium stress and barley seedlings treated with Cu^{2+} show enhanced LOX activity (Sakouhi et al., 2020; Ben et al., 2022).

11.5.6 ROLE OF LOX IN SEED VIGOR & VIABILITY, QUALITY OF SEED/FRUIT UNDER STORAGE

Several species of plants produce large amounts of LOXs in their seeds, like in soybean and other legumes, where there are a lot of proteins, and they may also serve as storage proteins, perhaps instead of only defense proteins (Porta and Sosa, 2002). Different oxylipin forms, including jasmonic acid, green leaf volatiles (GLVs), and the recently found death acid, are crucial for defense against infections and pests during storage. Both jasmonates and GLVs, which depend on 13-LOX, are essential for direct and indirect defenses, respectively (Viswanath et al., 2020). Hazelnut seeds, *Corylusavellana organoleptic* characteristics, and shelf life were found to be affected by LOXs (Santino et al., 2003). *Pennisetum glaucum* (Pearl millet) grows more quickly and is effectively protected against downy mildew disease as a result of high LOX expression induced by thiamine seed inoculation (Pushpalatha et al., 2012). As already discussed, LOXs help in various phases of crop growth, development, and harvesting. However, the generation of volatile components affects the flavor and scent of vegetable and fruit products. In freshly cut lettuce, the free radicals and hyperoxides so produced disrupt the cell membrane integrity by lipid peroxidation followed by phenolic substrate oxidation that leads to brown coloration (Ma et al., 2022). The discovery of two genes, MdLOX1a and MdLOX5e, linked to the production of fruit fragrance volatiles in apples suggested that a LOX action may be related to *Malus aphid* resilience responses (Vogt et al., 2013). Similarly, in melon, 13-LOX CmLOX18 produced C6 volatiles that enhanced fruit ripening during storage, indicating that LOX may change flavor (Zhang et al., 2017). But the storage quality of the crop is somehow directly or indirectly linked to LOX activity. It is the key enzyme in lipid metabolism. Since LOX activity includes the catalyzation of PUFA oxygenation with cis-1, cis, and 4-pentadiene to form conjugated double bonds consisting of polyunsaturated hydroperoxides (Zheng et al., 2022). The storage protein LOX can assist in the breakdown of the lipid bodies that have been accumulated. However, metabolites formed by lipid degradation, such as acetaldehyde and MDA, can cause cell damage and decline seed quality and longevity during storage. Therefore, loss of *LOX* genes can also enhance the quality of seeds (Wang et al., 2023). The *LOX* gene was removed from soybean seeds to reduce the off-flavoring of soy products (Kumar et al., 2022). Similar behavior was observed in young compared to the control showing low expression of SiLOX protein during storage (Ma et al., 2022). A study on wheat seeds revealed that higher *LOX* gene expressions shortened seeds' storage time period (Zhou et al., 2019). Rice seed storage was enhanced by down-regulating OsLOX by RNAi (Zhang et al., 2019). Therefore, it may be said that LOX can be an important storage protein for the defense of seeds, fruits, or vegetables, but the quality and flavor may not always be preserved.

11.5.7 ROLE OF LOXS AGAINST BIOTIC STRESS

Under stress, a plethora of molecular and cellular processes are activated in plants (Rejeb et al., 2014; Lamers et al., 2020), which in turn, causes a cross-wired web of physiological, morphological, and biochemical systems (Nejat and Mantri 2017; Saijo and Loo, 2020). In both abiotic as well as biotic environmental defense response challenges such as nematodes, fungi, bacteria, viruses, birds, insects, and rodents, LOXs are essential. However, healthy and developing plant tissues exhibit considerable LOX activity. LOXs also play a role in host-pathogen communications.

Plant biotic stressors include injury by herbivorous insects or animals and bacterial or fungal pathogen infection (Umar et al., 2021). Linoleic and linolenic acids, which are used as substrates by LOXs, are released from membrane phospholipids as a result of phospholipase-A being activated by Ca^{2+} and calmodulin. Subsequently, LOXs hydroperoxidize fatty acids and activate protein kinases that cause the expression of defense-related genes to be activated. A summary of the different roles of LOXs in biotic stress mitigation is presented in Table 11.2.

TABLE 11.1 Application of LOXs for Crop Storage and Stress Management

Plant species	Gene/Enzyme	Crop part	Function or Effect	References
Potato	LOX	Tuber	Improve tuber defense	Isayenka et al., 2022
Rice	*OsRCI-1*	Seedlings	Enhance the growth and the defensive system of plant	Liao et al., 2022
Peach	*PpLOX1*	Fruit	Maintain quality, usable-life of stored fruit	Yang et al., 2020
Rice	*$OsLOX_2$*	Seedling	Safe storage of rice	Huang et al., 2014
Avocado	LOX	Fruit	Prevent stem end rot and anthracnose	Obianom et al., 2019
Rice	*LOX3*	Grains	Enhanced storage tolerance	Ma et al., 2015
Damask rose	LOX	Flowers	Preserves aroma and quality of flowers	Ali et al., 2022
Agaricus bisporus	LOX	Fruit	Prevent deterioration	Zhou et al., 2022

TABLE 11.1 *(Continued)*

Plant species	Gene/Enzyme	Crop part	Function or Effect	References
Maize	*LOX10, LOX2*	Grains	Prevent quality deterioration	Dutta et al., 2019
Apples	LOX	Fruit	Improve color, and quality and reduce storage disorders	Sharma et al., 2020
Pointed guard	LOX	Fruit	Improved shelf life and quality	Siddiqui et al., 2021
Rose apple	LOX	Fruit	Prevent chilling injury	Duong et al., 2022
Wheat	LOX	Grains	Prevents lutein and lutein ester degradation	Mares et al., 2021
Rice	*LOX1, LOX2, LOX3*	Grains	Long-term safe storage	Cai et al., 2022
Sweet potato	*IbLOX1*	Tuber roots	Better storability of tubers	Tang et al., 2019
Pears	*PcLOX*	Fruit	Delay senescence	Li et al., 2020
Tomato	*lLOXD*	Fruit	Resistance toward fungal pathogen	Fu et al., 2022
Cucumber	*CsLOX*	Fruit	Regulates food aroma during storage	Yang et al., 2020
Peaches	*PpLOX*	Fruit	Prevent chilling injury	Chen et al., 2019
Peanut	*LOX3*	Seeds	Seed life prolonged with respect to temperature, moisture, and humidity	Zhang et al., 2020
Sapota	LOX	Fruit	Prevent chilling injury	Mirshekari et al., 2020
Longan	*DlLOX1,DlLOX2*	Fruit	Prevent pulp breakdown	Liu et al., 2022
Kiwi fruit	*GLOX*, LOX6	Fruit	Long term storage	Choi et al., 2022
Banana	LOX	Fruit	Helps in ripening and maintaining the aroma and quality of fruit	Zhu et al., 2020
Litchi	LOX	Fruit	Delays the browning of the pericarp	Wang et al., 2020
Goji berry	LOX	Fruits	Regulate healthy storage of fruit	Wang et al.
Citrus fruit	LOX	Fruit	Prevent fungal decay	Chen et al., 2019
Peach	*PpaLOX2.1, PpaLOX7.1, PpaLOX7.2, PpaLOX2.2*	Fruit	Fruit ripening while in the preservation process	Guo et al., 2017
Sweet cherry	*PavLOX*	Fruit	Prevent fungal growth	Pan et al., 2022
Foxtail millet	LOX	Seedling	Protection against saltiness and water deficit stress	Zhang et al., 2021
Lemon	LOX	Fruit	Prevent quality and injury during cold storage	Liao et al., 2022
Wheat	LOX	Grains	Processes storage quality of grains	Zheng et al., 2022
Carrot	LOX5, *LOX2*	Roots	Abiotic stress management	Surjadinata et al., 2021

11.5.8 LOX AGAINST PLANT PATHOGENS

The 9-LOX pathway plays a fundamental role in providing resistance against pathogen attacks, such as those caused by pathogens like bacteria and fungi (Park et al., 2010; Vellosillo et al., 2013; Kaya et al., 2023). In *A. thaliana*, the 9-LOX pathway modifies oxidative stressors, plant defense, and lipid peroxidation (Lopez et al., 2011). In *A. thaliana*, pepper-derived CaLOX1 over-expression led to greater resistance against *Alternaria brassicicola, Hyaloperonospora arabidopsidis,* and *Pseudomonas syringae*. Likewise, pathogen onslaught increases CaLOX1 expression even in pepper leaves (Hwang and Hwang, 2010). In maize, the 10-OPEA with 12 and 14-C cyclopentane (a)nones (together known as "death acids") have a critical role in increasing JA biosynthesis when *Cochliobolus heterostrophus* is present (Christensen et al., 2015). This infection elevates the production of genes related to defense activities, the activity of phytoalexins, and the potential for cytotoxicity eventually causing cell death.

Maize metamerically duplicated 9-*LOX* genes – *ZmLOX4* and *ZmLOX5*, were activated by the fungal attack, i.e., *Fusarium verticillioides* and *Cochliobolus carbonum*, and associated with a unique resistance mechanism (Park et al., 2010). The role of the 9-LOX route in the defense reaction to pathogenic warfare was examined in *Arabidopsis* against coconut fruit in kernel tissue and water during cold storage by reducing LOX activity (Meethaworn et al., 2019). LOX also degraded the carotenoid content during storage in foxtail millet, where the variety of plants having high LOX activity discolored the millet as *Pseudomonas syringae* and revealed that the 9-LOX causes improvements to the cell wall to prevent pathogen entrance (Kaur et al., 2014).

Additionally, *A. thaliana* roots have been found to have greater 9-LOX activity, where the enzyme's products modify the structure of the roots (Vellosillo et al., 2007). The 9-LOX oxylipins are involved in defense activities against *Phytophthora infestans* in early stages of potatoes (Vishwanath et al., 2020). Upon injection of *Aspergillus flavus* and *Fusarium verticillioides*, the *Spodoptera exigua* larvae feeding on Zea mays increase the 9-LOX expression to a slightly higher degree than 13-LOX or LOX3 (Woldemariam et al., 2018). Notably, Magnaporthe grisea infection increased LOX3 activity in rice leaves. By interacting with 13-LOX, the Osr9-LOX1 contributes to the pathway that offers resistance to herbivores that pierce and eat through their mouths. The 9-hydro(peroxy)fatty acids (HFAs) are produced by the interaction of almond and *Aspergillus carbonarius*, 9-LOX, and are then further converted into the C9 aldehydes HPL-branch route. When

TABLE 11.2 Role of Lipoxygenase Enzyme in Biotic Stress Mitigation in Crops

Tested Crop Species	LOX Gene or Enzyme	Biotic Stressor	Reported Effect	References
Pea	LOX	Nematode *(Heteroderagoettingiana)*	Upregulated the nematode resistance via genetic variations	Leone et al., 2001
Tobacco	9-LOX	Fungi (*Botrytis cinerea*)	LOX enhanced the susceptibility of the crop against the necrotrophic effect of susceptible pathogen	Seo et al., 2001
Tobacco	*LOX-1A*	Fungi *(Phytophthora parasiticanicotianae)*	The survival rate of roots and seedlings inoculated was increased inferencing that overexpression of Lox enzyme enhanced the susceptibility of the crop to fight the stress.	Mène-Saffrané et al., 2003
Pea	*LOXN2*	Nematode(*Heteroderagoettingiana*)	Mutants show tolerance to nematode	Veronico et al., 2006
Rice	9- LOX encoding – *OsMLOX1*	Insect *(Nilaparvatalugens*)	Higher levels of OsLOX1 induced JA production that induced resistance to herbivore in the plants rice.	Wang et al., 2008
A. thaliana	Pepper-9- LOX encoding– (*CaLOX1*)	Bacteria *(P. syringae, X. campestris) and* Fungi *(Colletotrichum spp., Alternaria brassicicola)*	Enhanced the phenotype modulation and stress improvement in the plants	Hwang and Hwang 2010
Passion fruit	LOXs	Insect (*Spodoptera frugiperda, Agraulisvanillae vanilla*)	LOX activity induced the wound and MeJA enzymes for plant growth	Jardim et al., 2010
Maize	*ZmLOX10*	Fungi *(Aspergillus flavus, Colletotrichum graminicola)* and Insects	It may promote aflatoxin buildup and induce anthracnose leaf blight and stalk rot resistance.	Christense 2011
A. thaliana	*LOX5*	Insect *(*Green peach aphid-*Myzuspersicae*sp.)	The infestation of aphids induced the expression of LOX5 and other oxylipins to combat the stress caused by aphids.	Louis et al., 2013

TABLE 11.2 *(Continued)*

Tested Crop Species	LOX Gene or Enzyme	Biotic Stressor	Reported Effect	References
Tomato	13-LOX encoding – *TomloxD*	Fungi *(Cladosporium fulvum)*	Upregulated the JA production and provided resistance against the thermal and pathogenic stress.	Hu et al., 2013
Tomato	13-LOX encoding *TomLoxD*	Insect and wound	*Spr8/TomLoxD* increased wound-induced JA synthesis, and wound-responsive gene expression and improved the tolerance to insect herbivory and necrotrophic pathogen infection.	Yan et al., 2013
Rice	*OsHI-LOX*	Insect (*Chilosuppressalis*)	Provided resistance against the chewing stress of the herbivore	Wang et al., 2013
Rice	9-LOX encoding-*Osr9LOX1*	Insect (Striped stem borer– *Chilosuppressalis*)	Regulated the JA burst and the interaction between the JA and SA	Zhou et al., 2014
A. thaliana	13-LOX encoding – *LOX4* and *LOX3*	Nematode (*Meloidogyne javanica, Heteroderaschachtii*)	The induction of the LOXs gene in seedlings led to a remarkable lessening of nematode galls and mass number of eggs and also upregulated the expression of antioxidant enzymes along with JA.	Ozalvo et al., 2014
Maize	9-LOX encoding – *ZmLOX3*	Nematode *(Meloidogyne incognita*)	Elevated JA and, thus, increased susceptibility towards root-knot nematode	Ozalvo et al., 2014
A. thaliana	*LOX3* and *LOX4*	Insect (*Spodoptera littoralis*)	Role in phytohormone synthesis that helps in combating the stress	Chauvin et al., 2016
Barley	*LOX 2.2*	Insects (Rhopalosiphumpadi, *Myzuspersicae)*	*LOX2.2* is involved in fundamental defense responses and contribution to the activation of JA-mediated reflexes.	Losvik 2017

TABLE 11.2 *(Continued)*

Tested Crop Species	LOX Gene or Enzyme	Biotic Stressor	Reported Effect	References
A. thaliana	*9-LOX encoding – DkLOX3*	Bacteria (*P.syringae*) and Fungi (*Botrytis cinerea*)	It enhanced the behavior of antioxidative enzymes and regulated the stress response	Hou et al., 2018
Maize	*LOX4–9*	Insect (Beet armyworm – *Spodoptera exigua*)	Gene expression modifications caused by caterpillars and enhanced survival of S. exigua on lox4 mutants.	Woldemariam et al., 2018
Melon	*CmLOX09*	Fungi *(Podosphaeraxanthii)*	It modulated the signaling, hormonal, and stress for the plants	Ju et al., 2018
Sweet Pepper	*CaLOX2*	Insect *(Frankliniella occidentalis*)	Defense against thrips	Sarde 2019
Maize	*LOX3*	Fungi *(Ustilago maydis*)	Developed the resistance against the smut disease caused by fungi via altering the genes	Pathi et al., 2020
Maize	9-LOX encoding – *ZmLOX5*	Fungi *(Fusarium graminearum)*	Oxylipin with JA production caused resistance to *Gibberella* stalk rot disease caused by fungi.	Wang et al., 2021
Sorghum	13-LOX encoding– *ZmLOX12*	Fungi *(Fusarium verticillioides*)	Enhanced resistance and defense response against the aphids.	Shrestha et al., 2021
Sorghum	9-LOXs encoding *ZmLOX1*, *ZmLOX3*, *ZmLOX4*, and *ZMLOX5*	Insect (Corn leaf aphid- *Rhopalosiphummaidis*)	Enhanced resistance and defense response against the aphids	Shrestha et al., 2021
Sorghum	*ZmLOX10*	Insect (*Spodoptera exigua*)	Enhanced resistance and defense response in crops against aphids.	Shrestha et al., 2021
Tomato	9-LOX	Nematode *(Meloidogyne incognita)*	The LOXs enzyme specially provided the chemicals that enhanced the resistance against the nematodes converting 9-hydroperoxides to 9-divinyl ethers that helped in the nematode.	Sanadhya et al., 2021

TABLE 11.2 *(Continued)*

Tested Crop Species	LOX Gene or Enzyme	Biotic Stressor	Reported Effect	References
Pepper	*LOX1*	Fungi (*Colletotrichum coccodes*)	9-Lox-derived oxylipins helped in preventing infection in solanaceous plants.	Sanadhya et al., 2021
Rice	13-LOX encoding – *OsRCI-1*	Insect (Brown plant hopper – *Nilaparvatalugens*)	BPHs had lesser colonization, fecundity, and biomass.	Liao et al., 2022
Maize	*ZmLOX3*	Fungi *(Ustilago maydis)*	The ROS burst caused by fungi was suppressed by the mutant plants via the protein Rip1 that bounds to lipoxygenase.	Saado et al., 2022
Rice and Wheat	*MnLOXandFeLOX*	Fungi (*Fusarium graminearum, Magnaportheoryzae, Geaumannomyces graminis)*	Iron and Manganese bound lipoxygenases helped in the natural synthesis of JA.	Oliw 2022
Maize	*Lox 9, ZmLOX4*	Fungi (*Fusarium verticillioides,Aspergillus flavus)*	Development of resistance to mycotoxins and, also limiting the buildup of accumulation of pathogens	Guche et al., 2022
A. thaliana	9-LOX encoding – *LOX3*	Nematode (*Meloidogyne incognita*)	Upregulated the expression of JA for the management of nematode stress	Wiśniewska et al., 2022

Sclerospora graminicola and downy mildew interact with pearl millet, the level of LOX activity increases and is associated with the degree of host resistance. Similarly, *Aspergillus flavus* and *Aspergillus parasitic* infections in peanuts cause LOX, and as a result, LOX feedback is used to create a variety of antimicrobial compounds (Song et al., 2013).

Two plastidial 13-*LOX* genes in maize, i.e., *ZmLOX6* and *ZmLOX10*, are activated by the fungus *Cochliobolus carbonum* when there are favorable contacts (Nemchenko et al., 2006; Pokotylo et al., 2015). Papaya seedlings showed anti-fungal activity by producing hydroperoxides from 13-LOX that inhibited the growth of the mycelium and sporangia in *Phytophthora palmivora* (Sujatha et al., 2012). As a result of wounding, late blight, and other pathogen infestations, several more isoforms of LOX were discovered in potatoes (Vishwanath et al., 2020). By preventing ROS formation and cell apoptosis in *A. thaliana,* overexpression of DkLOX3 displayed high resistance to *Botrytis cinerea* and *Pseudomonas syringae* (Hou et al., 2018). The LOX1 gene in guard cells was found to have a greater impact on the stomatal defense response than ABA (Montillet et al., 2013).

Mutations in LOX1 have been shown to decrease the stomatal closure process in tomatoes, increasing their susceptibility to *Pseudomonas syringae* (Vishwanath et al., 2020). Only the cyanobacteria Anabaena/Nostoc sp. PCC 7120, Cyanothece sp. PCC8801, Nostoc punctiforme PCC 73102, and *Acaryochloris marina* were shown to produce LOXs. These bacteria are Gram-negative. *Pseudomonas aeruginosa* and *Myxococcus xanthus,* Gram-positive Thermoactinomycetes. By altering the lipid signaling seen between the host and the invader during host airway epithelial colonization, studies demonstrated the biological involvement of LOX in biofilm formation. The oxidation of membrane lipids by LOX during infections can make it a virulence factor (hemolysis). The efficient expression of LOXs by the microbial hosts also contributes to the advancement of novel biocatalytic pathways for the generation of green leaf volatiles (Stolterfoht et al., 2019). Milk fermented by a *Lactococcus lactis* strain generating 15-LOX-1 was proven to possess both helpful and aggressive properties, i.e., it lowered inflammation-related responses and boosted immune function (Saraiva et al., 2015). In *A. thaliana*, actions of LOX3 and LOX4 to *Heterodera schachtii* and *Meloidogyne javanica* were noted. They behave differently and have unique metabolic and signaling mechanisms. An essential function of LOX4 is to direct plant defense against nematode invasion (Ozalvo et al., 2014). Even the administration of the pea seed-borne mosaic virus (PSbMV) also results in a rise in LOX activity (Cerna et al., 2017).

11.6 INSECT AND HERBIVOROUS PESTS

Higher plants defend themselves by triggering the production of genes involved in deterring herbivores, repairing wounds, and other defensive mechanisms during insect herbivory stress (Ryan et al., 2000; Walling et al., 2000; Li et al., 2002; Howe et al., 2008; Wu et al., 2010; Dicke et al., 2010; Sun et al., 2011). A panoramic spectrum of metabolic changes in plants is brought on by insect herbivory. While some induced metabolites act as poisons or repellents to herbivores, other induced metabolites behave as signaling molecules to engage defense mechanisms in plant sections that haven't yet been assaulted (War et al., 2018). Based on studies on tomato, Green and Ryan (1972) postulated that particular signals produced at the wound site go through the plant and switch on the production of various metabolites such as proteinase inhibitors and additional genes involved in defense distantly anticipating leaves in their seminal investigation of wounds. Such accumulation of strong inhibitors of animal's proteinases is definitely a defensive response of plants in their wound-healing property of leaves. Plants begin to release metabolites with harmful and preventative properties in response to pathogen attacks, while other metabolites serve as signaling molecules to activate defense-related reactions.

The LOX pathway, which is primarily used by oxylipins like 10-OPDA (a positional isomer of 12-OPDA), 12-OPDA through the action of 13-LOXs, and 10-oxo-11-phytoenoic acid (10-OPEA) through the action of 9-LOXs, serves an essential function in defense systems for herbivores that bite and chew vegetation (Pokotylo et al., 2015). Herbivory causes an anti-herbivore oxidative shift, damaging herbivores directly or indirectly and sending signals for controlling plant defense mechanisms (Kaur et al., 2014). While *Toxoptera aurantii* (tea aphid) feeding caused a pattern of induction and repression in the tea plant, feeding by *Empoasca vitis* (tea green leafhopper) caused induction of CsLOX1 to occur quickly (Liu et al., 2010).

Some insects and bacteria have genetic modifications that allow them to resist or abide by the inherent as well as triggered defenses of plants. These microbes and insect infestations can then establish and expand by consuming nutrients from plants, which cause disease and harm to the host plant. The performance of constitutive defenses and specialized defenses used against induced defenses greatly influences how plants and pathogens interact. Signal cross-talk in plants affected by diseases and insects amount of lipoxygenase mRNA is measured. Based on the type of biologic inducer (insect vs. disease), studies of LOX reveal that their induction in some plants does not adhere to a precise pattern, but each treatment elicits a unique pattern of gene expression.

Certain flora answered the pest attack through the LOX pathway by boosting LOX function and suppressing protease activity in their leaflets. Jasmonates (JAs) are a group of oxylipin regulatory proteins having overlapping functions in controlling plant development and stress responses. Defense against microbial infections, defense against insects, and reactions to stressful situations include Ultraviolet rays, ozone, and other abiotic stressors that rely on JA signaling (Glazebrook 2005; Wasternack 2007; Balbi and Devoto 2008; Howe and Jander, 2008). This critical requirement of JA signaling for induction of plant protection from insect pests was founded using the explorations into JA mutations of tomato plants (Li et al., 2005). Ribosome-inactivating proteins (RIPs) characteristics contribute to defense through antifungal, antibacterial, antiviral, and insecticidal actions. Different RIPs have been identified to exhibit antimicrobial activities: antitumor, antibacterial, antifungal, and broad-spectrum antiviral activities (Puri et al., 2009; Bian et al., 2010).

11.7 CONCLUSION

This chapter provides the economics of agricultural production globally and numerous biotic and abiotic conflicts influencing crop losses. Plants (or crops) react to infections caused by pathogens by releasing metabolites, some of which operate as signaling molecules to trigger defense responses and many others which have unfavorable or preventative effects. The various biological controls discussed, among which catalytic lipoxygenase is highly responsible for the alleviation of biotic stress and crop storage tolerance. Lipoxygenase is a dioxygenase enzyme in plants, animals, and fungi that helps hydroperoxide formation from polyunsaturated fatty acid. The oxylipin diversity and pathways are majorly responsible for the LOXs activity. The LOX enzyme confers tolerance against various biotic and abiotic stress conditions. Our chapter provides insight into the functioning of many LOXs with a special focus on storage and against biotic stress like pathogen attacks, insects, weeds, and herbivores. Potential applications of LOX in seed storage and feasibility, fruit ripening and usability, phytohormones availability for *LOX* gene, seed germination, and positive growth are majorly responsible for its effectiveness. Thus, in order to prevent the crops from biotic and abiotic stress, LOXs play an important role in mitigation to a larger extent which ultimately translates into better yield and future availability. This analysis provides up-to-date details on the plant LOXs. We also discussed the functional characteristics of the LOXs that aid in overall crop storage management, enabling it for longer shelf life. Therefore, because of their crucial function in plant physiology, LOXs produced by plants require additional consideration. The discovery of new attributes involved in LOX synthesis, usage, and administration will provide fresh insight into the world of crop management.

KEYWORDS

- **biotic stress,**
- **crop seed storage**
- **lipoxygenase**
- **seed viability**
- **senescence**

REFERENCES

Abass, A. B., Ndunguru, G., Mamiro, P., Alenkhe, B., Mlingi, N., & Bekunda, M. (2014). Post-harvest food losses in a maize-based farming system of semi-arid savannah area of Tanzania. *Journal of Stored Products Research, 57*, 49–57. https://doi.org/10.1016/j.jspr.2013.12.004.

Ahmad, S. S., & Tahir, I. (2018). Putrescine and jasmonates outplay conventional growth regulators in improving postharvest performance of *Iris germanica* L. cut scapes. *Proceedings of the National Academy of Sciences India Section B–Biological Sciences, 88*, 391–402. https://doi.org/10.1007/s40011-016-0767-2.

Ali, E. F., Issa, A. A., Al-Yasi, H. M., Hessini, K., & Hassan, F. A. (2022). The efficacies of 1-methylcyclopropene and chitosan nanoparticles in preserving the postharvest quality of damask rose and their underlying biochemical and physiological mechanisms. *Biology, 11*(2), 242. https://doi.org/10.3390/biology11020242.

Andre, E. M. I. L. E., & Hou, K. W. (1932). The presence of a lipoid oxidase in soybean, *Glycine soya* Lieb. *Comptes Rendus d'Academie des Sciences, 194*, 645–647.

Andreou, A., & Feussner, I. (2009). Lipoxygenases–Structure and reaction mechanism. *Phytochemistry, 70*, 1504–1510. https://doi.org/10.1016/j.phytochem.2009.05.008.

Arif, Y., Sami, F., Siddiqui, H., Bajguz, A., & Hayat, S. (2020). Salicylic acid in relation to other phytohormones in plant: A study towards physiology and signal transduction under challenging environment. *Environmental and Experimental Botany, 175*, 104040. https://doi.org/10.1016/j.envexpbot.2020.104040.

Babenko, L. M., Shcherbatiuk, M. M., Skaterna, T. D., & Kosakivska, I. V. (2017). Lipoxygenases and their metabolites in formation of plant stress tolerance. *Ukrainian Biochemical Journal, 89*, 5–21. https://doi.org/10.15407/ubj89.01.005.

Babenko, L. M., Kosakivska, I. V., & Skaterna, T. D. (2015). Jasmonic acid: Role in biotechnology and the regulation of plants biochemical processes. *Biotechnologia Acta, 8*(2), 36–51. https://doi.org/10.15407/biotech8.02.036.

Balbi, V., & Devoto, A. (2008). Jasmonate signaling network in *Arabidopsis thaliana*: Crucial regulatory nodes and new physiological scenarios. *New Phytologist, 177*, 301–318. https://doi.org/10.1111/j.1469-8137.2007.02292.x.

Banothu, V., & Uma, A. (2022). Effect of biotic and abiotic stresses on plant metabolic pathways. In F. A. Badria (Ed.), *Phenolic Compounds—Chemistry, Synthesis, Diversity, Non-Conventional Industrial Pharmaceutical and Therapeutic Applications* (pp. 119–130). IntechOpen. https://www.doi.org/10.54574/RJPP.14.05.

Barbaś, P., Skiba, D., Pszczółkowski, P., & Sawicka, B. (2022). Mechanisms of plant natural immunity and the role of selected oxylipins as molecular mediators in plant protection. *Agronomy, 12*(11), 2619. https://doi.org/10.3390/agronomy12112619.

Barbu, L. D. N., & Boiu-Sicuia, O. A. (2021). Plant-beneficial microbial inoculants and their formulation–A review. *Romanian Journal of Plant Protection, 14*, 32–43. http://www.doi.org/10.54574/RJPP.14.05.

Bardin, M., Ajouz, S., Comby, M., Lopez-Ferber, M., Graillot, B., Siegwart, M., & Nicot, P. C. (2015). Is the efficacy of biological control against plant diseases likely to be more durable than that of chemical pesticides? *Frontiers in Plant Science, 6*, 566. https://doi.org/10.3389/fpls.2015.00566.

Baysal, T., & Demirdöven, A. (2007). Lipoxygenase in fruits and vegetables: A review. *Enzyme and Microbial Technology, 40*, 491–496. https://doi.org/10.1016/j.enzmictec.2006.11.025.

Ben Massoud, M., Kharbech, O., Mahjoubi, Y., Chaoui, A., & Wingler, A. (2022). Effect of exogenous treatment with nitric oxide (NO) on redox homeostasis in barley seedlings (*Hordeum vulgare* L.) under copper stress. *Journal of Soil Science and Plant Nutrition, 22*, 1604–1617. https://doi.org/10.1007/s42729-021-00757-w.

Ben Rejeb, I., Pastor, V., & Mauch-Mani, B. (2014). Plant responses to simultaneous biotic and abiotic stress: Molecular mechanisms. *Plants, 3*, 458–475. https://doi.org/10.3390/plants3040458.

Bian, X., Shen, F., Chen, Y., Wang, B., Deng, M., & Meng, Y. (2010). PEGylation of alpha-momorcharin: Synthesis and characterization of novel anti-tumor conjugates with therapeutic potential. *Biotechnology Letters, 32*, 883–890. https://doi.org/10.1007/s10529-010-0242-8.

Bittner, A., Cieśla, A., Gruden, K., Lukan, T., Mahmud, S., Teige, M., Vothknecht, U. C., & Wurzinger, B. (2022). Organelles and phytohormones: A network of interactions in plant stress responses. *Journal of Experimental Botany, 73*(22), 7165–7181. https://doi.org/10.1093/jxb/erac384.

Blée, E. (2002). Impact of phyto-oxylipins in plant defense. *Trends in Plant Science, 7*, 315–322. https://doi.org/10.1016/S1360-1385(02)02290-2.

Boxall, R. A. (2001). Post-harvest losses to insects—A world overview. *International Biodeterioration & Biodegradation, 48*, 137–152. https://doi.org/10.1016/S0964-8305(01)00076-2.

Brash, A. R. (1999). Lipoxygenases: Occurrence, functions, catalysis, and acquisition of substrate. *Journal of Biological Chemistry, 274*, 23679–23682. https://doi.org/10.1074/jbc.274.34.23679.

Cai, C., Zhao, Z., Zhang, Y., Li, M., Li, L., Cheng, P., & Shen, W. (2022). Molecular hydrogen improves rice storage quality via alleviating lipid deterioration and maintaining nutritional values. *Plants, 11*(9), 2588. https://doi.org/10.3390/plants11192588.

Cerna, H., Černý, M., Habánová, H., Šafářová, D., Abushamsiya, K., Navrátil, M., & Brzobohatý, B. (2017). Proteomics offers insight to the mechanism behind *Pisum sativum* L. response to pea seed-borne mosaic virus (PSbMV). *Journal of Proteomics, 153*, 78–88. https://doi.org/10.1016/j.jprot.2016.05.018.

Chauvin, A., Lenglet, A., Wolfender, J. L., & Farmer, E. E. (2016). Paired hierarchical organization of 13-lipoxygenases in *Arabidopsis*. *Plants, 5*(2), 16. https://doi.org/10.3390/plants5020016.

Chen, C., Cai, N., Chen, J., & Wan, C. (2019). Clove essential oil as an alternative approach to control postharvest blue mold caused by *Penicillium italicum* in citrus fruit. *Biomolecules, 9*(5), 197. https://doi.org/10.3390/biom9050197.

Chen, M., Guo, H., Chen, S., Li, T., Li, M., Rashid, A., Xu, C., & Wang, K. (2019). Methyl jasmonate promotes phospholipid remodeling and jasmonic acid signaling to alleviate chilling injury in peach fruit. *Journal of Agricultural and Food Chemistry, 67*, 9958–9966. https://doi.org/10.1021/acs.jafc.9b03853.

Chen, X., Wang, D., Zhang, C., Wang, X., Yang, K., Wang, Y., Yang, X., & You, C. (2022). The apple lipoxygenase MdLOX3 regulates salt tolerance and ABA sensitivity. *Horticulturae, 8*(7), 651. https://doi.org/10.3390/horticulturae8070651.

Chen, Z., Chen, X., Yan, H., Li, W., Li, Y., Cai, R., & Xiang, Y. (2015). The lipoxygenase gene family in poplar: Identification, classification, and expression in response to MeJA treatment. *PLOS ONE, 10*(4), e0125526. https://doi.org/10.1371/journal.pone.0125526.

Choi, H. R., Baek, M. W., Tilahun, S., & Jeong, C. S. (2022). Long-term cold storage affects metabolites, antioxidant activities, and ripening and stress-related genes of kiwifruit cultivars. *Postharvest Biology and Technology, 189*, 111912. https://doi.org/10.1016/j.postharvbio.2022.111912.

Christensen, S. A. (2011). The function of the lipoxygenase ZmLOX10 in maize interactions with insects and pathogens. *Doctoral dissertation, Texas A&M University*. https://oaktrust.library.tamu.edu/bitstream/handle/1969.1/ETD-TAMU-2009-12-7459/CHRISTENSEN-DISSERTATION.pdf?sequence=3&isAllowed=y.

Christensen, S. A., Huffaker, A., Kaplan, F., Sims, J., Ziemann, S., Doehlemann, G., & Schmelz, E. A. (2015). Maize death acids, 9-lipoxygenase–derived cyclopente (a) nones, display activity as cytotoxic phytoalexins and transcriptional mediators. *Proceedings of the National Academy of Sciences, 112*(36), 11407–11412. https://doi.org/10.1073/pnas.1511131112.

Cole, B. K., Kuhn, N. S., Green-Mitchell, S. M., Leone, K. A., Raab, R. M., Nadler, J. L., & Chakrabarti, S. K. (2012). 12/15-Lipoxygenase signaling in the endoplasmic reticulum stress response. *American Journal of Physiology-Endocrinology and Metabolism, 302*(6), E654–E665. https://doi.org/10.1152/ajpendo.00373.2011.

Contreras, C., Tjellström, H., & Beaudry, R. M. (2016). Relationships between free and esterified fatty acids and LOX-derived volatiles during ripening in apple. *Postharvest Biology and Technology, 112*, 105–113. https://doi.org/10.1016/j.postharvbio.2015.10.009.

Craig, F. N. (1936). A fat oxidation system in *Lupinus albus*. *Journal of Biological Chemistry, 114*(3), 727–746. https://doi.org/10.1016/S0021-9258(18)74802-3.

Crozier, A. (2000). Biosynthesis of hormones and elicitor molecules. *Biochemistry and Molecular Biology of Plants, 46*, 1674–1681. https://cir.nii.ac.jp/crid/1574231875137724032.

Deluc, L. G., Quilici, D. R., Decendit, A., Grimplet, J., Wheatley, M. D., Schlauch, K. A., & Cramer, G. R. (2009). Water deficit alters differentially metabolic pathways affecting important flavor and quality traits in grape berries of cabernet sauvignon and chardonnay. *BMC Genomics, 10*, 212. https://doi.org/10.1186/1471-2164-10-212.

Dicke, M., & Baldwin, I. T. (2010). The evolutionary context for herbivore-induced plant volatiles: Beyond the 'cry for help.' *Trends in Plant Science, 15*(3), 167–175. https://doi.org/10.1016/j.tplants.2009.12.002.

Dubey, N. K., Singh, P., Prakash, B., & Mishra, P. K. (2012). Plant chemicals in post-harvest technology for management of fungal, mycotoxin and insect contamination of food commodities. In J. Mérillon & K. Ramawat (Eds.), *Plant* Defense*: Biological Control* (Progress in Biological Control, vol. 12, pp. 195–208). Springer. https://doi.org/10.1007/978-94-007-1933-0_11.

Duong, N. T. C., Uthairatanakij, A., Laohakunjit, N., Jitareerat, P., & Kaisangsri, N. (2022). An innovative single step of cross-linked alginate-based edible coating for maintaining postharvest quality and reducing chilling injury in rose apple cv. 'Tabtimchan' (*Syzygium samarangense*). *Scientia Horticulturae, 292*, 110648. https://doi.org/10.1016/j.scienta.2021.110648.

Dutta, S., Muthusamy, V., Chhabra, R., Zunjare, R. U., & Hossain, F. (2019). Characterization of diverse provitamin: A rich inbreds for lipoxygenase (LOXs) gene governing oxidative deterioration in maize. *International Journal of Chemical Studies, 7*, 841–846. Retrieved from: https://www.researchgate.net/profile/Suman-Dutta-7/publication/356910320 (accessed on 9 July 2024).

Ehrhardt-Brocardo, N. C. M., Coelho, C. M. M., & Souza, C. A. (2022). Storage protein composition during germination and its association with physiological seed quality in common bean. *Acta Scientiarum. Agronomy, 44*. https://doi.org/10.4025/actasciagron.v44i1.53434.

El Hadi, M. A. M., Zhang, F. J., Wu, F. F., Zhou, C. H., & Tao, J. (2013). Advances in fruit aroma volatile research. *Molecules, 18*, 8200–8229. https://doi.org/10.3390/molecules18078200.

El-Mogy, M. M., Ali, M. R., Darwish, O. S., & Rogers, H. J. (2019). Impact of salicylic acid, abscisic acid, and methyl jasmonate on postharvest quality and bioactive compounds of cultivated strawberry fruit. *Journal of Berry Research, 9*, 333–348. https://doi.org/10.3233/JBR-180349.

Elufisan, T. O., Oyedara, O. O., & Oluyide, O. O. (2023). Transcriptional regulation of phyto-oxylipins signaling pathway. In S. M. Shafi, C. Egbuna, & C. O. Adetunji (Eds.), *Phyto-Oxylipins: Metabolism, Physiological Roles, and Profiling Techniques* (pp. 22–41). CRC Press.

Farhangi-Abriz, S., Alaee, T., & Tavasolee, A. (2019). Salicylic acid but not jasmonic acid improved canola root response to salinity stress. *Rhizosphere, 9*, 69–71. https://doi.org/10.1016/j.rhisph.2018.11.009.

Farooq, M., Wahid, A., Kobayashi, N., Fujita, D., & Basra, S. M. A. (2009). Plant Drought Stress: Effects, Mechanisms and Management. In E. Lichtfouse, M. Navarrete, P. Debaeke, V. Souchère, & C. Alberola (Eds.), *Sustainable Agriculture* (pp. 153–188). Springer. https://doi.org/10.1007/978-90-481-2666-8_12.

Fox, T., & Fimeche, C. (2013). Global food waste not, want not. Institute of Mechanical Engineers. England and Wales. Retrieved from: https://catalogue.unccd.int/1482_IME-Global-Food-Report.pdf (accessed on 9 July 2024).

Gao, W., Feng, Z., Bai, Q., He, J., & Wang, Y. (2019). Melatonin-mediated regulation of growth and antioxidant capacity in salt-tolerant naked oat under salt stress. *International Journal of Molecular Sciences, 20*, 1176. https://doi.org/10.3390/ijms20051176.

Glazebrook, J. (2005). Contrasting mechanisms of defense against biotrophic and necrotrophic pathogens. *Annual Review of Phytopathology, 43*, 205–227. https://doi.org/10.1146/annurev.phyto.43.040204.135923.

Godfray, H. C. J., Beddington, J. R., Crute, I. R., Haddad, L., Lawrence, D., Muir, J. F., & Toulmin, C. (2010). Food security: The challenge of feeding 9 billion people. *Science, 327*(5967), 812–818. https://doi.org/10.1126/science.1185383.

Green, T. R., & Ryan, C. A. (1972). Wound-induced proteinase inhibitor in plant leaves: A possible defense mechanism against insects. *Science, 175*(4023), 776–777. https://doi.org/10.1126/science.175.4023.776.

Grover, D. K., & Singh, J. M. (2013). Post-harvest losses in wheat crop in Punjab: Past and Present. *Agricultural Economics Research Review, 26*(2), 293–297. https://www.indianjournals.com/ijor.aspx?target=ijor:aerr&volume=26&issue=2&article=016 (accessed on 9 July 2024).

Guche, M. D., Pilati, S., Trenti, F., Dalla Costa, L., Giorni, P., Guella, G., & Lanubile, A. (2022). Functional study of lipoxygenase-mediated resistance against *Fusarium verticillioides* and *Aspergillus flavus* infection in maize. *International Journal of Molecular Sciences, 23*(18), Article 10894. https://doi.org/10.3390/ijms231810894.

Guo, S., Song, Z., Ma, R., Yang, Y., & Yu, M. (2017). Genome-wide identification and expression analysis of the lipoxygenase gene family during peach fruit ripening under different postharvest treatments. *Acta Physiologiae Plantarum, 39*(5), Article 111. https://doi.org/10.1007/s11738-017-2409-6.

Haider, J., & Rai, A. B. (2021). Emergence of new insect pests on vegetables during the last decade: A case study. *Current Horticulture, 9*, 20–26. http://dx.doi.org/10.5958/2455-7560.2021.00003.0.

Han, C., Yin, X., He, D., & Yang, P. (2013). Analysis of proteome profile in germinating soybean seed, and its comparison with rice showing the styles of reserves mobilization in different crops. *PloS One, 8*(2), Article e56947. https://doi.org/10.1371/journal.pone.0056947.

Hayward, S., Cilliers, T., & Swart, P. (2017). Lipoxygenases: From isolation to application. *Comprehensive Reviews in Food Science and Food Safety, 16*(2), 199–211. https://doi.org/10.1111/1541-4337.12239.

Hell, K., Cardwell, K. F., Setamou, M., & Poehling, H. M. (2000). The influence of storage practices on aflatoxin contamination in maize in four agroecological zones of Benin, West Africa. *Journal of Stored Products Research, 36*(4), 365–382. https://doi.org/10.1016/S0022-474X(99)00056-9.

Horn, T., Adel, S., Schumann, R., Sur, S., Kakularam, K. R., Polamarasetty, A., & Heydeck, D. (2015). Evolutionary aspects of lipoxygenases and genetic diversity of human leukotriene signaling. *Progress in Lipid Research, 57*, 13–39. https://doi.org/10.1016/j.plipres.2014.11.001.

Hou, S., Lin, L., Lv, Y., Xu, N., & Sun, X. (2018). Responses of lipoxygenase, jasmonic acid, and salicylic acid to temperature and exogenous phytohormone treatments in *Gracilariopsis lemaneiformis* (Rhodophyta). *Journal of Applied Phycology, 30*(6), 3387–3394. https://doi.org/10.1007/s10811-018-1514-2.

Hou, Y., Ban, Q., Meng, K., He, Y., Han, S., Jin, M., & Rao, J. (2018). Overexpression of persimmon 9-lipoxygenase DkLOX3 confers resistance to *Pseudomonas syringae* pv. *tomato* DC3000 and *Botrytis cinerea* in Arabidopsis. *Plant Growth Regulation, 84*(2), 179–189. https://doi.org/10.1007/s10725-017-0331-y.

Howe, G. A., & Jander, G. (2008). Plant immunity to insect herbivores. *Annual Review of Plant Biology, 59*, 41–66. https://doi.org/10.1146/annurev.arplant.59.032607.092825.

Hu, T., Hu, Z., Zeng, H., Qv, X., & Chen, G. (2015). Tomato lipoxygenase D involved in the biosynthesis of jasmonic acid and tolerance to abiotic and biotic stress in tomato. *Plant Biotechnology Reports, 9*(1), 37–45. https://doi.org/10.1007/s11816-015-0341-z.

Hu, T., Zeng, H., Hu, Z., Qv, X., & Chen, G. (2013). Overexpression of the tomato 13-lipoxygenase gene TomloxD increases generation of endogenous jasmonic acid and resistance to *Cladosporium fulvum* and high temperature. *Plant Molecular Biology Reporter, 31*(5), 1141–1149. https://doi.org/10.1007/s11105-013-0581-4.

Huan, C., Yang, X., Wang, L., Kebbeh, M., Wang, Y., Dai, B., & Zhou, H. (2022). Methyl jasmonate treatment regulates α-linolenic acid metabolism and jasmonate acid signaling pathway to improve chilling tolerance in both stony hard and melting flesh peaches. *Postharvest Biology and Technology, 190*, Article 111960. https://doi.org/10.1016/j.postharvbio.2022.111960.

Huang, J., Cai, M., Long, Q., Liu, L., Lin, Q., Jiang, L., & Wan, J. (2014). *OsLOX2*, a rice type I lipoxygenase, confers opposite effects on seed germination and longevity. *Transgenic Research, 23*(4), 643–655. https://doi.org/10.1007/s11248-014-9803-2.

Hwang, I. S., & Hwang, B. K. (2010). The pepper 9-lipoxygenase gene *CaLOX1* functions in defense and cell death responses to microbial pathogens. *Plant Physiology, 152*(2), 948–967. https://doi.org/10.1104/pp.109.147827.

Isayenka, I., Duque-Yate, J., Goulet, M. C., Michaud, D., Beaulieu, C., & Beaudoin, N. (2023). Increased abundance of patatins, lipoxygenase and miraculins in a thaxtomin A-habituated potato Russet *Burbank somaclone* with enhanced resistance to common scab. *Plant Pathology, 72*(1), 100–111. https://doi.org/10.1111/ppa.13650.

Jacobo-Velázquez, D. A., González-Agüero, M., & Cisneros-Zevallos, L. (2015). Cross-talk between signaling pathways: The link between plant secondary metabolite production and wounding stress response. *Scientific Reports, 5*, Article 8608. https://doi.org/10.1038/srep08608.

Jardim, B. C., Perdizio, V. A., Berbert-Molina, M. A., Rodrigues, D. C., Botelho-Junior, S., Vicente, A. C., & Jacinto, T. (2010). Herbivore response in passion fruit (*Passiflora edulis* Sims) plants: Induction of lipoxygenase activity in leaf tissue in response to generalist and specialist insect attack. *Protein & Peptide Letters, 17*(4), 480–484. https://doi.org/10.2174/092986610790963744.

Jiao, C., Li, K., Zuo, Y., Gong, J., Guo, Z., & Shen, Y. (2022). Calmodulin1 and WRKY53 function in plant defense by negatively regulating the jasmonic acid biosynthesis pathway in Arabidopsis. *International Journal of Molecular Sciences, 23*(14), Article 7718. https://doi.org/10.3390/ijms23147718.

Ju, L. J., Zhang, C., Liao, J. J., Li, Y. P., & Qi, H. Y. (2018). An oriental melon 9-lipoxygenase gene *CmLOX09* response to stresses, hormones, and signal substances. *Journal of Zhejiang University Science B, 19*(8), 596–607. https://doi.org/10.1631/jzus.B1700388.

Kang, Y., Liu, W., Guan, C., Guan, M., & He, X. (2021). Evolution and functional diversity of lipoxygenase (LOX) genes in allotetraploid rapeseed (*Brassica napus* L.). *International Journal of Biological Macromolecules, 188*, 844–854. https://doi.org/10.1016/j.ijbiomac.2021.08.082.

Kaur, K. D., Jha, A., Sabikhi, L., & Singh, A. K. (2014). Significance of coarse cereals in health and nutrition: A review. *Journal of Food Science and Technology, 51*(7), 1429–1441. https://doi.org/10.1007/s13197-011-0612-9.

Kaya, C., Ugurlar, F., Ashraf, M., & Ahmad, P. (2023). Salicylic acid interacts with other plant growth regulators and signal molecules in response to stressful environments in plants. *Plant Physiology and Biochemistry, 196*, 431–443. https://doi.org/10.1016/j.plaphy.2023.02.006.

Kiaya, V. (2014). Post-harvest losses and strategies to reduce them. *Technical Paper on Postharvest Losses, Action Contre la Faim (ACF), 25*, 1–25.

Kotapati, K. V., Palaka, B. K., & Ampasala, D. R. (2017). Alleviation of nickel toxicity in finger millet (*Eleusine coracana* L.) germinating seedlings by exogenous application of salicylic acid and nitric oxide. *Crop J., 5*, 240–250. https://doi.org/10.1016/j.cj.2016.09.002.

Kućko, A., de Dios Alché, J., Tranbarger, T. J., & Wilmowicz, E. (2022). The acceleration of yellow lupine flower abscission by jasmonates is accompanied by lipid-related events in abscission zone cells. *Plant Sci., 316*, 111173. https://doi.org/10.1016/j.plantsci.2021.111173.

Kuhn, H., Banthiya, S., & Van Leyen, K. (2015). Mammalian lipoxygenases and their biological relevance. *Biochim. Biophys. Acta, Mol. Cell Biol. Lipids, 1851*, 308–330. https://doi.org/10.1016/j.bbalip.2014.10.002.

Kumar, D., & Kalita, P. (2017). Reducing postharvest losses during storage of grain crops to strengthen food security in developing countries. *Foods, 6*, 8. https://doi.org/10.3390/foods6010008.

Kumar, D., & Ohri, P. (2022). Say "NO" to plant stresses: Unravelling the role of nitric oxide under abiotic and biotic stress. *Nitric Oxide, 130*, 36–57. https://doi.org/10.1016/j.niox.2022.11.004.

Kumar, S., Shah, S. H., Vimala, Y., Jatav, H. S., Ahmad, P., Chen, Y., & Siddique, K. H. (2022). Abscisic acid: Metabolism, transport, crosstalk with other plant growth regulators, and its role in heavy metal stress mitigation. *Front. Plant Sci., 13*, 972856. https://doi.org/10.3389/fpls.2022.972856.

Kumar, V., Rani, A., Tayalkar, T., Jha, P., Mittal, P., Anshu, A. K., & Rawal, R. (2022). Genetic elimination of lipoxygenase-2 improves sprouting and tocopherols, and removal of Kunitz trypsin inhibitor enhances in vitro protein digestibility in soybean (*Glycine max*). *Plant Breed., 141*, 63–70. https://doi.org/10.1111/pbr.12998.

Lamers, J., Van Der Meer, T., & Testerink, C. (2020). How plants sense and respond to stressful environments. *Plant Physiol.*, 182, 1624–1635. https://doi.org/10.1104/pp.19.01464.

Lemeza, O. V., Zubo, Y. O., & Kusnetsov, V. V. (2010). Regulation of lipoxygenase gene expression in potato mini-tubers by phytohormones. *Russ. J. Plant Physiol., 57*, 715–719. https://doi.org/10.1134/S1021443710050158.

Leone, A., Melillo, M. T., & Bleve-Zacheo, T. (2001). Lipoxygenase in pea roots subjected to biotic stress. *Plant Sci., 16*1, 703–717. https://doi.org/10.1016/S0168-9452(01)00458-7.

León-García, E., Vela-Gutiérrez, G., Del Ángel-Coronel, O. A., Torres-Palacios, C., La Cruz-Medina, J. D., Gómez-Lim, M. A., & García, H. S. (2017). Increased postharvest life of *TomLoxB* silenced mutants of tomato (*Solanum lycopersicum*) var. TA234. *Plant Foods Hum. Nutr., 72*, 380–387. https://doi.org/10.1007/s11130-017-0629-y.

Li, C., Schilmiller, A. L., Liu, G., Lee, G. I., Jayanty, S., Sageman, C., & Howe, G. A. (2005). Role of β-oxidation in jasmonate biosynthesis and systemic wound signaling in tomato. *Plant Cell, 17*, 971–986. https://doi.org/10.1105/tpc.104.029108.

Li, L., Li, C., Lee, G. I., & Howe, G. A. (2002). Distinct roles for jasmonate synthesis and action in the systemic wound response of tomato. *Proc. Natl. Acad. Sci., 99*, 6416–6421. https://doi.org/10.1073/pnas.072072599.

Li, M., & Kim, C. (2021). Chloroplast ROS and stress signaling. *Plant Communications, 3*(1), 100264. https://doi.org/10.1016/j.xplc.2021.100264.

Li, X., Li, C., Cheng, Y., Hou, J., Zhang, J., & Ge, Y. (2020). Postharvest application of acibenzolar-S-methyl delays the senescence of pear fruit by regulating reactive oxygen species and fatty acid metabolism. *J. Agric. Food Chem., 68*, 4991–4999. https://doi.org/10.1021/acs.jafc.0c01031.

Li, Z. Q., Qian, X., Yan, J. H., Chen, J. Q., & Chen, Q. X. (2022). Genome-wide identification and characterization of the abiotic-stress-responsive lipoxygenase gene family in diploid woodland strawberry (*Fragaria vesca*). *J. Integr. Agric., 21*, 1982–1996. https://doi.org/10.1016/S2095-3119(21)63819-4.

Liang, X., Qian, R., Wang, D., Liu, L., Sun, C., & Lin, X. (2022). Lipid-derived aldehydes: New key mediators of plant growth and stress responses. *Biology, 11*, 1590. https://doi.org/10.3390/biology11111590.

Liao, L., Li, S., Li, Y., Huang, Z., Li, J., Xiong, B., & Wang, Z. (2022). Pre- or post-harvest treatment with MeJA improves post-harvest storage of lemon fruit by stimulating the antioxidant system and alleviating chilling injury. *Plants, 11*, 2840. https://doi.org/10.3390/plants11212840.

Liao, Z., Wang, L., Li, C., Cao, M., Wang, J., Yao, Z., & Lou, Y. (2022). The lipoxygenase gene OsRCI-1 is involved in the biosynthesis of herbivore-induced JAs and regulates plant defense and growth in rice. *Plant Cell Environ., 45*, 2827–2840. https://doi.org/10.1111/pce.14341.

Liavonchanka, A., & Feussner, I. (2006). Lipoxygenases: Occurrence, functions and catalysis. *J. Plant Physiol., 163*, 348–357. https://doi.org/10.1016/j.jplph.2005.11.006.

Liburkin-Dan, T., Nir-Zvi, I., Razon, H., Kessler, O., & Neufeld, G. (2022). Knock-Out of the five lysyl-oxidase family genes enables identification of lysyl-oxidase pro-enzyme regulated genes. *Int. J. Mol. Sci., 23*, 11322. https://doi.org/10.3390/ijms231911322.

Lim, C. W., Han, S. W., Hwang, I. S., Kim, D. S., Hwang, B. K., & Lee, S. C. (2015). The pepper lipoxygenase CaLOX1 plays a role in osmotic, drought and high salinity stress response. *Plant Cell Physiol., 56*, 930–942. https://doi.org/10.1093/pcp/pcv020.

Liu, C., Chen, L., Zhao, R., Li, R., Zhang, S., Yu, W., & Shen, L. (2019). Melatonin induces disease resistance to *Botrytis cinerea* in tomato fruit by activating jasmonic acid signaling pathway. *J. Agric. Food Chem., 67*, 6116–6124. https://doi.org/10.1021/acs.jafc.9b00058.

Liu, Q., Xie, H., Chen, Y., Lin, M., Hung, Y. C., & Lin, H. (2022). Alleviation of pulp breakdown in harvested longan fruit by acidic electrolyzed water in relation to membrane lipid metabolism. *Sci. Hortic., 304*, 111288. https://doi.org/10.1016/j.scienta.2022.111288.

Liu, S., & Han, B. (2010). Differential expression pattern of an acidic 9/13-lipoxygenase in flower opening and senescence and in leaf response to phloem feeders in the tea plant. *BMC Plant Biol., 10*, 228. https://doi.org/10.1186/1471-2229-10-228.

Lõhelaid, H., Teder, T., & Samel, N. (2015). Lipoxygenase-allene oxide synthase pathway in octocoral thermal stress response. *Coral Reefs, 34*, 143–154. https://doi.org/10.1007/s00338-014-1238-y.

López, M. A., Vicente, J., Kulasekaran, S., Vellosillo, T., Martínez, M., Irigoyen, M. L., & Castresana, C. (2011). Antagonistic role of 9-lipoxygenase-derived oxylipins and ethylene in the control of oxidative stress, lipid peroxidation and plant defense. *Plant J., 67*, 447–458. https://doi.org/10.1111/j.1365-313X.2011.04608.x.

Losvik, A., Beste, L., Glinwood, R., Ivarson, E., Stephens, J., Zhu, L. H., & Jonsson, L. (2017). Overexpression and down-regulation of barley lipoxygenase LOX2.2 affects jasmonate-regulated genes and aphid fecundity. *Int. J. Mol. Sci., 18*, 2765. https://doi.org/10.3390/ijms18122765.

Louis, J., & Shah, J. (2013). *Arabidopsis thaliana—Myzus persicae* interaction: Shaping the understanding of plant defense against phloem-feeding aphids. *Front. Plant Sci., 4*, 213. https://doi.org/10.3389/fpls.2013.00213.

Lu, J., Xu, Y., Wang, J., Singer, S. D., & Chen, G. (2020). The role of triacylglycerol in plant stress response. *Plants, 9*, 472. https://doi.org/10.3390/plants9040472.

Ma, L., Zhu, F., Li, Z., Zhang, J., Li, X., Dong, J., & Wang, T. (2015). TALEN-based mutagenesis of lipoxygenase LOX3 enhances the storage tolerance of rice (*Oryza sativa*) seeds. *PLoS One, 10*, e0143877. https://doi.org/10.1371/journal.pone.0143877.

Ma, Q., Wang, J., Cheng, L., Li, Y., Zhang, Q., Li, H., & Zhang, B. (2022). The potential function of SiLOX4 on millet discoloration during storage in foxtail millet. *Agriculture, 12*, 1283. https://doi.org/10.3390/agriculture12081283.

Ma, W., Li, J., Murtaza, A., Iqbal, A., Zhang, J., Zhu, L., & Hu, W. (2022). High-pressure carbon dioxide treatment alleviates browning development by regulating membrane lipid metabolism in fresh-cut lettuce. *Food Control, 134*, 108749. https://doi.org/10.1016/j.foodcont.2021.108749.

Maccarrone, M., Salucci, M. L., van Zadelhoff, G., Malatesta, F., Veldink, G., Vliegenthart, J. F., & Finazzi-Agro, A. (2001). Tryptic digestion of soybean lipoxygenase-1 generates a 60 kDa fragment with improved activity and membrane binding ability. *Biochemistry, 40*, 6819–6827. https://doi.org/10.1021/bi010187m.

Majumder, S., Bala, B. K., Arshad, F. M., Haque, M. A., & Hossain, M. A. (2016). Food security through increasing technical efficiency and reducing postharvest losses of rice production systems in Bangladesh. *Food Security, 8*, 361–374. https://doi.org/10.1007/s12571-016-0558-x.

Mangrauthia, S. K., Agarwal, S., Sailaja, B., Sarla, N., & Voleti, S. R. (2016). Transcriptome analysis of *Oryza sativa* (rice) seed germination at high temperature shows dynamics of genome expression associated with hormones signaling and abiotic stress pathways. *Trop. Plant Biol., 9*, 215–228. https://doi.org/10.1007/s12042-016-9170-7.

Mares, D. J., Cheong, J., Goonetilleke, S. N., & Mather, D. E. (2021). Lipoxygenase in wheat: Genetic control and impact on stability of lutein and lutein esters. *Foods, 10*, 1149. https://doi.org/10.3390/foods10051149.

Mazur, R., Trzcinska-Danielewicz, J., Kozlowski, P., Kowalewska, Ł., Rumak, I., Shiell, B. J., & Garstka, M. (2018). Dark-chilling and subsequent photo-activation modulate expression and induce reversible association of chloroplast lipoxygenase with thylakoid membrane in runner bean (*Phaseolus coccineus* L.). *Plant Physiol. Biochem., 122*, 102–112. https://doi.org/10.1016/j.plaphy.2017.11.015.

Meethaworn, K., Luckanatinwong, V., Zhang, B., Chen, K., & Siriphanich, J. (2019). Off-flavor caused by cold storage is related to induced activity of LOX and HPL in young coconut fruit. *LWT, 114*, 108329. https://doi.org/10.1016/j.lwt.2019.108329.

Mène-Saffrané, L., Esquerré-Tugayé, M. T., & Fournier, J. (2003). Constitutive expression of an inducible lipoxygenase in transgenic tobacco decreases susceptibility to *Phytophthora parasitica* var. *nicotianae*. *Mol. Breed., 12*, 271–282. https://doi.org/10.1023/B:MOLB.0000006754.19398.d4.

Menga, V., & Trono, D. (2020). The molecular and functional characterization of the durum wheat Lipoxygenase TdLOX2 suggests its role in hyperosmotic stress response. *Plants, 9*, 1233. https://doi.org/10.3390/plants9091233.

Mirshekari, A., Madani, B., Yahia, E. M., Golding, J. B., & Vand, S. H. (2020). Postharvest melatonin treatment reduces chilling injury in sapota fruit. *J. Sci. Food Agric., 100*, 1897–1903. https://doi.org/10.1002/jsfa.10198.

Montillet, J. L., Leonhardt, N., Mondy, S., Tranchimand, S., Rumeau, D., Boudsocq, M., & Hirt, H. (2013). An abscisic acid-independent oxylipin pathway controls stomatal closure and immune defense in Arabidopsis. *PLoS Biol., 11*, e1001513. https://doi.org/10.1371/journal.pbio.1001513.

Nejat, N., & Mantri, N. (2017). Plant immune system: Crosstalk between responses to biotic and abiotic stresses the missing link in understanding plant defense. *Curr. Issues Mol. Biol., 23*, 1–16. https://doi.org/10.21775/cimb.023.001.

Nemchenko, A., Kunze, S., Feussner, I., & Kolomiets, M. (2006). Duplicate maize 13-lipoxygenase genes are differentially regulated by circadian rhythm, cold stress, wounding, pathogen infection, and hormonal treatments. *J. Exp. Bot., 57*, 3767–3779. https://doi.org/10.1093/jxb/erl137.

Nene, Y. L., Reddy, M. V., Haware, M. P., Ghanekar, A. M., Amin, K. S., Pande, S., & Sharma, M. (2012). Field diagnosis of chickpea diseases and their control. Information bulletin no. 28 (revised). *International Crops Research Institute for the Semi-Arid Tropics.* http://oar.icrisat.org/id/eprint/6601 (accessed on 9 July 2024).

Obianom, C., Romanazzi, G., & Sivakumar, D. (2019). Effects of chitosan treatment on avocado postharvest diseases and expression of phenylalanine ammonia-lyase, chitinase and lipoxygenase genes. *Postharvest Biol. Technol., 147*, 214–221. https://doi.org/10.1016/j.postharvbio.2018.10.004.

Oliw, E. H. (2022). Iron and manganese lipoxygenases of plant pathogenic fungi and their role in biosynthesis of jasmonates. *Arch. Biochem. Biophys., 722*, 109169. https://doi.org/10.1016/j.abb.2022.109169.

Ozalvo, R., Cabrera, J., Escobar, C., Christensen, S. A., Borrego, E. J., Kolomiets, M. V., & Brown Horowitz, S. (2014). Two closely related members of *Arabidopsis* 13-lipoxygenases (13-LOXs), LOX3 and LOX4, reveal distinct functions in response to plant-parasitic nematode infection. *Mol. Plant Pathol., 15*, 319–332. https://doi.org/10.1111/mpp.12094.

Pan, L., Chen, X., Xu, W., Fan, S., Wan, T., Zhang, J., & Cai, Y. (2022). Methyl jasmonate induces postharvest disease resistance to decay caused by *Alternaria alternata* in sweet cherry fruit. *Sci. Hortic., 292*, 110624. https://doi.org/10.1016/j.scienta.2021.110624.

Pandey, P., Irulappan, V., Bagavathiannan, M. V., & Senthil-Kumar, M. (2017). Impact of combined abiotic and biotic stresses on plant growth and avenues for crop improvement by exploiting physio-morphological traits. *Front. Plant Sci., 8*, 537. https://doi.org/10.3389/fpls.2017.00537.

Park, Y. S., Kunze, S., Ni, X., Feussner, I., & Kolomiets, M. V. (2010). Comparative molecular and biochemical characterization of segmentally duplicated 9-lipoxygenase genes ZmLOX4 and ZmLOX5 of maize. *Planta, 231*, 1425–1437. https://doi.org/10.1007/s00425-010-1143-8.

Pathi, K. M., Rink, P., Budhagatapalli, N., Betz, R., Saado, I., Hiekel, S., & Kumlehn, J. (2020). Engineering smut resistance in maize by site-directed mutagenesis of LIPOXYGENASE 3. *Front. Plant Sci., 11*, 543895. https://doi.org/10.3389/fpls.2020.543895.

Permyakova, M. D., Permyakov, A. V., Osipova, S. V., Pshenichnikova, T. A., Shishparenok, A. A., Rudikovskaya, E. G., Rudikovsky, A. V., Verkhoturov, V. V., & Börner, A. (2017). Chromosome regions associated with the activity of lipoxygenase in the genome D of

Triticum aestivum L. under water deficit. *Russian Journal of Plant Physiology, 64*, 28–40. https://doi.org/10.1134/S1021443717010113.

Pinkerton, J. N., Ivors, K. L., Miller, M. L., & Moore, L. W. (2000). Effect of soil solarization and cover crops on populations of selected soilborne plant pathogens in western Oregon. *Plant Disease, 84*, 952–960. https://doi.org/10.1094/PDIS.2000.84.9.952.

Pokotylo, I. V., Kolesnikov, Y. S., Derevyanchuk, M. V., Kharitonenko, A. I., & Kravets, V. S. (2015). Lipoxygenases and plant cell metabolism regulation. *Ukrainian Biochemical Journal, 87*, 41–55. https://europepmc.org/article/med/26255338.

Porta, H., & Rocha-Sosa, M. (2002). Plant lipoxygenases: Physiological and molecular features. *Plant Physiology, 130*, 15–21. https://doi.org/10.1104/pp.010787.

Puri, M., Kaur, I., Kanwar, R. K., Gupta, R. C., Chauhan, A., & Kanwar, J. R. (2009). Ribosome inactivating proteins (RIPs) from *Momordica charantia* for antiviral therapy. *Current Molecular Medicine, 9*, 1080–1094. https://doi.org/10.2174/156652409789839071.

Pushpalatha, H. G., Sudisha, J., Geetha, N. P., Amruthesh, K. N., & Shekar Shetty, H. (2011). Thiamine seed treatment enhances LOX expression, promotes growth and induces downy mildew disease resistance in pearl millet. *Biologia Plantarum, 55*, 522–527. https://www.bp.ueb.cas.cz/pdfs/bpl/2011/03/15.pdf.

Roychoudhury, A., Basu, S., & Sengupta, D. N. (2009). Effects of exogenous abscisic acid on some physiological responses in a popular aromatic indica rice compared with those from two traditional non-aromatic indica rice cultivars. *Acta Physiologiae Plantarum, 31*, 915–926. https://doi.org/10.1007/s11738-009-0305-4.

RoyChowdhury, M., Li, X., Qi, H., Li, W., Sun, J., Huang, C., & Wu, D. (2016). Functional characterization of 9-/13-LOXs in rice and silencing their expressions to improve grain qualities. *Biomedical Research International, 2016*. https://doi.org/10.1155/2016/4275904.

Ryan, C. A. (2000). The systemin signaling pathway: Differential activation of plant defensive genes. *Biochimica et Biophysica Acta–Protein Structure and Molecular Enzymology, 1477*, 112–121. https://doi.org/10.1016/S0167-4838(99)00269-1.

Saado, I., Chia, K. S., Betz, R., Alcântara, A., Pettkó-Szandtner, A., Navarrete, F., D'Auria, J. C., Kolomiets, M. V., Melzer, M., Feussner, I., & Djamei, A. (2022). Effector-mediated relocalization of a maize lipoxygenase protein triggers susceptibility to *Ustilago maydis*. *The Plant Cell, 34*, 2785–2805. https://doi.org/10.1093/plcell/koac105.

Saha, I., Hasanuzzaman, M., & Adak, M. K. (2021). Abscisic acid priming regulates arsenite toxicity in two contrasting rice (*Oryza sativa* L.) genotypes through differential functioning of sub1A quantitative trait loci. *Environmental Pollution, 287*, 117586. https://doi.org/10.1016/j.envpol.2021.117586.

Saijo, Y., & Loo, E. P. I. (2020). Plant immunity in signal integration between biotic and abiotic stress responses. *New Phytologist, 225*, 87–104. https://doi.org/10.1111/nph.15989.

Sakouhi, L., Mahjoubi, Y., Labben, A., Kharbech, O., Chaoui, A., & Djebali, W. (2022). Effects of cadmium–selenium interaction on glyoxalase and antioxidant systems of *Pisum sativum* germinating seeds. *Journal of Plant Growth Regulation, 42*, 3084–3099. https://doi.org/10.1007/s00344-022-10772-5.

Sanadhya, P., Kumar, A., Bucki, P., Fitoussi, N., Carmeli-Weissberg, M., Borenstein, M., & Brown-Miyara, S. (2021). Tomato divinyl ether-biosynthesis pathway is implicated in modulating of root-knot nematode *Meloidogyne javanica's* parasitic ability. *Frontiers in Plant Science, 12*, 670772. https://doi.org/10.3389/fpls.2021.670772.

Santino, A., De Paolis, A., Gallo, A., Quarta, A., Casey, R., & Mita, G. (2003). Biochemical and molecular characterization of hazelnut (*Corylus avellana*) seed lipoxygenases. *European Journal of Biochemistry, 270*, 4365–4375. https://doi.org/10.1046/j.1432-1033.2003.03831.x.

Saraiva, T. D., Morais, K., Pereira, V. B., de Azevedo, M., Rocha, C. S., Prosperi, C. C., Gomes-Santos, A. C., Bermudez-Humaran, L., Faria, A. M. C., Blottiere, H. M., & Azevedo, V. (2015). Milk fermented with a 15-lipoxygenase-1-producing Lactococcus lactis alleviates symptoms of colitis in a murine model. *Current Pharmaceutical Biotechnology, 16*, 424–429.

Sarde, S. J., Bouwmeester, K., Venegas-Molina, J., David, A., Boland, W., & Dicke, M. (2019). Involvement of sweet pepper CaLOX2 in jasmonate-dependent induced defense against Western flower thrips. *Journal of Integrative Plant Biology, 61*, 1085–1098. https://doi.org/10.1111/jipb.12742.

Sarker, A., Nandi, R., Kim, J. E., & Islam, T. (2021). Remediation of chemical pesticides from contaminated sites through potential microorganisms and their functional enzymes: Prospects and challenges. *Environmental Technology & Innovation, 23*, 101777. https://doi.org/10.1016/j.eti.2021.101777.

Schiller, D., Contreras, C., Vogt, J., Dunemann, F., Defilippi, B. G., Beaudry, R., & Schwab, W. (2015). A dual positional specific lipoxygenase functions in the generation of flavor compounds during climacteric ripening of apple. *Horticulture Research, 2*, 15003. https://doi.org/10.1038/hortres.2015.3.

Seo, H. S., Song, J. T., Cheong, J. J., Lee, Y. H., Lee, Y. W., Hwang, I., Lee, J. S., & Choi, Y. D. (2001). Jasmonic acid carboxyl methyltransferase: A key enzyme for jasmonate-regulated plant responses. *Proceedings of the National Academy of Sciences, 98*, 4788–4793. https://doi.org/10.1073/pnas.081557298.

Shaban, M., Ahmed, M. M., Sun, H., Ullah, A., & Zhu, L. (2018). Genome-wide identification of lipoxygenase gene family in cotton and functional characterization in response to abiotic stresses. *BMC Genomics, 19*, 599. https://doi.org/10.1186/s12864-018-4985-2.

Shahzad, A., Ullah, S., Dar, A. A., Sardar, M. F., Mehmood, T., Tufail, M. A., & Haris, M. (2021). Nexus on climate change: Agriculture and possible solution to cope future climate change stresses. *Environmental Science and Pollution Research, 28*, 14211–14232. https://doi.org/10.1007/s11356-021-12649-8.

Sharma, N., Khanna, K., Jasrotia, S., Kumar, D., Bhardwaj, R., & Ohri, P. (2023). Metabolites and chemical agents in the plant roots: An overview of their use in plant-parasitic nematode management. *Nematology*. https://doi.org/10.1163/15685411-bja10220.

Sharma, N., Kour, S., Kumar, D., Kaur, R., Khajuria, A., & Ohri, P. (2022). Role of brassinosteroids (BRs) in modulating antioxidative defense mechanism in plants growing under abiotic and biotic stress conditions. In T. Aftab & K. R. Hakeem (Eds.), *Antioxidant Defense in Plants* (pp. 325–367). Springer. https://doi.org/10.1007/978-981-16-7981-0_15.

Sharma, R. R., Datta, S. C., & Varghese, E. (2020). Kaolin-based particle film sprays reduce the incidence of pests, diseases and storage disorders and improve postharvest quality of 'Delicious' apples. *Crop Protection, 127*, 104950. https://doi.org/10.1016/j.cropro.2019.104950.

Shi, Y., Mandal, R., Singh, A., & Pratap Singh, A. (2020). Legume lipoxygenase: Strategies for application in food industry. *Legume Science, 2*, e44. https://doi.org/10.1002/leg3.44.

Shibata, D., Steczkoli, J., Dixon, J. E., Hermodson, M., Yazdanparast, R., & Axelrod, B. (1987). Primary structure of soybean lipoxygenase-1. *Journal of Biological Chemistry, 262*, 10080–10085. https://doi.org/10.1016/S0021-9258(18)61079-8.

Shimada, T. L., & Hara-Nishimura, I. (2015). Leaf oil bodies are subcellular factories producing antifungal oxylipins. *Current Opinion in Plant Biology, 25*, 145–150. https://doi.org/10.1016/j.pbi.2015.05.019.

Shrestha, K., Pant, S., & Huang, Y. (2021). Genome-wide identification and classification of lipoxygenase gene family and their roles in sorghum-aphid interaction. *Plant Molecular Biology, 105*, 527–541. https://doi.org/10.1007/s11103-020-01107-7.

Siddiqui, M. W., Homa, F., Lata, D., & Ahmad, M. S. (2021). Exogenous nitric oxide delays ripening and maintains postharvest quality of pointed gourd during storage. *Journal of Plant Growth Regulation, 40*, 2371–2378. https://doi.org/10.1007/s00344-020-10270-6.

Singh, P., Arif, Y., Miszczuk, E., Bajguz, A., & Hayat, S. (2022). Specific roles of lipoxygenases in development and responses to stress in plants. *Plants, 11*, 979. https://doi.org/10.3390/plants11070979.

Smith, W. L., & Murphy, R. C. (2016). The eicosanoids: Cyclooxygenase, lipoxygenase and epoxygenase pathways. In N. D. Ridgway & R. S. McLeod (Eds.), *Biochemistry of Lipids, Lipoproteins and Membranes* (pp. 259–296). Elsevier. https://doi.org/10.1016/B978-0-444-63438-2.00009-2.

Song, H., Wang, P., Li, C., Han, S., Lopez-Baltazar, J., Zhang, X., & Wang, X. (2016). Identification of lipoxygenase (LOX) genes from legumes and their responses in wild type and cultivated peanut upon *Aspergillus flavus* infection. *Scientific Reports, 6*, 35245. https://doi.org/10.1038/srep35245.

Srivastava, C., & Subramanian, S. (2016). Storage insect pests and their damage symptoms: An overview. *Indian Journal of Entomology, 78*, 53–58. http://dx.doi.org/10.5958/0974-8172.2016.00025.0.

Stathers, T., Holcroft, D., Kitinoja, L., Mvumi, B. M., English, A., Omotilewa, O., & Torero, M. (2020). A scoping review of interventions for crop postharvest loss reduction in sub-Saharan Africa and South Asia. *Nature Sustainability, 3*, 821–835. https://doi.org/10.1038/s41893-020-00622-1.

Steele, P., & Gibbs, M. (2018). Crop diseases and their control. *Scientific e-Resources*. ISBN 1839471859, 9781839471858 Publisher: Scientific e-Resources, 2018 (EDTECH). https://books.google.co.in/books?id=ZeLEDwAAQBAJ.

Stolterfoht, H., Rinnofner, C., Winkler, M., & Pichler, H. (2019). Recombinant lipoxygenases and hydroperoxide lyases for the synthesis of green leaf volatiles. *Journal of Agricultural and Food Chemistry, 67*, 13367–13392. https://doi.org/10.1021/acs.jafc.9b02690.

Sujatha, B., & Devi, P. U. M. (2012). Antifungal potential of papaya lipoxygenase metabolites against *Phytophthora palmivora*. *Journal of Pure and Applied Microbiology, 6*, 433–438. https://www.cabdirect.org/cabdirect/abstract/20123220065.

Surjadinata, B. B., Jacobo-Velázquez, D. A., & Cisneros-Zevallos, L. (2021). Physiological role of reactive oxygen species, ethylene, and jasmonic acid on UV light induced phenolic biosynthesis in wounded carrot tissue. *Postharvest Biology and Technology, 172*, 111388. https://doi.org/10.1016/j.postharvbio.2020.111388.

Szeliga, M., Ciura, J., Grzesik, M., & Tyrka, M. (2019). Identification of candidate genes involved in steroidal alkaloids biosynthesis in organ-specific transcriptomes of *Veratrum nigrum* L. *Gene, 712*, 143962. https://doi.org/10.1016/j.gene.2019.143962.

Tang, J., Wang, S. Q., Hu, K. D., Huang, Z. Q., Li, Y. H., Han, Z., & Zhang, H. (2019). Antioxidative capacity is highly associated with the storage property of tuberous roots in different sweet potato cultivars. *Scientific Reports, 9*, 1–10. https://doi.org/10.1038/s41598-019-47604-8.

Tayyab, N., Naz, R., Yasmin, H., Nosheen, A., Keyani, R., Sajjad, M., & Roberts, T. H. (2020). Combined seed and foliar pre-treatments with exogenous methyl jasmonate and salicylic acid mitigate drought-induced stress in maize. *PLOS ONE, 15*, e0232269. https://doi.org/10.1371/journal.pone.0232269.

Thakur, M., & Udayashankar, A. C. (2019). Lipoxygenases and their function in plant innate mechanism. In S. Jogaiah & M. Abdelrahman (Eds.), *Bioactive Molecules in Plant Defense: Signaling in Growth and Stress* (pp. 133–143). Springer. https://doi.org/10.1007/978-3-030-27165-7_8.

Ul Hassan, M. N., Zainal, Z., & Ismail, I. (2015). Green leaf volatiles: Biosynthesis, biological functions and their applications in biotechnology. *Plant Biotechnology Journal, 13*, 727–739. https://doi.org/10.1111/pbi.12368.

Umar, O. B., Ranti, L. A., Abdulbaki, A. S., Bola, A. L., Abdulhamid, A. K., Biola, M. R., & Victor, K. O. (2021). Stresses in plants: Biotic and abiotic. In M. R. Amsari (Ed.), *Current Trends in Wheat Research*. IntechOpen.

Umate, P. (2011). Genome-wide analysis of lipoxygenase gene family in Arabidopsis and rice. *Plant Signaling & Behavior, 6*, 335–338. https://doi.org/10.4161/psb.6.3.13546.

Upadhyay, R. K., & Mattoo, A. K. (2018). Genome-wide identification of tomato (*Solanum lycopersicum* L.) lipoxygenases coupled with expression profiles during plant development and in response to methyl-jasmonate and wounding. *Journal of Plant Physiology, 231*, 318–328. https://doi.org/10.1016/j.jplph.2018.10.001.

Upadhyay, R. K., Handa, A. K., & Mattoo, A. K. (2019). Transcript abundance patterns of 9- and 13-lipoxygenase subfamily gene members in response to abiotic stresses (heat, cold, drought or salt) in tomato (*Solanum lycopersicum* L.) highlights member-specific dynamics relevant to each stress. *Genes, 10*, 683. https://doi.org/10.3390/genes10090683.

Vellosillo, T., Aguilera, V., Marcos, R., Bartsch, M., Vicente, J., Cascón, T., Hamberg, M., & Castresana, C. (2013). Defense activated by 9-lipoxygenase-derived oxylipins requires specific mitochondrial proteins. *Plant Physiology, 161*, 617–627. https://doi.org/10.1104/pp.112.207514.

Vellosillo, T., Martínez, M., Lopez, M. A., Vicente, J., Cascon, T., Dolan, L., Hamberg, M., & Castresana, C. (2007). Oxylipins produced by the 9-lipoxygenase pathway in Arabidopsis regulate lateral root development and defense responses through a specific signaling cascade. *Plant Cell, 19*, 831–846. https://doi.org/10.1105/tpc.106.046052.

Veronico, P., Giannino, D., Melillo, M. T., Leone, A., Reyes, A., Kennedy, M. W., & Bleve-Zacheo, T. (2006). A novel lipoxygenase in pea roots. Its function in wounding and biotic stress. *Plant Physiology, 141*, 1045-–1055. https://doi.org/10.1104/pp.106.081679.

Viswanath, K. K., Varakumar, P., Pamuru, R. R., Basha, S. J., Mehta, S., & Rao, A. D. (2020). Plant lipoxygenases and their role in plant physiology. *Journal of Plant Biology, 63*, 83–95. https://doi.org/10.1007/s12374-020-09241-x.

Vogt, J., Schiller, D., Ulrich, D., Schwab, W., & Dunemann, F. (2013). Identification of lipoxygenase (LOX) genes putatively involved in fruit flavor formation in apple (*Malus* × *domestica*). *Tree Genetics & Genomes, 9*, 1493–1511. https://doi.org/10.1007/s11295-013-0653-5.

Walling, L. L. (2000). The myriad plant responses to herbivores. *Journal of Plant Growth Regulation, 19*(2), 195–216. https://doi.org/10.1007/s003440000026.

Wang, F., Xu, H., Zhang, L., Shi, Y., Song, Y., Wang, X., Cai, Q., He, W., Xie, H., & Zhang, J. (2023). The lipoxygenase OsLOX10 affects seed longevity and resistance to saline-alkaline stress during rice seedlings. *Plant Molecular Biology, 111*, 415–428. https://doi.org/10.1007/s11103-023-01334-8.

Wang, J., Hu, T., Wang, W., Hu, H., Wei, Q., Wei, X., & Bao, C. (2019). Bioinformatics analysis of the lipoxygenase gene family in radish (*Raphanus sativus*) and functional characterization in response to abiotic and biotic stresses. *International Journal of Molecular Sciences, 20*(23), 6095. https://doi.org/10.3390/ijms20236095.

Wang, Q., Li, J., Hu, L., Zhang, T., Zhang, G., & Lou, Y. (2013). OsMPK3 positively regulates the JA signaling pathway and plant resistance to a chewing herbivore in rice. *Plant Cell Reports, 32*(7), 1075–1084. https://doi.org/10.1007/s00299-013-1389-2.

Wang, Q., Sun, Y., Wang, F., Huang, P. C., Wang, Y., Ruan, X., Ma, L., Li, X., Kolomiets, M. V., & Gao, X. (2021). Transcriptome and oxylipin profiling joint analysis reveal opposite roles of 9-oxylipins and jasmonic acid in maize resistance to gibberella stalk rot. *Frontiers in Plant Science, 12*, 699146. https://doi.org/10.3389/fpls.2021.699146.

Wang, R., Shen, W., Liu, L., Jiang, L., Liu, Y., Su, N., & Wan, J. (2008). A novel lipoxygenase gene from developing rice seeds confers dual position specificity and responds to wounding and insect attack. *Plant Molecular Biology, 66*(4), 401–414. https://doi.org/10.1007/s11103-007-9278-0.

Wang, T., Hu, M., Yuan, D., Yun, Z., Gao, Z., Su, Z., & Zhang, Z. (2020). Melatonin alleviates pericarp browning in litchi fruit by regulating membrane lipid and energy metabolisms. *Postharvest Biology and Technology, 160*, 111066. https://doi.org/10.1016/j.postharvbio.2019.111066.

Wang, W., Ni, Z. J., Song, C. B., Ma, W. P., Cao, S. Q., & Wei, Z. J. (2023). Hydrogen sulfide treatment improves quality attributes via regulating the antioxidant system in goji berry (*Lycium barbarum* L.). *Food Chemistry, 405*, 134858. https://doi.org/10.1016/j.foodchem.2022.134858.

War, A. R., Taggar, G. K., Hussain, B., Taggar, M. S., Nair, R. M., & Sharma, H. C. (2018). Plant defense against herbivory and insect adaptations. *AoB Plants, 10*, ply037. https://doi.org/10.1093/aobpla/ply037.

Wari, D., Aboshi, T., Shinya, T., & Galis, I. (2022). Integrated view of plant metabolic defense with particular focus on chewing herbivores. *Journal of Integrative Plant Biology, 64*(3), 449–475. https://doi.org/10.1111/jipb.13204.

Wasternack, C. (2007). Jasmonates: An update on biosynthesis, signal transduction and action in plant stress response, growth and development. *Annals of Botany, 100*(4), 681–697. https://doi.org/10.1093/aob/mcm079.

Wei, X., Wei, X., Guan, W., Nong, W., Chen, R., Tao, X., & Mao, L. (2022). ABA-responsive transcription factor ABF1-1 promotes JA biosynthesis to accelerate suberin polyphenolic formation in wounded kiwifruit (*Actinidia chinensis*). *Postharvest Biology and Technology, 187*, 111850. https://doi.org/10.1016/j.postharvbio.2022.111850.

Weichert, H., Stenzel, I., Berndt, E., Wasternack, C., & Feussner, I. (1999). Metabolic profiling of oxylipins upon salicylate treatment in barley leaves—Preferential induction of the reductase pathway by salicylate. *FEBS Letters, 464*(1–2), 133–137. https://doi.org/10.1016/S0014-5793(99)01697-X.

Wiśniewska, A., Wojszko, K., Różańska, E., Lenarczyk, K., & Sobczak, M. (2022). *Arabidopsis thaliana* AtHRS1 gene is involved in the response to *Heterodera schachtii* infection and its overexpression hampers development of syncytia and involves a jasmonic acid-dependent mechanism. *Journal of Plant Physiology, 272*, 153680. https://doi.org/10.1016/j.jplph.2022.153680.

Woldemariam, M. G., Ahern, K., Jander, G., & Tzin, V. (2018). A role for 9-lipoxygenases in maize defense against insect herbivory. *Plant Signaling & Behavior, 13*(3), 4709–4723. https://doi.org/10.1080/15592324.2017.1422462.

Wu, J., & Baldwin, I. T. (2010). New insights into plant responses to the attack from insect herbivores. *Annual Review of Genetics, 44*, 1–24. https://doi.org/10.1146/annurev-genet-102209-163500.

Wu, L., Wang, S., Tian, L., Wu, L., Li, M., Zhang, J., Li, P., Zhang, W., & Chen, Y. (2018). Comparative proteomic analysis of the maize responses to early leaf senescence induced by preventing pollination. *Journal of Proteomics, 177*, 75–87. https://doi.org/10.1016/j.jprot.2018.02.017.

Xi, Y., Cheng, D., Zeng, X., Cao, J., & Jiang, W. (2016). Evidences for chlorogenic acid—A major endogenous polyphenol involved in regulation of ripening and senescence of apple fruit. *PLoS One, 11*(2), e0146940. https://doi.org/10.1371/journal.pone.0146940.

Xing, Q., Liao, J., Cao, S., Li, M., Lv, T., & Qi, H. (2020). *CmLOX10* positively regulates drought tolerance through jasmonic acid-mediated stomatal closure in oriental melon (*Cucumis melo* var. *makuwa* Makino). *Scientific Reports, 10*(1), 17452. https://doi.org/10.1038/s41598-020-74550-7.

Xu, H., Wei, Y., Zhu, Y., Lian, L., Xie, H., Cai, Q., Chen, Q., Lin, Z., Wang, Z., Xie, H., & Zhang, J. (2015). Antisense suppression of *LOX3* gene expression in rice endosperm enhances seed longevity. *Plant Biotechnology Journal, 13*(4), 526–539. https://doi.org/10.1111/pbi.12277.

Yan, L., Zhai, Q., Wei, J., Li, S., Wang, B., Huang, T., Du, M., Sun, J., Kang, L., Li, C. B., & Li, C. (2013). Role of tomato lipoxygenase D in wound-induced jasmonate biosynthesis and plant immunity to insect herbivores. *PLoS Genetics, 9*(11), e1003964. https://doi.org/10.1371/journal.pgen.1003964.

Yang, C., Duan, W., Xie, K., Ren, C., Zhu, C., Chen, K., & Zhang, B. (2020). Effect of salicylic acid treatment on sensory quality, flavor-related chemicals and gene expression in peach fruit after cold storage. *Postharvest Biology and Technology, 161*, 111089. https://doi.org/10.1016/j.postharvbio.2019.111089.

Yang, X. Y., Jiang, W. J., & Yu, H. J. (2012). The expression profiling of the lipoxygenase (LOX) family genes during fruit development, abiotic stress and hormonal treatments in cucumber (*Cucumis sativus*). *International Journal of Molecular Sciences, 13*(2), 2481–2500. https://doi.org/10.3390/ijms13022481.

Yang, Y., Zhang, T., Wang, X., Wu, M., Li, X., Guo, X., Fang, Y., & Chen, S. (2020). Relationship between *CsLOX* gene expression and C6 and C9 aldehydes during cucumber fruit storage. *Postharvest Biology and Technology, 161*, 111085. https://doi.org/10.1016/j.postharvbio.2019.111085.

Yao, M., Zhou, X., Ji, Y., Luo, M., Sun, Y., Zhou, Q., & Ji, S. (2022). Potential of ethylene in alleviating cold-induced volatile esters loss of 'Nanguo' pears by regulating the lipoxygenase pathway. *Environmental and Experimental Botany, 194*, 104723. https://doi.org/10.1016/j.envexpbot.2021.104723.

Zhang, B., Shen, J. Y., Wei, W. W., Xi, W. P., Xu, C. J., Ferguson, I., & Chen, K. (2010). Expression of genes associated with aroma formation derived from the fatty acid pathway during peach fruit ripening. *Journal of Agricultural and Food Chemistry, 58*(11), 6157–6165. https://doi.org/10.1021/jf100172e.

Zhang, C., Cao, S., Jin, Y., Ju, L., Chen, Q., Xing, Q., & Qi, H. (2017). *Melon13-lipoxygenase CmLOX18* may be involved in C6 volatiles biosynthesis in fruit. *Scientific Reports, 7*, 2816. https://doi.org/10.1038/s41598-017-02559-6.

Zhang, P., Wu, H., Zhang, F., Cao, P., Cai, M., Song, W., & Jiang, L. (2019). Improving the storage tolerance of rice seeds by down-regulating *OsLOX* by RNAi. *Journal of Nanjing Agricultural University, 42*(5), 996–1005. https://doi.org/10.7685/jnau.201904043.

Zhang, Q., Zhao, Y., Zhang, J., Li, X., Ma, F., Duan, M., Zhang, B., & Li, H. (2021). The responses of the lipoxygenase gene family to salt and drought stress in foxtail millet (*Setaria italica*). *Life, 11*(11), 1169. https://doi.org/10.3390/life11111169.

Zhang, Y. T., Zhang, Y. L., Chen, S. X., Yin, G. H., Yang, Z. Z., Lee, S., Liu, C. G., Zhao, D. D., Ma, Y. K., Song, F. Q., & Yang, F. S. (2015). Proteomics of methyl jasmonate induced defense response in maize leaves against Asian corn borer. *BMC Genomics, 16*, 1–16. https://doi.org/10.1186/s12864-015-1363-1.

Zhang, Y., & Zhang, Y. (2020). Effect of lipoxygenase-3 on storage characteristics of peanut seeds. *Journal of Stored Products Research, 87*, 101589. https://doi.org/10.1016/j.jspr.2020.101589.

Zheng, W. Y., Zhao, L., Li, Y. M., Li, J., Zhu, Z. H., & Yao, D. N. (2022). Association analysis between Lipoxygenase activity and SSR markers in wheat grains. *Cereal Research Communications, 50*(2), 297–303. https://doi.org/10.1007/s42976-021-00168-8.

Zhou, G., Ren, N., Qi, J., Lu, J., Xiang, C., Ju, H., Cheng, J., & Lou, Y. (2014). The 9-lipoxygenase *Osr9-LOX1* interacts with the 13-lipoxygenase-mediated pathway to regulate resistance to chewing and piercing-sucking herbivores in rice. *Physiologia Plantarum, 152*(1), 59–69. https://doi.org/10.1111/ppl.12148.

Zhou, G., Zhang, F., Long, W., Chen, D., Wu, X., Cai, Q., & Wu, S. (2019). Changes in viability of Lincang hulled wheat seeds under different storage temperatures and changes in lipoxygenase (*lox*) gene activity and expression levels during germination. *International Journal of Agricultural Biology, 21*(5), 979–988. https://doi.org/10.17957/IJAB/15.0983.

Zhou, Z., Han, P., Bai, S., Ma, N., Fang, D., Yang, W., Hu, Q., & Pei, F. (2022). Caffeic acid-grafted-chitosan/polylactic acid film packaging enhances the postharvest quality of *Agaricus bisporus* by regulating membrane lipid metabolism. *Food Research International, 158*, 111557. https://doi.org/10.1016/j.foodres.2022.111557.

Zhu, F. Y., Bao, Y. J., Chen, M. X., Zhang, Y., Fernie, A. R., Zhang, J., Huang, B. X., & Cao, F. L. (2023). Emerging role of jasmonic acid in woody plant development. *Authorea*. https://doi.org/10.22541/au.167420536.65415896/v1.

Zhu, J., Wang, X., Guo, L., Xu, Q., Zhao, S., Li, F., Yan, X., Liu, S., & Wei, C. (2018). Characterization and alternative splicing profiles of the lipoxygenase gene family in tea plant (*Camellia sinensis*). *Plant, Cell & Environment, 41*(8), 1765–1781. https://doi.org/10.1093/pcp/pcy091.

Zorya, S., Morgan, N., Diaz Rios, L., Hodges, R., Bennett, B., Stathers, T., & Lamb, J. (2011). Missing food: The case of postharvest grain losses in sub-Saharan Africa. Retrieved from: http://siteresources.worldbank.org/INTARD/Resources/MissingFoods10_web_final1.pdf (accessed on 9 July 2024).

CHAPTER 12

DNA Helicase: The Futuristic Instrument to Increase Biotic Crop Stress Tolerance

SWATI RAY

Department of Food Technology, Haldia Institute of Technology, ICARE Complex, Hatiberia, Haldia, West Bengal, India

ABSTRACT

Technological development using cutting-edge molecular tools is gaining importance in expanding agricultural produce to meet the increasing demand of population expansion. Biological stress-induced responses of crop plants have emerged as a new area of research to increase crop productivity. Biological stress can be defined as any condition which may hamper the normal functioning of the plant. It includes abiotic environmental stresses like high salinity of the soil, drought conditions, cold exposure, etc., or biotic stresses like interaction with several other deleterious microorganisms, insects, pests, or endogenous metabolism that may harm the overall growth and survival of plants which also affect its yield. Lesser yield leads to huge losses that have a direct impact on the economic backbone of the country. Helicases are motor proteins that respond under stressed conditions by unwinding the nucleic acid's primary and secondary structures. Though the basic functions of these plant helicases are related to DNA repair, recombination, meiotic control, gene silencing, and replication progress with respect to genomic stabilization; they also perform some secondary functions in plants. DNA helicases, through their control and maintenance of genetic stability, act as a natural immune barrier for the plant for maintaining its innate resistivity to any foreign biological agents or their toxins. This chapter discusses the basic plant helicases that were identified in some of the plants. It also scans the

Climate Change and Biotic Factors: A Molecular Approach.
Ashutosh Singh, Saurabh Pandey, and Amarjeet Kumar, (Eds.)

probabilities of understanding the mode of action of these helicase proteins to manage various biotic stress responses of crop plants. It evaluates their potential as genetic tools for modifying crop plants to make them more resistant to biotic stresses.

12.1 INTRODUCTION

All genomes are stable in their natural form and are mostly associated with normal cellular functions. However, to perform genetic functions such as replication, transcription, recombination, or any kind of cellular repair, the naturally stable DNA double helical structure has to reach a higher energy state and unwind a portion of it. This transition to a higher energy state is fueled by the hydrolysis of nucleoside triphosphates or NTPs, which aids in the breakdown of hydrogen bonds and other covalent or non-covalent interactions in the duplex DNA so that it can unwind and become available to be used for various activities of the cell. These DNA-dependent NTPase activities generally occur in three stages:

- Binding of the helicase to the Duplex DNA strand
- Nucleoside triphosphate (NTP) or Di-nucleoside triphosphates (dNTP) binding and simultaneous hydrolysis of the molecules
- The unwinding of the DNA double-stranded helical structure.

The unwinding of the energetically stable DNA structure is only possible with the help of some motor proteins or enzymes called helicases that utilize the hydrolytic energy of NTP or dNTP to open up a portion of the DNA strand for necessary metabolic functions (Tuteja, 2003). The first ever unwinding of DNA with the help of proteins was reported by Hoffmann-Berling and his co-workers in 1976, where they isolated the first prokaryotic DNA helicase from *Escherichia coli.* Later in 1978, DNA helicase was isolated from a Lily plant and given the term "Helicase" by Hoffmann-Berling. Over time, eight more helicases were isolated from plants that share some common characteristics (Brosh et al., 2003). Biotic and abiotic stresses surrounding the plant may lead to the generation of pro-oxidants like reactive oxygen species or ROS. Biotic stresses such as insects, pests, and bacterial or fungal toxins may even produce hydrogen peroxide, superoxide, hydroxyl ion, and singlet oxygen from ROS (Figure 12.1). ROS can also activate signal transduction pathways that may cause irreversible damage to plant cells (Huang et al., 2019).

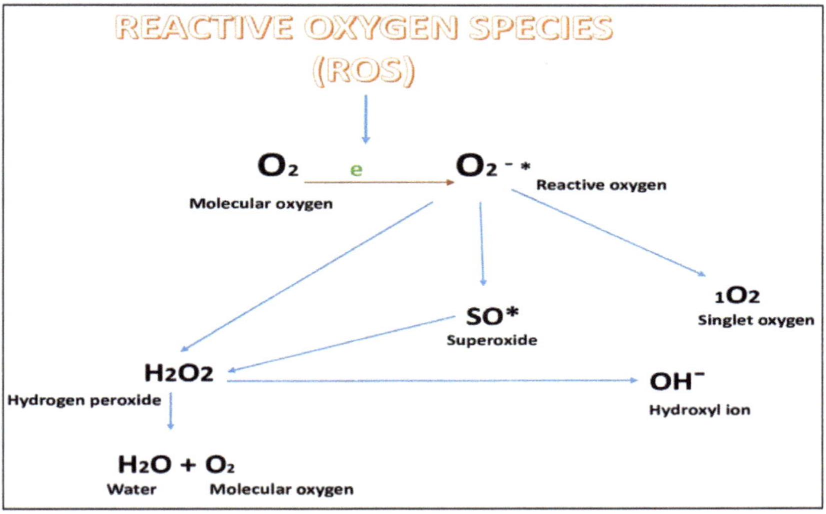

FIGURE 12.1 Conversion of reactive oxygen species to its various high-energy components in response to biotic stress.

Generation of this oxidative stress may cause retardation in plant growth and development, delayed growth, reduction in the amount of seed production, etc. (Gilroy et al., 2014; Loiacono et al., 2012; Miller et al., 2010; Roldan-Arjona and Ariza, 2009; Bray and West, 2005).

DNA Helicases have shown evidence of involvement in biological processes such as replication, transcription of rDNA, initiation of translation, genetic recombination, pre-translational rRNA processing, repair of DNA, telomere repair, excision repair, cell proliferation during flower development, maintenance of pattern in genomic methylation, a continuation of the plant cell cycle, etc., mainly in the cell organelles like nucleus, mitochondria, and chloroplast (Tuteja, 2003). Though the *in vivo* tests of these DNA helicases are still not well reported especially for eukaryotic cells; clear evidence indicates their role in the overall cellular function, growth, and development of plants which makes them an ideal tool for biotechnological engineering of DNA. This chapter will discuss in general different plant helicases and the kinds of biotic stresses any plant encounters. Then specifically it will focus on the role of some of these helicases in the management of biotic stresses in plants.

12.2 BIOTIC STRESSES IN PLANTS

Living organisms are gifted with various natural strategies to deal with stress. Animals have the advantage that they can simply avoid some dangerous exposure by motion. The non-motile nature of plants has made it tricky to survive under these challenging conditions (Krishna and Yan, 2015). Plant growth is severely hampered by a number of stress factors such as abiotic, biotic, and endogenous factors. The environmental or external factor is considered abiotic that may create abiotic stresses hampering the overall growth of the plant. Examples of abiotic stresses may include high salinity, severe drought, extreme heat or cold, the presence of heavy metals, exposure to ozone or UV radiation, and nutritional deficiencies. These factors are also involved in affecting productivity as a direct or indirect consequence of global climate change (Mareri et al., 2022). Biotic stress occurs mostly as a result of the presence of deleterious microorganisms such as bacteria, viruses, fungi, or their toxins. It may also include biological agents like insects, arachnids, nematodes, or weeds. These agents may actually kill the host plant or are responsible for depriving the host of its nutrients, which may lead to stunted growth and vigor of plants (Figure 12.2). This causes a large amount of pre and postharvest losses, which have direct consequences on the economy of the country. Naturally, plants do not have any adaptive nature of an immune system or memory of past infection, which makes them extremely vulnerable to any kind of diseases. In order to cope with the disease breakdown, plants have developed certain biotic stress resistance genes that act as their defense system (Singla and Krattinger, 2016).

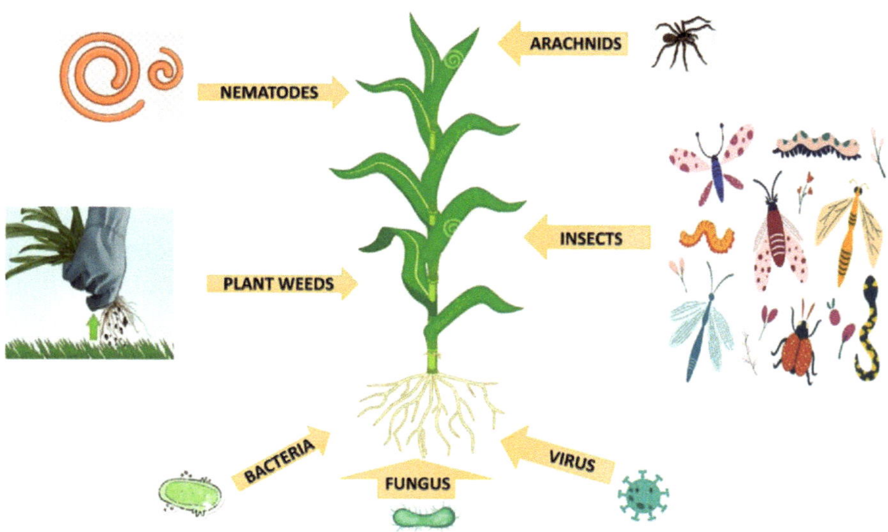

FIGURE 12.2 The biotic factors that affect any plant growth.

The endogenous factors are also known as internal or genetic factors that often give different characteristics in the same types of plants under the same environmental conditions. Phytohormones or regulators that are present in trace amounts in plants can act both as growth enhancers (Auxin, gibberellin, cytokinin) or inhibitors (Abscisic acid) or may have dual functions (Ethylene). They may also include the metabolic systems of the plant which determine the ultimate growth system and morphology of plants.

These factors have cumulative, synergistic, or potentiating effects that imbalance the amount of ROS that should be present in the cell, causing DNA lesions and peroxidation of lipids. However, plants have developed a set of repair mechanisms to maintain the integrity and uniformity of the membrane as well as its genetic framework (de Medeiros et al., 2015).

Some studies reported in Australia studied on foliar diseases of barley such as net blotch (*Drechslera teres* f.sp. *teres*), spotty net blotch (*Drechslera teres* f.sp. *maculata*), powdery mildew (*Blumeria graminis*), barley scald (*Rhynchosporium secalis*), leaf rust (*Puccinia hordei*), yellow dwarf virus, etc. These biotic elements or stresses may affect overall yield and crop quality by growing on or inside the plants causing stunted growth, chlorosis, rot, or local injury (Balodi et al., 2017). An average of 32% loss can be perceived in yield for grains, vegetables, or other plantation crops which not only inhibits production but also a threat to food safety and nutrition (Wang et al., 2013; Eicher-Miller et al., 2009).

12.3 TYPES AND FUNCTIONS OF PLANT DNA HELICASE

DNA is the self-replicating carrier of genetic information, which must be converted to RNA and ultimately to protein in order to engage in the daily activity of any living organism. The unwinding of the DNA double-stranded duplex structure is obligatory to continue with this train of activities performed by the molecule. DNA helicases are a group of protein molecules, typically enzymes, that perform the process of DNA unwinding to make it available for the synthesis of RNA and protein. They have a higher affinity towards single-stranded DNA than towards double-stranded DNA, which is essential for NTP binding NTPase activity needed to unwind duplex DNA. The binding results in a conformational change in the helicases that causes a change in protein structure, though the presence of single-stranded DNA enhances the function. It also helps to safeguard the genomic stability of plants, which may help in their survival. Though the nature of functions performed in different species or cell types is different, it is highly conserved due to their ability to stabilize the genome in all circumstances. Functional variations along with numbers in different kingdoms may vary with their presence in different genetic orthologs (Dorn and Puchta, 2019). The different kinds of helicases that can bring forth stress-induced open reading frames available for plants can be summarized below.

12.3.1 RECQ HELICASES

RecQ is one of the most ubiquitous helicases discovered so far, as it is involved in almost all domains. Its number in different organisms increases with the increase in genetic complexity, but the presence of at least one homolog is evident. The RecQ helicase family in plants consists of six subfamilies (Hartung and Puchta, 2006), which can be illustrated in Figure 12.3.

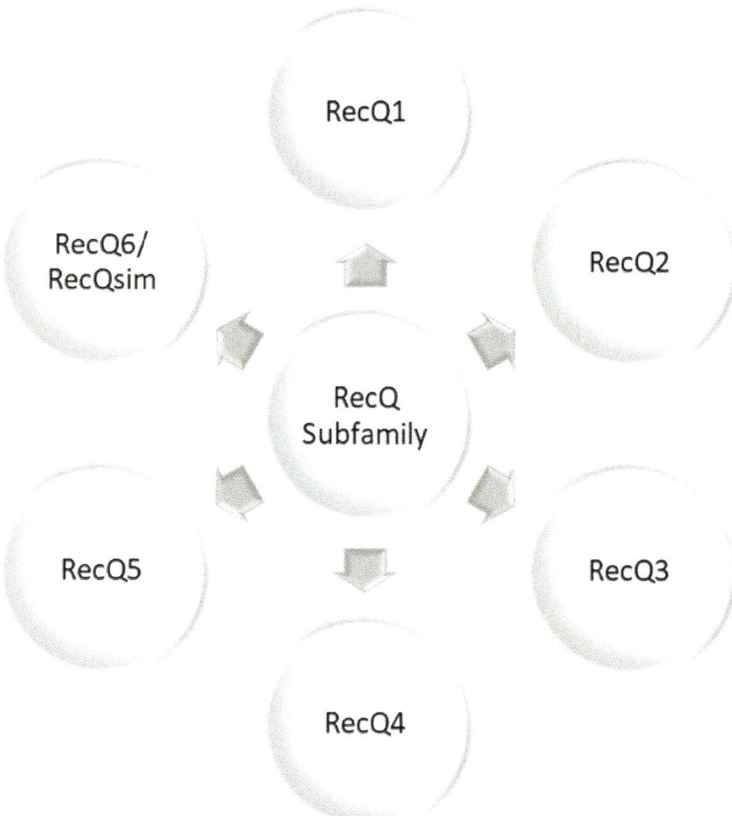

FIGURE 12.3 The RecQ helicase subfamilies.

The RecQ4 subfamily is the most abundant one, which is proven to have the most contribution in maintaining genomic stability (Ellis et al., 1995). RecQ4A and RecQ4B are the RecQ4 homologs that have been isolated from the plant Arabidopsis, of which RecQ4A was found to be a functional homolog of the similar BLM in humans (Hartung et al., 2000; Hartung et al., 2007). In a tomato hybrid plant between *Solanum lycopersicum* and *S. pimpinellifolium,* a 1.8-fold increase in CO was found due to a biallelic recq4 knockout (de Maagd et al., 2020). Very strong inhibitory effects of RecQ4 found in moss. Physcomitrium patens with similar conformation result in various pea and rice hybrid plants (Wiedemann et al., 2018; Mieulet et al., 2018). Both RecQ2 and RecQ3 have specificities towards DNA on a single molecular level as observed in *Arabidopsis thaliana*. While RecQ2 prefers to unwind DNA and RecQ3 has typical rewinding activity after unwinding for a short stretch of DNA. The 3'–5' helicase activity for each protein is found to be ATP-dependent (Klaue et al., 2013). In *Oryza sativa*, simultaneous expression of RecQ1, RecQ2, RecQ6, or RecQsim was observed stimulated by MMS, bleomycin, and crosslinker MMC (Saotome et al., 2006). Plants and some fungi may possess an additional helicase that has similar characteristics to the classical RecQ homologs. Hrq1, identified in yeast, can perform 3'–5' helicase activity along with cross-linked repair with nucleotide excision repair (Groocock et al., 2012; Kwon et al., 2012; Choi et al., 2014).

12.3.2 FE-S CLUSTER HELICASES

This family is characterized by the presence of bound metal in its core structure. Mutation in these proteins may cause hereditary diseases that may hamper the overall genomic stability (Rudolf et al., 2006; Suhasini and Brosh, 2013). The homologs of human helicases XPD, CHL1, FANCJ, and RTEL1 were identified in plants as well, of which FANCJ and RTEL1 are related to genomic stability in plants (Liu et al., 2003). The RTEL1 (regulator of telomere elongation helicase 1) is one of the major FE-S cluster helicases that are significant both in plants and animal cells with respect to genomic stability. The homolog of RTEL1 present in *Arabidopsis thaliana* was found to be inhibitory to homologous recombination and repairs crosslinks and DNA damage associated with replication in the meristem of the root system of the plant. Severe developmental defects such as degenerated root development, fused, flattened shoot system, stunted growth, and infertility set off if there is a parallel loss of helicases RTEL1 and RecQ4A (Recker et al., 2014).

Physcomitrium patens RTEL1 homolog proves that the basic DNA repair functions are associated with this helicase as the absence of it shows extreme growth deficiencies and sensitivity to bleomycin and irradiation by ultraviolet light (Kami Sugi et al., 2016; Goffova et al., 2019). RTEL1 and telomerase TERT deficient *Arabidopsis* and *Physcomitrium* plants may have shorter telomeres resulting in developmental defects as they have stabilizing effects in case of repetitive DNA sequences in plants. In both plants, RTEL1 affects the stability of 45S rDNA as there is evidence of a reduction in copy number found in the respective mutant. This is significant as it shows conserved functions of these helicases in the stabilization of repetitive sequences in the plant genome (Recker et al., 2014).

Intercrosslinked (ICL) repair of DNA is possible through the use of FANCJ helicase homolog, which is a member of the Fanconi anemia (FA) pathway (Kim et al., 2012). *Arabidopsis* contains this FANCJ homolog named AtFANCJA and AtFANCJB, which are separated by only two genes on chromosome 1 and have diverse functions. Though AtFANCJB does not perform any function related to *HR* gene suppression, as it is more associated with mutant lines of wild-type plants (Barber et al., 2008). Both helicases have specific autonomous functions that could be identified in root meristems of plants. The main function includes replication-based DNA repair and stabilizing rDNA repeats, as is evident from the reduction to one-third copy number in 45S rDNA in the double mutant compared to the wild-type plant variant (Dorn et al., 2019).

12.3.3 OTHER SIGNIFICANT HELICASES

Besides the basic helicases mentioned above, the multiple helicase family plays important roles in stabilizing genomes in plants through DNA repair, recombination, and replication. Due to the limited number of studies available on plants, only some isolated reports suggest their roles (Niu and Klein, 2016). The SRS2 helicase homolog in *Arabidopsis* exhibits 3'–5' helicase activity in branched DNA structures by unwinding Holliday junctions partially and performing synthesis-dependent strand annealing. It can also perform the exceptional work of re-annealing the displaced DNA strand after elongation is complete (Blanck et al., 2009).

FANCM is another FA-associated helicase that stabilizes genomes in plants by supporting them on multiple levels. Its function in somatic HR is inimical, which makes all naturally occurring HR mutant lines unregulated (Knoll et al., 2012). It repairs replicative DNA damage without MUS81 endonuclease in the lethal variant of the double mutant line (Dangel et al., 2014). The absence of FANCM causes an anomaly in meiotic recombination with an increase in crosslinking three times normal (Crismani et al., 2012; Blary et al., 2018). In *Brassica rapa* and *Brassica napus*, the CO formation is subdued due to FANCM activity (Tuteja et al., 2011).

The MCM2–7 helicase performs authorization and replication initiation, which is highly regulated and can conduct systematic unwinding of DNA (Di et al., 2009). MCM proteins are very important in the DNA replication of *Arabidopsis* since their over- or under-expression may have some inhibitory effects that could even lead to embryo lethality (Herridge et al., 2014). The MCM2–7 helicase also plays a role in seed development, where each subunit has its own unique function. A defective MCM2–7 complex may accumulate damaged DNA and prevent DNA repair genes from functioning (Rizvi et al., 2016). MCM3 and MCM6 from *Arabidopsis* and pea show ATP-dependent unwinding and 3’–5’ helicase activity (Tran et al., 2010; Dang et al., 2011). The overexpression of MCM6 in tobacco plants helps in salinity resistance, which is actually a type of abiotic stress (Dang et al., 2011). The MCM8 of *Arabidopsis* shows meiotic chromosomal fragmentation without affecting the number of crosslinks (Crismani et al., 2013).

The C-terminal DNA polymerase and the N-terminal DNA helicase are combined in a single genetic product called POLQ, which has typical similarities with the Rec Q family of helicases. The homologs for POLQ were identified in plants and named TEBICHI (TEB) (Inagaki et al., 2006). This helicase is chiefly involved in DNA repair, showing hypersensitive mutant lines (Klemm et al., 2017; Gelvin S, 2017). In *Physcomitrium patens,* the POLQ homolog shows prominent expression after a double-strand break is induced, confirming its role in the double-strand repair mechanism (Goffova et al., 2019). T-DNA integrated plant transformation *via* Agrobacterium into the genomes of plants is a crucial step for applied biotechnology that is mediated by POLQ homolog in plants (Gelvin, 2017).

So, it becomes very clear that the function of helicases in plants with respect to genetic stabilization is highly conserved. The overall role of the above-mentioned plant helicases can be summarized in Figure 12.4. The plant helicases that are mainly identified in some plant species include Arabidopsis, rice, pea, soybean, lily, etc. (Tuteja and Tuteja, 2004; Pandey et al., 2020).

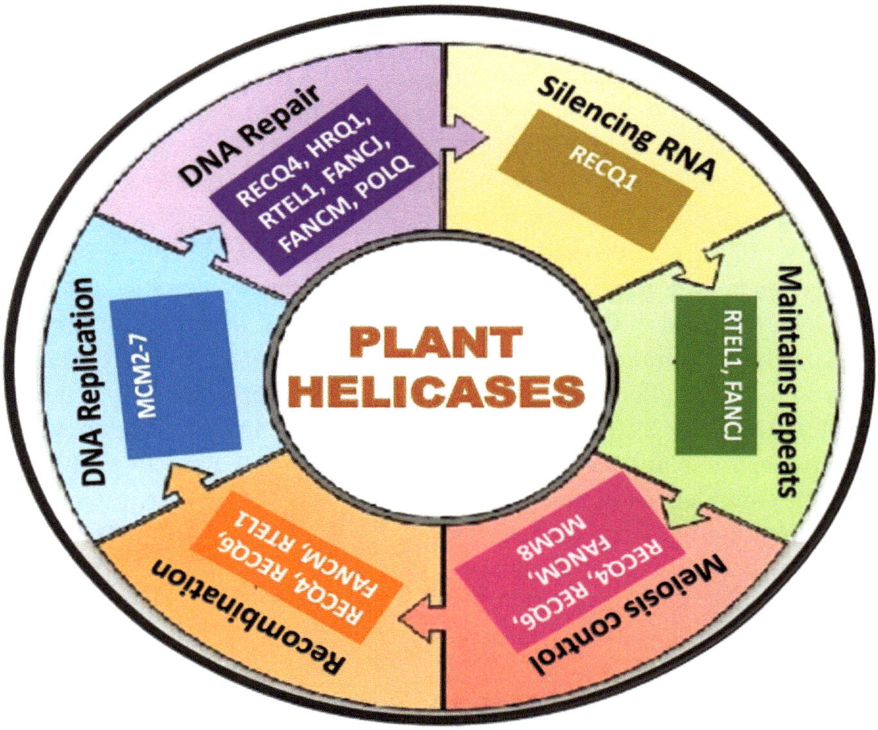

FIGURE 12.4 The summation of the functions of different plant helicases.

12.4 ROLE OF DNA HELICASE IN BIOTIC STRESS MANAGEMENT

Various plant diseases caused by microorganisms, pests, insects, arachnids, nematodes, and weeds compel the inclusion of advanced technologies in crop production in order to prevent large annual crop production loss not only locally but on a worldwide scale. High demand for the consumer-preferred but disease-labile crops leads to their larger production, which makes it more difficult to control and eradicate diseases altogether. This also creates environmental conditions that fuel the growth of more and more disease-producing agents.

The population explosion worldwide requires a continued supply of staple food with higher yields since the amount of cultivable land is also shrinking.

This is not possible without the abolishment of any kind of plant diseases that may hamper yield or crop sustainability. Therefore, the use of traditional as well as novel management techniques to avert crop disease has become an absolute necessity. Plant helicase enzymes have the capacity to reduce or inhibit diseases in crops which is often considered a natural defense mechanism that the plant uses to protect themselves to make up for their sessile nature (Sundin et al., 2016). In coping with the biotic stress factors highlighted above, helicases are considered quite efficient. Table 12.1 enlists some of these incidences of helicase action towards genomic maintenance which acts as a natural immune system for prevention of plant diseases.

TABLE 12.1 Actions of Different Plant Helicases in Genomic Stability in Various Crop Plants

Name of the Helicase	Function	Origin or Source	Reference
AtRECQ2	• Cures semi-mobility of Holliday Junction (HJ) • Helps in regression of replication fork		Kobbe et al., 2008
AtRECQ3	• In vitro stabilization of replication fork		Kobbe et al., 2009
AtRECQ4A	• Associated with DNA damage repair • During prophase, regulates telomeric interaction among non-homologous chromosomes		Hartung et al., 2007
AtRECQsim	• Suppresses methyl methane sulphonate (MMS) sensitivity in mutant variants	Thale cress (*Arabidopsis thaliana*)	Bagherieh-Najjar, M.B. 2003
MER3	• Discourages mutant generation by reduced fertility and crossover in mutants		Chen et al., 2005
HRQ1	• DNA crosslinking and excision repair		Rohrig et al., 2018
MCM2-MCM7	• Involved in stabilizing replication process in plants		Herridge et al., 2014
RTEL1	• Plant DNA replication, repair, and recombination • Maintenance of telomere by stabilizing DNA repeats		Hu et al., 2015
AtKu70 & AtKu80	• Participates in repair of double strand breakage.		Tamura et al., 2002
AtXPB1 & AtXPD/UVH6	• Associated with pathway of nucleotide excision repair.		Costa et al., 2002; Tuteja and Tuteja, 2013
CTF7	• Maintains stability of genes in plants through control of meiosis.		Mannuss et al., 2018

TABLE 12.1 *(Continued)*

Name of the Helicase	Function	Origin or Source	Reference
RAD5A	• Direct role in DNA repair and genetic recombination to maintain stability		Mannuss et al., 2018
AtFANCJB	• Intrastrand crosslink repair of plant DNA		Knoll et al., 2012
FANCM	• Deals with meiotic crossovers		Yelina et al., 2015
OsRECQ1, OsRECQ2, OsRECQsim & OsRECQ886	• Gene expression is stimulated by toxic agents in meristematic tissue • Helps in the control and silencing of genes.	Rice (*Oryza sativa*)	Knoll and Puchta, 2011
PsMCM6	• Helps in the plant replication process	*Pea* (*Pisum sativum)*	Dangel et al., 2014
DNA helicase	• 3'–5' helicase activity with ATP binding • Hydrolase activity • Nucleic acid binding • Aids in DNA repair, recombination, and replication	Wild Soybean (*Glycine soja*)	Uniprot, 2023a
DNA helicase	• ATP-dependent and independent DNA unwinding and maintenance of replication stability	White Lily (*Lilium candidum*)	Hotta and Stern, 1978
Rep Protein	• Enhances virus-induced gene silencing through Sardinia virus	Tomato (*Solanum lycopersicum*)	Clerot and Bernardi, 2006
AC3 protein	• Enhances virus-induced gene silencing through the Gemini virus	Tomato (*Solanum lycopersicum*)	Pasumarthy et al., 20111
DNA helicase	• ATP dependent helicase activity similar to RecQ.	Field mustard (*Brassica rapa*)	String Consortium, 2023
DNA helicase	• Associated with nucleotide excision repair • 3'–5' helicase activity with ATP binding • Hydrolase activity • DNA binding and incision	Canola or Rapeseed (*Brassica napus*)	Uniprot, 2023b
RecQ4	• Embryogenesis and development • Homologous recombination	Spreading Earthmoss (*Physcomitrium patens*	Wiedemann et al., 2018; Zhu L, 2021
RTEL1	• Basic DNA repair function • Stabilization of repetitive sequences	Spreading Earthmoss (*Physcomitrium patens*	Kamisugi et al., 2016; Goffova et al., 2019

Though stresses are undesirable and affect yield and overall development of crops, the silver lining is that they actually induce the capacity of the repair pathways, giving rise to mutations and genetic variations. New and improved morphological characters and genotypes help in evolving the plant through adaptation in different adverse ecosystems (Wielgoss et al., 2013). Plant breeders and scientists have utilized this to their advantage to develop new plant variants which naturally resist biotic stresses from germination to maturity (Oladosu et al., 2016; Chaudhary et al., 2019; Raina et al., 2020). The pathways at different cell cycle stages through which the stabilization or damage repair takes place can be one of these – Nucleotide excision repair, mismatch repair, base excision repair, homologous recombination repair, non-homologous end-joining repair, interstrand crosslink repair, direct chemical reversal, etc. (Chatterjee and Walker, 2017). The maneuver for gene stabilization and biotic stress management in plants can be illustrated in Figure 12.5.

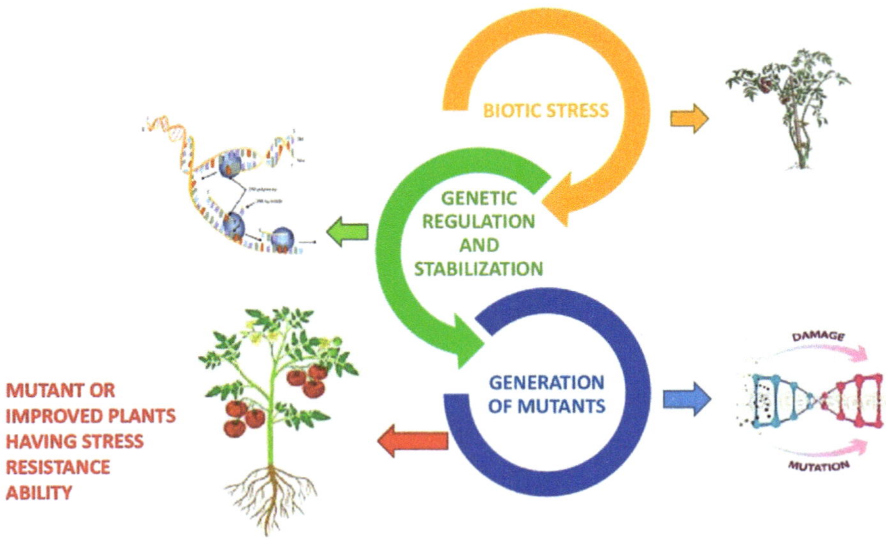

FIGURE 12.5 The overall schema of helicase performance in biotic stress management.

12.4 SUMMARY

The vehicle of genetic information transfer is entrusted with deoxyribonucleic acid or DNA. In order to perform and regulate genetic functions, a large number of enzymes are assigned the task of maintaining stability in the regular functioning of genes. Through processes like DNA replication, transcription, or various other repair mechanisms, the helicase group of enzymes prevents genetic instability. Due to the non-motile nature of plants, they are subjected to more stress than animals or human beings. So this set of helicase enzymes is their way to mitigate damage in their cells. These inherent abilities of plants not only help them to survive but also eliminate repetitive genes which are otherwise very commonly known to hamper genetic stability in plants. The correct interplay of these helicases is necessary to alleviate various biotic stresses that may influence crop quality and yield. This also facilitates innovation in the production of new varieties of plants through crossing over or mutant generation that can be more resistant to any form of biotic stress. Plant breeders often utilize this trait of helicases to gain control over plant genes through genetic engineering. Though the basic enzyme function is conserved in most plants, with species variation their actions become slightly different. The knowledge about these wonderful genetic tools, helicases, not only will help researchers to upgrade agricultural practices but also it will become the main weapon in the eradication of food insecurity worldwide.

KEYWORDS

- **biotic stress**
- **helicase**
- **molecular tool**
- **plant stress tolerance**

REFERENCES

Bagherieh-Najjar, M. B. (2003). Arabidopsis RecQsim, a plant-specific member of the RecQ helicase family, can suppress the MMS hypersensitivity of the yeast sgs1 mutant. *Plant Molecular Biology, 52*(3), 273–284.

Balodi, R., Bisht, S., Ghatak, A., & Rao, K. H. (2017). Plant disease diagnosis: Technological advancements and challenges. *Indian Phytopathology, 70*(2), 275–281.

Barber, L. J., Youds, J. L., Ward, J. D., McIlwraith, M. J., O'Neil, N. J., Petalcorin, M. I. R., Martin, J. S., Collis, S. J., Cantor, S. B., Auclair, M., et al. (2008). RTEL1 maintains genomic stability by suppressing homologous recombination. *Cell, 135*(2), 261–271.

Blanck, S., Kobbe, D., Hartung, F., Fengler, K., Focke, M., & Puchta, H. (2009). A SRS2 homolog from Arabidopsis thaliana disrupts recombinogenic DNA intermediates and facilitates single strand annealing. *Nucleic Acids Research, 37*(21), 7163–7176.

Blary, A., Gonzalo, A., Eber, F., Bérard, A., Bergès, H., Bessoltane, N., Charif, D., Charpentier, C., Cromer, L., Fourment, J., et al. (2018). FANCM limits meiotic crossovers in Brassica crops. *Frontiers in Plant Science, 9*, Article 368. https://doi.org/10.3389/fpls.2018.00368.

Bray, C. M., & West, C. E. (2005). DNA repair mechanisms in plants: Crucial sensors and effectors for the maintenance of genome integrity. *New Phytologist, 168*(3), 511–528. https://doi.org/10.1111/j.1469-8137.2005.01548.x.

Brosh, R. M. Jr., & Matson, S. W. (2020). History of DNA Helicases. *Genes, 11*(3), Article 255. https://doi.org/10.3390/genes11030255.

Chatterjee, N., & Walker, G. C. (2017). Mechanisms of DNA damage, repair, and mutagenesis. *Environmental and Molecular Mutagenesis, 58*(4), 235–263. https://doi.org/10.1002/em.22087.

Chaudhary, J., Deshmukh, R., & Sonah, H. (2019). Mutagenesis approaches and their role in crop improvement. *Plants, 8*(11), Article 467. https://doi.org/10.3390/plants8110467.

Chen, C., Zhang, W., Timofejeva, L., Gerardin, Y., & Ma, H. (2005). The Arabidopsis ROCK-N-ROLLERS gene encodes a homolog of the yeast ATP-dependent DNA helicase MER3 and is required for normal meiotic crossover formation: Arabidopsis crossover formation requires an MER3 homolog. *The Plant Journal, 43*(3), 321–334.

Choi, D. H., Min, M. H., Kim, M. J., Lee, R., Kwon, S. H., & Bae, S. H. (2014). Hrq1 facilitates nucleotide excision repair of DNA damage induced by 4-nitroquinoline-1-oxide and cisplatin in Saccharomyces cerevisiae. *Journal of Microbiology, 52*(4), 292–298.

Clérot, D., & Bernardi, F. (2006). DNA helicase activity is associated with the replication initiator protein rep of tomato yellow leaf curl geminivirus. *Journal of Virology, 80*(22), 11322–11330. https://doi.org/10.1128/JVI.00924-06.

Costa, R. M. A., Morgante, P. G., Berra, C. M., Nakabashi, M., Bruneau, D., Bouchez, D., Sweder, K. S., Van Sluys, M. A., & Menck, C. F. M. (2002). The participation of AtXPB1, the XPB/RAD25 homologue gene from Arabidopsis thaliana, in DNA repair and plant development: AtXPB1 functions in DNA repair and development. *The Plant Journal, 28*(4), 385–395.

Crismani, W., Girard, C., Froger, N., Pradillo, M., Santos, J. L., Chelysheva, L., Copenhaver, G. P., Horlow, C., & Mercier, R. (2012). FANCM limits meiotic crossovers. *Science, 336*(6088), 1588–1590.

Crismani, W., Portemer, V., Froger, N., Chelysheva, L., Horlow, C., Vrielynck, N., & Mercier, R. (2013). MCM8 is required for a pathway of meiotic double-strand break repair independent of DMC1 in Arabidopsis thaliana. *PLoS Genetics, 9*(12), e1003165.

Dang, H. Q., Tran, N. Q., Gill, S. S., Tuteja, R., & Tuteja, N. (2011). A single subunit MCM6 from pea promotes salinity stress tolerance without affecting yield. *Plant Molecular Biology, 76*(1–2), 19–34.

Dangel, N. J., Knoll, A., & Puchta, H. (2014). MHF1 plays Fanconi anaemia complementation group M protein (FANCM)-dependent and FANCM-independent roles in DNA repair and homologous recombination in plants. *The Plant Journal, 78*(5), 822–833.

de Maagd, R. A., Loonen, A., Chouaref, J., Pele, A., Meijer-Dekens, F., Fransz, P., & Bai, Y. (2020). CRISPR/Cas inactivation of RECQ4 increases homeologous crossovers in an interspecific tomato hybrid. *Plant Biotechnology Journal, 18*(4), 805–813. https://doi.org/10.1111/pbi.13248.

de Medeiros, N. M. C., de Medeiros, A. L. M., Silva, H. C., & Scortecci, K. C. (2015). Recent advances in plant DNA repair. In C. C. Clarke (Ed.), *Advances in DNA Repair*. IntechOpen. https://doi.org/10.5772/59998.

Di Ni, A., Sozzani, R., Blanchet, S., Domenichini, S., Reuzeau, C., Cella, R., Bergounioux, C., & Raynaud, C. (2009). The Arabidopsis MCM2 gene is essential to embryo development and its over-expression alters root meristem function. *New Phytologist, 184*(2), 311–322.

Dorn, A., & Puchta, H. (2019). DNA helicases as safekeepers of genome stability in plants. *Genes, 10*(12), Article 1028. https://doi.org/10.3390/genes10121028.

Dorn, A., Feller, L., Castri, D., Röhrig, S., Enderle, J., Herrmann, N. J., Block-Schmidt, A., Trapp, O., Köhler, L., & Puchta, H. (2019). An Arabidopsis FANCJ helicase homologue is required for DNA crosslink repair and rDNA repeat stability. *PLoS Genetics, 15*(6), e1008174.

Eicher-Miller, H. A., Mason, A. C., Abbott, A. R., McCabe, G. P., & Boushey, C. J. (2009). The effect of Food Stamp Nutrition Education on the food insecurity of low-income women participants. *Journal of Nutrition Education and Behavior, 41*(3), 161–168. https://doi.org/10.1016/j.jneb.2008.06.004.

Ellis, N. A., Groden, J., Ye, T. Z., Straughen, J., Lennon, D. J., Ciocci, S., Proytcheva, M., & German, J. (1995). The Bloom's syndrome gene product is homologous to RecQ helicases. *Cell, 83*(4), 655–666.

Gelvin, S. B. (2017). Integration of Agrobacterium T-DNA into the plant genome. *Annual Review of Genetics, 51*, 195–217.

Gilroy, S., Suzuki, N., Miller, G., Choi, W., Toyota, M., Devireddy, A. R., & Mittler, R. (2014). A tidal wave of signals: Calcium and ROS at the forefront of rapid systemic signaling. *Trends in Plant Science, 19*(10), 623–630. https://doi.org/10.1016/j.tplants.2014.06.013.

Goffová, I., Vágnerová, R., Peška, V., Franek, M., Havlová, K., Holá, M., Zachová, D., Fojtová, M., Cuming, A., Kamisugi, Y., et al. (2019). Roles of RAD51 and RTEL1 in telomere and rDNA stability in Physcomitrella patens. *The Plant Journal, 98*(6), 1090–1105.

Groocock, L. M., Prudden, J., Perry, J. J. P., & Boddy, M. N. (2012). The RecQ4 orthologue Hrq1 is critical for DNA interstrand cross-link repair and genome stability in fission yeast. *Molecular and Cellular Biology, 32*(2), 276–287.

Hartung, F., & Puchta, H. (2006). The RecQ gene family in plants. *Journal of Plant Physiology, 163*(3), 287–296. https://doi.org/10.1016/j.jplph.2005.10.013.

Hartung, F., Plchová, H., & Puchta, H. (2000). Molecular characterisation of RecQ homologues in Arabidopsis thaliana. *Nucleic Acids Research, 28*(21), 4275–4282.

Hartung, F., Suer, S., & Puchta, H. (2007). Two closely related RecQ helicases have antagonistic roles in homologous recombination and DNA repair in Arabidopsis thaliana. *Proceedings of the National Academy of Sciences, 104*(47), 18836–18841.

Hartung, F., Suer, S., & Puchta, H. (2007). Two closely related RecQ helicases have antagonistic roles in homologous recombination and DNA repair in Arabidopsis thaliana. *Proceedings of the National Academy of Sciences, 104*(47), 18836–18841.

Herridge, R. P., Day, R. C., & Macknight, R. C. (2014). The role of the MCM2–7 helicase complex during Arabidopsis seed development. *Plant Molecular Biology, 86*(1–2), 69–84.

Hotta, Y., & Stern, H. (1978). DNA unwinding protein from meiotic cells of Lilium. *Biochemistry, 17*(9), 1872–1880. https://doi.org/10.1021/bi00603a011.

Hu, Z., Cools, T., Kalhorzadeh, P., Heyman, J., & De Veylder, L. (2015). Deficiency of the Arabidopsis helicase RTEL1 triggers a SOG1-dependent replication checkpoint in response to DNA cross-links. *The Plant Cell, 27*(1), 149–161.

Huang, H., Ullah, F., Zhou, D. X., Yi, M., & Zhao, Y. (2019). Mechanisms of ROS regulation of plant development and stress responses. *Frontiers in Plant Science, 10,* Article 800. https://doi.org/10.3389/fpls.2019.00800.

Inagaki, S., Suzuki, T., Ohto, M. A., Urawa, H., Horiuchi, T., Nakamura, K., & Morikami, A. (2006). Arabidopsis TEBICHI, with helicase and DNA polymerase domains, is required for regulated cell division and differentiation in meristems. *The Plant Cell, 18*(4), 879–892.

Kamisugi, Y., Whitaker, J. W., & Cuming, A. C. (2016). The transcriptional response to DNA-double-strand breaks in Physcomitrella patens. *PLoS ONE, 11*(9), e0161204.

Kim, H., & D'Andrea, A. D. (2012). Regulation of DNA cross-link repair by the Fanconi anemia/BRCA pathway. *Genes & Development, 26*(13), 1393–1408.

Klaue, D., Kobbe, D., Kemmerich, F., Kozikowska, A., Puchta, H., & Seidel, R. (2013). Fork sensing and strand switching control antagonistic activities of RecQ helicases. *Nature Communications, 4,* Article 2024. https://doi.org/10.1038/ncomms3024.

Klemm, T., Mannuß, A., Kobbe, D., Knoll, A., Trapp, O., Dorn, A., & Puchta, H. (2017). The DNA translocase RAD5A acts independently of the other main DNA repair pathways, and requires both its ATPase and RING domain for activity in Arabidopsis thaliana. *The Plant Journal, 91*(4), 725–-740.

Knoll, A., & Puchta, H. (2011). The role of DNA helicases and their interaction partners in genome stability and meiotic recombination in plants. *Journal of Experimental Botany, 62*(5), 1565–1579.

Knoll, A., Higgins, J. D., Seeliger, K., Reha, S. J., Dangel, N. J., Bauknecht, M., & Puchta, H. (2012). The Fanconi anemia ortholog FANCM ensures ordered homologous recombination in both somatic and meiotic cells in Arabidopsis. *The Plant Cell, 24*(4), 1448–1464.

Kobbe, D., Blanck, S., Demand, K., Focke, M., & Puchta, H. (2008). AtRECQ2, a RecQ-helicase homologue from Arabidopsis thaliana, is able to disrupt different recombinogenic DNA-structures in vitro. *The Plant Journal, 55*(3), 397–405. https://doi.org/10.1111/j.0960-7412.2008.03511.x.

Kobbe, D., Blanck, S., Focke, M., & Puchta, H. (2009). Biochemical characterization of AtRecQ3 reveals significant differences relative to other RecQ helicases. *Plant Physiology, 151*(3), 1658–1666.

Krishna, N., & Yan, L. (2015). A paradigm of reactive oxygen species and programmed cell death in plants. *Journal of Cell Science & Therapy, 6*(2), 1–2. https://doi.org/10.4172/2157-7013.1000202.

Kwon, S. H., Choi, D. H., Lee, R., & Bae, S. H. (2012). Saccharomyces cerevisiae Hrq1 requires a long 3'-tailed DNA substrate for helicase activity. *Biochemical and Biophysical Research Communications, 427*(4), 623–628.

Liu, Z., Hong, S. W., Escobar, M., Vierling, E., Mitchell, D. L., Mount, D. W., & Hall, J. D. (2003). Arabidopsis UVH6, a homolog of human XPD and yeast RAD3 DNA repair genes, functions in DNA repair and is essential for plant growth. *Plant Physiology, 132*(3), 1405–1414.

Loiacono, F., & De Tullio, M. (2012). Why we should stop inferring simple correlations between antioxidants and plant stress resistance: Towards the antioxidomic era. *OMICS: A Journal of Integrative Biology, 16*(3), 160–167. https://doi.org/10.1089/omi.2011.0099.

Mannuss, A., Dukowic-Schulze, S., Suer, S., Hartung, F., Pacher, M., & Puchta, H. (2010). RAD5A, RECQ4A, and MUS81 have specific functions in homologous recombination and define different pathways of DNA repair in Arabidopsis thaliana. *The Plant Cell, 22*(10), 3318–3330.

Mareri, L., Parrotta, L., & Cai, G. (2022). Environmental stress and plants. *International Journal of Molecular Sciences, 23*(10), Article 5416. https://doi.org/10.3390/ijms23105416.

Mieulet, D., Aubert, G., Bres, C., Klein, A., Droc, G., Vieille, E., & Guiderdoni, E. (2018). Unleashing meiotic crossovers in crops. *Nature Plants, 4*(12), 1010–1016.

Miller, B., Suzuki, N., Ciftci-Yilmaz, S., & Mittler, R. (2010). Reactive oxygen species homeostasis and signaling during drought and salinity stresses. *Plant, Cell & Environment, 33*(4), 453–467. https://doi.org/10.1111/j.1365-3040.2009.02041.x.

Niu, H., & Klein, H. L. (2016). Multifunctional roles of Saccharomyces cerevisiae Srs2 protein in replication, recombination and repair. *FEMS Yeast Research, 17*(7), fow111.

Oladosu, Y., Rafii, M. Y., Abdullah, N., Hussin, G., Ramli, A., Rahim, H. A., & Latif, M. A. (2016). Principle and application of plant mutagenesis in crop improvement: A review. *Biotechnology & Biotechnological Equipment, 30*(1), 1–16. https://doi.org/10.1080/13102818.2015.1087333.

Pandey, S., Prasad, A., Sharma, N., & Prasad, M. (2020). Linking the plant stress responses with RNA helicases. *Plant Science, 299*, Article 110607.

Pasumarthy, K. K., Mukherjee, S. K., & Choudhury, N. R. (2011). The presence of tomato leaf curl Kerala virus AC3 protein enhances viral DNA replication and modulates virus induced gene-silencing mechanism in tomato plants. *Virology Journal, 8*, Article 178. https://doi.org/10.1186/1743-422X-8-178.

Raina, A., Laskar, R. A., Tantray, Y. R., Khursheed, S., Wani, M. R., Khan, S., & Rai, M. (2020). Characterization of induced high yielding cowpea mutant lines using physiological, biochemical and molecular markers. *Scientific Reports, 10*(1), Article 1–22.

Recker, J., Knoll, A., & Puchta, H. (2014). The Arabidopsis thaliana homolog of the helicase RTEL1 plays multiple roles in preserving genome stability. *The Plant Cell, 26*(12), 4889–4902.

Rizvi, I., Choudhury, N. R., & Tuteja, N. (2016). Arabidopsis thaliana MCM3 single subunit of MCM2–7 complex functions as 3' to 5' DNA helicase. *Protoplasma, 253*(2), 467–475. https://doi.org/10.1007/s00709-015-0803-5.

Röhrig, S., Dorn, A., Enderle, J., Schindele, A., Herrmann, N. J., Knoll, A., & Puchta, H. (2018). The RecQ-like helicase HRQ1 is involved in DNA crosslink repair in Arabidopsis in a common pathway with the Fanconi anemia-associated nuclease FAN1 and the post-replicative repair ATPase RAD5A. *New Phytologist, 218*(4), 1478–1490.

Roldan-Arjona, T., & Ariza, R. R. (2009). Repair and tolerance of oxidative DNA damage in plants. *Mutation Research/Reviews in Mutation Research, 681*(2–3), 169–179. https://doi.org/10.1016/j.mrrev.2008.07.003.

Rudolf, J., Makrantoni, V., Ingledew, W. J., Stark, M. J. R., & White, M. F. (2006). The DNA repair helicases XPD and FancJ have essential iron-sulfur domains. *Molecular Cell, 23*(6), 801–808. https://doi.org/10.1016/j.molcel.2006.07.026.

Saotome, A., Kimura, S., Mori, Y., Uchiyama, Y., Morohashi, K., & Sakaguchi, K. (2006). Characterization of four RecQ homologues from rice (*Oryza sativa* L. cv. Nipponbare). *Biochemical and Biophysical Research Communications, 345*(3), 1283–1291. https://doi.org/10.1016/j.bbrc.2006.05.004.

Singla, J., & Krattinger, S. G. (2016). Biotic stress resistance genes in wheat. In C. Wrigley, H. Corke, K. Seetharaman, & J. Faubion (Eds.), *Encyclopedia of Food Grains* (2nd ed., pp. 388–392). Academic Press. https://doi.org/10.1016/B978-0-12-394437-5.00229-1.

String Consortium. (2023). Global Core Biodata Resource. Retrieved from: https://string-db.org/.

Suhasini, A. N., & Brosh, R. M. (2013). DNA helicases associated with genetic instability, cancer, and aging. *Advances in Experimental Medicine and Biology, 767*, 123–144.

Sundin, G. W., Castiblanco, L. F., Yuan, X., Zeng, Q., & Yang, C. H. (2016). Bacterial disease management: Challenges, experience, innovation and future prospects: Challenges in Bacterial Molecular Plant Pathology. *Molecular Plant Pathology, 17*(10), 1506–1518. https://doi.org/10.1111/mpp.12436.

Tamura, K., Adachi, Y., Chiba, K., Oguchi, K., & Takahashi, H. (2002). Identification of Ku70 and Ku80 homologues in *Arabidopsis thaliana*: Evidence for a role in the repair of DNA double-strand breaks. *The Plant Journal, 29*(6), 771–781.

Tran, N. Q., Dang, H. Q., Tuteja, R., & Tuteja, N. (2010). A single subunit MCM6 from pea forms homohexamer and functions as DNA helicase. *Plant Molecular Biology, 74*(3), 327–336. https://doi.org/10.1007/s11103-010-9668-0.

Tuteja, N. (2003). Plant DNA helicases: The long unwinding road. *Journal of Experimental Botany, 54*(392), 2201–2214. Retrieved from: http://www.jstor.org/stable/23697928 (accessed on 9 July 2024).

Tuteja, N., & Tuteja, R. (2004). Prokaryotic and eukaryotic DNA helicases: Essential molecular motor proteins for cellular machinery. *European Journal of Biochemistry, 271*(10), 1835–1848.

Tuteja, N., Tran, N. Q., Dang, H. Q., & Tuteja, R. (2011). Plant MCM proteins: Role in DNA replication and beyond. *Plant Molecular Biology, 77*(6), 537–545.

Tuteja, R., & Tuteja, N. (2013). Analysis of DNA repair helicase UvrD from *Arabidopsis thaliana* and *Oryza sativa*. *Plant Physiology and Biochemistry, 71*, 254–260.

UniProt. (2023a). *Glycine Soja* (Wild soybean). Retrieved from: https://www.uniprot.org/uniprotkb/A0A445HRF0/entry (accessed on 9 July 2024).

UniProt. (2023b). *Brassica Napus* (Rape). Retrieved from: https://www.uniprot.org/uniprotkb/A0A078J769/entry (accessed on 9 July 2024).

Wang, C., Deng, P., Chen, L., Wang, X., Ma, H., et al. (2013). A wheat WRKY transcription factor TaWRKY10 confers tolerance to multiple abiotic stresses in transgenic tobacco. *PLOS ONE, 8*(6), Article e65120. https://doi.org/10.1371/journal.pone.0065120.

Wiedemann, G., van Gessel, N., Köchl, F., Hunn, L., Schulze, K., Maloukh, L., & Reski, R. (2018). RecQ helicases function in development, DNA repair, and gene targeting in *Physcomitrella patens*. *The Plant Cell, 30*(4), 717–736. https://doi.org/10.1105/tpc.17.00894.

Wielgoss, S., Barrick, J. E., Tenaillon, O., Wiser, M. J., Dittmar, J., Cruveiller, S., & MacLean, R. C. (2013). Mutation rate dynamics in a bacterial population reflect tension between adaptation and genetic load. *Proceedings of the National Academy of Sciences, 110*(1), 222–227. https://doi.org/10.1073/pnas.1219574110.

Yelina, N. E., Lambing, C., Hardcastle, T. J., Zhao, X., Santos, B., & Henderson, I. R. (2015). DNA methylation epigenetically silences crossover hot spots and controls chromosomal domains of meiotic recombination in *Arabidopsis*. *Genes & Development, 29*(20), 2183–2202.

Zhu, L. (2021). Targeted gene knockouts by protoplast transformation in the moss *Physcomitrella patens*. *Frontiers in Genome Editing, 3*, Article 719087. https://doi.org/10.3389/fgeed.2021.719087.

CHAPTER 13

Molecular Strategies for Climate Resilient Agriculture in Cereal Crops and Its Tolerance Mechanism

AJAY KUMAR CHANDRA,[1] PRIYANKA AGARWAL,[1] ALKA BHARATI,[2] RADHESHYAM KUMAWAT,[3] PRAMOD KUMAR PANDEY,[4] and AMARJEET KUMAR[5]

[1]*ICAR–Indian Agricultural Research Institute, New Delhi, India*

[2]*Scientist, ICAR–Central Agroforestry Research Institute, Jhansi, Uttar Pradesh, India*

[3]*Research Scholar, College of Post-Graduate Studies in Agricultural Sciences, Central Agricultural University (Imphal), Umiam, Meghalaya, India*

[4]*Assistant Maize Breeder, AICRP on Maize, College of Agriculture, Central Agricultural University, Kyrdemkulai, Meghalaya, India*

[5]*Assistant Professor, Genetics and Plant Breeding, College of Agriculture, Central Agricultural University (Imphal), Kyrdemkulai, Meghalaya, India*

ABSTRACT

As a result of global climate alterations, fueled by the rapidly growing population, agriculture being highly vulnerable and vital to humanity, has experienced enormous stresses. Given the prevalence of many stressors and the high frequency of such conditions, stress-resilient cultivars are advocated as a very reliable alternative to integrate existing in the era of climate change. Focusing effort and resources on innovative breakthroughs in plant breeding is crucial since the challenges posed by the global warming

Climate Change and Biotic Factors: A Molecular Approach.
Ashutosh Singh, Saurabh Pandey, and Amarjeet Kumar, (Eds.)

repercussions and the urgency of food security can no longer be solved by conventional breeding techniques. Consequently, agricultural biotechnology assisted with newly evolving molecular biology toolkits is reaching its age, and it has made incredible progress in this area. Because of these, the scientists were able to identify the genes responsible for underlying resilience mechanisms to concerned stresses using different strategies, which are often equipped with precise, cost-effective, and eco-friendly tools. These genes have been then validated and pyramided into cultivars with concurrent durable multi-stress resistance. Additionally, modern biotechnological tools like Multi-Omics-Led breeding, genomic selection (GS), GWAS studies, TILLING, gene editing technologies (CRISPR-Cas9 system), plant noncoding RNAs, Pan-Genomics, etc. have opened up a new prospect for precise plant genome engineering without off-target introgression. Though these emerging technologies are encouraging, the road ahead is very challenging for the targeted improvement of cereal crops. Thus, the cross-talk and effects of stresses make breeding initiatives more difficult, demanding more vigorous research efforts. The strategy to design crops/cultivars that can sustain different climatic stresses while delivering a reasonable yield and, ultimately, strengthening food security is of more importance. These efforts can also improve our knowledge concerning plant genomes to develop high-throughput innovative breeding methods.

13.1 INTRODUCTION

The world's biggest challenges facing humanity today are mitigating altered global climate, addressing human nutritional demands, and securing a sustainable energy supply. The primary sources of food energy for humans and animals are cereal crops, which are often grasses that are grown for their edible grains. As a result, cereal crops are grown in greater quantities compared to all other crop species. In terms of production, the most important cereal grains are wheat, rice, and maize (Agarwal et al., 2021; Nyamgoud et al., 2022). These cereals, which require higher resources than climate-smart "underutilized" cereal crops, contribute to nearly 40% of our daily caloric intake. Barley, sorghum, oat, finger millet, pearl millet, kodo millet, foxtail millet, and barnyard millet comprise additional cereal crops commonly known as "minor or underutilized" (Mohanta et al., 2018; Chandra et al., 2019; Zandalinas et al., 2021). Yet, changing climate jeopardizes the continued production of important cereal crops, predominantly the top cereals, threatening global food and nutritional security. Agriculture has been effective in feeding the world's population since the past century through varietal crossing, breeder selection, and novel agronomic innovations from the 1960s, including the expanded supply of irrigation and fertilizers. This resulted in substantial increases in agricultural productivity, resulting in significant crop production rises, especially rice, wheat, and maize (Kumar et al., 2018; Hsu et al., 2021). However, the advanced homogenized agricultural system, defined by resource-intensive crop species with high-calorie contents but confined nutritional diversity, had already lowered genetic diversity among most cultivated crops, limiting their potential for adaptation and thus raising malnutrition concerns. The fact that larger yield increases in recent decades are now difficult to maintain is even more unsettling. This is largely because "elite" crop varieties have reached their genetic potential, leading to full-scale genetic erosion and a shortage of fertile land mainly due to climate alteration, urbanization, and land degradation (Chandra et al., 2020a; Chandra et al., 2021). Diversification of the world's food supply, raising agricultural productivity, and addressing nutritional challenges while adapting to changing climate would necessitate a radical shift from a conventional to a more wide-ranging transformative approach. This entails developing climate-resilient cultivars that can sustain productivity under climate-aggravated conditions, thus expanding our focus on crop wild relatives (CWRs) and better exploiting underutilized crop species. This is made possible by recent developments in plant genomics that have transformed our potential to mine key genes that regulate important agronomic traits. Genomics, which is the study of an organism's whole genomic sequence, especially in plants, entails the study of all attributes of plant genomes and target genes at the sequence level, including various polymorphisms, mutations, and sequential phylogenetic interactions. It is composed of three components: structural, functional, and comparative genomics. Structural genomics is concerned with the characterization of genome structure, specifically the identification and comprehension of the physical architecture of the genome, as well as the position and attribution of genomic features to produce agronomically important traits (Zenda et al., 2021). Functional genomics, on the other hand, entails explaining gene function using copious information that includes sequence and map-based data. On the other hand, transcriptomics is a subset of functional genomics that is concerned with including both quantitative and qualitative transcription patterns. Correspondingly, comparative genetics and genomics in crop science are predominantly focused on uncovering allelic variations that control the improved phenotype. Plant gene functions and their pertaining governing networks have also been studied using several functional genomics tools, and functional genomics investigations may deliver the most readily available data for crop breeding (Agarwal et al., 2021; Nyamgoud et al., 2022).

Genomic technologies, when combined with novel methods such as gene/genome editing and genomic selection (GS), have the potential to speed up the rate of genetic advancement in crop breeding (Sood et al., 2019). As a result, employing the latest technology will abbreviate breeding cycles and reduce breeding expenses while boosting crop traits for mitigation of climate change and nutritional value. Climate change is fascinating for plant breeders to search for useful traits across all potential genetic sources (Hsu et al., 2021). As a result, integrating underutilized crop species into crop breeding enables breeders to access diversification for specific traits that are discovered in orphaned crop species yet not considered elite crop genotypes. Despite their proven agro-economic importance, orphan crops keep lagging in overall breeding programs. But even if the need to cope with climate change and resolve global food and nutritional insecurity barriers through nutritious, sustainable, and resilient agri-food systems, along with scientific breakthroughs through functional genomics, genome editing (CRISPR-Cas9 system), and genome-wide association study (GWAS) tools, have reignited scientific communities towards such orphan crops (Chandra et al., 2020b; Zenda et al., 2021; Rivero et al., 2022).

In this chapter, we tried to recapitulate recent advances in plant stress genomics and their possible applications in expanding stress-response mechanisms, with a focus on cereal crops. We mainly discussed widely available genetic and genomics tools such as marker-assisted breeding, Multi-Omics-Led breeding, genomic selection (GS), GWAS studies, TILLING, gene editing technologies (CRISPR-Cas9 system), plant noncoding RNAs, and Pan-Genomics as a complement to other crop genotyping platforms. Finally, as a concluding remark, we have provided a viewpoint for future plant stress genomics research concerning stress genomics to develop climate-smart "resilience crops" having nutritional superiority.

13.2 PHYSIOLOGICAL AND METABOLIC EFFECT OF STRESS ON CEREALS

Stress disrupts plant homeostasis, interfering with key physiological and metabolic processes, restricting energy production, and jeopardizing cellular integrity. To mitigate the effects of stress, plants must quickly adjust their physiological and metabolic responses, establishing a new homeostasis mechanism. Through acclimation, plants can modify their physiology and/or growth and reproduction strategies over time. Because distinct stresses have different impacts on plants, the acclimation responses of plants toward each stress condition may necessitate a different response strategy. Plants, for example, close their stomata opening during drought to prevent excess water loss but open their stomata during heat to increase transpiration and cool down their leaves' surface (Zandalinas et al., 2021; Hsu et al., 2021). Each stress may therefore stimulate a different acclimation response in plants. The biggest concern that plants usually face during stress responses is that the two different stresses influencing the plant at a given time may demand different, or sometimes resistant, metabolic and physiological responses. Using drought and heat as instances, these two different stresses necessarily require dissenting stomatal responses at the same morpho-physiological state (Table 13.1). Under such stress conditions, the plant may prioritize one adaptation approach over the other or use an entirely new response strategy. The specific acclimation strategy to employ for various stresses is likely to be influenced by the severity of each stress experienced impacting the plant, along with the timing by which they affect the plant (Rivero et al., 2022). Scientists have started to dissect plants' metabolic and physiological responses to uncover different stress responses in recent decades.

TABLE 13.1 Physiological and Metabolic Effects of Different Biotic and Stress on Cereal Crops

S. No.	Target Crops	Growth Stage	Stress Description	Physiological And Metabolic Effects	References
1.	Maize	Emergence to maturity	Salinity (1, 50, and 100 mM NaCl)	Stunted growth, reduced chlorophyll, and enhanced ROS level	Forieri et al., 2015
2.	Wheat	Emergence to maturity	~40% water deficit	Up to 20–25% grain yield loss	Daryanto et al., 2017
3.	Rice	Emergence to maturity	Drought stress	>40% grain yield loss	Zhang et al., 2018
4.	Barley	Anthesis and grain-filling stage	20–60% water deficit	Up to 50–60% total yield loss	Alghabari and Ihsan, 2018
5.	Wheat	Emergence to maturity	High temperature (30–45 °C)	Reduced LAI, productive tillers, photosynthesis, seed setting and number, and lower seed protein content	Sun et al., 2018

TABLE 13.1 *(Continued)*

S. No.	Target Crops	Growth Stage	Stress Description	Physiological And Metabolic Effects	References
6.	Pearl millet	Early seedling stage	Drought stress	>50% biomass loss	Debieu et al., 2018
7.	Maize	Tasselling stage	High temperature (> 38 °C) for 15 days	7–17 % decrease in yield	Hussain et al., 2019
8.	Maize	Pollination stage	Short-duration water deficit	Reduced seed setting, 28–32% yield loss, and down-regulation of photosynthetic genes	Wang et al., 2019
9.	Wheat	Vegetative to reproductive stage	Early or late water lodging	Up to 14–39.4% total yield loss	Farkas et al., 2020
10.	Rice	Seedling emergence	High temperature (> 40 °C) for 15 days	Delayed and reduced seedling germination	Rane et al., 2021
11.	Rice	Emergence to maturity	Salinity (50 mM and 75 mM NaCl)	Up to 40–75% total yield loss	Dramalis et al., 2021
12.	Wheat	Maturity to reproductive stage	Salinity (100 and 175 mM NaCl)	Delayed spike emergence, reduced fertility, and reduction in spikelet per spike	Rane et al., 2021
13.	Wheat	Seedling to maturity stage	Fungal (rust pathogen)	Up to 90–100% total yield loss	Raghunandan et al., 2022

13.3 GENETIC AND MOLECULAR RESPONSES OF PLANTS TO REGULATE DIFFERENT STRESS

As a consequence of the stimulation of various stress-response signaling cascades, molecular sensors, and other regulatory networks, the co-occurrence of diverse stresses might indeed result in the activation of additive, conflicting, or even irrelevant molecular responses. Besides this, during stress, the synthesis of unique and sometimes antagonistic hormonal signals/cross-talk may increase simultaneously over time, resulting in complementary molecular responses.

Since different stresses may necessitate different stress-tolerant strategies, plants may favor another sort of response over another or implement a combination of genetic and molecular strategies (Yang et al., 2019; Zandalinas et al., 2021). Genome-wide association studies (GWAS) or Transcriptomics of plants confined to different stress that categorize both as widely known and distinct responses. Previously, prevalent signaling cascades, molecular regulators, proteins, and genes that function during various stress were widely studied (Zandalinas et al., 2021). Such regulatory mechanisms, modulators, and genes could have been associated with ubiquitous stress responses, comprising signaling, or phytohormonal cross-talk. In Arabidopsis plants, for instance, 29 transcripts were prevalently up-regulated in expression in responses to heat and drought. These included various heat shock proteins (HSPs) as well as genes/transcripts associated with the ethylene and ABA signaling pathways.

A significant overlap in transcript feedback was also discovered in wheat during heat and drought stress. Furthermore, *G-BOX BINDING FACTOR3 (GBF3)* was articulated during all drought, salt, heat, osmotic, and biotic stress stresses, suggesting that *GBF3* may be associated with broad sense stress responses. A transcriptomics survey of *A. thaliana* subjected to multifaceted stress such as heat, light, salt, cadmium, acidity, and pesticides revealed that the relatively stable threshold of 127 and 136 transcript levels have been significantly over-expressed in response to studied stress. Some of the over-expressed transcripts throughout this group were entailed in redox control, ABA responses, stress responses, and transcription regulation, whereas others have been associated with carbohydrate and amino acid metabolism, glutathione transferase, peroxidase activities, and heme-binding (Zandalinas et al., 2021). Aside from responses to these stresses, several transcriptome-wide investigations of plants under different stresses have revealed varied responses unique to different stress. For instance, blending drought and heat stress transformed the expression levels of over 770 genes/transcripts that were not affected by either heat or drought stress (Yang et al., 2019).

Intriguingly, the transcriptome-wide study identified various stress pathways activated in response to many abiotic stresses (Figure 13.1). Classical stress-response pathways including unfolded protein response, osmoregulation, heat shock factors (HSFs), and autophagy, act as a major defining factor during stress responses (Yang et al., 2019; Zandalinas et al., 2021). It was also postulated that some of these factors could be supplemented by yet undiscovered pathways since the same set of stress or stress combinations contributed to the increased expression of transcripts having unknown functions. GWAS studies could identify novel genic loci that can be uniquely associated with the different or same set of stresses (Chandra et al., 2020b).

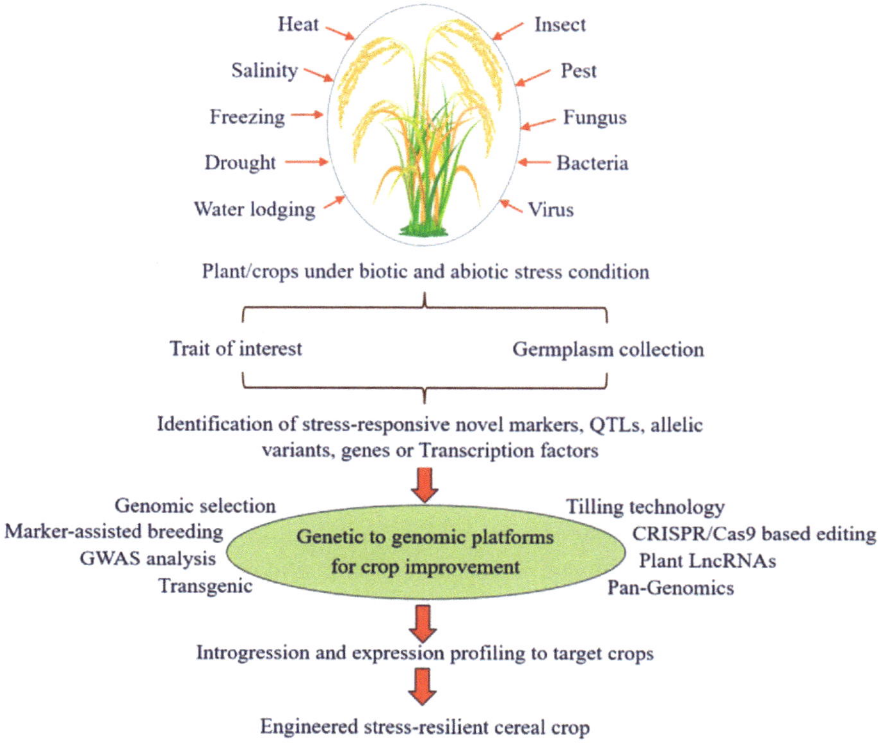

FIGURE 13.1 Systematic Scheme Showing Genetics to Genomics Approaches for Improvement of Stress-Resilient Cereal Crops.

Furthermore, maize lines with greater tolerance to heat and drought stress were discovered in a study involving 300 inbred lines. Tolerance to a combined effect of heat and drought was linked to genetic markers that deviated from all those attributed with tolerance to specific stresses (Table 13.2). Additionally, a negative impact of alkalinity on powdery mildew resistance was discovered in a tomato introgression line of 31 lines confined to a combination of salt content and powdery mildew. Moreover, Davila Olivas et al. (2017) investigated the natural variation influencing tolerance to various variations of biotic and abiotic stresses in 350 *Arabidopsis* lines.

13.4 STRESS SENSING AND SIGNALING DURING DIFFERENT BIOTIC AND ABIOTIC STRESS

Stress sensing and signaling in plants involve a variety of signal transduction mechanisms. These include various stress-specific sensors, signaling

TABLE 13.2 Details of Explored Genetic and Genomic Tools for Improvement of Climate-Resilient Cereal Crops

S. No.	Candidate Gene/Tfs/ Modulators	Target Crop	Stress	Description of Improved Traits	Modern Tools Under Investigation	References
1.	*mtHsp70, Hsp101, OsWRKY11, FAD7, OsGSK1, OsHsfA2e, SBPase,* and *Sp17*	Rice	High temperature, heat,	Increased chlorophyll content, enhanced photosynthesis and CO_2 assimilation, and growth rate	MAS, expression profiling, and other functional genomics tools	Wei et al., 2013; Zafar et al., 2018
2.	*SUB1A-1, AP2/ERF, OsTPP7, SK1–2, SD1, LGF1,* and *OsHSD1*	Rice	Submergence tolerance	Intermodal elongation, enhanced dormancy, and high chlorophyll retention	Multi-omics led breeding	Kuroha et al., 2018; Rane et al., 2021
3.	*SOS1–3, HKT, PUT, SPD, SPM, HVP1–10, TIP1, NCP1, GLP1, SAPK8,* and *OsTPS8*	Wheat, rice, maize, and barley	Salinity, and drought	Minimize uptake of toxin, increased ion homeostasis, protect cells from damage, maintain membrane integrity, and regulate gene expression	LC-MS/MS analysis, qRT-PCR profiling, and cloning	Vishal et al., 2018; Chandra et al., 2022a
4.	*SBEIIb, ERF922, DEP1, EDR1, MLO1, DREB1–2, ERF3,* and *ALS2*	Wheat, rice, and maize	Drought, salt, powdery mildew, blast, herbicide,	Enhanced grain yield, increased grain weight and GPC, erect panicle, and male sterility	CRISPR/Cas9 system	Ansari et al., 2020; Zenda et al., 2021
5.	*EcFER1, EcIRT2, EcYSL2, EcZIP1* and *EcZTP29*	Finger millet	Drought tolerance	Enhanced iron and Zinc acquisition and enrichment in grains	GWAS, Multi-omics led breeding, nano modulation, and qRT-PCR profiling	Chandra et al., 2020c; Chandra et al., 2021
6.	*Ppd, Ppd-D1,* and *VRN1*	Wheat, and barley	Drought tolerance	Early flowering	Functional genomics	Rane et al., 2021
7.	*LrTrk/YrTrk*	Wheat	Stripe rust resistance	27 NILs resistant to both leaf and rust pathotypes	Marker-assisted backcross	Mallick et al., 2022
8.	*TaZHD1* and *TaZHD10*	Wheat	Drought tolerance	RILs tolerant to leaf rolling	Multi-omics led breeding, qRT-PCR profiling, miRNA studies, and epigenomics	Chandra et al., 2022b

cascades, and pumps that generate different stress-signature molecules, such as scavengers, ROS sensors, and producers that can generate diverse ROS signatures all over cellular compartments, and different kinases and phosphatases that decipher the different signals produced and trigger differently, or networks of gene/protein regulatory pathways (Mohanta et al., 2018). A variety of stress-related enzymes, channels, and proteins are also thought to co-localize, resulting in the generation of stress-specific signalosomes. The fact that several stress-sensing and signaling events take place in cell organelles and entail a multifaceted mode of interaction among various organelles that many other hormones, signaling molecules, and processes, such as myristylation, ubiquitination, and S-nitrosylation, are involved in the regulation of response processes (Mohanta et al., 2018). Given this level of complexity, it is not unexpected that relatively little is known about how divergent stress-specific signals are assimilated when two or even more stresses influence the plant at the same time. Some examples of the integration of various stress-sensing and signaling events during these stresses have been documented in recent years (Table 13.2). These also include members of various transcription factor (TF) families.

Certain MYB TFs, for example, were discovered to be relevant for drought and heat stress in *Arabidopsis.* The regulation of various MYB TFs in *Arabidopsis* in response to a stress combination like heat and drought, heat and high light, and heat and salt demonstrated that the relative expression of *MYB7* and *MYB32* was relatively higher while the *MYB30* and *MYB51* were repressed during these stresses (Zandalinas et al., 2021). Members of the WRKY TF family, in addition to MYBs, have also been revealed to respond to a variety of stresses. For example, *AtWRKY40* works in dissent to *AtWRKY18* and *AtWRKY60* to improve *Arabidopsis* endurance to osmotic and salt stress through ABA signaling (Iyer et al., 2013). The AP2/ERF family is another TF family that plays key roles in stress responses in plants. DREBs were discovered to positively regulate salt, cold, heat, and drought tolerance, in addition to various stress combinations, by regulating stress-responsive genes/TFs. Furthermore, different ERF subfamilies such as *ERF-VII, ERF-X, ERF-VI,* and *ERF-XI* have emerged as key players in *Arabidopsis* in response to different metabolic changes, phytohormones (ABA and ethylene), and abiotic stresses such as cold, heat, drought, flooding, and salinity (Xie et al., 2019; Zandalinas et al., 2021). Furthermore, numerous AP2/ERF TFs, including *ERF4, ERF25, ERF88, ERF109, ERF57, ERF99*, and *DREB1D*, are significantly up-regulated in response to a combination of heat and high light stress in Arabidopsis (Zandalinas et al., 2021; Rivero et al., 2022).

13.5 MODERN TOOLS FOR UNDERSTANDING CROP RESPONSES TO BIOTIC AND ABIOTIC STRESSES

13.5.1 MARKER-ASSISTED SELECTION (MAS) AND MULTI-OMICS-LED BREEDING

To deliver optimal altitude in marker-assisted breeding and to screen large accessions having a multi-trait assembly, many plant breeding programs are paired with various molecular/genomic approaches. With the introduction of phenotypic plasticity, Multi-omic-led breeding, and high-throughput NGS provided a framework for identifying stress-responsive loci/QTLs/genes that are predicted to adversely affect total crop yield. QTL analysis of yield-related attributes in crops under various stress conditions facilitates the creation of novel cultivars with improved abiotic stress adaptability. Merchuk-Ovant et al. (2016) used marker-assisted selection (MAS) to detect novel genic loci/QTLs related to drought stress in bread and durum wheat cultivars. Under drought stress, three stable QTLs were identified: *qDTY3.1*, *qDTY6.1*, and *qDTY6.2*, all of which significantly impact total grain yield. They also executed QTL mapping using wheat recombinant inbred lines (RILs) under various stress conditions like heat, drought, flooding, and a combination of drought and heat at the same time (Table 13.2).

13.5.2 GENOME-WIDE ASSOCIATION STUDIES

Genome-wide association studies (GWAS) effectively comprehend the entire array of genetic variants in panels/cultivars to identify allelic variants associated with target traits, particularly stress. GWAS has been used in many crops to uncover the genetic workflow implicated in genetic and molecular resistance under varied climates. GWAS has a wide range of applications in plants, including biotic and abiotic stresses. Heat, drought, and salt tolerance have also been described employing GWAS. Qin et al. (2016) examined 13 drought-responsive traits in 373 different *A. tauschii* varieties. Salt tolerance QTLs were discovered on chromosomes 1, 4, 6, and 7. A novel QTL called "Saltol" was identified on chromosome 1 and is attributed to salt tolerance at the early-seedling stage (Raza et al., 2019). Chopra et al. (2017) identified several cold and heat-stress-tolerant QTLs/genes in *Sorghum bicolor*. In another survey, Chen et al. (2017) searched for heat-tolerant traits like leaf blotching (LB) and leaf firing (LF) in Sorghum during the seedling growth. GWAS was also used to determine the relationship between SNPs and genotype in terms of heat tolerance (Table 13.2). Five SNPs were found to be associated with LB traits, while nine SNPs were found to be associated with LF traits. Besides, 14 putative genes associated with target SNPs that have stress-responsive expression against abiotic stresses were discovered (Raza et al., 2019).

13.5.3 GENOME SELECTION (GS)

Genomic selection (GS) is a fascinating tool that has the potential to revolutionize crop improvement by screening elite germplasm with high marker densities and high-throughput phenotyping, boosting polygenic traits, and developing economically viable breeding lines. Currently, the potential of GS to accelerate the pace of genetic improvements in major crops has driven the creation of multi-environment models for key genomic estimation (Sood et al., 2019). Cuevas et al. (2017) proposed a modern multi-environment model based on the Bayesian model for genomic estimation. These methods were applied to four wheat cultivars and one maize cultivar, and the CIMMYT data bank demonstrated that the GxE model has relatively high significance rates as well as improved genomic predictions than other reported models. So far, approximately 40 studies based on GS have already been published. Wheat is the crop that has received the most attention, with 29 GS studies. Furthermore, durum wheat, oat, and barley have 1, 2, and 5 scientific studies, respectively. The most promising marker used in GS was Diversity Array Technology (DArT), followed by genotyping by sequencing (GBS) and single nucleotide polymorphism (SNP). These experiments indicated that GS could be used effectively in crop breeding programs. GS designs for wheat were employed extensively to identify germplasm with greater ability to deal with climate change. Crain et al. (2018) investigated various GS techniques for detecting phenotypic datasets from high-throughput phenotyping. Drought and heat stresses were investigated in 1000 elite wheat cultivars at CIMMYT using a high-throughput screening approach.

13.5.4 TILLING TECHNOLOGY

TILLING (targeting induced local lesions in genomes) technology has transformed mutagenesis-based plant breeding efforts. As a reverse genetics tool, TILLING uses chemical mutagenesis in conjunction with a mismatch detection method to detect a string of single-base paired allelic variants within a desired gene. This method produces a diverse set of mutant alleles, is quick and easy to implement, and can be applied to any organism capable of chemical mutagenesis. TILLING has been used successfully in a variety of cereal crop species, including rice, wheat, maize, and sorghum (Raza et al., 2019). Even though TILLING populations have traditionally been employed in reverse genetic perspectives, recent advances in whole-genome sequence analysis have provided novel possibilities for identifying mutations in putative candidate genes. The accessibility of sequence information from the mutant population would therefore allow TILLING populations to be used for forward genetic strategies such as starting from an interesting phenotype detected in specific mutant lines and cloning the implicit genes through association mapping (Qin et al., 2016). The utility of TILLING innovation, for example, has been demonstrated in the cloning of Stb6, a wheat gene. Stb6 encodes a highly conserved wall-associated receptor kinase (WAK) that induces pathogen resistance (PR) to *Septoria tritici* blotch (STB). Raza et al. (2019) discovered missense and nonsense mutations in two of these candidate genes via genome mapping. TILLING technology might also offer new opportunities for cloning genes responsive to biotic and abiotic stress in cereal crops in the context of a changing global climate.

13.5.5 TRANSGENIC

The discovery of stress-responsive TFs is a substantial step towards enhancing stress-resilient cultivars. These TFs can influence the morphology of genes in crops that are genetically modified under different stresses. Many transgenic plants have been developed through genetic engineering to combat biotic and abiotic stresses (Chandra et al., 2022a). Contrasting with control, genetically modified plants show significantly higher resistance to climate variability. Several transcription factors (TFs) have been identified as plant-specific TFs, including *AP2/ERFBP* group. This *AP2/ERFBP* TF family is involved across many plant growth promoters as well as stress responses. AP2/EREBP TFs are classified into four sub-groups based on their similarity and number. AP2, RAV, ERF, and DREB TFs are the groups in the class. DREB and ERF are two main subfamilies that have attracted considerable interest because of their functions in biotic and abiotic stress responses in the plant. In cold and drought stress, the DREB TFs have played an important role in gene regulation (Table 13.2). *DREB* TFs have widely been studied in plants such as *Arabidopsis*, rice, wheat, maize, barley, soybean, and tomato, for stress response. *DREB1* has also been studied extensively in *Arabidopsis* and rice for its cold stress regulation, whereas *DREB2* is also reported to play a crucial role in high temperature, freezing, salinity, and drought stresses (Raza et al., 2019).

The MYB TF family, also known as the myeloblastosis oncogene, is a large group of TFs explored in eukaryotes and widely distributed across plant species. These TFs are also associated with biotic and abiotic stress responses. Li et al. (2015) compiled numerous MYB TFs' resilience to abiotic stresses in *Arabidopsis* and other plants. Some MYB TFs, like AtMYB44, AtMYB60, and AtMYB61, have been identified to improve tolerance against drought in *Arabidopsis* transgenic plants by monitoring stomata movement. Yang et al. (2019) generated transgenic rice by over-expressing OsMYB2 to improve rice tolerance to dehydration, chilling, and salinity. Chen et al. (2017) revealed that the *ZmMYB30* gene was engineered into *Arabidopsis to* improve tolerance to salinity in maize.

Another major TF family is WRKY, which is distributed widely in plants in response to abiotic and biotic stresses. To enhance tolerance to abiotic stress in transgenic plants, *OsWRKY11* genes were overexpressed in transgenic rice to improve tolerance to drought and heat stresses (Raza et al., 2019). Similarly, Niu et al. (2012) studied the *TaWRKY19* gene overexpression level in transgenic wheat in response to drought, salinity, and freezing stresses. Maize ZmWRKY33 genes have likely been subjected to response to drought, freeze, salinity, and ABA-induced stresses (Chandra et al., 2022a). Under salt stress conditions, transgenic *Arabidopsis* with *ZmWRKY33* gene overexpression showed substantial tolerance (Li et al., 2015). NAC TFs play an important role in many mechanisms in the plant, including cell division, flower growth, and stress responses. Myriad NAC TFs have been identified and attributed to close links in both abiotic and biotic stresses in a variety of plants, including rice, *Arabidopsis,* maize, and soybean NAC TFs. The effects of *OsNAC1* overexpression in transgenic rice on drought and salinity tolerance were investigated (Raza et al., 2019).

13.5.6 CRISPR/CAS9 SYSTEM

CRISPR/Cas9 is a technologically advanced genome editing strategy that relies on the prokaryotic defensive response triggered by the type II RNA mechanism, which safeguards prokaryotes from invading viruses. CRISPR-Cas9 has modernized genome editing by generating candidate gene mutations and knocking down SNPs in a genome. In contrast to other editing tools such as ZFNs, MNs, and TALENs, CRISPR-editing methods have been extensively used in crop improvement (Dongariyal et al., 2023). Furthermore, it has a high potential to help crop breeding in the pursuit of high-yielding stress-resilient cultivars. Most importantly, the CRISPR/Cas9 tool is evolving into an entirely reasonable and eco-friendly strategy for the emergence of edited non-transgenic crops to combat extreme climate and ensure global food security (Table 13.2). CRISPR/Cas9 has been employed extensively in plant genome editing to deal with both biotic and abiotic stresses (Sood et al., 2019). Using the CRISPR genome editing tool, investigators silenced the TaDREB2 and TaERF3 genes to generate abiotic resistance (Raza et al., 2019). The CRISPR/Cas9 technique was used to generate tomato mutants with significantly higher concentration levels of malondialdehyde, H_2O_2, and proline. CRISPR/Cas9 technology has been employed on rice mutant lines, leading to improved tolerance to cold stress. The potential of a gene like *OsANN3* in cold tolerance was demonstrated and proposed as a potential candidate for developing transgenic rice with enhanced cold resilience (Shen et al., 2017).

13.5.7 PLANT LONG NONCODING RNAS (LNCRNAS)

In the ongoing decades, studies suggest that plant lncRNAs function as "key biological regulators" in a wide range of cellular processes such as genomic imprinting, cell differentiation, stress response, and epigenetic modification (Yu et al., 2019). Several plant lncRNAs function as primary transcribed miRNAs and siRNAs, like small regulatory RNAs proposed to have a role in protein re-localization and phosphate homeostasis. Furthermore, several stress response events in the plant have already been characterized using lncRNAs. For example, using intense transcriptomic sequencing, researchers have found 584 lncRNAs in foxtail millet that respond to modeled drought stress (Zenda et al., 2021). Similarly, in rice, drought-responsive 98 lncRNAs were found to modulate drought-responsive genes associated with various metabolic and gene-regulatory networks, whereas in wheat, heat-responsive 77 lncRNAs were found to respond to heat stress (Chung et al., 2016). Various lncRNAs also play critical roles in stress response modulation by target mimics of diverse miRNAs that can regulate the expression level of key stress-responsive genes/TFs at transcriptional, post-transcriptional, and epigenomic levels (Chandra et al., 2022b). Since field crops are constantly exposed to a wide range of stresses, they can reprogram their gene expression at the exposed stage, even if they are at transcriptional, post-transcriptional, and/or posttranslational levels. Uncovering the detailed roles of lncRNAs in particular abiotic and biotic stresses will thus aid in the development of lncRNA biomarkers pertinent to climate-resilient crops/cultivars.

13.5.8 PAN-GENOMICS

A pan-genome is defined as the total number of genes in a biological taxon, such as a species, which is composed of core genes distributed across all individuals and a set of interchangeable genes that are either individual-specific or partially shared. Meanwhile, genome structural variation (SV) like copy number variants (CNVs) and presence-absence variants (PAVs) can play critical roles in affecting important traits that are climate-relevant. The dispensable genome, in particular, has been found to contain genes involved in adaptation to biotic and abiotic conditions, such as phosphorus starvation, and head smut resistance in rice and maize, respectively. As a result, pan-genomic studies will aid in dissecting the genetic basis of agronomically important traits, allowing for the integration of genetic variation with improved traits via GWAS or QTL studies. Furthermore, expanding pan-genomics will also allow for rapid exploitation of CWRs to promote diversity within available gene pools, expanding the toolkit accessible for crop improvement initiatives (Zenda et al., 2021). Pan-genomics, in conjunction with NGS techniques, will further allow for a more thorough comprehension of crop diversity and the attribution of novel alleles. Consequently, detailed pan-genomic studies can offer to broaden the gene pool for expedited genetic improvement for better yields and climate-resilient cultivars.

13.6 CHALLENGES AND FUTURE PROSPECTS FOR THE DEVELOPMENT OF STRESS-RESILIENT CROPS

Improving resilience to abiotic stresses in crops demands an understanding of the underlying intricacies as well as a restraint to avoid grain yield losses caused by the introgression of specific traits. A viable strategy for tackling this challenging task must include greater yield and trait improvement unique to particular agroecologies. Even though transgenics have demonstrated repeatedly the possibility of improving stress tolerance, breeding abiotic stress-tolerant crops is difficult. This is because of the implicit machinery that influences the important traits and genes involved.

As marker-assisted selections (MAS) targeting stress-responsive genes and QTLs become more common in crop breeding, there is ample opportunity to integrate various omics strategies to substantially increase plant stress tolerance.

CRISPR/Cas editing has grown rapidly for genetic improvement of tolerance to biotic and abiotic stresses in plants. Recent developments in multi-omic approaches, such as genomics, transcriptomics, proteomics, metabolomics, and phenomics, have opened up new avenues for realizing abiotic stress responses from plants. With detailed insight into the methodologies of stress tolerance that have been derived through systems biology, NGS sequencing, functional, and other multi-omics approaches, breeding for stress-tolerant crops can indeed be significantly improved (Figure 13.1, Table 13.2). The molecular insights from epigenetics can help us understand the genetic and molecular mechanisms of stress response memory. Technological advancements in molecular breeding may hasten the use of functional markers in marker-assisted selection (MAS).

Furthermore, genomics can aid in the interpretation of the gene networks related to plant stress. Genetic mapping, GWAS, functional genomics, and NGS sequencing, in combination with recently developed phenotyping strategies, are currently being used to examine the genetic variability of plants at the single-base level. The integration of large-scale genome sequencing, CRISPR-Cas technologies, metabolomics, pan-genomics, and machine learning software has also greatly accelerated research into the temporal and spatial regulation of gene expression regulating important gene-metabolic networks. The success of emerging technologies, however, heavily relies on the discovery of mechanisms that control growth and development, grain yield, and stress resilience under field conditions. Realizing the cellular, metabolic, and developmental pathways of plants adapt in response to pollinators, pathogens, and dynamic environmental stresses will be critical (Table 13.1). New and improved cultivars must have "stable" genetic information to mitigate losses resulting from individual and combined effects of stresses. Changing stomatal state, photosynthetic efficiency, effective use of water and nutrients, and incentivizing plant interactions with beneficial microflora can be of great importance. Among the potential avenues for improving plant yields and productivity in the face of the expected adverse conditions resulting from climate change are:

1. Enhancing endurance to the combined effect of stress concerning changing climatic conditions includes the synchronous exposure of

plants to heat with flooding, salinity, drought, high CO_2 levels, and insect-pest attacks. Realizing the key metabolic, physiological, and signaling mechanisms underlying stress responses will be required. Consistent collaboration and cross-disciplinary cross-talk between research groups will be necessary to achieve this monumental task.

2. Escalating NUE and WUE via vascular tissue, stomata, roots, biochemical engineering, and enhancing plant-microbiome interactions within its endosphere, rhizosphere, and phyllosphere. Genome-wide analysis of plants/crops and microflora, combined with extensive metagenomic, sensing, and imaging technology, might be critical.
3. Enhancing plant procreation and grain-filling mechanisms under stress-induced conditions. This objective could be met by increasing heat and dehydration tolerance and improving photoassimilate allocation from source (leaf) to sink (flower) tissues.
4. Improving photosynthetic efficiency by modifying the abundant supply of photochemicals and reducing photorespiration contributes to increased CO_2 assimilation and photosynthetic efficiency. Alteration of the stomatal aperture and stomatal density control machinery and optimizing light acquisition by plant canopy are additional targets to achieve this goal.
5. Use green chemistry, nanoparticle technology, synthetic biology, and artificial intelligence to generate and incorporate novel acclimation approaches for mitigation of climate change strategies that are not currently evolved in crop genomes.
6. Developing, implementing, and continuously improving precision farming will allow farmers to ameliorate particular climate conditions in real time. Optimized mapping technologies, robotics, drone technologies, imaging, artificial intelligence, and effective use of bio-chemicals with irrigation systems are of great importance.

Plant breeders, physiologists, plant biochemists, ecologists, system biologists, molecular biologists, and metabolic engineers, along with other disciplines, will thus need to work in collaboration more actively in the progression of these technological breakthroughs. The journey ahead is challenging and lengthy, but "climate time" has already arrived.

KEYWORDS

- **climate resilient**
- **cross-talk**
- **food security**
- **multi-omics**
- **multi-stress resistance**

REFERENCES

Agarwal, P., Jha, S.K., Sharma, N.K., Raghunanadan, K., et al. (2021). Identification of the improved genotypes with 2NS/2AS translocation through molecular markers for imparting resistance to multiple biotic stresses in wheat. *Indian Journal of Genetics and Plant Breeding, 81*, 28–34.

Alghabari, F., & Ihsan, M. (2018). Effects of drought stress on growth, grain filling duration, yield and quality attributes of barley (*Hordeum vulgare* L.). *Bangladesh Journal of Botany, 47*, 421–428.

Ansari, W.A., Chandanshive, S.U., Bhatt, V., Nadaf, A.B., et al. (2020). Genome editing in cereals: Approaches, applications and challenges. *International Journal of Molecular Sciences, 21*, 4040.

Chandra, A.K., Chandora, R., Sood, S., & Malhotra, N. (2020b). Global production, demand, and supply. In M. Singh & S. Sood (Eds.), *Millets and Pseudo Cereals-Genetic Resources and Breeding* (pp. 7–18). Woodhead Publishing.

Chandra, A.K., Gururani, K., Gupta, A., & Kumar, A. (2019). Biotechnological and OMICS approaches to Study the Mechanisms of Stress Responses in Agricultural Crop. In S.K. Suman, A.K. Chandra, & P. Kumar (Eds.), *Plant Stress Biology* (pp. 01–15). New Delhi Publishers.

Chandra, A.K., Jha, S.K., Agarwal, P., Mallick, N., Murukan, N., & Vinod. (2022b). Leaf Rolling in Bread Wheat (*Triticum aestivum* L.) is controlled by the Upregulation of a Pair of closely linked/duplicate Zinc Finger Homeodomain class transcription factors during moisture stress conditions. *Frontiers in Plant Science, 13*, 1038881.

Chandra, A.K., Joshi, A., Tripathi, A., Kumar, A., Pandey, S., Singh, A., Lal, D., Bharati, A., Adhikari, S., & Dinkar, V. (2022a). Climate resilience maize: Heat stress, signaling, and molecular interventions. *Journal of Plant Growth Regulation*. doi.org/10.1007/s00344-022-10844-6.

Chandra, A.K., Pandey, D., Tiwari, A., Gururani, K., Agarwal, A., Dhasmana, A., & Kumar, A. (2021). Metal-based nanoparticles trigger the differential expression of key regulatory genes which regulate iron and zinc homeostasis mechanisms in Finger Millet. *Journal of Cereal Science, 100*, 103235.

Chandra, A.K., Pandey, D., Tiwari, A., Sharma, D., Agarwal, A., Sood, S., & Kumar, A. (2020a). An omics study of iron and zinc homeostasis in finger millet: Biofortified foods for micronutrient deficiency in an era of climate change? *OMICS: A Journal of Integrative Biology, 24*(12), 688–705.

Chen, J., Chopra, R., Hayes, C., Morris, G., Marla, S., Burke, J., Xin, Z., & Burow, G. (2017). Genome-wide association study of developing leaves' heat tolerance during vegetative growth stages in a sorghum association panel. *Plant Genome, 10*, 1–15.

Chopra, R., Burow, G., Burke, J.J., Gladman, N., & Xin, Z. (2017). Genome-wide association analysis of seedling traits in diverse Sorghum germplasm under thermal stress. *BMC Plant Biology, 17*, 1–12.

Chung, P.J., Jung, H., Jeong, D.H., Ha, S.H., Choi, Y.D., & Kim, J.K. (2016). Transcriptome profiling of drought responsive noncoding RNAs and their target genes in rice. *BMC Genomics, 17*, 563.

Crain, J., Mondal, S., Rutkoski, J., Singh, R.P., & Poland, J. (2018). Combining high-throughput phenotyping and genomic information to increase prediction and selection accuracy in wheat breeding. *Plant Genome, 11*, 1–14.

Cuevas, J., Crossa, J., Montesinos-López, O.A., Burgueño, J., Pérez-Rodríguez, P., & de los Campos, G. (2017). Bayesian genomic prediction with genotype × environment interaction kernel models. *G3: Genes, Genomes, Genetics, 7*, 41–53.

Daryanto, S., Wang, L., & Jacinthe, P.A. (2017). Global synthesis of drought effects on cereal, legume, tuber and root crops production: A review. *Agricultural Water Management, 179*, 18–33.

Davila Olivas, N.H., Kruijer, W., Gort, G., Wijnen, C.L., van Loon, J.J.A., & Dicke, M. (2017). Genome-wide association analysis reveals distinct genetic architectures for single and combined stress responses in Arabidopsis thaliana. *New Phytologist, 213*, 838–851.

Debieu, M., Sine, B., Passot, S., Grondin, A., et al. (2018). Response to early drought stress and identification of qtls controlling biomass production under drought in pearl millet. *PLoS ONE, 13*, e0201635.

Dongariyal, A., Chandra, A.K., Dongriyal, A., Kumar, A., & Sharma, P. (2023). Tending genome editing via crispr/cas9-induced mutagenesis: Opportunity and challenges for yield, quality and nutritional improvement of fruit crops. *Scientia Horticulturae, 311*, 111790.

Dramalis, C., Katsantonis, D., & Koutroubas, S.D. (2021). Rice growth, assimilate translocation, and grain quality in response to salinity under mediterranean conditions. *AIMS Agriculture and Food, 6*, 255–272.

Farkas, Z., Varga-Laszlo, E., Anda, A., Veisz, O., & Varga, B. (2020). Effects of waterlogging, drought and their combination on yield and water-use efficiency of five Hungarian winter wheat varieties. *Water, 12*, 1318.

Forieri, I., Hildebrandt, U., & Rostas, M. (2015). Salinity stress effects on direct and indirect defense metabolites in maize. *Environmental and Experimental Botany, 122*, 68–75.

Hsu, P.K., Dubeaux, G., Takahashi, Y., & Schroeder, J.I. (2021). Signaling mechanisms in abscisic acid-mediated stomatal closure. *The Plant Journal, 105*, 307–332.

Hussain, H.A., Men, S., Hussain, S., et al. (2019). Interactive effects of drought and heat stresses on morpho-physiological attributes, yield, nutrient uptake and oxidative status in maize hybrids. *Scientific Reports, 9*, 3890.

Iyer, N.J., Tang, Y., & Mahalingam, R. (2013). Physiological, biochemical and molecular responses to a combination of drought and ozone in *Medicago truncatula*. Plant, Cell & Environment, 36, 706–72.

Kumar, A., Gupta, S., Abrol, G., & Chandra, A. (2018). Deriving innovations through intervention of newer technologies in integrated manner for value addition and development of value-added products in agri-food-nutrition and health sectors. *Acta Scientiae Agricolae, 2*(11), 78–79.

Kuroha, T., Nagai, K., Gamuyao, R., et al. (2018). Ethylene-gibberellin signaling underlies adaptation of rice to periodic flooding. *Science, 361*, 181–186.

Li, C., Ng, C.K.Y., & Fan, L.M. (2015). MYB transcription factors, active players in abiotic stress signaling. *Environmental and Experimental Botany, 114*, 80–91.

Mallick, N., Jha, S.K., Agarwal, P., Kumar, S., et al. (2022). Marker-assisted transfer of leaf and stripe rust resistance from *T. turgidum* var. durum cv. Trinakria to wheat variety HD2932. *Frontiers in Genetics, 13*, 941287.

Merchuk-Ovnat, L., Barak, V., Fahima, T., Ordon, F., et al. (2016). Ancestral QTL alleles from wild emmer wheat improve drought resistance and productivity in modern wheat cultivars. *Frontiers in Plant Science, 7*, 452.

Mohanta, T.K., Bashir, T., Hashem, A., & Abd_Allah, E.F. (2017). Systems biology approach in plant abiotic stresses. *Plant Physiology and Biochemistry, 121*, 58–73.

Niu, C.F., Wei, W., Zhou, Q.Y., Tian, A.G., et al. (2012). Wheat WRKY genes TaWRKY2 and TaWRKY19 regulate abiotic stress tolerance in transgenic Arabidopsis plants. *Plant Cell and Environment, 35*, 1156–1170.

Nyamgoud, S., Tyagi, S., Chandra, A.K., et al. (2022). Development and characterization of leaf rust resistant *Triticum timopheevii* derived introgression lines in hexaploidy wheat. *Indian Journal of Genetics and Plant Breeding, 82*(4), 1–10.

Qin, P., Lin, Y., Hu, Y., Liu, K., et al. (2016). Genome-wide association study of drought-related resistance traits in *Aegilops tauschii. Genetics and Molecular Biology, 39*, 398–407.

Raghunandan, K., Tanwar, J., Patil, S.N., Chandra, A.K., et al. (2022). Identification of novel broad-spectrum leaf rust resistance sources from Khapli wheat landraces. *Plants, 11*, 1965.

Rane, J., Singh, A.K., Kumar, M., Boraiah, K.M., et al. (2021). The adaptation and tolerance of major cereals and legumes to important abiotic stresses. *International Journal of Molecular Sciences, 22*, 12970.

Raza, A., Razzaq, A., Mehmood, S.S., Zou, X., et al. (2019). Impact of climate change on crops adaptation and strategies to tackle its outcome: A review. *Plants, 8*(2), 34.

Rivero, R.M., Mittler, R., Blumwald, E., & Zandalinas, S.I. (2022). Developing climate-resilient crops: Improving plant tolerance to stress combination. *The Plant Journal, 109*(2), 373–389.

Shen, C., Que, Z., Xia, Y., Tang, N., et al. (2017). Knock out of the annexin gene OsAnn3 via CRISPR/Cas9-mediated genome editing decreased cold tolerance in rice. *Journal of Plant Biology, 60*, 539–547.

Sood, S., Joshi, D.C., Chandra, A.K., & Kumar, A. (2019). Phenomics and genomics of finger millet: Current status and future prospects. *Planta, 250*(3), 731–751.

Sun, A., Somayananda, I., Sunoj, V.S.J., Singh, K., et al. (2018). Heat stress during flowering affects time of day of flowering, seed-set and grain quality in spring wheat (*Triticum aestivum* L.). *Crop Science, 58*, 380–392.

Vishal, B., Krishnamurthy, P., Rengasamy, R., & Prakash, P.K. (2018). OsTPS8 controls yield-related traits and confers salt stress tolerance in rice by enhancing suberin deposition. *New Phytologist, 221*, 1369–1386.

Wang, B., Liu, C., Zhang, D., He, C., Zhang, J., & Li, Z. (2019). Effects of maize organ-specific drought stress response on yields from transcriptome analysis. *BMC Plant Biology, 19*, 335.

Wei, H., Liu, J., Wang, Y., et al. (2013). A dominant major locus in chromosome 9 of rice (*Oryza sativa* L.) confers tolerance to 48°C high temperature at seedling stage. *Journal of Heredity, 104*, 287–294.

Xie, Z., Nolan, T.M., Jiang, H., & Yin, Y. (2019). AP2/ERF transcription factor regulatory networks in hormone and abiotic stress responses in Arabidopsis. *Frontiers in Plant Science, 10*, 228.

Yang, J., Duan, G., Li, C., Liu, L., Han, G., Zhang, Y., et al. (2019). The cross-talks between jasmonic acid and other plant hormone signaling highlight the involvement of jasmonic acid as a core component in plant response to biotic and abiotic stresses. *Frontiers in Plant Science, 10*, 1349.

Yu, Y., Zhang, Y., Chen, X., & Chen, Y. (2019). Plant noncoding RNAs: Hidden players in development and stress responses. *Annual Review of Cell and Developmental Biology, 35*, 407–431.

Zafar, S.A., Hameed, A., Nawaz, M.A., Wei, M.A., et al. (2018). Mechanisms and molecular approaches for heat tolerance in rice (*Oryza sativa* L.) under climate change scenario. *Journal of International Agriculture, 17*(4), 726–738.

Zandalinas, S.I., Sengupta, S., Fritschi, F.B., Azad, R.K., et al. (2021). The impact of multifactorial stress combination on plant growth and survival. *New Phytologist, 230*, 1034–1048.

Zenda, T., Liu, S., Dong, A., & Duan, H. (2021). Advances in cereal crop genomics for resilience under climate change. *Life, 11*, 502.

Zhang, J., Zhang, S., Cheng, M., Jiang, H., et al. (2018). Effect of drought on agronomic traits of rice and wheat: A meta-analysis. *International Journal of Environmental Research and Public Health, 15*, 839.

Index

C

D

E

F

H

I

M

P

Q

R

S

T